de Gruyter Series in Logic and Its Applications 5

Martin Zeman

Inner Models and Large Cardinals

Walter de Gruyter
Berlin · New York 2002

Author

Martin Zeman
Department of Mathematics
University of California at Irvine
261 Multipurpose Science & Technology Bldg.
Irvine, CA 92697
USA

Mathematics Subject Classification 2000: 03-02; 03E35; 03E45; 03E55

Keywords: Inner model, Extender, Fine structure.

Printed on acid-free paper which falls within the guidelines of the ANSI to ensure permanence and durability

Library of Congress – Cataloging-in-Publication Data

Zeman, Martin, 1964–
Inner models and large cardinals / Martin Zeman.
p. cm. – (De Gruyter series in logic and its applications ; 5)
Includes bibliographical references and index.
ISBN 3-11-016368-3
1. Constructibility (Set theory) 2. Large cardinals (Mathematics) I. Title. II. Series.
QA248 .Z46 2001
511.3'22–dc21 2001047562

Die Deutsche Bibliothek – Cataloging-in-Publication Data

Zeman, Martin:
Inner models and large cardinals / Martin Zeman. – Berlin ; New York : de Gruyter, 2001
(De Gruyter series in logic and its applications ; 5)
ISBN 3-11-016368-3

ISSN 1438–1893

Printed in Germany.
Typesetting using the Authors' TEX files: I. Zimmermann, Freiburg – Printing and binding: Hubert & Co. GmbH & Co. KG, Göttingen – Cover design: Rainer Engel, Berlin.

Preface

The purpose of this book is to give an introduction to the theory of fine structural inner models. These models are constructible relative to coherent extender sequences and admit a fine structure theory analogous to that of Gödel's constructible universe. The theory of such models originates in Jensen's analysis of the constructible universe and in the work of Dodd and Jensen on the core model, and was later pursued by Jensen, Mitchell, Steel and others.

It is not possible to explain the point of inner model theory on just a few lines, but I would like to draw attention to at least one of its aspects. In many cases, this theory can be viewed as a counterpart to the theory of forcing in the following sense. While forcing allows us to construct various kinds of set-theoretical objects from large cardinals, inner model theory enables us to analyze the complexity of such objects in terms of large cardinals. Such analysis can be then used to determine lower bounds for the consistency strengths of various set-theoretic principles. Let me illustrate this with the following example. It is well-known that Prikry forcing makes a measurable cardinal ω-cofinal. On the other hand, inner model theory can be used to show that measurability is necessary for this task: If a regular cardinal can be turned into an ω-cofinal one, then this cardinal must be measurable in an inner model. Other examples involve, for instance, the consistency strength of the failure of the singular cardinal hypothesis, that of the failure of the combinatorial principle square or the consistency strength of projective uniformization. There is also a body of applications to descriptive set theory, that are of a slightly different nature. Here, various definable subsets of reals can be characterized in terms of inner models. For instance, the set C_{2n+2} is precisely the set of reals of the minimal inner model for n Woodin cardinals, and scales for various pointclasses can be directly constructed using inner models, avoiding any reference to determinancy.

My intention is to give an account of inner model theory that would serve as an introductory text for those not familiar with the field and that could also serve as a text book for graduate students. At the same time, I attempt to provide the reader with a glimpse of some current developments. The book can informally be divided into three parts. The first part comprises Chapters 1 through 3 and gives a detailed introduction to the general fine structure theory of acceptable structures. This theory can be developed abstractly, without any reference to inner models or large cardinals. The second part, consisting of Chapters 4 through 8, contains both the full core model theory for measures of order 0 as well as instructions for generalizing this theory for models that can contain up to one strong cardinal (Chapter 8). The third part of the book is just Chapter 9 and provides an introduction to the theory of Jensen extender models that are beyond one strong cardinal. The main difference between the theories below and beyond one strong cardinal rests in the complexity of the comparison process. In the former case, comparison arguments can be carried out using mere *linear* iterability of the models in question, which considerably simplifies the whole theory. One can say that inner model theory at this level has reached its definitive version. In the latter case, however, the comparison process is highly non-linear and the essence of the iterability problem is the choice of cofinal well-founded branches through iteration trees. This problem belongs to one of the central open questions in inner model theory, as all iterability proofs known today can be carried out only under smallness conditions ranging at the level of (several) Woodin cardinals. In this book, however, we will not touch this problem at all, and merely derive various consequences of iterability that determine the inner structure of extender models.

This exposition is based on Jensen's handwritten notes [Je88], [Je1] and [Je97]. The presentation of the material, however, often diverges from that in Jensen's notes, as it takes into account later developments. Most of the results presented here (and all relevant ones) are due to Dodd, Jensen, Mitchell, Steel and others, and only a few techniques from Chapter 6 and 9 are due to me; a more detailed account of credits can be found in notes at the end of each chapter. Also, many proofs that I present can be replaced by much simpler proofs if we work with measures of order 0, so the reader might get an impression that I made the things more complicated than they really are. My intention, however, is to present methods that easily generalize to higher models.

I should also include some words on how to use this book. In first three chapters I give a very general account of fine structure theory. Knowledge of this theory helps to understand fine structural aspects of inner models in general; however, this generality is superfluous if we work at the level of measures of order 0. This might confuse those readers who are beginners in fine structure theory. For this reason, I recommend first reading the beginning of Chapter 1 that precedes the exposition of Σ^*-relations, then go directly to Chapter 2 and after that learn only the basic facts about fine ultrapowers from the beginning of Chapter 3. To make the first reading of Chapter 3 easier, it is helpful to assume that $n = 0$, i.e. that $\omega\varrho_M^1 \leq \kappa$ where κ is the critical point of the ultrapower in question. The above assumption enables the reader to avoid the use of Σ^*-theory (or any substitute for it), as fine ultrapowers are nothing more than coarse ultrapowers in this case. Having accumulated the knowledge of the above-mentioned material, one can proceed directly to Chapter 4 where the theory of extender models for measures of order 0 begins. Later in the course of the reading, one can return back to the first three chapters to learn new facts about general fine structure theory and fine ultrapowers; this should be done precisely at those points where deeper knowledge of this theory is necessary for understanding the arguments. Chapters 4 through 7 constitute a compact exposition of the core model theory for measures of order 0. Although it is possible to skip over various parts of them, the full understanding of the theory requires the knowledge of the methods used in the proofs of the basic lemmata on the comparison process, that of the solidity theorem, condensation lemma and the weak covering theorem, as well as the construction of $\boldsymbol{K}$ and the verification of its basic properties. These topics, however, constitute the bulk of the four chapters mentioned above. Chapter 8 is intended for those readers who would like to work with inner models up to one strong cardinal and avoid non-linear iterations. As we have already mentioned, it contains a recipe for generalizing the theory for measures of order 0 to the core model theory up to one strong cardinal. This generalization is merely an elaboration on the previous theory and Chapter 8 provides the reader with only those lemmata that do not follow from the methods developed previously in a straightforward way. Thus, the reading of this chapter requires a reasonable knowledge of the previous theory. Chapter 9 is an introduction to Jensen fine structure theory for higher extender models; this theory is based on the so-called λ-indexing. The chapter builds on the fine structure theory developed in the first three chapters of the book in its full generality, but is relatively independent of the remaining chapters. This should enable the reader familiar with the basics of inner model theory who is interested in Jensen extender models to proceed to Chapter 9 directly. I choose to present Jensen extender models based on λ-indexing and $*$-iterability that is also based on λ-indexing, as this considerably simplifies the general iterability theory including the Neeman–Steel lemma, thus somewhat facilitating the reader's first contact with such models.

Acknowledgments. I would like to thank all those people who encouraged, helped and supported me during the preparation of this book. Since the list of these people would be too long, I mention just three of them, to whom I am particularly indebted: Ronald Jensen, from whom I learned set theory, Ralf Schindler, who shared the office with me during the last two years and who was always free to discuss the topic with me and Michael Bruening, who proofread the whole text. I am also grateful to the following institutions, where various parts of this book were prepared and/or which provided me with a financial support: Humboldt University Berlin (Mathematical Institute), Friedrich-Wilhelm University Bonn (Mathematical Institute), Vienna University (Institute of Formal Logic), Slovak Academy of Sciences (Mathematical Institute) and Deutsche Forschungsgemeinschaft (DFG).

Bratislava and Vienna, September 2001 *Martin Zeman*

Contents

Chapter 1

Fine Structure

In this section we shall describe the fine structure theory for acceptable structures of the form $\langle J_\alpha^A, B\rangle$ which is general enough for the applications in core model theory. We assume some familiarity with the theory as described in [Je72] or [De84] and shall automatically use various properties of the above-mentioned structures without further referring to these sources.

1.1 Acceptable J-Structures

Definition. A J*-structure* is an amenable structure of the form $\langle J_\alpha^A, B\rangle$ for an ordinal α and predicates A, B (or, mutatis mutandis, $\langle J_\alpha^{A_1,\dots,A_n}, B_1, \dots, B_n\rangle$).

Hence $M = \langle J_\alpha^A, B\rangle$ is of the form $\langle |J_\alpha^A|, A, B\rangle$, where $|J_\alpha^A|$ is the underlying set. We shall often abuse the notation (if there is no danger of confusion) and write J_α^A or M instead of $|J_\alpha^A|$. The same also applies to the symbols for the predicates A_i, B_j and the predicates themselves – we will usually write A_i for both the predicate and the corresponding predicate symbol, as the meaning will be clear from the context.

Sometimes it is convenient to work with the auxiliary hierarchy S_α^A of [Je72], which is defined as follows:

$$\begin{aligned} S_0^A &= \emptyset \\ S_{\alpha+1}^A &= S^A(S_\alpha^A) \\ S_\lambda^A &= \bigcup_{\xi<\lambda} S_\xi^A \quad \text{for limit } \lambda \end{aligned}$$

where S^A is an operator which, applied to a set x, adds images of $x \cup \{x\}$ under a certain carefully chosen fixed *finite* list of rudimentary functions. The following equality is often useful:

$$J_\alpha^A = S_{\omega\alpha}^A$$

for all ordinals α. This is a consequence of the fact that the class of all functions rudimentary in A has a finite basis. An example of such a basis is the following list of functions:

$$\begin{aligned}
F_0(x,y) &= \{x,y\}\\
F_1(x,y) &= x-y\\
F_2(x,y) &= x\times y\\
F_3(x,y) &= \{\langle u,z,v\rangle;\ z\in x\ \&\ \langle u,v\rangle\in y\}\\
F_4(x,y) &= \{\langle u,v,z\rangle;\ z\in x\ \&\ \langle u,v\rangle\in y\}\\
F_5(x,y) &= \bigcup x\\
F_6(x,y) &= \mathrm{dom}(x)\\
F_7(x,y) &= \in\cap(x\times x)\\
F_8(x,y) &= \{x''\{z\};\ z\in y\}\\
F_9(x,y) &= A\cap x,
\end{aligned}$$

where $\langle x_1, x_2, \dots x_n\rangle = \langle x_1, \langle x_2, \dots, x_n\rangle\rangle$. It is useful to define the hierarchy S_α so that each of its levels is transitive. For this purpose we need a collection of functions such that given any transitive set U, the images of elements of U under these functions constitute a transitive set. One can easily see that this fails for the above basis. However, this shortcoming can be remedied by expanding our list of rudimentary functions in order to guarantee transitivity. One such possibility is to add the following functions:

$$\begin{aligned}
F_{10}(x,y) &= \langle x,y\rangle\\
F_{11}(x,y) &= x''\{y\}\\
F_{12}(x,y) &= \langle \mathrm{left}(y), x, \mathrm{right}(y)\rangle\\
F_{13}(x,y) &= \langle \mathrm{left}(y), \mathrm{right}(y), x\rangle\\
F_{14}(x,y) &= \{\mathrm{left}(y), \langle \mathrm{right}(y), x\rangle\}\\
F_{15}(x,y) &= \{\mathrm{left}(y), \langle x, \mathrm{right}(y)\rangle\},
\end{aligned}$$

where $\mathrm{left}(y) = u$ and $\mathrm{right}(y) = v$ if $y = \langle u, v\rangle$ and $\mathrm{left}(y) = 0 = \mathrm{right}(y)$ if y is not an ordered pair. We can then set

$$S^A(U) = \bigcup_{i=0}^{15} F_i''(U\cup\{U\})^2.$$

Definition. Let $\alpha, \alpha_0, \dots, \alpha_{\ell-1}$ be ordinals and $\ell \in \omega$.

a) $\prec\alpha_0, \dots, \alpha_{\ell-1}\succ$ is the *Gödel ℓ-tuple*, i.e. an ordinal which is a result of an iterative application of the Gödel pairing function to the above ℓ-tuple of ordinals.

b) $(\alpha)_i^\ell$ is the *i-th Gödel projection* ($i < \ell$), i.e. if $\alpha = \prec\alpha_0, \dots, \alpha_{\ell-1}\succ$ then $(\alpha)_i^\ell = \alpha_i$.

Recall that the functions introduced in the above definition are primitive recursive. One of the useful consequences of this fact is that they have a canonical Σ_1 definition, which is uniform over all transitive primitive recursively closed structures.

We shall now state two important facts about J-structures, which are proved in [Je72].

Convention. Throughout this book we shall work with a fixed recursive enumeration of Σ_1-formulae with one free variable $\langle \varphi_i ; i \in \omega \rangle$. We usually write $\varphi_i(x_1, \ldots, x_\ell)$ instead of $\varphi_i(\langle x_1, \ldots, x_\ell \rangle)$.

Theorem 1.1.1. *Let M be a J-structure. The Σ_1-satisfaction relation $\models_M^{\Sigma_1}$ is then uniformly $\Sigma_1(M)$.*

Theorem 1.1.2. *Let M be a J-structure. The Σ_1-Skolem function h_M is then uniformly $\Sigma_1(M)$.*

Remark. The above two theorems are to be understood as follows. There are Σ_1-formulae Φ, Ψ such that whenever M is a J-structure,

a) Φ defines $\models_M^{\Sigma_1}$, i.e. $M \models \varphi_i(x) \leftrightarrow M \models \Phi(i, x)$

b) Ψ defines h_M, i.e. $y = h_M(i, x) \leftrightarrow M \models \Psi(i, x, y)$

where $\langle \varphi_i ; i \in \omega \rangle$ is the enumeration of Σ_1-formulae we fixed above.

In the following we describe a useful class of formulae, which lies somewhere between Σ_1 and Π_2. It turns out that many notions can be expressed by statements belonging to this class.

Definition. We say that φ is a *Q-formula* iff φ is of the form $(\forall u)(\exists v \supset u)\varphi(v)$, where φ is Σ_1 and does not contain u as a free variable. Instead of $(\forall u)(\exists v \supset u)$ we write briefly Qv. The above formula then has the form $(Qv)\varphi(v)$ and we read "for cofinally many v, $\varphi(v)$". A map π which preserves Q formulae is called *Q-preserving* and we write

$$\pi : \bar{M} \underset{Q}{\longrightarrow} M.$$

A property which is characterized by a Q-formula is also called a *Q-condition.*

Definition. Let U, V be transitive structures. A map $\sigma : U \to V$ is *cofinal* iff every $y \in V$ is contained as a subset in some $\sigma(x)$ where $x \in U$.

Let $\sigma : U \to V$ be a Σ_1-preserving map, where U, V are transitive structures, and let φ be a Q-formula. It is easy to see that

a) φ is preserved downwards,

b) if σ is cofinal, then φ is preserved upwards.

Note also that Q-formulae are closed under & and $\vee$.

We now introduce the notion of acceptability which is fundamental for the general fine structure theory presented later in this section. As will follow from the definition, acceptability can be considered as a strong version of GCH.

Definition. A J-structure $M = \langle J_\alpha^A, B\rangle$ is *acceptable* iff the following holds: Whenever $\xi < \alpha$ and $\mathcal{P}(\tau) \cap J_{\xi+1}^A \not\subset J_\xi^A$ for some $\tau < \omega\xi$, there is a surjective map $f : \tau \to \omega\xi$ in $J_{\xi+1}^A$. (This means that $\operatorname{card}(\omega\xi) \leq \tau$ in $J_{\xi+1}^A$.)

Lemma 1.1.3. *Being an acceptable J-structure is a Q-property. More precisely: There is a fixed Q-sentence Ψ such that for any $M = \langle |M|, A, B\rangle$ which is transitive and closed under unordered pairing, M is an acceptable J-structure iff $M \models \Psi$.*

Proof. By [Je72], the statement $\langle U, A\rangle = J_\alpha^A$ is a Q-condition for $\langle |M|, A\rangle$ as above. Amenability can be expressed by

$$(Qu)(\exists z)\,(z = B \cap u).$$

It only remains to prove that the fact that we collapse $\omega\xi$ whenever we add a new bounded subset is expressible in a Q-fashion. We note first that a J-structure M is acceptable iff the following holds in M:

$$\begin{aligned}&(\forall\,\omega\xi)(\exists\, n \in \omega)(\forall\, m \geq n)(\forall\, \tau < \omega\xi)\\ &[\mathcal{P}(\tau) \cap S_{\omega\xi+m}^A \not\subset S_{\omega\xi}^A \longrightarrow (\exists\, f \in S_{\omega\xi+m}^A)(f : \tau \overset{\text{onto}}{\longrightarrow} \omega\xi)].\end{aligned} \tag{1.1.1}$$

Denote the sentence from (1.1.1) by ψ. It is easy to see that if M satisfies ψ then M is acceptable. To see the converse, fix ξ and suppose there is a $\tau < \omega\xi$ such that $\mathcal{P}(\tau) \cap S_{\omega(\xi+1)}^A \not\subset S_{\omega\xi}^A$. Let τ be minimal with this property. By acceptability of M, there is an $n \in \omega$ such that $S_{\omega\xi+n}^A$ contains a function f mapping τ onto $\omega\xi$. Using f, it is easy to construct a surjective map $f_{\tau'} : \tau' \to \omega\xi$ for any $\tau' < \omega\xi$ whose power set in $S_{\omega(\xi+1)}^A$ is larger than that in $S_{\omega\xi}^A$ (since $\tau' \geq \tau$) and it follows immediately that $f_{\tau'} \in S_{\omega\xi+n+2}^A$; so we have a uniform bound for all such functions.

If the height of M is $\omega\alpha$ for some limit α, then acceptability is equivalent to the statement

$$(Q\xi)(S_{\omega\xi}^A \models \psi).$$

Otherwise we have to state (1.1.1) explicitly for the last level. Hence, the desired condition is then

$$\begin{aligned}&\langle U, A, B\rangle \text{ is an amenable } J\text{-structure}\\ &\&\ (Q\zeta)(\lim(\zeta) \longrightarrow S_\zeta^A \models \psi)\ \&\ [(Q\zeta)(\zeta \text{ is limit}) \vee \varphi]\end{aligned} \tag{1.1.2}$$

where φ is the sentence

$$(Q\zeta)(\exists\,\beta < \zeta)[\lim(\beta)\ \&\ (\forall\,\eta < \zeta)(\eta > \beta \to \operatorname{succ}(\eta))\ \&\ \varphi'(\beta, \zeta)]$$

and $\varphi'(\beta, \zeta)$ the formula

$$(\forall \tau \leq \beta)[(\exists u \in S_\zeta^A)(u \notin S_\beta^A \;\&\; u \subset \tau) \longrightarrow (\exists f \in S_\zeta^A)(f : \tau \xrightarrow{\text{onto}} \omega\beta)].$$

φ is clearly a Q-sentence, hence (1.1.2) is a Q-sentence as well. Denote this formula by Ψ. Then M is an acceptable J-structure iff $M \models \Psi$. □(Lemma 1.1.3)

Corollary 1.1.4. a) *If $\pi : \bar{M} \underset{\Sigma_1}{\longrightarrow} M$ and M is acceptable, then so is $\bar{M}$.*

b) *If $\pi : \bar{M} \underset{Q}{\longrightarrow} M$ and $\bar{M}$ is acceptable, then so is M. This holds in particular if π is a Σ_0 cofinal map.*

Lemma 1.1.5. *Let $M = J_\alpha^A$ be acceptable and let $\omega\varrho \in M$ be a cardinal in M. Given $u \in J_\varrho^A$, any $a \in M$ which is a subset of u is in fact an element of J_ϱ^A.*

Proof. Since $u \in J_\varrho^A$, there is a surjective map $g : \tau \to u$ in J_ϱ^A for some $\tau < \varrho$. Set $\bar{a} = g^{-1}{}''a$. Then $\bar{a} \subset \tau$ and $a \in J_\varrho^A \longleftrightarrow \bar{a} \in J_\varrho^A$. But if $\bar{a} \notin J_\varrho^A$, then there is an $f : \tau \xrightarrow{\text{onto}} \omega\xi$, where ξ is such that $\bar{a} \in J_{\xi+1}^A - J_\xi^A$; notice that $\xi \geq \varrho$. This contradicts the fact that $\omega\varrho$ is a cardinal in J_α^A. Consequently, $a \in J_\varrho^A$. □(Lemma 1.1.5)

Lemma 1.1.6. *Let M be as above and $\omega\varrho$ be a successor cardinal in M. Let $a \subset J_\varrho^A$ be such that* card$(a) < \omega\varrho$ *in M. Then $a \in J_\varrho^A$.*

Proof. Let $\gamma < \omega\varrho$ and $g \in M$ be such that $g : \gamma \xrightarrow{\text{onto}} a$. Define $f : \gamma \to \omega\varrho$ by

$$\zeta = f(\xi) \longleftrightarrow g(\xi) \in S_{\zeta+1}^A - S_\zeta^A.$$

Then $f \in M$.

Claim. f is bounded in $\omega\varrho$.

Let $\delta < \omega\varrho$ be such that $\text{rng}(f) \subset \delta$. Then $a \subset S_\delta^A \in J_\varrho^A$ and, by Lemma 1.1.5, $a \in J_\varrho^A$. Hence it suffices to prove the above claim.

Proof of the claim. Suppose f is unbounded in $\omega\varrho$. Define $G : \omega\varrho \to M$ by

$$G(\eta) = \text{ the } <^A\text{-least function of } \gamma \text{ onto } \eta.$$

This is possible since $\omega\varrho$ is a successor cardinal in M, so we can pick γ large enough that all we have done so far goes through. Clearly G is definable over J_ϱ^A, hence $G \in M$.

Now define $F : \gamma \times \gamma \xrightarrow{\text{onto}} \omega\varrho$ by

$$F(\xi, \eta) = G(f(\xi))(\eta).$$

Then $F \in M$. To obtain a contradiction, it suffices to find a map $h : \omega\delta \xrightarrow{\text{onto}} \gamma \times \gamma$ in M for some $\delta < \varrho$, since $F \circ h$ will then collapse $\omega\varrho$ inside M. Pick any $\delta < \varrho$ such that $\gamma \leq \omega\delta$. This is possible by the assumption on $\omega\varrho$. Then there is a map $h : \omega\delta \xrightarrow{\text{onto}} J_\delta^A \supset \gamma \times \gamma$ such that $h \in M$. □(Claim) □(Lemma 1.1.6)

Corollary 1.1.7. *Let M, ϱ be as in Lemma 1.1.6. Then $J_\varrho^A \models \mathrm{ZFC}^-$. (Hence $\omega\varrho = \varrho$.)*

Proof. The point here is to verify the separation and replacement schemata in J_ϱ^A, since the rest of the axioms hold in J_ϱ^A automatically. The former follows from Lemma 1.1.5 and the latter from Lemma 1.1.6 in a straightforward way. □(Corollary 1.1.7)

Corollary 1.1.8. *Let M, ϱ be as above where $\omega\varrho$ is a limit cardinal in M. (Hence $\omega\varrho = \varrho$.) Then $J_\varrho^A \models \mathrm{Z} + \mathrm{AC}$ (where Z is Zermelo set theory, i.e. ZF without replacement).*

Proof. We only have to verify the power set axiom; the rest goes through as before. Let $a \in J_\varrho^A$. Pick $\gamma < \varrho$ such that γ is a cardinal in J_ϱ^A and $a \in J_\gamma^A$. Then for every $x \in \mathcal{P}(a) \cap J_\varrho^A$, $x \in J_\gamma^A$. Hence $\mathcal{P}(a) \cap J_\varrho^A \in J_\varrho^A$. □(Corollary 1.1.8)

Corollary 1.1.9. *Let M, ϱ be as above (hence $\omega\varrho = \varrho$). Then*

$$|J_\varrho^A| = H_\varrho^M \stackrel{\text{def}}{=} \{x\,;\ \mathrm{card}\,(\mathrm{trcl}(x)) < \varrho \ \text{ in } M\}.$$

Proof. Clearly $|J_\varrho^A| \subset H_\varrho^M$. So it is sufficient to prove the converse. Suppose it fails. Let ϱ be the least counterexample. Then ϱ is a successor cardinal in M. Let $x \in H_\varrho^M$ be $\in$-minimal such that $x \notin J_\varrho^A$. Then $x \subset J_\varrho^A$. Since $\mathrm{card}(x) < \varrho$ in M, $x \in J_\varrho^A$ by Lemma 1.1.6. Contradiction. □(Corollary 1.1.9)

1.2 The Σ_1-Projectum

We now introduce the central notions of fine structure theory – the notions *projectum, standard code* and *good parameter.* We stress that we are working with arbitrary J-structures and that these structures have, in general, very few closure properties. This means that there might be bounded definable subsets of these structures (in a precise sense) failing to be *elements*. However, each J-structure has an initial segment which is "firm" in the sense that it does contain all sets reasonably definable over the whole structure. The height of this "firm" segment is called the *projectum. Standard codes* are (boldface) definable relations computing truth and *good parameters* are parameters which occur in the definitions of standard codes.

Definition. The Σ_1-*projectum* ϱ_M of an acceptable J-structure M is defined by

$$\varrho_M = \text{ the least } \varrho \in \mathbf{On} \text{ such that } \mathcal{P}(\omega\varrho) \cap \mathbf{\Sigma}_1(M) \not\subset M.$$

Lemma 1.2.1. *Let M be as above. If $\omega\varrho \in M$, then $\omega\varrho$ is a cardinal in M.*

Proof. Suppose not. Set $\varrho = \varrho_M$. Let $f \in M$ be such that $f : \gamma \xrightarrow{\text{onto}} \omega\varrho$ for some $\gamma < \omega\varrho$ and $A \in \mathbf{\Sigma}_1(M)$ be such that $a = A \cap \omega\varrho \notin M$. Let $\bar{a} = f^{-1\prime\prime}a$. Then $\bar{a} \notin M$, since otherwise $a = f''\bar{a} \in M$. On the other hand, $\bar{a} \in M$ by the definition of ϱ, since $\bar{a} \subset \gamma$ and $\bar{a} \in \mathbf{\Sigma}_1(M)$. Contradiction. □(Lemma 1.2.1)

Lemma 1.2.2. *Let M be as above and $\varrho = \varrho_M$. Then $\omega\varrho$ is a Σ_1-cardinal in M (i.e. there is no $\mathbf{\Sigma}_1(M)$ partial map from a $\gamma < \omega\varrho$ onto $\omega\varrho$).*

Proof. Suppose there is such a map, say $f : \gamma \overset{\text{onto}}{\longrightarrow} \omega\varrho$. We know that there is a $\mathbf{\Sigma}_1(J^A_\varrho)$ map of $\omega\varrho$ onto J^A_ϱ. Hence there is a $\mathbf{\Sigma}_1(M)$ map $g : \gamma \overset{\text{onto}}{\longrightarrow} J^A_\varrho$. Define a set b by

$$\xi \in b \longleftrightarrow \xi \notin g(\xi).$$

Then b is clearly $\mathbf{\Sigma}_1(M)$ and $b \subset \gamma$. Moreover, $b \notin J^A_\varrho$ by a diagonal argument. Hence $b \notin M$: this follows from Lemma 1.1.5 if $\omega\varrho \in M$ and immediately otherwise. On the other hand, $\gamma < \omega\varrho$, and therefore $b \in M$. Contradiction. □(Lemma 1.2.2)

Corollary 1.2.3. *Let M be acceptable and $\varrho = \varrho_M$.*

a) *If $B \subset J^A_\varrho$ is $\mathbf{\Sigma}_1(M)$, then $\langle J^A_\varrho, B\rangle$ is amenable.*

b) $|J^A_\varrho| = H^M_{\omega\varrho}$.

Definition. Let $M = \langle J^B_\alpha, D\rangle$ be an acceptable J-structure, $\varrho = \varrho_M$ and $p \in M$. We define

$$A^p_M = \{\langle i, x\rangle \in \omega \times H^M_{\omega\varrho};\ M \models \varphi_i(x, p)\}$$

where $\langle \varphi_i;\ i \in \omega\rangle$ is the recursive enumeration of Σ_1-formulae fixed at the beginning of this book. A^p_M is called the *standard code* determined by p. Let us stress that A^p_M is the intersection of $\omega \times H^M_{\omega\varrho}$ with a set $\tilde{A}^p_M$ (defined in an obvious way) which is $\Sigma_1(M)$ in p. We shall often write $A^p_M(i, x)$ instead of $\langle i, x\rangle \in A^p_M$. The structure

$$M^p = \langle J^B_\varrho, A^p_M\rangle$$

is called the *reduct* determined by p. If $\delta = \varrho$ or $\delta < \varrho$ where $\omega\delta$ is a cardinal in M, we also set

$$\begin{aligned} A^{p,\delta}_M &= A^p \cap J^B_\delta \\ M^{p,\delta} &= \langle J^B_\delta, A^{p,\delta}_M\rangle. \end{aligned}$$

We shall omit the subscript $_M$ whenever possible (i.e. whenever there is no danger of confusion).

Definition. Let M be acceptable and $\varrho = \varrho_M$.

P_M = the set of all $p \in M$ for which there is a $B \subset \mathbf{On} \cap M$ such that

a) B is $\Sigma_1(M)$ in p

b) $B \cap \omega\varrho \notin M$.

The elements of P_M are called *good parameters*.

Lemma 1.2.4. *Let M be as before, $p \in M$ and $A = A_M^p$. Then*

$$p \in P_M \leftrightarrow A \cap (\omega \times \omega\varrho_M) \notin M.$$

Proof. ($\rightarrow$) Pick B which witnesses that $p \in P_M$. Suppose B is defined by φ_i. Hence $B(\xi) \leftrightarrow \langle i, \xi \rangle \in A$, which means that if $A \cap (\omega \times \omega\varrho_M)$ is in M, then so is $B \cap \omega\varrho_M$. Hence the former is not an element of M.

($\leftarrow$) Now suppose $A \cap (\omega \times \omega\varrho_M) \notin M$. Let $f : \omega \times \omega\varrho_M \longrightarrow \omega\varrho_M$ be defined by $f(i, \omega\xi + j) = \omega\xi + 2^i \cdot 3^j$. Clearly f is $\Sigma_1(M)$ uniformly and if $\omega\varrho \in M$ then $f \in M$. Let $B = f''A$. Then B is $\Sigma_1(M)$ in p and $B \cap \omega\varrho_M \notin M$: if $\omega\varrho_M \in M$, this follows from the fact that $f \in M$ and $A \cap (\omega \times \omega\varrho_M) = f^{-1}{}''(B \cap \omega\varrho_M)$; if $\omega\varrho_M \notin M$, this follows from the fact that B is cofinal in $\omega\varrho_M$. □(Lemma 1.2.4)

Definition. Let M be acceptable and $\varrho = \varrho_M$. We set

$$R_M = \text{the set of all } r \in M \text{ such that } h_M(\omega\varrho \cup \{r\}) = |M|.$$

The elements of R_M are called *very good parameters*.

Lemma 1.2.5. $R_M \subset P_M \neq \emptyset$.

Proof. The inequality follows immediately from the definition of P_M. As to the inclusion, define the set $a \subset \omega \times \mathbf{On} \cap M$ by

$$\langle i, \xi \rangle \in a \longleftrightarrow \langle i, \xi \rangle \notin h_M(i, \langle \xi, p \rangle)$$

for a $p \in R_M$. By a diagonal argument, $a \cap \omega \times \omega\varrho_M \notin M$ and a is $\Sigma_1(M)$ in p. Using the map f from the proof of Lemma 1.2.4 it is easy to turn the set a into some $b \subset \mathbf{On} \cap M$ such that b is $\Sigma_1(M)$ in p and $b \cap \omega\varrho_M \notin M$. □(Lemma 1.2.5)

1.3 Downward Extension of Embeddings Lemmata

Given a Σ_0 preserving map between the reducts of two acceptable structures, the question naturally arises whether the map can be extended to a map between the original structures. It turns out that this is possible and we shall study this in detail, since it will provide us with a useful technique for later applications. The conjunction of the following three lemmas is called the *downward extension of embeddings lemma.*

Lemma 1.3.1. *Let $\pi : \bar{M}^{\bar{p}} \xrightarrow[\Sigma_0]{} M^p$, where $\bar{p} \in R_{\bar{M}}$. Then there is a unique $\tilde{\pi} : \bar{M} \xrightarrow[\Sigma_0]{} M$ such that $\tilde{\pi} \supset \pi$ and $\tilde{\pi}(\bar{p}) = p$. Moreover, $\tilde{\pi} : \bar{M} \xrightarrow[\Sigma_1]{} M$.*

Proof. *Uniqueness.* Assume that $\tilde{\pi}$ has the above properties. Let $x \in \bar{M}$. Then $x = h_{\bar{M}}(i, \langle \xi, \bar{p} \rangle)$ for some $i \in \omega$ and $\xi < \omega\varrho_M$. Let $\bar{H}$ be $\Sigma_0(\bar{M})$ such that

$(\exists z)\bar{H}(z, x, i, \xi, \bar{p})$ defines the Skolem function $h_{\bar{M}}$ (this involves a slight abuse of notation). $\bar{H}$ has a uniform definition, so let H have the same definition over M. Pick z such that $\bar{H}(z, x, i, \xi, \bar{p})$. Since $\tilde{\pi}$ is Σ_0-preserving, we have $H(\tilde{\pi}(z), \tilde{\pi}(x), i, \tilde{\pi}(\xi), p)$, i.e. $\tilde{\pi}(x) = h_M(i, \langle \tilde{\pi}(\xi), p \rangle) = h_M(i, \langle \pi(\xi), p \rangle)$. Hence, there is at most one such $\tilde{\pi}$.

Existence. The above proof of uniqueness suggests how to define the extension $\tilde{\pi}$. Here we show that such a definition is correct. We first observe:

Claim. Suppose that $\varphi(v_1, \dots, v_\ell)$ is a Σ_1-formula. Let $\bar{x}_i = h_{\bar{M}}(j_i, \langle \bar{\xi}_i, \bar{p} \rangle)$ for some $j_i \in \omega$, $\bar{\xi}_i < \omega \varrho_M$ and $x_i = h_M(j_i, \langle \xi_i, p \rangle)$ where $\xi_i = \pi(\bar{\xi}_i)$ $(i = 1, \dots, \ell)$. Then

$$\bar{M} \models \varphi(\bar{x}_1, \dots, \bar{x}_\ell) \quad \text{iff} \quad M \models \varphi(x_1, \dots, x_\ell).$$

Proof. $\bar{M} \models \varphi(\bar{x}_1, \dots, \bar{x}_\ell)$ is equivalent to

$$\bar{M} \models \varphi(h_{\bar{M}}(j_1, \langle \bar{\xi}_1, \bar{p} \rangle), \dots, h_{\bar{M}}(j_\ell, \langle \bar{\xi}_\ell, \bar{p} \rangle)).$$

Since $h_{\bar{M}}$ has a uniform Σ_1-definition over $\bar{M}$, there is a Σ_1-formula ψ such that the above is equivalent to

$$\bar{M} \models \psi(\bar{\xi}_1, \dots, \bar{\xi}_\ell, \bar{p}). \tag{1.3.1}$$

The formula ψ clearly does not depend on the actual structure in question, so the statement $M \models \varphi(x_1, \dots, x_\ell)$ is equivalent to

$$M \models \psi(\xi_1, \dots, \xi_\ell, p). \tag{1.3.2}$$

Now suppose $\psi(\xi_1, \dots, x_\ell, p) \longleftrightarrow \varphi_k(\langle \xi_1, \dots, x_\ell \rangle, p)$ in our fixed recursive enumeration of Σ_1-formulae, hence (1.3.1) is equivalent to

$$A^{\bar{p}}_{\bar{M}}(k, \langle \bar{\xi}_1, \dots, \bar{\xi}_\ell \rangle) \tag{1.3.3}$$

and (1.3.2) is equivalent to

$$A^{p}_{M}(k, \langle \xi_1, \dots, \xi_\ell \rangle). \tag{1.3.4}$$

Since π is Σ_0-preserving, (1.3.3) and (1.3.4) are equivalent. □(Claim)

Define $\tilde{\pi}$ by

$$\tilde{\pi}(h_{\bar{M}}(i, \langle \xi, \bar{p} \rangle)) \simeq h_M(i, \langle \pi(\xi), p \rangle)$$

for $i \in \omega$ and $\xi \subset \omega \varrho_{\bar{M}}$. We have to verify several facts:

- $\tilde{\pi}$ is well defined.
  Let $h_{\bar{M}}(j_1, \langle \bar{\xi}_1, \bar{p} \rangle) = h_{\bar{M}}(j_2, \langle \bar{\xi}_2, \bar{p} \rangle)$ for some $j_1, j_2 \in \omega$ and $\bar{\xi}_1, \bar{\xi}_2 < \omega \varrho_{\bar{M}}$. We have to show that $h_M(j_1, \langle \xi_1, p \rangle) = h_M(j_2, \langle \xi_2, p \rangle)$, where $\xi_i = \pi(\bar{\xi}_i)$ $(i = 1, 2)$. By the above claim, this follows immediately.
- $\tilde{\pi}$ is Σ_1-preserving.
  Since $\tilde{\pi}$ is a well defined map this follows immediately from the above claim.

- $\tilde{\pi} \supset \pi$.
There is an $i \in \omega$ such that the equality $x = h_M(i, \langle x, q \rangle)$ holds uniformly and independently of q. In particular, for $\xi < \omega\varrho_{\bar{M}}$ we have $\xi = h_{\bar{M}}(i, \langle \xi, \bar{p} \rangle)$, so

$$\tilde{\pi}(\xi) = h_M(i, \langle \pi(\xi), p \rangle) = \pi(\xi).$$

- $\tilde{\pi}(\bar{p}) = p$.
Similarly as above, there is an $i \in \omega$ such that $q = h(i, \langle x, q \rangle)$ uniformly. Hence $\bar{p} = h_{\bar{M}}(i, \langle 0, \bar{p} \rangle)$ and $\tilde{\pi}(\bar{p}) = h_M(i, \langle 0, p \rangle) = p$. □(Lemma 1.3.1)

Lemma 1.3.2. *Let $\bar{M}, M, \bar{p}, p, \pi, \tilde{\pi}$ be as above. Suppose moreover that $p \in R_M$. Let $\pi : \bar{M}^{\bar{p}} \underset{\Sigma_n}{\longrightarrow} M^p$. Then*

$$\tilde{\pi} : \bar{M} \underset{\Sigma_{n+1}}{\longrightarrow} M.$$

Proof. We shall proceed by induction. Suppose the lemma holds for n. Suppose we have a Σ_{n+1} formula φ which is of the form

$$(\exists z_1)(\forall z_2) \ldots (\exists z_{n+1})\bar{\varphi}(z_1, \ldots, z_{n+1}, x_1, \ldots, x_\ell), \tag{1.3.5}$$

where $\bar{\varphi}$ is Σ_0 (so n is even). For simplicity, assume that $n = 2$. Given a J-structure N and an arbitrary $q \in R_N$, the structure N satisfies (1.3.5) iff it satisfies the formula

$$(\exists \xi_1 < \omega\varrho)(\forall \xi_2 < \omega\varrho)(\psi_1(\xi_1, \xi_2, \zeta_1, \ldots, \zeta_\ell, q) \longrightarrow \psi_1(\xi_1, \xi_2, \zeta_1, \ldots, \zeta_\ell, q))$$

where $\psi_1(\xi_1, \xi_2, \zeta_1, \ldots, \zeta_\ell, q)$ is the Σ_1-formula

$$(\exists z)(z = k_N(\langle \xi_2, q \rangle)),$$

$\psi_2(\xi_1, \xi_2, \zeta_1, \ldots, \zeta_\ell, q)$ is the Σ_1-formula

$$(\exists z_3)\bar{\varphi}(k_N(\langle \xi_1, q \rangle), k_N(\langle \xi_2, q \rangle), z_3, k_N(\langle \zeta_1, q \rangle), \ldots, k_N(\langle \zeta_\ell, q \rangle)),$$

$k_N(\langle \xi, q \rangle) \simeq h_N((\xi)_1, \langle (\xi)_2, q \rangle)$ and $x_i = k_N(\langle \zeta_i, q \rangle)$ for some suitable $\zeta_i < \omega\varrho_N$; here $i = 1, \ldots, \ell$ (recall that $(\ )_i$ is the i-th coordinate in Gödel pairing). This can be seen by a straightforward induction on n using the fact that q is a very good parameter. Notice that this transformation is uniform, i.e. the formulae ψ_1 and ψ_2 depend merely on $\bar{\varphi}$ and *not* on N. Hence there are $k_1, k_2 \in \omega$ such that, setting

$$B^q_N(k_1, k_2, z) \longleftrightarrow (A^q_N(k_1, z) \longrightarrow A^q_N(k_2, z)),$$

the above statement can be expressed in the Σ_2-fashion over N^q in the following way:

$$(\exists \xi_1)(\forall \xi_2) B^q_N(k_1, k_2, \langle \xi_1, \xi_2, \zeta_1, \ldots, \zeta_\ell \rangle).$$

Then

$$\begin{aligned}\bar{M} \models \varphi(x_1, \ldots, x_\ell) &\longleftrightarrow (\exists \xi_1) \ldots (\forall \xi_n) B^{\bar{p}}_{\bar{M}}(k_1, k_2, \langle \xi_1, \ldots, \xi_n, \zeta_1, \ldots, \zeta_\ell \rangle) \\ &\longleftrightarrow (\exists \xi_1) \ldots (\forall \xi_n) B^{p}_{M}(k_1, k_2, \langle \xi_1, \ldots, \xi_n, \pi(\zeta_1), \ldots, \pi(\zeta_\ell) \rangle) \\ &\longleftrightarrow M \models \varphi(\tilde{\pi}(x_1), \ldots, \tilde{\pi}(x_\ell)).\end{aligned}$$

Similar, but slightly more complicated reductions can be done for arbitrary n; we leave this to the reader. □(Lemma 1.3.2)

Lemma 1.3.3. *Let $\pi : N \underset{\Sigma_0}{\longrightarrow} M^p$, where N is a J-structure. Then there are unique $\bar{M}$, $\bar{p}$ such that $\bar{p} \in R_{\bar{M}}$ and $N = \bar{M}^{\bar{p}}$.*

Proof. Let $M = \langle J^B_\alpha, D \rangle$, $M^p = \langle J^B_\varrho, A \rangle$ and $N = \langle J^{\bar{B}'}_{\bar{\varrho}}, \bar{A} \rangle$. Let further $X = \operatorname{rng}(\pi)$, $Y = h_M(X \cup \{p\})$, $\bar{M}$ be the transitive collapse of Y and $\tilde{\pi} : \bar{M} \to Y$ the associated uncollapsing map (i.e. the inverse to the Mostowski collapse). The map $\tilde{\pi}$ is clearly Σ_1-preserving, hence $\bar{M}$ is of the form $\langle J^{\bar{B}}_{\bar{\alpha}}, \bar{D} \rangle$. We shall show that $\bar{M}$ is the desired structure. We prove first

$$\tilde{\pi} \supset \pi \quad \text{and} \quad J^{\bar{B}'}_{\bar{\varrho}} = J^{\bar{B}}_{\bar{\varrho}}, \tag{1.3.6}$$

which follows immediately from

$$\text{if } x \in X,\ y \in Y \text{ and } y \in x \text{ then } y \in X \tag{1.3.7}$$

since the latter tells us that the collapsing map for Y restricted to X coincides with π^{-1}.

So suppose $y = h_M(i, \langle z, p \rangle)$ for some $z \in X$. Since both $y, z \in M^p$, this can be equivalently expressed in the form $A(k, \langle y, z \rangle)$ for some $k \in \omega$. Thus we have

$$(\exists v \in x) A(k, \langle v, z \rangle)$$

which is preserved by a Σ_0 map, so

$$(\exists v \in \bar{x}) \bar{A}(k, \langle v, \bar{z} \rangle)$$

where $(\bar{x}, \bar{z}) = \pi^{-1}(x, z)$. Let $\bar{y} \in \bar{x}$ be such that $A(k, \langle \bar{y}, \bar{z} \rangle)$. Such an $\bar{y}$ is uniquely determined: if $\bar{y}_1$ were another one, we would have

$$A(k, \langle \pi(\bar{y}), z \rangle) \ \&\ A(k, \langle \pi(\bar{y}_1), z \rangle),$$

which means that $\pi(\bar{y}) = h_M(i, \langle z, p \rangle) = \pi(\bar{y}_1)$, hence $\bar{y} = \bar{y}_1$. This argument also shows $\pi(\bar{y}) = y$. Hence $y \in X$. □(1.3.6), (1.3.7)

Let

$$\bar{p} = \tilde{\pi}^{-1}(p).$$

Note that there is a $\mathbf{\Sigma}_1(\bar{M})$ map of $\omega\bar{\varrho}$ onto $\bar{M}$; this follows immediately from the fact that $\bar{M} = h_{\bar{M}}(J^{\bar{B}}_{\bar{\varrho}} \cup \{\bar{p}\})$ and that there is a $\mathbf{\Sigma}_1(\bar{M})$ map of $\omega\bar{\varrho}$ onto $J^{\bar{B}}_{\bar{\varrho}}$. So we have

$$\varrho_{\bar{M}} \leq \bar{\varrho}. \tag{1.3.8}$$

Note also that if $i \in \omega$ and $x \in N$, then

$$\bar{A}(i,x) \longleftrightarrow A(i,\tilde{\pi}(x)) \longleftrightarrow M \models \varphi_i(\tilde{\pi}(x),p) \longleftrightarrow \bar{M} \models \varphi_i(x,\bar{p}) \tag{1.3.9}$$

where $\langle\varphi_i\rangle_i$ is the fixed recursive enumeration of the Σ_1-formulae. Our aim is to show

$$\varrho_{\bar{M}} = \bar{\varrho} \tag{1.3.10}$$

which reduces to proving the inequality $\bar{\varrho} \leq \bar{\varrho}_M$. Let P be $\Sigma_1(\bar{M})$ in $\bar{q}$. By the fact that $h_M(J^{\bar{B}}_{\bar{\varrho}} \cup \{\bar{p}\}) = \bar{M}$ we can find an R which is $\Sigma_1(\bar{M})$ and such that

$$P(z) \longleftrightarrow R(z,x,\bar{p})$$

for some fixed $x \in N$. By (1.3.9), there is an $i \in \omega$ such that

$$P(z) \longleftrightarrow \bar{A}(i,\langle z,x\rangle).$$

Hence if $\gamma < \omega\bar{\varrho}$ then $P \cap \gamma$ is a projection of $\bar{A} \cap \{i\} \times \gamma \times \{x\} \in N \subset M$, thus it is itself in M. This proves (1.3.10).

As an immediate consequence of (1.3.9) and (1.3.10) we get

$$\bar{A} = A^{\bar{p}}_{\bar{M}}. \tag{1.3.11}$$

It only remains to prove that

$$\bar{p} \in R_{\bar{M}}. \tag{1.3.12}$$

If $\bar{M} = N$ this is trivial. Otherwise $\omega\bar{\varrho}$ is closed under the Gödel pairing function, hence there is a $\Sigma_1(M)$ map of $\omega\bar{\varrho}$ onto $J^{\bar{B}}_{\bar{\varrho}}$. Since $h_{\bar{M}}(J^{\bar{B}}_{\bar{\varrho}} \cup \{\bar{p}\}) = \bar{M}$, the proof is complete. □(Lemma 1.3.3)

1.4 Upward Extension of Embeddings Lemma

We now present a method which gives a solution to the problem dual to that from the previous subsection, namely the problem of extending an embedding from the reduct of a J-structure to the whole structure, that is, extending an embedding to a larger *domain*. This problem is a bit more delicate than the previous one, since such an extension need not always exist; therefore we have to strengthen our requirements on the embeddings we intend to extend. The difference between forming downward and upward extensions lies in the fact that downward extensions are related to taking hulls and collapsing them, which is always possible, whilst upward extensions are related to forming ultrapowers (as will become clear later), which have transitive isomorphs only if they are well-founded. The method presented in the next lemma is included mainly for the sake of completeness and motivation. In Chapter 3 we shall develop a more powerful method for building upwards extensions, that does not even require the domains of extended maps to be precisely the reducts of structures.

Definition. $\pi : \bar{M} \to M$ is a *strong embedding* iff

a) $\pi : \bar{M} \underset{\Sigma_1}{\longrightarrow} M$

b) For any $\bar{R}$, R such that $\bar{R}$ is rudimentary over $\bar{M}$ and R is rudimentary over M by the same definition, the following holds:

If $\bar{R}$ is well-founded, then so is R.

The *upward extension of embeddings lemma* is the conjunction of the following lemma together with Lemmata 1.3.1 and 1.3.2.

Lemma 1.4.1. *Let $\pi : \bar{M}^{\bar{p}} \to N$ be a strong embedding, where N is acceptable and $\bar{p} \in R_{\bar{M}}$. Then there are unique M, p such that $N = M^p$ and $p \in R_M$. Moreover, $\tilde{\pi}$ is strong, where $\tilde{\pi} \supset \pi$, $\tilde{\pi} : \bar{M} \underset{\Sigma_1}{\longrightarrow} M$ and $\tilde{\pi}(\bar{p}) = p$.*

Proof. Uniqueness. Suppose $\tilde{\pi}_1 : \bar{M} \to M_1$ and $\tilde{\pi}_2 : \bar{M} \to M_2$ are two extensions of π satisfying the conclusions of the lemma and that p_1, p_2 are the corresponding parameters. Then $A_{M_1}^{p_1} = A = A_{M_2}^{p_2}$ and every $x \in M_k$ is of the form $h_{M_k}(i, \langle \xi, p_k \rangle)$ for some $i \in \omega$ and $\xi \in \mathbf{On} \cap M$ $(k = 1, 2)$. Let $\sigma : M_1 \to M_2$ be the map sending $h_{M_1}(i, \langle \xi, p_1 \rangle)$ to $h_{M_2}(i, \langle \xi, p_2 \rangle)$. Then σ is well defined since

$$(\exists z)(z = h_{M_1}(i, \langle \xi, p_1 \rangle)) \longleftrightarrow A(j, \langle i, \xi \rangle) \longleftrightarrow (\exists z)(z = h_{M_2}(i, \langle \xi, p_2 \rangle)) \quad (1.4.1)$$

for an appropriate j (i.e. j is such that $(\exists z)(z = h_{M_k}(i, \langle \xi, p_k \rangle))$ can be expressed as $M_k \models \varphi_j(\langle i, \xi \rangle)$ for $k = 1, 2$). Also, σ is Σ_1-preserving, since given any Σ_1-formula $\psi(v_1, \ldots, v_\ell)$ and $x_s = h_{M_1}(i_s, \langle \xi_s, p_1 \rangle)$ $(s = 1, \ldots, \ell)$,

$$\begin{aligned} M_1 \models \psi(x_1, \ldots, x_\ell) &\longleftrightarrow M_1 \models \psi(h_{M_1}(i_1, \langle \xi_1, p_1 \rangle), \ldots, h_{M_1}(i_\ell, \langle \xi_\ell, p_1 \rangle)) \\ &\longleftrightarrow A(j, \langle \langle i_1, x_1 \rangle, \ldots \langle i_\ell, \xi_\ell \rangle \rangle) \\ &\longleftrightarrow M_2 \models \psi(h_{M_2}(i_1, \langle \xi_1, p_1 \rangle), \ldots, h_{M_2}(i_\ell, \langle \xi_\ell, p_1 \rangle)) \\ &\longleftrightarrow M_2 \models \psi(\sigma(x_1), \ldots, \sigma(x_\ell)) \end{aligned}$$

for a suitable j (which is, of course, independent of the x_s's). It is then easy to see that σ is structure preserving and $\sigma \circ \tilde{\pi}_1 = \tilde{\pi}_2$. Furthermore, (1.4.1) implies that $\operatorname{rng}(\sigma) = M_2$. Thus, $M_1 = M_2$, $\tilde{\pi}_1 = \tilde{\pi}_2$ and $p_1 = p_2$.

Existence. The idea of the construction is simple: using the fact that $\bar{p} \in R_{\bar{M}}$, we encode the whole structure $\bar{M}$ and its satisfaction relation in a rudimentary fashion over $\bar{M}^{\bar{p}}$. The preservation qualities of π then guarantee that the corresponding relations with the same rudimentary definitions over N encode the required structure M; the process of decoding will also yield the extension $\tilde{\pi}$. However, the verification of all details is somewhat technical.

Suppose $\bar{M} = \langle J_{\bar{\alpha}}^{\bar{B}}, \bar{D} \rangle$. Let $\bar{k}(\langle i, z \rangle) \simeq h_{\bar{M}}(i, \langle z, \bar{p} \rangle)$ and $\bar{d} = \operatorname{dom}(\bar{k})$. Then $\bar{d}$-membership is expressible by a Σ_1-statement over $\bar{M}$ in $\bar{p}$ as

$$x \in \bar{d} \longleftrightarrow (\exists i, z)(i \in \omega \;\&\; x = \langle i, z \rangle \;\&\; (\exists y)(y = h_{\bar{M}}(i, \langle \xi, \bar{p} \rangle))).$$

So there is an $i_0 \in \omega$ such that for every $x \in \bar{M}^{\bar{p}}$ we have $x \in \bar{d}$ iff $A^{\bar{p}}_{\bar{M}}(i_0, x)$. Note that the latter is a rudimentary relation over $\bar{M}^{\bar{p}}$. Similarly, the identity and membership relations as well as the membership of $\bar{B}$ and $\bar{D}$ can be expressed in a Σ_1-fashion over $\bar{M}$ in $\bar{p}$, and therefore in a rudimentary fashion over $\bar{M}^{\bar{p}}$. More precisely, we introduce relations $\bar{I}$, $\bar{E}$, $\bar{B}^*$ and $\bar{D}^*$ over $\bar{d}$ as follows

$$\begin{aligned} x\bar{I}y &\longleftrightarrow \bar{k}(x) = \bar{k}(y) \\ x\bar{E}y &\longleftrightarrow \bar{k}(x) \in \bar{k}(y) \\ \bar{B}^*(x) &\longleftrightarrow \bar{B}(\bar{k}(x)) \\ \bar{D}^*(x) &\longleftrightarrow \bar{D}(\bar{k}(x)) \end{aligned}$$

and set

$$\bar{\mathbb{D}} = \langle \bar{d}, \bar{I}, \bar{E}, \bar{B}^*, \bar{D}^* \rangle.$$

Thus $\bar{\mathbb{D}}$ encodes the structure $\bar{M}$: $\bar{I}$ is a congruence relation on $\bar{\mathbb{D}}$, $\bar{E}$ represents the membership relation and $\bar{k}$ is the Mostowski collapsing isomorphism between $\bar{\mathbb{D}}/\bar{I}$ and $\bar{M}$. We denote the Σ_1-satisfaction relation for $\bar{\mathbb{D}}$ by $\bar{T}$. (The symbol for $=$ is interpreted in $\bar{\mathbb{D}}$ as $\bar{I}$, the symbol for $\in$ as $\bar{E}$, and the symbols for $\bar{B}$, $\bar{D}$ as $\bar{B}^*$ and $\bar{D}^*$, respectively.) More precisely, for $x_1, \ldots, x_\ell \in \bar{d}$ and $i \in \omega$ we have

$$\bar{T}(i, \langle x_1, \ldots, x_\ell \rangle) \longleftrightarrow \bar{\mathbb{D}} \models \varphi_i(x_1, \ldots, x_\ell)$$

where $\langle \varphi_i;\ i \in \omega \rangle$ is our fixed recursive enumeration of Σ_1-formulae (remember that $\langle x_1, \ldots, x_\ell \rangle = \langle x_1, \langle x_2 \ldots, x_\ell \rangle \rangle$ and that we write $\varphi_i(x_1, \ldots, x_\ell)$ instead of $\varphi_i(\langle x_1, \ldots, x_\ell \rangle)$). One can easily show the following equivalence by induction:

$$\bar{T}(i, \langle x_1, \ldots, x_\ell \rangle) \longleftrightarrow \bar{M} \models \varphi_i(\bar{k}(x_1), \ldots, \bar{k}(x_\ell)).$$

For atomic formulae this follows immediately from the definitions of the relations $\bar{I}, \bar{E}, \bar{B}^*$ and $\bar{D}^*$. Let us do the induction step for the formula $\varphi_j(v)$ of the form $(\exists z \in \text{left}(v))\varphi_i(z, v)$ and leave the rest to the reader (recall that $\text{left}(v) = v_1$ and $\text{right}(v) = v_2$ if $v = \langle v_1, v_2 \rangle$ and undefined otherwise). Then

$$\begin{aligned} \bar{\mathbb{D}} \models \varphi_j(x_1, \ldots, x_\ell) &\longleftrightarrow (\exists w \in \bar{d})[w\bar{E}x_1 \;\&\; \bar{\mathbb{D}} \models \varphi_i(w, x_1, x_2, \ldots, x_\ell)] \\ &\longleftrightarrow (\exists w)[\bar{k}(w) \in \bar{k}(x_1) \\ &\qquad\qquad \& \; \bar{M} \models \varphi_i(\bar{k}(w), \bar{k}(x_1), \bar{k}(x_2), \ldots, \bar{k}(x_\ell))] \\ &\longleftrightarrow \bar{M} \models \varphi_j(\bar{k}(x_1), \ldots, \bar{k}(x_\ell)). \end{aligned}$$

The middle equivalence follows by the induction hypothesis, the last one by the fact that if there is a z witnessing the bottom formula, then such a z is always of the form $\bar{k}(w)$ for some w.

Let d, I, E, B^* and D^* be rudimentary over N by the same rudimentary definitions as their counterparts over $\bar{M}^{\bar{p}}$. It follows from the above that $\bar{T}$ is $\Sigma_1(\bar{M})$ in $\bar{p}$ and therefore rudimentary over $\bar{M}^{\bar{p}}$. Let

$$\mathbb{D} = \langle d, I, E, B^*, D^* \rangle$$

and T be a relation which is rudimentary over N by the same rudimentary definition as $\bar{T}$. We show that T is the Σ_1-satisfaction predicate for $\mathbb{D}$. Strictly speaking, we must show that the following equivalences hold, where $\varphi_i(v)$ has the form indicated on the left hand side:

$$\begin{array}{ll}
\mathrm{left}(v) = \mathrm{right}(v) & T(i, \langle x, y\rangle) \longleftrightarrow xIy \\
\mathrm{left}(v) \in \mathrm{right}(v) & T(i, \langle x, y\rangle) \longleftrightarrow xEy \\
B(v) & T(i, x) \longleftrightarrow B^*(x) \\
D(v) & T(i, x) \longleftrightarrow D^*(x) \\
 & \\
\varphi_{i_1}(v)\ \&\ \varphi_{i_2}(v) & T(i, \vec{x}) \longleftrightarrow T(i_1, \vec{x})\ \&\ T(i_2, \vec{x}) \\
\neg\varphi_j(v) & T(i, \vec{x}) \longleftrightarrow \neg T(j, \vec{x}) \\
 & \\
(\exists z \in \mathrm{left}(v))\varphi_j(z, v) & T(i, \langle x, \vec{y}\rangle) \longleftrightarrow (\exists z \in d)(zEx\ \&\ T(j, \langle z, x, \vec{y}\rangle)) \\
(\forall z \in \mathrm{left}(v))\varphi_j(z, v) & T(i, \langle x, \vec{y}\rangle) \longleftrightarrow (\forall z \in d)(zEx\ \&\ T(j, \langle z, x, \vec{y}\rangle)) \\
 & \\
(\exists z)\varphi_j(z, v) & T(i, \vec{x}) \longleftrightarrow (\exists z \in d)T(j, \langle z, \vec{x}\rangle).
\end{array}$$

Here $\vec{x}$ stands for $\langle x_1, \ldots, x_\ell\rangle$. We again proceed by induction on formulae. Suppose first that φ_i is an atomic formula, say the formula $\mathrm{left}(v) = \mathrm{right}(v)$. Then

$$(\forall x, y \in \bar{d})(\bar{T}(i, \langle x, y\rangle) \longleftrightarrow x\bar{I}y).$$

This is a Π_1-statement over $\bar{M}^{\bar{p}}$, since the predicates $\bar{d}$, $\bar{T}$ and $\bar{I}$ are rudimentary. Since π is Σ_1-preserving, $T(i, \langle x, y\rangle)$ iff xIy iff $\mathbb{D} \models \varphi_i(x, y)$ for all $x, y \in d$.

Now suppose $\varphi_i(v)$ is the formula $\varphi_{i_1}(v)\ \&\ \varphi_{i_2}(v)$. Then

$$(\forall x_1, \ldots, x_\ell \in \bar{d})[\bar{T}(i, \langle x_1, \ldots, x_\ell\rangle) \longleftrightarrow \bar{T}(i_1, \langle x_1, \ldots, x_\ell\rangle)\ \&\ \bar{T}(i_2, \langle x_1, \ldots, x_\ell\rangle)].$$

This is again a Π_1-statement over $\bar{M}^{\bar{p}}$, so we obtain

$$\begin{aligned}
T(i, \langle x_1, \ldots, x_\ell\rangle) &\longleftrightarrow T(i_1, \langle x_1, \ldots, x_\ell\rangle)\ \&\ T(i_2, \langle x_1, \ldots, x_\ell\rangle) \\
&\longleftrightarrow \mathbb{D} \models \varphi_{i_1}(x_1, \ldots, x_\ell)\ \&\ \varphi_{i_2}(x_1, \ldots, x_\ell) \\
&\longleftrightarrow \mathbb{D} \models \varphi_i(x_1, \ldots, x_\ell);
\end{aligned}$$

the second equivalence follows from the induction hypothesis. We proceed similarly if $\varphi_i(v)$ is of the form $\neg\varphi_j(v)$.

Finally, suppose that $\varphi_i(v)$ introduces a quantifier; say $\varphi_i(v)$ is of the form $(\exists z \in \mathrm{left}(v))\varphi_j(z, v)$. The implication $(\longleftarrow)$ follows easily: Since $\bar{T}$ is a satisfaction relation for $\bar{\mathbb{D}}$, we have

$$(\forall x, y)[(\exists z)(z\bar{E}x\ \&\ \bar{T}(j, \langle z, x, y\rangle)) \longrightarrow \bar{T}(i, \langle x, y\rangle)].$$

This is a Π_1-statement over $\bar{M}^{\bar{p}}$ and is therefore preserved upwards by π. To see the converse, let $\bar{g}$ be a $\Sigma_1(\bar{M})$-function in $\bar{p}$ uniformizing the relation

$$z \in \bar{k}(x)\ \&\ \varphi_j(z, \bar{k}(x), \bar{k}(y)).$$

Then $\bar{g}(u) \simeq h_{\bar{M}}(m, \langle u, \bar{p}\rangle) \simeq \bar{k}(\langle m, u\rangle)$ for an appropriate $m \in \omega$, hence

$$\varphi_i(\bar{k}(x), \bar{k}(y)) \longrightarrow \langle m, x, y\rangle \in \mathrm{dom}(\bar{k}) \ \&\ \bar{k}(\langle m, x, y\rangle) \in \bar{k}(x) \ \& \\ \&\ \varphi_j(\bar{k}(\langle m, x, y\rangle), \bar{k}(x), \bar{k}(y))$$

holds in $\bar{\mathbb{D}}$ for all x, y in $\bar{M}^{\bar{p}}$. Translating this into the language of $\bar{\mathbb{D}}$ we obtain

$$(\forall x, y)[\bar{T}(i, \langle x, y\rangle) \longrightarrow \langle m, x, y\rangle \in \bar{d} \ \&\ \langle m, x, y\rangle \bar{E} x \ \&\ \bar{T}(j, \langle\langle m, x, y\rangle, x, y\rangle)],$$

which is again a Π_1-statement over $\bar{M}^{\bar{p}}$. The required implication for $\mathbb{D}$ then follows immediately.

One consequence of the fact that T is a satisfaction relation for $\mathbb{D}$ is:

$$\pi : \bar{\mathbb{D}} \underset{\Sigma_2}{\longrightarrow} \mathbb{D},$$

as follows immediately by the fact that $\bar{T}, T$ are rudimentary over $\bar{M}^{\bar{p}}, N$ respectively by the same rudimentary definition. This implies that I is a congruence relation on $\mathbb{D}$ and E is extensional modulo this congruence relation; the map π simply carries both properties from $\bar{\mathbb{D}}$ over to $\mathbb{D}$ (extensionality being Π_2). Note also that $\bar{E}$ is well-founded modulo the congruence relation $\bar{I}$ (in other words, the relation $x\bar{E}y \ \&\ \neg(x\bar{I}y)$ is well-founded). Hence E is well-founded modulo the congruence relation I by the strongness of π.

Let M be the transitive collapse of $\mathbb{D}$ and k be the collapsing map. Define $\tilde{\pi} : \bar{M} \to M$ by

$$\tilde{\pi}(\bar{k}(x)) = k(\pi(x)) \quad \text{for all } x \in \bar{d}.$$

It follows immediately that

$$\tilde{\pi} : \bar{M} \underset{\Sigma_2}{\longrightarrow} M.$$

In the following we show that M has all the required properties. Note first that M is a J-structure, say $M = \langle J_\alpha^B, D\rangle$. For the rest of the proof fix $i, i^* \in \omega$ so that

$$x = h_Q(i, \langle x, q\rangle) \quad \text{holds uniformly over any } J\text{-structure } Q$$
$$\bar{p} = h_{\bar{M}}(i^*, \langle 0, \bar{p}\rangle).$$

The former can be arranged if we choose i so that $\varphi_i(v)$ is the formula $\mathrm{left}(v) = \mathrm{left}(\mathrm{right}(v))$; the latter by taking $\varphi_{i^*}(v)$ to be $\mathrm{left}(v) = \mathrm{right}(\mathrm{right}(v))$ and using the fact that $h_{\bar{M}}$ is the corresponding uniformizing function. Then

$$x = \bar{k}(\langle i, x\rangle) \text{ for all } x \in \bar{M}^{\bar{p}} \text{ and } \bar{p} = \bar{k}(\langle i^*, 0\rangle).$$

We first observe that $|N| \subset M$. Given any $x \in \bar{M}^{\bar{p}}$, $\bar{k}(\langle j, z\rangle) \in x = \bar{k}(\langle i, x\rangle)$ iff $\bar{k}(\langle j, z\rangle) = w = \bar{k}(\langle i, w\rangle)$ for some $w \in x$; in other words,

$$\langle j, z\rangle \in \bar{d} \longrightarrow [\langle j, z\rangle \bar{E} \langle i, x\rangle \longleftrightarrow (\exists w \in x)(\langle j, z\rangle \bar{I} \langle i, w\rangle)]$$

holds for all $x, y \in \bar{M}^{\bar{p}}$ and $j \in \omega$. Quantifying over $\bar{M}^{\bar{p}}$ yields a Π_1-statement, which is preserved under π. This allows us to conclude: Whenever $x \in N$ (note: $\langle i, x\rangle \in d$ by the fact that π is Σ_1-preserving and the "same" holds for $\bar{M}^{\bar{p}}$),

$$k(\langle i, x\rangle) = \{k(\langle i, w\rangle);\ w \in x\}.$$

It is now straightforward to verify by $\in$-induction that $k(\langle i, x\rangle) = x$ for all $x \in N$. Furthermore, given any $x \in \bar{M}^{\bar{p}}$,

$$\tilde{\pi}(x) = \tilde{\pi}(\bar{k}(\langle i, x\rangle)) = k(\pi(\langle i, x\rangle)) = k(\langle i, \pi(x)\rangle) = \pi(x),$$

so $\tilde{\pi} \supset \pi$.

Let B', A and ϱ be such that $N = \langle J_\varrho^{B'}, A\rangle$ (recall that, by our assumption, N is an acceptable structure). We prove that A is the Σ_1-satisfaction relation of M restricted to N. Set $p = k(\langle i^*, 0\rangle)$. It is easy to check that $\tilde{\pi}(\bar{p}) = p$. Given an $x \in \bar{M}^{\bar{p}}$ and $j \in \omega$, we have $A_{\bar{M}}^{\bar{p}}(j, x) \longleftrightarrow \bar{M} \models \varphi_j(x, \bar{p}) \longleftrightarrow \bar{M} \models \varphi_j(\bar{k}(\langle i, x\rangle), \bar{k}(\langle i^*, 0\rangle)) \longleftrightarrow \bar{T}(j, \langle\langle i, x\rangle, \langle i^*, 0\rangle\rangle)$. Hence, the Π_1-statement

$$(\forall x)(\forall j \in \omega)[A_{\bar{M}}^{\bar{p}}(j, x) \longleftrightarrow \bar{T}(j, \langle\langle i, x\rangle, \langle i^*, 0\rangle\rangle)]$$

is preserved by π, which means that

$$A(j, x) \longleftrightarrow T(j, \langle\langle i, x\rangle, \langle i^*, 0\rangle\rangle) \longleftrightarrow M \models \varphi_j(x, p) \tag{1.4.2}$$

for every $x \in N$ and $j \in \omega$.

This equivalence easily yields that $B' = B \cap N$ and $N = \langle J_\varrho^B, A\rangle$ for ϱ as above. Pick a j such that $\varphi_j(u, v)$ is the formula "$B^*(u)$" for any acceptable structure of the form $\langle J_{\alpha^*}^{B^*}, D^*\rangle$. Since $\bar{B}(x) \longleftrightarrow A_{\bar{M}}^{\bar{p}}(j, x)$ holds for every $x \in \bar{M}^{\bar{p}}$, the preservation properties of π guarantee that $A(j, x) \longleftrightarrow B'(x)$. However, $A(j, x)$ is equivalent to $B(x)$ for all $x \in N$, as follows from (1.4.2).

To see that $A = A_M^p$ it suffices to show that $\varrho = \varrho_M$, as $|N| = H_\varrho^M$ then follows immediately (recall that $\omega\varrho = \mathbf{On} \cap N$). The computation below will also yield that $p \in R_M$. Notice that

$$\bar{k}(\langle j, x\rangle) = h_{\bar{M}}(j, \langle x, \bar{p}\rangle) = h_{\bar{M}}(\bar{k}(\langle i, j\rangle), \langle \bar{k}(\langle i, x\rangle), \bar{k}(\langle i^*, 0\rangle)\rangle),$$

where the left equality is simply the definition of $\bar{k}$. Leaving out the middle term we obtain a Σ_1-statement, so it can be represented by some $m \in \omega$. More precisely, $\bar{T}(m, \langle\langle j, x\rangle, \langle i, j\rangle, \langle i, x\rangle, \langle i^*, 0\rangle\rangle)$ holds for all $x \in \bar{M}^{\bar{p}}$ and $j \in \omega$. As above we apply π to obtain the corresponding statement for T and all $x \in N$. Using the fact that the Σ_1-Skolem functions have uniform definitions we infer that $k(\langle j, x\rangle) = h_M(k(\langle i, j\rangle), \langle k(\langle i, x\rangle), k(\langle i^*, 0\rangle)\rangle) = h_M(j, \langle x, p\rangle)$. Since the values $k(\langle j, x\rangle)$ range over all of M, we have (note that there is a *lightface* $\Sigma_1(M)$ map from $\omega\varrho$ onto $|N|$)

$$M = h_M(\omega\varrho \cup \{p\}) \quad \text{and} \quad \omega\varrho_M \leq \omega\varrho.$$

The latter is obviously a consequence of the former. On the other hand, given any $r \in M$ we can pick a $\xi \in \mathbf{On} \cap N$ such that $r = h_M(j, \langle \xi, p \rangle)$ for some j. For any Σ_1 formula ψ, $M \models \psi(\eta, r)$ iff $M \models \varphi_m(\eta, \xi, p)$ for a suitable m; the latter is equivalent to $A(\langle \eta, \xi \rangle)$ by (1.4.2). Taken together, $M \models \psi(\eta, r)$ can be expressed in a rudimentary fashion over N. Since N is amenable, every bounded $\mathbf{\Sigma}_1(M)$ subset of $\omega\varrho$ is an element of N, which means that $\omega\varrho \leq \omega\varrho_M$. Thus,

$$\omega\varrho = \omega\varrho_M \quad \text{and} \quad p \in R_M.$$

It only remains to show that $\tilde{\pi}$ is strong. Let $\bar{R}$, R be binary relations which are rudimentary over $\bar{M}$, M respectively by the same rudimentary definition. Define $\bar{R}^*$, R^* as follows

$$x\bar{R}^*y \longleftrightarrow x, y \in \bar{d} \;\&\; \bar{k}(x)\bar{R}\bar{k}(y)$$
$$xR^*y \longleftrightarrow x, y \in d \;\&\; k(x)Rk(y).$$

Then $\bar{R}^*$ is well-founded since $\bar{R}$ is – any decreasing chain $x_0, x_1, \ldots, x_n, \ldots$ in $\bar{R}^*$ yields a decreasing chain $\bar{k}(x_0), \bar{k}(x_1), \ldots, \bar{k}(x_n), \ldots$ in $\bar{R}$, hence no such chain can be infinite. Furthermore, $\bar{R}^*$, R^* are rudimentary over $\bar{M}^{\bar{p}}$, N respectively by the same rudimentary definition. As π is strong, R^* must be well-founded. Hence R must be well-founded as well, since every infinite decreasing chain $x_0, x_1, \ldots, x_n, \ldots$ in R is of the form $k(z_0), k(z_1), \ldots, k(z_n), \ldots$ for some $z_0, z_1, \ldots, z_n, \ldots$ and the latter would be an infinite decreasing chain in R^*.

1.5 Iterated Projecta

In the following we shall show how to iterate the process of defining a projectum and forming a standard code and introduce the notion of n-th projectum, n-th standard code and n-th reduct.

Definition. Let $M = \langle J_\beta^B, D \rangle$ be an acceptable J-structure. For $n \in \omega$ we inductively define the *n-th projectum* ϱ_M^n, the *n-th standard code* $A_M^{n,p}$ and the *n-th reduct* $M^{n,p}$ for $p \in \bigtimes_{i \in n} H^M_{\omega\varrho_M^i}$ as follows:

$$\varrho_M^0 = \beta, \qquad A_M^{0,\emptyset} = \emptyset, \qquad M^{0,\emptyset} = M,$$
$$\varrho_M^{n+1} = \min\{\varrho_{M^{n,p}};\ p \in \bigtimes_{i \in n} H^M_{\omega\varrho_M^i}\},$$
$$A^{n+1,p} = A^{p(n),\delta}_{M^{n,p\restriction n}}, \qquad M^{n+1,p} = (M^{n,p\restriction n})^{p(n),\delta} \quad \text{where } \delta = \varrho_M^{n+1}.$$

We also set $H_M^n = H^M_{\omega\varrho_M^n}$ and $\Gamma_M^n = \bigtimes_{i \in n} H_M^i$.

Definition. We define $P^n_M, R^n_M \subset \Gamma^n_M$ as follows:

$$P^0_M = \emptyset$$
$$P^{n+1}_M = \{p \in \Gamma^{n+1}_M;\ p \upharpoonright n \in P^n_M \ \&\ \varrho_{M^{n,p\upharpoonright n}} = \varrho^{n+1}_M \ \&\ p(n) \in P_{M^{n,p\upharpoonright n}}\}$$

R^n_M is defined in the same way but with R^n_M in place of P^n_M.

As before, we call the elements of P^n_M *good parameters* and the elements of R^n_M *very good parameters*. We shall again drop the subscript $_M$ whenever possible.

Remark. For M as above we have

a) $p \in R^n_M \longleftrightarrow p \in P^n_M \ \&\ (\forall i \in n)(p(i) \in R_{M^{i,p\upharpoonright i}})$

b) $R^n_M \subset P^n_M \neq \emptyset$

c) $p \in P^n_M \longrightarrow (\forall i \in n)(p(i) \in P_{M^{i,p\upharpoonright i}})$ and similarly for R^n_M. Moreover, if $p \upharpoonright (n-1) \in R^{n-1}_M$, then the equivalence holds.

Definition. M is *n-sound* iff $R^n_M = P^n_M$. M is *sound* iff M is n-sound for all $n \in \omega$.

We can now formulate a downward extension of embeddings lemma as the conjunction of the following three lemmas which, in turn, are immediate consequences of the corresponding lemmas for the first projectum.

Lemma 1.5.1. *Let $\bar{M}$, M be acceptable and $\pi : \bar{M}^{n,\bar{p}} \underset{\Sigma_0}{\longrightarrow} M^{n,p}$, where $\bar{p} \in R^n_{\bar{M}}$. Then there is a unique map $\tilde{\pi} \supset \pi$ such that $\mathrm{dom}(\tilde{\pi}) = \bar{M}$, $\tilde{\pi}(\bar{p}) = p$ and, setting $\tilde{\pi}_i = \tilde{\pi} \upharpoonright H^i_{\bar{M}}$,*

$$\tilde{\pi}_i : \bar{M}^{i,\bar{p}\upharpoonright i} \underset{\Sigma_0}{\longrightarrow} M^{i,p\upharpoonright i} \quad \text{for } i \leq n.$$

The map $\tilde{\pi}_i$ in fact Σ_1-preserving for $i \in n$.

Lemma 1.5.2. *Suppose that $\bar{M}$, M, $\bar{p}$, p, π, $\tilde{\pi}$ are as above and $p \in R^n_M$. Let $\pi : \bar{M}^{n,\bar{p}} \underset{\Sigma_\ell}{\longrightarrow} M^{n,p}$ where $\ell \in \omega$. Then*

$$\tilde{\pi}_i : \bar{M}^{i,\bar{p}\upharpoonright i} \underset{\Sigma_{\ell+n-i}}{\longrightarrow} M^{i,p\upharpoonright i} \quad \text{for } i \leq n.$$

Hence, $\tilde{\pi}_0 : \bar{M} \underset{\Sigma_{\ell+n}}{\longrightarrow} M$.

Lemma 1.5.3. *Let $\pi : N \underset{\Sigma_0}{\longrightarrow} M^{n,p}$, where M is as above. Then there are unique $\bar{M}$, $\bar{p}$ such that $\bar{p} \in R^n_{\bar{M}}$ and $N = \bar{M}^{n,\bar{p}}$.*

The upward extension of embeddings lemma is the conjunction of the following lemma together with Lemmata 1.5.1 and 1.5.2.

Lemma 1.5.4. *Let $\pi : \bar{M}^{n,\bar{p}} \to N$ be strong, where $\bar{M}$ is an acceptable J-structure and $\bar{p} \in R^n_{\bar{M}}$. Then there are unique M, p such that M is acceptable, $p \in R^n_M$ and $M^{n,p} = N$. Moreover, if $\tilde{\pi}$ is as in Lemma 1.5.1, then $\tilde{\pi}$ is strong.*

If we take $\pi : \bar{M}^{n,\bar{p}} \to M^{n,p}$ as in Lemma 1.5.1 and form the corresponding extension $\tilde{\pi}$, it is easy to see that for every $\bar{q} \in \Gamma^n_{\bar{M}}$ and $q = \tilde{\pi}(\bar{q})$,

$$\tilde{\pi}_i : \bar{M}^{i,\bar{q}\restriction i} \underset{\Sigma_1}{\longrightarrow} M^{i,q\restriction i} \quad \text{for } i \in n. \tag{1.5.1}$$

This suggests the general notion of a $\Sigma_\ell^{(n)}$-preserving embedding, where n indicates preservation at the n-th level, i.e. if $\tilde{\pi}(\bar{q}) = q$ then (1.5.1) and

$$\tilde{\pi}_n : \bar{M}^{n,\bar{q}} \underset{\Sigma_\ell}{\longrightarrow} M^{n,q}. \tag{1.5.2}$$

It turns out that there is a canonical class of formulae, the so called $\Sigma_\ell^{(n)}$-formulae, such that the above embeddings are exactly those which are elementary with respect to this class.

1.6 Σ^*-Relations

The idea of a $\Sigma_\ell^{(n)}$ relation is to capture definability over the n-th reduct *without* actually talking about the reduct. We have seen in the previous section that a Σ_ℓ relation over the first reduct of M is Σ_ℓ in a $\Sigma_1(M)$-relation, namely the first standard code. This leads to a hierarchy of formulae which captures the notion "Σ_ℓ in", where $\Sigma_1^{(n+1)}$ is inductively defined to be "Σ_1 in $\Sigma_1^{(n)}$". Such hierarchy is more convenient for the purposes of the fine structure theory than the Lévy hierarchy and it is not surprising that the two hierarchies give rise to different hierarchies of definable relations.

Definition. $\mathcal{L}^*$ is a language with predicate symbols $\in$, $=$, B, D (or, more generally, $B_1, \ldots, B_m, D_1, \ldots, D_m$) and variables v^i of type $i \in \omega$. (Recall that, by a slight abuse of notation, we use B to denote both the predicate and the corresponding predicate symbol; see the beginning of section 1.1.) $\mathcal{L}^*$ is interpreted in M in the obvious way, letting v^i range over H^i_M ($i \in \omega$). (Hence the larger the index, the smaller the domain.)

Given $k, n \in \omega$, the set of formulae $\Sigma_k^{(n)}$ is defined by induction on n as follows.

$\Sigma_0^{(n)}$ is the smallest set Σ of formulae satisfying

a) $x^i \in y^j$, $x^i = y^j$, $B(x^i)$, $D(x^i)$ are in Σ,

b) $\Sigma_1^{(m)} \subset \Sigma$ for $m < n$,

c) Σ is closed under $\&, \vee, \to, \leftrightarrow, \neg$,

d) if φ is in Σ then so are $(\forall x^n \in x^i)\varphi$, $(\exists x^n \in x^i)\varphi$ for $i \geq n$.

For $k > 0$ we then set

$\Sigma_k^{(n)}$ is the set of all formulae of the form $(\exists x_1^n)(\forall x_2^n)\dots(Qx_k^n)\varphi$
where $\varphi \in \Sigma_0^{(n)}$ and Q is $\exists$ or $\forall$ depending on whether k is odd or even.

$\Pi_k^{(n)}$ is the set of all formulae of the form $(\forall x_1^n)(\exists x_2^n)\dots(Qx_k^n)\varphi$
where $\varphi \in \Sigma_0^{(n)}$ and Q is $\exists$ or $\forall$ depending on whether k is even or odd.

We also set: $\Pi_0^{(m)} = \Sigma_0^{(m)}$

$$\Sigma^* = \Sigma^{(\omega)} = \bigcup_{n\in\omega} \Sigma_0^{(n)}.$$

When dealing with $\mathcal{L}^*$-model theory of M we informally use x^i as a variable for elements of H_M^i. Using this notation we define:

Definition. $R(x_1^{i_1}, \dots, x_\ell^{i_\ell})$ is a $\Sigma_k^{(n)}(M)$ *relation of type* $\langle i_1, \dots, i_\ell\rangle$ (in the parameters $q_1^{j_1}, \dots, q_m^{j_m}$) iff R is defined over M by a $\Sigma_k^{(n)}$-formula $\varphi(v_1^{i_1}, \dots, v_\ell^{i_\ell})$ (in the parameters $q_1^{j_1}, \dots, q_m^{j_m}$).

R is $\mathbf{\Sigma}_k^{(n)}(M)$ iff R is $\Sigma_k^{(n)}(M)$ in some parameters. Similarly we define $\Pi_k^{(n)}(M)$ and $\mathbf{\Pi}_k^{(n)}(M)$.

Finally we set $\Sigma^*(M) = \Sigma^{(\omega)}(M) = \bigcup_{n\in\omega} \Sigma_0^{(n)}(M)$ and similarly for $\mathbf{\Sigma}^*(M)$.

We shall often write $\Sigma_k^{(n)}$ instead of $\Sigma_k^{(n)}(M)$ if there is no danger of confusion.

Definition. We say that a relation f is a $\Sigma_k^{(n)}(M)$*-function to* H_M^i *of argument type* $\langle i_1, \dots, i_\ell\rangle$ iff $y^i = f(x_1^{i_1}, \dots, x_\ell^{i_\ell})$ is a $\Sigma_k^{(n)}(M)$-relation of type $\langle i, i_1, \dots, i_\ell\rangle$.

$\mathbf{\Sigma}_k^{(n)}(M)$-functions are then defined in a straightforward way.

A relation in the above sense is not characterized by its graph. It could be identified by a pair consisting of its graph and the argument type.

Lemma 1.6.1. *Let $n, \ell < \omega$. If $R(x^h, \vec{x})$ is $\Sigma_\ell^{(n)}$ then so is $R(x^k, \vec{x})$ for $k \geq h$.*

Proof. Immediate. □(Lemma 1.6.1)

Definition. $R(x^k, \vec{x})$ from the above lemma is called a *specialization* of R.

Lemma 1.6.2. *Let $n, \ell < \omega$. If $R(x^k, \vec{x})$ is $\Sigma_\ell^{(n)}$ and $k \geq h \geq n$, then R is a specialization of a relation $R(x^h, \vec{x})$, which is $\Sigma_\ell^{(n)}$.*

Proof. Immediate. □(Lemma 1.6.2)

Hence, every $\Sigma_\ell^{(n)}$-relation can be obtained by specialization from a $\Sigma_\ell^{(n)}$-relation whose arguments are of type $\leq n$. It is also apparent that $\Sigma_\ell^{(n)}$-relations are closed under permutation of arguments and insertion of dummy variables. Another consequence of the definition is:

Lemma 1.6.3. *Let $n, \ell < \omega$. $R(\vec{x}^{\,n+1}, \dots, \vec{x}^{\,0})$ is $\Sigma_\ell^{(n+1)}(M)$ iff the relation $R_{\vec{x}} = \{\langle \vec{x}^{\,n+1}\rangle;\ R(\vec{x}^{\,n+1}, \vec{x})\}$ is uniformly $\Sigma_\ell(\langle H_M^{n+1}, Q_{\vec{x}}^1, \dots, Q_{\vec{x}}^m\rangle)$ (here $\vec{x}$ stays for $\vec{x}^{\,n}, \dots, \vec{x}^{\,0}$), where each $Q_{\vec{x}}^i$ $(i = 1, \dots, m)$ has the form*

$$Q_{\vec{x}}^i = \{\langle \vec{z}^{\,n+1}\rangle;\ Q^i(\vec{z}^{\,n+1}, \vec{x})\}$$

and $Q^i(\vec{z}^{\,n+1}, \vec{x})$ is $\Sigma_1^{(n)}(M)$. (Here $\langle \vec{x}^{\,n+1}\rangle$ is the tuple $\langle x_1^{n+1}, \dots, x_h^{n+1}\rangle$ if $\vec{x}^{\,n+1}$ is the list $x_1^{n+1}, \dots x_h^{n+1}$.)

Proof. $(\rightarrow)$ If $\varphi(\vec{x}^{\,n+1}, \dots, \vec{x}^{\,0})$ is the defining formula for $R(\vec{x}^{\,n+1}, \dots, \vec{x}^{\,0})$ we can rewrite this formula into a prenex form

$$\begin{aligned}(\exists v_1^{n+1})(\forall v_2^{n+1}) \dots (\exists w_1^{n+1} \in u_{i_1}^{n+1})(\forall w_2^{n+1} \in u_{i_2}^{n+1}) \dots \\ \dots \psi(\vec{v}^{\,n+1}, \vec{w}^{\,n+1}, \vec{x}^{\,n+1}, \vec{x})\end{aligned} \tag{1.6.1}$$

where each $u_{i_j}^{n+1}$ is either some v_h^{n+1} or some x_h^{n+1} and ψ is a propositional composition of $\Sigma_1^{(n)}$-formulae. Let $Q^i(\vec{z}^{\,n+1}, \vec{x})$, $i = 1, \dots, m$, be the relations defined over M by these formulae and $Q(\vec{z}^{\,n+1}, \vec{x})$ be the propositional combination of the Q^i's corresponding to the structure of ψ. Then $R_{\vec{x}}$ can be expressed as

$$(\exists v_1)(\forall v_2) \dots (\exists w_1 \in u_1)(\forall w_2 \in u_2) \dots Q_{\vec{x}}(\vec{v}, \vec{w}, \vec{y}) \tag{1.6.2}$$

(where each u_i is either some v_h or some y_h), which is the desired form.

$(\leftarrow)$ If $R_{\vec{x}}$ is $\Sigma_1(\langle H_M^{n+1}, Q_{\vec{x}}^1, \dots, Q_{\vec{x}}^m\rangle)$, we can write its definition as in (1.6.2), hence $Q_{\vec{x}}$ is a propositional combination of $Q_{\vec{x}}^i$, $i = 1, \dots, m$. If we now replace $Q_{\vec{x}}^i$ by Q^i, $(\vec{v}, \vec{w})$ by $(\vec{v}^{\,n+1}, \vec{w}^{\,n+1})$ and $\vec{y}$ by $\vec{x}^{\,n+1}$, we get the form (1.6.1), which says that $R(\vec{x}^{\,n+1}, \dots, \vec{x}^{\,0})$ is $\Sigma_\ell^{(n+1)}$. □(Lemma 1.6.3)

Remark. The reduction in Lemma 1.6.3 is clearly uniform for all M; the $\Sigma_1^{(n)}$ definition of Q^i depends only on the $\Sigma_\ell^{(n+1)}$ definition of R.

Using Lemma 1.6.3, we get the following lemma by a straightforward induction.

Lemma 1.6.4. *Let $n \in \omega$, $1 \le \ell \in \omega$. Then*

$$R(\vec{x}^{\,n}, \dots, \vec{x}^{\,0}) \text{ is } \Sigma_\ell^{(n)}(M) \text{ iff } R_{\vec{x}} \text{ is uniformly } \Sigma_\ell(M^{n,p(\vec{x})})$$

where $R_{\vec{x}} = \{\langle \vec{x}^{\,n}\rangle;\ R(\vec{x}^{\,n}, \vec{x})\}$ and $p(\vec{x}) = \langle\langle \vec{x}^{\,0}\rangle, \dots, \langle \vec{x}^{\,n-1}\rangle\rangle$.

Lemma 1.6.5. *Let $n \in \omega$. If $R(\vec{x}^{\,n}, \dots, \vec{x}^{\,0})$ is $\Sigma_0^{(n)}(M)$, then $R_{\vec{x}}$ is uniformly rudimentary in $M^{n,p(\vec{x})}$. If $R_{\vec{x}}$ is $\Sigma_0(M^{n,p(\vec{x})})$, then $R(\vec{x}^{\,n}, \dots, \vec{x}^{\,0})$ is $\Sigma_0^{(n)}(M)$.*

Proof of Lemma 1.6.4. ($\rightarrow$) Let $R(\vec{x}^{\,n+1},\dots,\vec{x}^{\,0})$ be $\Sigma_1^{(n+1)}(M)$. As follows from Lemma 1.6.3, there are $Q^1(\vec{z}^{\,n+1},\vec{x}),\dots,Q^m(\vec{z}^{\,n+1},\vec{x})$ which are $\Sigma_1^{(n)}(M)$ such that $R_{\vec{x}}$ is $\Sigma_1(\langle H_M^{n+1}, Q_{\vec{x}}^1,\dots,Q_{\vec{x}}^m\rangle)$. Each Q^i ($i=1,\dots,m$) is a specialization of a $\tilde{Q}^i(\vec{z}^{\,n},\vec{x})$, hence by the induction hypothesis, $\tilde{Q}^i_{\vec{x}\restriction n}$ is $\Sigma_1(M^{n,p(\vec{x})\restriction n})$ and the same then obviously holds for $Q^i_{\vec{x}\restriction n}$. By the definition of standard code, $Q^i_{\vec{x}}$ is uniformly rudimentary $M^{n+1,p(\vec{x})}$, hence $R_{\vec{x}}$ is uniformly $\Sigma_1(\langle H_M^{n+1}, A_M^{n+1,p(\vec{x})}\rangle)$ and we are done.

($\leftarrow$) We first state a useful claim.

Claim. There is $A^*(\vec{x}^{\,n+1},\dots,\vec{x}^{\,0})$ which is $\Sigma_1^{(n)}(M)$ such that

$$\langle\vec{x}^{\,n+1}\rangle \in A_M^{n+1,p(\vec{x})} \longleftrightarrow A^*(\vec{x}^{\,n+1},\dots,\vec{x}^{\,0}), \tag{1.6.3}$$

i.e. $A_M^{n+1,p(\vec{x})} = A^*_{\vec{x}}$. The definition of $A^*(\vec{x}^{\,n+1},\dots,\vec{x}^{\,0})$ is uniform for all J-structures.

Proof. By induction on n. Note first that $A_M^{n+1,p(\vec{x})} = A^{\langle x^n\rangle}_{M^{n,p(\vec{x})\restriction n}}$. By the induction hypothesis, there is a $\Sigma_1^{(n-1)}(M)$ relation $\tilde{A}(\vec{y}^{\,n},\dots,\vec{y}^{\,0})$ such that

$$\vec{y}^{\,n} \in A_M^{n,p(\vec{y})} \longleftrightarrow \tilde{A}(\vec{y}^{\,n},\dots,\vec{y}^{\,0})$$

(i.e. $A_M^{n,p(\vec{y})} = \tilde{A}_{\vec{y}}$). Now $A_M^{n+1,p(\vec{x})}$ is $\Sigma_1(M^{n,p(\vec{x})\restriction n})$ in $\langle\vec{x}^{\,n}\rangle$, i.e. $\Sigma_1(\langle H_M^n, \bar{A}_{\vec{x}}\rangle)$, where $\bar{A}(\vec{x}^{\,n+1},\dots,\vec{x}^{\,0})$ is an obvious specialization of $\tilde{A}$. Thus $\bar{A}$ is $\Sigma_1^{(n-1)}(M)$. Now define A^* as in (1.6.3). By Lemma 1.6.3, A^* is $\Sigma_1^{(n)}$. □(Claim)

Now suppose $R_{\vec{x}}$ is $\Sigma_1(M^{n+1,p(\vec{x})})$. This means $R_{\vec{x}}$ is $\Sigma_1(\langle H_M^{n+1}, A^*_{\vec{x}}\rangle)$, where A^* is as in the above claim. By Lemma 1.6.3, R is $\Sigma_1^{(n+1)}(M)$. □(Lemma 1.6.4)

The proof of Lemma 1.6.5 is – up to some minor details – the same.

1.7 $\Sigma_\ell^{(n)}$-Embeddings

Definition. Let $\bar{M}, M$ be acceptable J-structures. $\pi : \bar{M} \underset{\Sigma_\ell^{(n)}}{\longrightarrow} M$ iff

a) $\pi : \bar{M} \to M$;

b) if $\varphi(v^{j_1},\dots,v^{j_m})$ is a $\Sigma_\ell^{(n)}$-formula such that $j_1,\dots,j_m \le n$ and $x_i \in H_{\bar{M}}^{j_i}$ for $i=1,\dots,m$, then $\pi(x_i) \in H_M^{j_i}$ and

$$\bar{M} \models \varphi(x_1,\dots,x_m) \longleftrightarrow M \models \varphi(\pi(x_1),\dots,\pi(x_m)).$$

Remark. If $\pi : \bar{M} \to M$ is $\Sigma_\ell^{(n)}$-preserving, then $\pi'' H^i_{\bar{M}} \subset H^i_M$ for $i \leq n$. If moreover $\ell > 0$ or $i < n$, we also have $\pi^{-1}{}'' H^i_M \subset H^i_{\bar{M}}$, since in this case the property $(\exists v^i)(v^i = x)$ is preserved downwards.

Lemma 1.7.1. *Let $n, \ell \in \omega$. Then $\pi : \bar{M} \underset{\Sigma_\ell^{(n)}}{\longrightarrow} M$ iff the following hold:*

a) $\pi : \bar{M} \to M$.

b) *If $\bar{p} \in \Gamma^n_{\bar{M}}$ and $p = \pi(\bar{p})$, then $p \in \Gamma^n_M$ and $\pi \upharpoonright H^n_{\bar{M}} : \bar{M}^{n,\bar{p}} \underset{\Sigma_\ell}{\longrightarrow} M^{n,p}$.*

Proof. The lemma is an immediate consequence of the two previous Lemmata. In case $\ell = 0$, the proof uses the fact that $\pi : \bar{N} \underset{\Sigma_0}{\longrightarrow} N$ implies $\pi : \bar{N} \underset{\text{rud}}{\longrightarrow} N$ for J-models. □(Lemma 1.7.1)

We are now ready to reformulate the downward extension of embeddings lemma in terms of $\Sigma_\ell^{(n)}$-embeddings. The relevant step towards this is

Lemma 1.7.2. *Let $\pi : \bar{M} \to M$ be such that $\pi \upharpoonright H^i_{\bar{M}} : \bar{M}^{i,\bar{p}\upharpoonright i} \underset{\Sigma_1}{\longrightarrow} M^{i,p\upharpoonright i}$ for $i < n$ and $\pi \upharpoonright H^n_{\bar{M}} : \bar{M}^{n,\bar{p}} \underset{\Sigma_\ell}{\longrightarrow} M^{n,p}$, where $\bar{p} \in R^n_{\bar{M}}$, $p = \pi(\bar{p})$ and $\ell \in \omega$. Then*

$$\pi : \bar{M} \underset{\Sigma_\ell^{(n)}}{\longrightarrow} M.$$

Proof. We shall apply Lemma 1.7.1 here. Let $\bar{q} \in \Gamma^n_{\bar{M}}$ and $q = \pi(\bar{q})$. Then clearly $q \in \Gamma^n_M$. Since $\bar{p} \in R^n_{\bar{M}}$, there are $\bar{z} \in H^1_{\bar{M}}$ and $j \in \omega$ such that $\bar{q}(0) = h_{\bar{M}}(j, \langle \bar{z}, \bar{p}(0)\rangle)$. Let $z = \pi(\bar{z})$. Then $q(0) = h_M(j, \langle z, p(0)\rangle)$. Now

$$A^{1,\bar{q}\upharpoonright 1}_{\bar{M}}(i, \bar{x}) \longleftrightarrow \models^{\Sigma_1}_{\bar{M}} \varphi_i(\bar{x}, h_{\bar{M}}(j, \langle \bar{z}, \bar{p}(0)\rangle)) \longleftrightarrow A^{1,\bar{p}\upharpoonright 1}_{\bar{M}}(k, \langle \bar{z}, \bar{x}\rangle)$$

for some $k \in \omega$ and similarly for $A^{1,q\upharpoonright 1}_M$ and $A^{1,p\upharpoonright 1}_M$. Hence by induction we can find a parameter $\bar{y} \in H^n_{\bar{M}}$ such that $A^{n,\bar{q}}_{\bar{M}}$ is rudimentary in $A^{n,\bar{p}}_{\bar{M}}$ in the parameter $\bar{y}$ and $A^{n,q}_M$ is rudimentary in $A^{n,p}_M$ in the parameter $y = \pi(\bar{y})$ by the same definition. By the preservation properties of the map $\pi \upharpoonright H^n_{\bar{M}}$ we conclude: $\pi \upharpoonright H^n_{\bar{M}} : \bar{M}^{n,\bar{q}} \underset{\Sigma_\ell}{\longrightarrow} M^{n,q}$. This holds for every $\bar{q} \in \Gamma^n_{\bar{M}}$, hence π is $\Sigma_\ell^{(n)}$-preserving. □(Lemma 1.7.2)

Lemma 1.3.1 can be then reformulated as

Lemma 1.7.3. *Let $\bar{M}$, M be J-structures, $\bar{p} \in R^n_{\bar{M}}$, and $\pi : \bar{M}^{n,\bar{p}} \underset{\Sigma_\ell}{\longrightarrow} M^{n,p}$ where $n, \ell \in \omega$. Then there is a unique map $\tilde{\pi} \supset \pi$ such that $\tilde{\pi}(\bar{p}) = p$ and $\tilde{\pi} : \bar{M} \underset{\Sigma_\ell^{(n)}}{\longrightarrow} M$.*

□(Lemma 1.7.3)

Lemma 1.7.4. *Let M be an acceptable J-structure and $p \in \Gamma^n_M$. The following are equivalent:*

a) $p \in P_M^n$.

b) $A_M^{n,p} \notin M$.

c) *There is a relation A which is $\Sigma_1^{(n-1)}(M)$ in p and $A \cap \omega\varrho_M^n \notin M$.*

Proof. a) → b) We proceed by induction. For $n = 1$ this follows from Lemma 1.2.4. Now suppose that it holds for n. We now consider the induction step. Let $p \in P_M^{n+1}$. Then $A_M^{n+1,p} = A^{p(n)}_{M^{n,p\restriction n}}$.

Case 1. $\varrho_M^{n+1} < \varrho_M^n$. Then $A_M^{n+1,p} \notin M^{n,p\restriction n}$ by Lemma 1.2.4, hence $A_M^{n+1,p} \notin M$ by acceptability (if $\omega\varrho_M^n \notin M$ this follows immediately without any use of acceptability).

Case 2. $\varrho_M^{n+1} = \varrho_M^n$. In this case $A_M^{n,p\restriction n}$ is rudimentary in $A_M^{n+1,p}$. Hence, if $A_M^{n+1,p}$ is an element of M, then so is $A_M^{n,p\restriction n}$, which contradicts with the induction hypothesis.

b) → c) If $\omega\varrho_M^n \notin M$, this is trivial. So suppose the contrary. Then there is a $\Sigma_1(M)$ function f mapping $\omega\varrho_M^n$ onto H_M^n. Set

$$A(\xi) \longleftrightarrow (\exists z^{n-1})(z^{n-1} = f(\xi) \ \& \ \models^{\Sigma_1}_{M^{n-1,p\restriction n-1}} \varphi_{(z^n)_1}((z^n)_2, p(n-1))$$

(remember that $(z)_i$ is the i-th Gödel component, if defined). A is clearly $\Sigma_1^{(n-1)}(M)$ in p, since the same holds for $\models^{\Sigma_1}_{M^{n-1,p\restriction n-1}}$, as follows from the claim in the proof of Lemma 1.6.4. $A \cap \omega\varrho_M^n \notin M$, since otherwise $A_M^{n,p} = f''(A \cap \omega\varrho_M^n) \in M$. This is a consequence of the fact that f has the same Σ_1 definition over $J^B_{\varrho_M^n}$ as over M, hence $f \in M$. Contradiction.

c) → b) Let A be $\Sigma_1^{(n-1)}$ in p such that $A \cap \omega\varrho_M^n \notin M$ and $A^*(x^n, \dots, x^0)$ $\Sigma_1^{(n-1)}(M)$ satisfying

$$A(x^n) \longleftrightarrow A^*(x^n, p(n-1), \dots, p(0)).$$

Then A^* is $\Sigma_0^{(n)}$, hence by Lemma 1.6.5, $A \cap \omega\varrho_M^n = A_p^*$ is (uniformly) rudimentary in $A_M^{n,p}$. Hence $A_M^{n,p} \notin M$ since otherwise $A \cap \omega\varrho_M^n \in M$.

b) → a) We proceed by induction. For $n = 1$ this is a part of the definition. So suppose it holds for n. Let $A_M^{n+1,p} \notin M$. Hence $A_M^{n+1,p} \notin M^{n,p\restriction n}$ which tells us that $p(n) \in P_{M^{n,p\restriction n}}$. Now we have to prove that $A_M^{n,p\restriction n} \notin M$, since this together with the induction hypothesis implies that $p \restriction n \in P_M^n$, which completes the proof. Suppose $A_M^{n,p\restriction n} \in M$. Then $M^{n,p\restriction n} \in M$. As $A_M^{n+1,p}$ is defined over $M^{n,p\restriction n}$, it must be an element of M. Contradiction. □(Lemma 1.7.4)

Lemma 1.7.5. *Let $\bar{M}$, M be acceptable structures and $\pi : \bar{M} \underset{\Sigma_1^{(n)}}{\longrightarrow} M$. Then $(\pi^{-1})'' P_M^n \subset P_{\bar{M}}^n$.*

Proof. Suppose not. Let $p \in P^n_M$ be such that $\bar{p} = \pi^{-1}(p) \notin P^n_{\bar{M}}$. Let A be a $\Sigma_1^{(n-1)}(M)$ relation in p satisfying $A \cap \omega\varrho^n_M \notin M$. Finally let $\bar{A}$ have the same $\Sigma_1^{(n-1)}$-definition over $\bar{M}$ in $\bar{p}$. Then $\bar{a} = \bar{A} \cap \omega\varrho^n_{\bar{M}} \in \bar{M}$. Hence

$$(\forall z^n)(z^n \in \bar{a} \longleftrightarrow \bar{A}(z^n)).$$

This is a $\Pi_1^{(n)}$ statement over $\bar{M}$, thus it is preserved by π. Set $a = \pi(\bar{a})$. Then $A \cap \omega\varrho^n_M = a \in M$. Contradiction. □(Lemma 1.7.5)

Definition. For an acceptable structure M we set $\varrho^\omega_M = \min\{\varrho^m_M;\ m \in \omega\}$.

Note that for $n \in \omega$, if $p \in P^n_M$, then the n-tuple $q = \langle p, 0, \ldots, 0\rangle \in P^n_M$, as $A^{n,p}_M$ is uniformly rudimentary in $A^{n,q}_M$. Since there is an $n \in \omega$ such that $\varrho^n_M = \varrho^\omega_M$, we can conclude that $P^*_M \neq \emptyset$, where P^*_M is defined as follows.

Definition. Let M be acceptable. Then

P^*_M = the set of all $p \in M$ such that $\langle p, 0, \ldots, 0\rangle \in P^n_M$ for all $n \in \omega$;

R^*_M is defined similarly.

Corollary 1.7.6. *Let $\bar{M}$, M be acceptable structures and $\pi : \bar{M} \underset{\Sigma^*}{\longrightarrow} M$. Then $(\pi^{-1})'' P^*_M \subset P^*_{\bar{M}}$.* □(Corollary 1.7.6)

Corollary 1.7.7. *For acceptable M and $p \in M$ the following are equivalent:*

a) $p \in P^*_M$.

b) $A^{n,p}_M \notin M$ *for all* $n \in \omega$.

c) *For every $n \in \omega$ there is a relation A which is $\Sigma_1^{(n)}(M)$ in p and such that $A \cap \omega\varrho^{n+1}_M \notin M$.* □(Corollary 1.7.7)

1.8 Substitution and Good Functions

$\Sigma_1^{(n)}$ relations are not, in general, closed under substitution of $\Sigma_1^{(n)}$-functions. There are, however, important cases, where such substitution is possible. In this section we shall define a class of functions which allow substitutions. We shall call these functions *good*. The first step towards the notion of a good function is the next lemma.

Lemma 1.8.1. *Let $m \leq n \in \omega$, $1 \leq \ell \in \omega$. Let $R(\vec{x}^n, \ldots, \vec{x}^0)$ be $\Sigma_\ell^{(n)}(M)$ for an acceptable M. Let $\vec{F}^n, \ldots, \vec{F}^0$ be such that each $F^i_j(\vec{z}^m, \ldots \vec{z}^0)$ is a $\Sigma_1^{(i)}(M)$ map to H^i_M. Then $R(\vec{F}(\vec{z}))$ is uniformly $\Sigma_\ell^{(n)}(M)$.*

Remark. The functions F_j^i may be partial on $\bigtimes_{h=1}^{m} H_M^h$. As usual, $R(\vec{F}(\vec{z}))$ holds iff the $\vec{F}(\vec{z})$ exist and R holds for $\vec{F}(\vec{z})$.

Proof of Lemma 1.8.1. We proceed by induction. The case $n = 1$ is straightforward. So suppose it holds for n. At this point we could use Lemma 1.6.3 as a black box, however we rather use its proof method in order to see more clearly that the result is uniform. It is also clear that it suffices to prove the lemma for $\ell = 1$, since the rest is just a matter of writing additional quantifiers. So let $R(\vec{x}^{\,n+1}, \ldots, \vec{x}^{\,0})$ be $\Sigma_1^{(n+1)}(M)$. By the proof of Lemma 1.6.3, R can be expressed as in (1.6.1), i.e. in the form

$$(\exists z^{n+1})B(\vec{v}^{\,n+1}, \vec{x}^{\,n+1})R^*(z^{n+1}, \vec{v}^{\,n+1}, \vec{x}^{\,n+1}, \ldots, \vec{x}^{\,0})$$

where $B(\vec{v}^{\,n+1}, \vec{x}^{\,n+1})$ is a block of bounded quantifiers and R^* is a propositional combination of $\Sigma_1^{(n)}(M)$ relations. By the induction hypothesis, we can substitute the functions $\vec{F}^n, \ldots, \vec{F}^0$ for $\vec{x}^{\,n}, \ldots, \vec{x}^{\,0}$ in R^* and obtain a $\Sigma_1^{(n)}(M)$-relation, since each of the $\Sigma_1^{(n)}(M)$ relations occurring in the propositional combination R^* is a specialization of a $\Sigma_1^{(n)}(M)$ relation with arguments of type $\leq n$. Hence we obtain a relation of the form

$$(\exists z^{n+1})\bar{R}(z^{n+1}, \vec{x}^{\,n+1}, \vec{F}^n(\vec{y}), \ldots, \vec{F}^0(\vec{y})),$$

where $\bar{R}$ is $B(\vec{v}^{\,n+1}, \vec{x}^{\,n+1})R^*$, hence $R(\vec{x}^{\,n+1}, \vec{F}^n(\vec{y}), \ldots, \vec{F}^0(\vec{y}))$ is $\Sigma_1^{(n)}(M)$. But then we can express $R(\vec{F}(\vec{y}))$ as

$$(\exists \vec{x}^{\,n+1})[(\mathop{\&}_i x_i^{n+1} = F_i^{n+1}(\vec{y})) \;\&\; R(\vec{x}^{\,n+1}, \vec{F}(\vec{y}), \ldots, \vec{F}^0(\vec{y}))],$$

which is clearly $\Sigma_1^{(n+1)}(M)$. It is clear from this argument that the transformation from $R(\vec{x})$ to $R(\vec{F}(\vec{y}))$ is uniform, i.e. the definition of the latter depends only on the definition of the former and the those of $F^0, \ldots, F^n$. □(Lemma 1.8.1)

One consequence of the above lemma is that $\Sigma_1^{(n)}$-relations are, essentially, characterized by their graphs after all. Let $R(x^{i_1}, \ldots, x^{i_n})$, $R(y^{j_1}, \ldots, y^{j_n})$ have the same graph, where $i_1, \ldots, i_n, j_1, \ldots, j_n \leq n$. Then the former is $\Sigma_1^{(n)}(M)$ iff the latter is. To see this, note that we can convert from one to the other by composition with the identity functions $y^{j_h} = x^{i_h}$. The same argument shows that a relation is $\Sigma_1^{(n)}(M)$ iff the relation with the same graph and arguments of type 0 is $\Sigma_1^{(n)}(M)$. We shall now extend the above result to a larger class of functions.

Definition. The *good* $\Sigma_1^{(n)}(M)$*-functions* comprise the smallest class Σ such that

a) Each partial $\Sigma_1^{(i)}(M)$ map $F(x_1^{j_1}, \ldots, x_k^{j_k})$ to H_M^i, where $i, j_1, \ldots, j_k \leq n$, is in Σ.

b) If $F(x_1^{j_1}, \ldots, x_k^{j_k})$ is in Σ and $G_i(\vec{z})$ is a map to $H_M^{j_i}$ which belongs to Σ for $i = 1, \ldots, k$ (the arguments of $G_i(\vec{z})$ all being of type $\leq n$), then the composition $F(G_1(\vec{z}), \ldots, G_k(\vec{z}))$ is in Σ.

Lemma 1.8.2. *Let $R(x_1^{j_1}, \ldots, x_k^{j_k})$ be a $\Sigma_\ell^{(n)}(M)$ relation, where $n \in \omega$, $1 \leq \ell \in \omega$ and $j_1, \ldots, j_k \leq n$. Let $F_i(\vec{z})$ be a good $\Sigma_1^{(n)}(M)$ map to $H_M^{j_i}$ for $i = 1, \ldots, k$. Then $R(F_1(\vec{z}), \ldots, F_k(\vec{z}))$ is uniformly $\Sigma_\ell^{(n)}(M)$.*

Proof. This follows in a straightforward way by the iterated application of Lemma 1.8.1. It is also clear that the definition of $R(\vec{F}(\vec{z}))$ depends only on the definition of R, the way F^i are composed from the basic functions (i.e. the functions as in a) of the previous definition) and on the definitions of these basic functions. □(Lemma 1.8.2)

Lemma 1.8.3. *($\Sigma_1^{(n)}$-uniformization). Let $R(y^n, \vec{x}^{\,n}, \ldots, \vec{x}^{\,0})$ be a $\Sigma_1^{(n)}(M)$ relation over an acceptable structure M. There is a $\Sigma_1^{(n)}(M)$-function F to H_M^n such that*

a) $\mathrm{dom}(F) = \{\vec{x}, (\exists\, y^n) R(y^n, \vec{x})\}$;

b) $(\exists\, y^n) R(y^n, \vec{x}) \longrightarrow R(F(\vec{x}), \vec{x})$.

Moreover, there is an F as above whose definition is uniform.

Proof. By Lemma 1.6.4, $R_{\vec{x}}(y^n, \vec{x}^{\,n})$ is uniformly $\Sigma_1(M^{n,p(\vec{x})})$ (see Lemma 1.6.3, 1.6.4 for the notation). Hence there is $i \in \omega$ which codes the definition of $R_{\vec{x}}$. Since i depends neither on M nor on $\vec{x}$, we can define F by

$$F(\vec{x}^{\,n}, \ldots, \vec{x}^{\,0}) \simeq h_{M^{n,p(\vec{x})}}(i, \langle \vec{x}^{\,n} \rangle).$$

By Lemma 1.6.4, the function $F(\vec{x}^{\,n}, \ldots, \vec{x}^{\,0})$ is $\Sigma_1^{(n)}(M)$ since $h_{M^{n,p(\vec{x})}}$ is uniformly $\Sigma_1(M^{n,p(\vec{x})})$. □(Lemma 1.8.3)

Definition. $\Sigma_\omega^{(n)} = \bigcup_{\ell \in \omega} \Sigma_\ell^{(n)}$.

Definition. $Q^{(n)}$ = the set of formulae of the form

$$(Qx^n)\varphi = (\forall z^n)(\exists x^n)(z^n \subset x^n \;\&\; \varphi(x^n)),$$

where φ is a $\Sigma_1^{(n)}$-formula. We also set

$$Q^* = \bigcup_{n \in \omega} Q^{(n)}.$$

Then

$$\pi : \bar{M} \underset{Q^{(n)}}{\longrightarrow} M, \qquad \pi : \bar{M} \underset{Q^{(*)}}{\longrightarrow} M$$

have the obvious meaning.

Lemma 1.8.4. *Let M be acceptable, $p \in R_M^n$. Then each $x \in M$ has the form $F(y, p)$, where $y \in H_M^{n+1}$ and F is a good $\Sigma_1^{(n)}(M)$ function. Moreover, there is an F as above such that its definition is uniform for all M.*

Proof. By induction. For $n = 1$ this follows from the definition of R_M. Now suppose the lemma holds for n. Let $p \in R_M^{n+1}$. Pick $x \in M$. By the definition, $p \restriction n \in R_M^n$ and $p(n) \in R_{M^{n,p\restriction n}}$. Hence there is a good uniformly $\Sigma_1^{(n-1)}(M)$ function G such that $x = G(z, p \restriction n)$ for some $z \in H_M^n$. Moreover, $p(n) \in R_{M^{n,p\restriction n}}$, hence there is $\langle i, y\rangle \in \omega \times H_M^{n+1}$ such that $z = h_{M^{n,p\restriction n}}(i, \langle y, p(n)\rangle)$. The map $h_{M^{n,p\restriction n}}$ is uniformly $\Sigma_1(M^{n,p\restriction n})$ and therefore uniformly $\Sigma_1^{(n)}(M)$. Thus, it suffices to set

$$F(w, p) \simeq G(h_{M^{n,p\restriction n}}((w)_1, \langle (w)_2, p(n)\rangle), p \restriction n)$$

as F is clearly good and uniformly $\Sigma_1^{(n)}(M)$. □(Lemma 1.8.4)

Definition. For M as above and $p \in \Gamma_M^n$ we set

a) $h_M^{n,p} = h_{M^{n,p}}$

b) $\tilde{h}_M^n =$ the good uniformly $\Sigma_1^{(n-1)}(M)$ function with two parameters which is the result of iterated composition of the Skolem functions of i-th reducts ($i = 1, \ldots, n-1$) as in the proof of Lemma 1.8.4.

Hence, if $p \in R_M^n$, then each $x \in M$ is of the form $\tilde{h}_M^n(z, p)$ for some $z \in H_M^n$.

Corollary 1.8.5. *If $\omega\varrho_M^n \in M$ then there is a good uniformly $\Sigma_1^{(n-1)}(M)$ function $F(x_1, x_2)$ such that for any $p \in R_M^n$, $F(-, p)$ maps $\omega\varrho_M^n$ onto M. In particular, $\tilde{h}_M^n$ is such a function.*

Proof. This follows immediately by the fact that $h_M^{k,p}(\omega\varrho_M^{k+1} \cup \{p(k)\}) = H_M^k$ for every $k < n$. □(Corollary 1.8.5)

Corollary 1.8.6. *Let M be acceptable and $p \in R_M^n$. Then every $A \subset H_M^n$ which is $\mathbf{\Sigma}_1^{(n)}(M)$ is $\mathbf{\Sigma}_1(M^{n,p})$.*

Proof. Let $A^*(v_1, v_2)$ be $\Sigma_1^{(n)}(M)$ such that $A(x) \leftrightarrow A^*(x, q)$ for some $q \in M$. Let $s \in H_M^n$ be such that $q = \tilde{h}_M^n(s, p)$. Then

$$A(x^n) \longleftrightarrow A^*(x^n, \tilde{h}_M^n(s, p)).$$

The right side is clearly $\Sigma_1^{(n)}(M)$ in s, p, hence A is $\mathbf{\Sigma}_1(M^{n,p})$ in s. □(Corollary 1.8.6)

Corollary 1.8.7. *Let M be acceptable and $p \in R_M^n$. Then $\varrho_M^{n+1} = \varrho_{M^{n,p}}$.*

Proof. Clearly $\varrho_M^{n+1} \leq \varrho_{M^{n,p}}$, so it suffices to prove the converse. Let $q \in P_M^{n+1}$ and A be $\Sigma_1^{(n)}(M)$ in q such that $A \cap \omega\varrho_M^{n+1} \notin M$. By the previous corollary, $A \cap H_M^n$ is $\mathbf{\Sigma}_1(M^{n,p})$, hence $\varrho_{M^{n,p}} \leq \varrho_M^{n+1}$. □(Corollary 1.8.7)

Corollary 1.8.8. *Let p be acceptable, $p \in R^n_M$ and $m \leq n$. Then*

a) $\varrho^n_M = \varrho^{n-m}_{M^{m,p\restriction m}}$

b) $M^{n,p} = (M^{m,p\restriction m})^{n-m,q}$,

where $q \in \bigtimes_{i<n-m} H^{m+i}_M$ is such that $q(i) = p(m+i)$.

Proof. By a straightforward induction using the definition of reducts and the previous corollary. □(Corollary 1.8.8)

Corollary 1.8.9. *Let M be acceptable and $p, q \in R^n_M$. Then*

$$\mathbf{\Sigma}_1(M^{n,p}) = \mathbf{\Sigma}_1(M^{n,q}).$$

Proof. By a straightforward use of Lemma 1.6.4 and Corollary 1.8.6. □(Corollary 1.8.9)

Definition. Let M be an acceptable J-structure. Let f be a $\Sigma^{(n)}_1(M)$ function. A formula φ is a *functionally absolute* $\Sigma^{(n)}_1$ *definition* for f iff

a) φ defines f over M (i.e. $y = f(x_1, \ldots, x_\ell) \leftrightarrow M \models \varphi(y, x_1, \ldots, x_\ell)$),

b) φ is $\Sigma^{(n)}_1$,

c) for every acceptable structure N, φ defines a function over N.

Definition. Let M be as above and $f(v^{i_1}_1, \ldots, v^{i_\ell}_\ell) \in \Sigma^{(n)}_1(M)$ in p. We say that f has a functionally absolute $\Sigma^{(n)}_1$ definition in p iff there is a function $F(v^{i_1}_1, \ldots, v^{i_\ell}_\ell, w)$ with a functionally absolute $\Sigma^{(n)}_1$ definition such that

$$f(x_1, \ldots, x_\ell) \simeq F(x_1, \ldots, x_\ell, p)$$

for all $x_1, \ldots, x_\ell$.

Functionally absolute definitions are useful when we want to "transfer" definable functions from one structure to another via an embedding. If the embedding is sufficiently preserving, we can define a relation over the target structure using the same definition, but such a relation can, in general, fail to be a function. However, it will be a function, if we use a functionally absolute definition, provided the function has one.

Note that the Skolem functions do have functionally absolute definitions, in fact, their canonical definitions are functionally absolute. Another important example of functions which possess functionally absolute definitions, are the functions $\tilde{h}^n_M$, which are composed from Skolem functions in a uniform way. It turns out that there is a fairly large class of functions, which have a functionally absolute definition.

Lemma 1.8.10. *Every good $\Sigma^{(n)}_1(M)$ function has a functionally absolute $\Sigma^{(n)}_1$ definition.*

Proof. By induction on the defining schema of good functions. The induction step is straightforward since if a collection of functions has such a definition, then any composition has one as well – one just has to put these definitions together in a straightforward way (as in the proof of Lemma 1.8.10. Hence it suffices to prove

Claim. Every $\Sigma_1^{(n)}(M)$ function to H_M^n has a functionally absolute definition.

Proof. Note first that this follows immediately if $n = 0$. In this case we have $f(x) \simeq h_M(i, x)$ for an appropriate $i \in \omega$ and the Skolem function, as mentioned above, does have such a definition.

For $n > 0$ we have to do a little bit more work. Let $f(\vec{x})$ be as required, where $\vec{x} = \vec{x}^{\,n}, \ldots, \vec{x}^{\,0}$. Fix a definition of f, say φ. By Lemma 1.6.4, $g_{\vec{x}}(\vec{x}^{\,n})$ is $\Sigma_1(M^{n,p(\vec{x})})$, where $g_{\vec{x}}(\vec{x}^{\,n}) \simeq f(\vec{x}^{\,n}, \ldots, \vec{x}^{\,0})$, $p(\vec{x}) = \langle \vec{x}^{\,n-1}, \ldots, \vec{x}^{\,0} \rangle$. Moreover, the definition of $g_{\vec{x}}$ over $M^{n,p(\vec{x})}$ depends only on φ and is independent of $\vec{x}$. Hence there is a fixed $i \in \omega$ such that for all $\vec{x}$ we have

$$g_{\vec{x}}(\vec{x}^{\,n}) \simeq h_M^{n,p(\vec{x})}(i, \vec{x}^{\,n}). \tag{1.8.1}$$

Since $h_M^{n,p(\vec{x})}$ has a uniform definition independent of $\vec{x}$, the right side of (1.8.1) gives us the required functionally absolute $\Sigma_1^{(n)}$ definition. □(Lemma 1.8.10)

This together with Lemma 1.8.2 has a significant consequence.

Lemma 1.8.11. *Let $\bar{M}$, M be acceptable and $\pi : \bar{M} \to M$ a $\Sigma_1^{(n)}$-preserving map. Let $\bar{R}(\vec{v}^{\,n}, \ldots, \vec{v}^{\,0})$ be a $\Sigma_1^{(n)}(\bar{M})$ relation and $R(\vec{v}^{\,n}, \ldots, \vec{v}^{\,0})$ the $\Sigma_1^{(n)}(M)$ relation by the same definition. Finally let $\vec{\bar{F}}^n(z_1^{i_1}, \ldots, z_\ell^{i_\ell}), \ldots, \vec{\bar{F}}^0(z_1^{i_1}, \ldots, z_\ell^{i_\ell})$ be good $\Sigma_1^{(n)}(\bar{M})$ functions and $\vec{F}^n(z_1^{i_1}, \ldots, z_\ell^{i_\ell}), \ldots, \vec{F}^0(z_1^{i_1}, \ldots, z_\ell^{i_\ell})$ the good $\Sigma_1^{(n)}(M)$ functions by the same functionally absolute definitions. Let $\bar{a}_j \in H_{\bar{M}}^{i_j}$ and $a_j = \pi(\bar{a}_j)$ $(i = 1, \ldots, \ell)$. Then*

$$\bar{R}(\vec{\bar{F}}^n(\bar{a}_1, \ldots, \bar{a}_\ell), \ldots, \vec{\bar{F}}^0(\bar{a}_1, \ldots, \bar{a}_\ell)) \text{ iff } R(\vec{F}^n(a_1, \ldots, a_\ell), \ldots, \vec{F}^0(a_1, \ldots, a_\ell)).$$

Proof. By induction on the defining schema of $\vec{F}$. The first step is to prove this for $\vec{F}$ such that each $\vec{F}^i$ is a $\Sigma_1^{(i)}(M)$ function to H_M^i. By Lemma 1.8.1, there is an effective method which, given a definition of R, say φ_R, and functionally absolute $\Sigma_1^{(i)}$ definitions of F_j^i, say φ_j^i, produces a $\Sigma_1^{(n)}$ formula φ satisfying

$$\varphi(z_1^{i_1}, \ldots, z_\ell^{i_\ell}) \longleftrightarrow (\exists \vec{y})\Big(\mathop{\&}_{i,j} \varphi_j^i(y_j^i, z_1^{i_1}, \ldots, z_\ell^{i_\ell}) \;\&\; \varphi_R(\vec{y})\Big). \tag{1.8.2}$$

On the other hand, given φ, we can effectively reconstruct the formula on the right side. Hence, since φ is $\Sigma_1^{(n)}$,

$$\bar{M} \models \varphi(\bar{a}, \ldots, \bar{a}_\ell) \longleftrightarrow M \models \varphi(a_1, \ldots, a_\ell),$$

and the rest follows from (1.8.2).

Once we have this, the induction step is straightforward, since it consists of applying the same argument at each level of the defining schema of the F^i_j's. □(Lemma 1.8.11)

Corollary 1.8.12. *Let* $\pi : \bar{M} \underset{\Sigma_1^{(n)}}{\longrightarrow} M$, $f(z_1^{i_1}, \ldots, z_\ell^{i_\ell})$ *be a good* $\Sigma_1^{(n)}(M)$ *function and* $\bar{f}(z_1^{i_1}, \ldots, z_\ell^{i_\ell})$ *be a good* $\Sigma_1^{(n)}(\bar{M})$ *function by the same functionally absolute definition. Let* $\bar{a}_i \in H^i_{\bar{M}}$ *and* $a_i = \pi(\bar{a}_i)$ *for* $i = 1, \ldots, \ell$. *Then*

$$f(a_1, \ldots, a_\ell) \quad \text{is defined iff} \quad \bar{f}(\bar{a}_1, \ldots, \bar{a}_\ell) \ \text{is.}$$

Proof. The only non-trivial direction is (→), which we now prove. The fact that $f(a_1, \ldots, a_\ell)$ is defined can be expressed as

$$(\exists y^0)\,(y^0 = f(a_1, \ldots, a_\ell))$$

which is a $\Sigma_1^{(n)}$ statement, since it is the result of substituting a good $\Sigma_1^{(n)}$ function into a Σ_1 relation. By Lemma 1.8.11,

$$(\exists y^0)\,(y^0 = \bar{f}(\bar{a}_1, \ldots, \bar{a}_\ell)).$$

□(Corollary 1.8.12)

Lemma 1.8.13. *Let M be acceptable and* $p \in R^n_M$. *Then* $\mathbf{\Sigma}_\ell(M) \subset \mathbf{\Sigma}_\ell^{(n)}(M)$ *for every* $\ell \geq 1$.

Proof. Let $A(\vec{y})$ be $\Sigma_\ell(M)$ in q, i.e. let

$$A(\vec{y}) \longleftrightarrow (\exists x_1)(\forall x_2)\ldots(\exists x_\ell)A^*(x_1, \ldots, x_\ell, \vec{y}, q),$$

where we w.l.o.g. suppose that ℓ is odd (if ℓ is even then do the usual trick $\neg(\exists x_\ell)\neg A^*$) and A^* is $\Sigma_0(M)$. By straightforward induction on ℓ we then show that A can be expressed in the following form

$$(\exists z_1^n)(\forall z_2^n)\ldots(\exists z_\ell^n)A^*(\tilde{h}^n_M(z_1^n, p), \ldots, \tilde{h}^n_M(z_\ell^n, p), \vec{y}, q),$$

which is clearly $\Sigma_\ell^{(n)}(M)$ in p, q. □(Lemma 1.8.13)

Lemma 1.8.14. *Let M be acceptable and* $p \in R^n_M$. *Then* $\mathbf{\Sigma}_\omega(M) = \mathbf{\Sigma}_\omega^{(n)}(M)$. *In particular, this holds if M is sound.*

Proof. The essential part is the previous lemma. To see the inclusion ⊃, note that each quantifier of the sort $\exists x^n$, $\forall x^n$ can be replaced by a corresponding ordinary quantifier if $\omega\varrho^n_M \notin M$ or by a corresponding bounded ordinary quantifier if $\omega\varrho^n_M \in M$.

□(Lemma 1.8.14)

Definition. We write

$$\pi : \bar{M} \underset{\Sigma_0^{(n)}}{\longrightarrow} M \text{ cofinally}$$

to express the following:

$$\pi : \bar{M} \underset{\Sigma_0^{(n)}}{\longrightarrow} M \quad \text{and} \quad \bigcup \pi'' H^n_{\bar{M}} = H^n_M.$$

Remark. This clearly implies $\pi \restriction H^n_{\bar{M}} : \bar{M}^{n,\bar{p}} \underset{\Sigma_0}{\longrightarrow} M^{n,p}$ cofinally for all $\bar{p}$, hence $\pi : \bar{M} \underset{\Sigma_1^{(n)}}{\longrightarrow} M$.

For $\mathbf{\Sigma}_n(M)$ relations which are subsets of H^n_M we can in fact do much better:

Lemma 1.8.15. *Let M be acceptable and $p \in R^n_M$. Let $A \subset H^n_M$ be a $\mathbf{\Sigma}_{n+1}(M)$-relation. Then A is $\mathbf{\Sigma}_1^{(n)}(M)$.*

Proof. We proceed by induction. The case $n = 0$ is trivial. Let us suppose now that the statement holds for n. Suppose also w.l.o.g. that n is odd. Let $p \in R^{n+1}$ and $A \subset H^{n+1}_M$ be a $\Sigma_{n+2}(M)$ relation in q, i.e.

$$A(\vec{x}) \longleftrightarrow (\exists z_1)(\forall z_2)\ldots(\exists z_{n+2})A^*(z_{n+2},\ldots,z_1,\vec{x},q)$$

for some $\Sigma_0(M)$ relation A^*. As $p(0) \in R_M$, we get, exactly as in Lemma 1.8.13, that $A(\vec{x})$ holds iff

$$(\exists z^1_1)(\forall z^1_2)\ldots,(\forall z^1_{n+1})B(z^1_{n+1},\ldots,z^1_1,\vec{x},q^*,p(0)), \tag{1.8.3}$$

where $q^* \in H^1_M$ is such that $q = \tilde{h}^1_M(q^*, p(0))$ and

$$\begin{aligned} &B(z^1_{n+1},\ldots,z^1_1,\vec{x},q^*,p(0)) \\ &\quad\longleftrightarrow (\exists z^0_{n+2})A^*(z^1_{n+2},\tilde{h}^1_M(z^1_{n+1},p(0)),\ldots,\tilde{h}^1_M(z^1_1,p(0)),\vec{x},\tilde{h}^1_M(q^*,p(0))). \end{aligned}$$

The relation $B(z^1_{n+1},\ldots,z^1_1,\vec{x},q^*,p(0))$ is $\Sigma_1(M)$ in $q^*, p(0)$ and, hence, rudimentary over $M^{p(0)}$. Consequently, (1.8.3) is $\mathbf{\Sigma}_{n+1}(M^{p(0)})$.

We now apply the induction hypothesis to $M^{p(0)}$. This is possible by Corollary 1.8.8 which tells us that $H^{n+1}_M = H^n_{M^{p(0)}}$ and that $M^{p(0)}$ has a very good parameter $r \in R^n_{M^{p(0)}}$ (such that $r(i) = p(i+1)$ for $i \in n$). Hence (1.8.3) holds and therefore A is $\mathbf{\Sigma}_1^{(n)}(M^{p(0)})$ and hence $\mathbf{\Sigma}_1^{(n+1)}(M)$. □(Lemma 1.8.15)

1.9 Standard Parameters

It is easy to see that $h_M([\mathbf{On} \cap M]^{<\omega}) = M$ for any J-structure M. In fact, we have used this several times already. Given an acceptable structure M, any relation A which is $\Sigma_1^{(n)}(M)$ in some p is therefore $\Sigma_1^{(n)}(M)$ in some $p' \in [\mathbf{On} \cap M]^{<\omega}$. In other words, we can w.l.o.g. restrict ourselves to those good parameters which are finite sets of ordinals. Such parameters are well-ordered in a simple canonical way.

Definition. Let $a, b \in [\mathbf{On}]^{<\omega}$. Set

$$a <^* b \longleftrightarrow (\exists \alpha \in b)(a - (\alpha + 1) = b - (\alpha + 1) \ \&\ \alpha \notin a).$$

The ordering $<^*$ has a rudimentary definition, therefore it is absolute for transitive rudimentarily closed structures and also is preserved under Σ_0-embeddings. If we view finite sets of ordinals as finite decreasing sequences, $a <^* b$ precisely when a precedes b lexicographically.

Lemma 1.9.1. *$[\mathbf{On}]^{<\omega}$ is well-ordered by $<^*$.*

Proof. Straightforward. □(Lemma 1.9.1)

Definition. Let M acceptable. The $<^*$-least $p \in P_M^* \cap [\mathbf{On}]^{<\omega}$ is called the *standard parameter* of M and is denoted by p_M.

The $<^*$-least $p \in P_M^n \cap [\mathbf{On}]^{<\omega}$ is called the *standard parameter above* $\omega\varrho_M^n$ and is denoted by $p_{M,n}$.

Notation. a) If $a \in [\mathbf{On} \cap M]^{<\omega}$ we set

$$a^i = a \cap [\omega\varrho_M^{i+1}, \omega\varrho_M^i]$$
$$a \restriction i = a - \omega\varrho_M^i.$$

b) We often write briefly A_M^n, M^n, h_M^n instead of A_M^{n,p_M}, $M^{n,p_M \restriction n}$, $h_M^{n,p_M \restriction n}$.

Corollary 1.9.2. *Let $p \in [\mathbf{On} \cap M]^{<\omega}$. Then:*

a) $p \in P_M^n \longrightarrow p \restriction (n-1) \in P_M^{n-1} \ \&\ p^{n-1} \in P_{M^{n-1,p\restriction n-1}}$

b) $p \in R_M^n \longleftrightarrow (\forall i < n)(p^i \in R_{M^{i,p\restriction i}})$.

If $\omega\varrho_{M^{n-1,p\restriction(n-1)}} = \omega\varrho_M^n$, then the equivalence holds.

Proof. Straightforward. □(Corollary 1.9.2)

Lemma 1.9.3. *Let $p \in R_M^n$. Then p can be lengthened to some $p' \in P_M^{n+1}$. If $p \in [\mathbf{On}]^{<\omega}$, then p can be lengthened to some $p' \in [\mathbf{On}]^{<\omega}$.*

Proof. The first part follows immediately from Corollary 1.8.7. The second follows then by the fact that each $z \in H_M^{n+1}$ can be replaced by a finite set of ordinals in a $\Sigma_1(M^{n,p})$ fashion. □(Lemma 1.9.3)

Lemma 1.9.4. *Let M be sound and $p \in P_M^n$. Then p can be lengthened to a $p' \in P_M^*$. Moreover, if $p \in [\mathbf{On}]^{<\omega}$, there is p' as above of the same form.*

Proof. By iterated application of Lemma 1.9.3 □(Lemma 1.9.4)

Corollary 1.9.5. *Let M be n-sound. Then* $p_{M,n-1} = p_{M,n} \restriction (n-1)$.

Proof. By Lemma 1.9.4. □(Corollary 1.9.5)

Lemma 1.9.6. *M is sound iff* $p_{M,n} \in R^n_M$ *for all* $n \in \omega$.

Proof. We shall prove the non-trivial direction ($\leftarrow$). Note first that it is sufficient to prove

$$P^n_M \cap [\mathbf{On}]^{<\omega} = R^n_M \cap [\mathbf{On}]^{<\omega} \quad \text{for all } n \in \omega,$$

since we can replace any parameter by a finite set of ordinals in a $\Sigma_1(M)$ fashion and retain its goodness, respectively very goodness. For each n we now prove

$$p_{M,n} \in R^n_M \longrightarrow R^n_M = P^n_M. \tag{1.9.1}$$

This holds trivially for $n = 0$. Now suppose $n > 0$ is the least such that the statement (1.9.1) fails. Hence $P^n_M - R^n_M \neq \emptyset$. Let q be the $<^*$-least element of $P^n_M - R^n_M$. This means $p <^* q$, where $p = p_{M,n}$. Let i be the first such that $p^i <^* q^i$.

Claim. $i = n - 1$.

Proof. Suppose $i < n-1$. Let $\bar{M}, \bar{q}, \pi$ be unique such that

$$\left\{\begin{array}{l} \bar{q} \in R^n_{\bar{M}} \\ \pi : \bar{M} \to M \text{ is } \Sigma^{(n-1)}_1\text{-preserving} \\ \pi(\bar{q}) = q \\ \pi \restriction H^n_{\bar{M}} = \mathrm{id}. \end{array}\right. \tag{1.9.2}$$

By the induction hypothesis, $p \restriction (i+1) \in R^{i+1}_M$. Hence there is a good $\Sigma^{(i)}_1(M)$ function f such that $q^i = f(z, p \restriction i, p^i)$ for some $z \in [\omega\varrho^{i+1}_M]^{<\omega}$. So

$$(\exists z^{i+1})(\exists r^i <^* q^i)(z^{i+1}, r^i \in [\mathbf{On}]^{<\omega} \;\&\; q^i = f(z^{i+1}, p \restriction i, r^i). \tag{1.9.3}$$

This is a $\Sigma^{(i+1)}_1(M)$ statement and hence is preserved downwards. We thus obtain finite sets of ordinals $\bar{z} \subset H^{i+1}_{\bar{M}}$ and $\bar{r} \subset H^i_{\bar{M}}$ such that $\bar{q}^i = \bar{f}(\bar{z}, \bar{p} \restriction i, \bar{r})$, where $\bar{f}$ has the same functionally absolute $\Sigma^{(i)}_1$ definition over $\bar{M}$ as f over M. Setting $r = \pi(\bar{r})$, $z = \pi(\bar{z})$ we get $q^i = f(z, p \restriction i, r)$, so letting $q_1 = q \restriction i \cup r \cup z \cup (q \cap \omega\varrho^{i+1}_M)$ we have $q_1 \in P^n_M$. Moreover, $q_1 <^* q$ since $r <^* q^i$ and $q_1 \notin R^n_M$, as $q_1 \cup \omega\varrho^{n+1}_M \subset \mathrm{rng}(\pi) \neq M$. But this is a contradiction to our choice of q. □(Claim)

The proof in case $i = n-1$ is – up to one little amendment – exactly the same. Now we cannot quantify over H^n_M as in (1.9.3) since we do not have enough preservation. But π is constant on $H^n_{\bar{M}}$, so we in fact do not need to quantify; the rest follows from the fact that $\bar{z} = z$ in this case. □(Lemma 1.9.6)

Remark. By the above proof, an obvious reformulation of the previous lemma gives us a characterization of n-soundness.

There is a class of structures, for which the above characterization of soundness has a particularly nice form. This class comprises all of the structures J_α, $\alpha \in \mathbf{On}$ and, as will turn out later, also the mice, which are the building blocks of the core model.

Definition. Let M be acceptable. We say that the good parameters of M can be *lengthened* iff for every $n \in \omega$ and every $p \in P_M^n$, p can be lengthened to some $p' \in P_M^{n+1}$.

Lemma 1.9.7. *Let M be as in the above definition. Then*

a) $p_{M,n} = p_{M,n+1} \upharpoonright n$ *for every* $n \in \omega$,

b) $p_{M,n} = p_M \upharpoonright n$ *for every* $n \in \omega$.

Proof. a) Clearly $p_{M,n} \leq^* p_{M,n+1} \upharpoonright n$ since the latter is an element of P_M^n. To see the converse note that $p_{M,n}$ can be lengthened to a $q \in P_M^{n+1}$, hence $p_{M,n+1} \leq^* q$ and the same holds for the restrictions of both to n.

The proof of b) is similar, but in this case we iteratively lengthen $p_{M,n}$ to a $q \in P_M^*$.

□(Lemma 1.9.7)

Lemma 1.9.8. *Let M be as above. Then*

a) *M is n-sound iff* $p_{M,n} \in R_M^n$;

b) *M is sound iff* $p_M \in R_M^*$.

Proof. It clearly suffices to prove a). But this is an immediate consequence of Lemma 1.9.6, since in this case $p_{M,n} \in R_M^n$ implies $p_{M,i} \in R_M^i$ for all $i \leq n$.

□(Lemma 1.9.8)

1.10 Two Applications to L

We shall now illustrate how to use the above machinery in the simplest case – in the constructible universe L. We shall in fact prove two important lemmata.

Lemma 1.10.1. *For each* $\alpha \in \mathbf{On}$, J_α *is acceptable.*

Lemma 1.10.2. *For each* $\alpha \in \mathbf{On}$, J_α *is sound.*

We shall prove these two lemmata simultaneously. The proof goes by induction on α in a zig-zag way in the sense that we use soundness of J_α to prove the acceptability of $J_{\alpha+1}$ and then, knowing this, its soundness.

Proof. The case $\alpha = 1$ is trivial. Now suppose both lemmata hold for all $\beta < \alpha$.

Claim 1. J_α is acceptable. This is trivial for α limit; for $\alpha = \beta + 1$ it is clear that the only thing we have to prove is the following:

$$\text{If there are } \tau < \beta \text{ and } a \subset \tau \text{ such that } a \in J_{\beta+1} - J_\beta, \text{ then there is an } f \in J_{\beta+1} \text{ such that } f : \tau \xrightarrow{\text{onto}} \omega\beta. \tag{1.10.1}$$

We prove (1.10.1). Suppose there are such τ, a and take τ to be the least one such that there is a as above. Then

$$\tau = \omega\varrho^\omega_{J_\beta}. \tag{1.10.2}$$

To see this note first that if n is such that $\omega\varrho^n_{J_\beta} = \omega\varrho^\omega_{J_\beta}$, then we have a new $\mathbf{\Sigma}_1^{(n)}(J_\beta)$ subset of $\omega\varrho^\omega_{J_\beta}$. Such a set is $\mathbf{\Sigma}_\omega(J_\beta)$ by Lemma 1.8.14, hence $\tau \leq \omega\varrho^\omega_{J_\beta}$.

Once we know this we can use Lemma 1.8.15 to get the converse. Let $a \subset \tau$ such that $a \in J_{\beta+1} - J_\beta$. Then $a \in \mathbf{\Sigma}_n(J_\beta)$ for some $n > 1$. By the above inequality, $a \subset H^n_{J_\beta}$. The above-mentioned lemma then yields that a is $\mathbf{\Sigma}_1^{(n-1)}(J_\beta)$, since, by the induction hypothesis, J_β is sound (and hence $R^*_{J_\beta} \neq \emptyset$). Consequently, $\omega\varrho^\omega_{J_\beta} \leq \tau$. □(1.10.2)

Now we use the induction hypothesis once again. By soundness of J_β, there is a $\mathbf{\Sigma}^*(J_\beta)$ function $f(z, x)$ such that $f(-, p_{J_\beta})$ maps $\omega\varrho^\omega_{J_\beta} = \tau$ onto J_β. But such an f is then $\mathbf{\Sigma}_\omega(J_\beta)$ and hence $f \in J_{\beta+1}$. □(1.10.1) □(Claim 1)

Claim 2. J_α is sound.

Proof. We shall make use of Lemma 1.9.6 here. Hence, for $n \in \omega$ we prove

$$p_{J_\alpha,n} \in R^n_{J_\alpha}. \tag{1.10.3}$$

Suppose this is false. Pick the first n such that $p = p_{J_\alpha,n} \notin R^n_{J_\alpha}$. Let a be $\Sigma_1^{(n-1)}(J_\alpha)$ in p such that $a \cap \omega\varrho^n_{J_\alpha} \notin J_\alpha$. Using the downward extension of embeddings lemma we construct the unique $J_{\bar\alpha}$, $\bar p$, π such that the same conditions as in (1.9.2) hold but with p and $\bar p$ in place of q and $\bar q$, respectively, J_α and $J_{\bar\alpha}$ in place of M and $\bar M$, respectively. Hence $a \cap \omega\varrho^n_{J_\alpha} = \bar a \cap \omega\varrho^n_{J_{\bar\alpha}}$ where $\bar a$ is $\Sigma_1^{(n-1)}(J_{\bar\alpha})$ by the same definition. Hence $\bar\alpha$ cannot be less than α, since otherwise $a \cap \omega\varrho^n_{J_\alpha} \in J_\alpha$. Consequently, $\bar\alpha = \alpha$ and $\bar p \in P^n_{J_\alpha}$. It is also clear by the construction that $\bar p \leq^* p$. But $p \leq^* \bar p$ since p is the standard parameter. Hence, $p = \bar p$. But this means $p \in R^n_{J_\alpha}$. Contradiction.

□(Claim 2) □(Lemma 1.10.1, 1.10.2)

1.11 More on Downward Extensions of Embeddings

Let M be acceptable, $p \in \Gamma^n_M$ and $\pi : N \to M^{n,p}$ be Σ_1-preserving. Set $X = \text{rng}(\pi)$. Now let $\bar{M}, \bar{p}, \tilde{\pi}$ be the unique objects given by the downward extension of embeddings lemma and $Y = \text{rng}(\tilde{\pi})$. Then:

a) If $f(x_1^{i_1}, \ldots, x_\ell^{i_\ell})$ is a good $\Sigma_1^{(n)}(M)$-function and $a_j \in H^{i_j}_M \cap Y$ for $j = 1, \ldots, \ell$, then $f(a_1, \ldots, a_\ell) \in Y$ as well, since it is equal to $\tilde{\pi}(\bar{f}(\bar{a}_1, \ldots, \bar{a}_\ell))$, where $\bar{a}_j = \tilde{\pi}^{-1}(a_j)$ for $j = 1, \ldots, \ell$ and $\bar{f}$ is $\Sigma_1^{(n)}(\bar{M})$ by the same functionally absolute definition. Hence Y is closed under the good $\Sigma_1^{(n)}(M)$-functions.

b) Y is the smallest set such that $X \subset Y$ and a) holds.

This follows from the fact that $\tilde{h}^n_{\bar{M}}(H^n_{\bar{M}} \cup \{\bar{p}\}) = \bar{M}$, hence $\tilde{h}^n_M(X \cup \{p\}) = Y$ (that is, Y cannot be smaller).

In other words, Y is the closure of $X \cup \{p\}$ under the good $\Sigma_1^{(n)}(M)$ functions (we also call it the "$\Sigma_1^{(n)}(M)$-hull of $X \cup \{p\}$") and is also the least $\Sigma_1^{(n)}$-elementary substructure of M containing $X \cup \{p\}$. $\bar{M}$ is then the transitive collapse of Y. Since this terminology corresponds to that used for Σ_n substructures, we shall often use it instead of formally referring to the downward extension of embeddings lemma. Another reason for this is, that we, in practice, do not start from a map $\pi : N \to M^{n,p}$, but rather from a subset Z of $M^{n,p}$, being interested in the least $\Sigma_1^{(n)}$-preserving map $\tilde{\pi} : \bar{M} \to M$ whose range contains $Z \cup \{p\}$. Such a map is clearly uniquely determined by the $\Sigma_1^{(n)}(M)$-hull of $Z \cup \{p\}$.

The following lemma, which will be often useful, justifies the heuristics described above.

Definition. Let $M = \langle |M|, A_1, \ldots, A_n \rangle$ and $X \subset |M|$. Then

$$M|X = \langle |M| \cap X, A_1 \cap X, \ldots, A_n \cap X \rangle.$$

Lemma 1.11.1. *Let $M = \langle J^B_\beta, D \rangle$ be an acceptable structure and $X \subset M$ be closed under good $\mathbf{\Sigma}_1^{(n)}(M)$ functions and such that $X \cap P^n_M \neq \emptyset$. Let $\bar{M}$ be the transitive collapse of $M|X$ and σ be the inverse of the collapsing map. Then $\sigma : \bar{M} \underset{\Sigma_1^{(n)}}{\longrightarrow} M$.*

Proof. Note first that for every $i \leq n$ and every $\Sigma_1^{(i)}$ formula $\varphi(z^i, v_1, \ldots, v_\ell)$ the following holds: if $x_1, \ldots x_\ell \in X$ then

$$M \models (\exists z^i)\varphi(z^i, x_1, \ldots, x_\ell) \quad \text{iff} \quad M|X \models (\exists z^i)\varphi(z^i, x_1, \ldots, x_\ell) \tag{1.11.1}$$

where the variables of i-th type are interpreted over $H^i_M \cap X$. This follows by a straightforward induction on the $\Sigma_1^{(n)}$ formulae using Lemma 1.8.3 and the fact that X is closed under good $\mathbf{\Sigma}_1^{(n)}(M)$ functions: if the left side of (1.11.1) holds, then there is a good $\Sigma_1^{(i)}(M)$ function f such that $z = f(x_1, \ldots, x_\ell)$ witnesses this. Consequently, $z \in X \cap H_M^i$.

Collapse $M|X$ to some $\bar{M}$. Let σ be the inverse of the collapsing map and $H_i = \sigma^{-1}{}'' H_M^i \cap X$ and $\omega\varrho_i = \mathbf{On} \cap H_i$ for $i \leq n$. We now interpret the variables of type i as elements of H_i for $i \leq n$. In this pseudointerpretation the following holds

$$\bar{M} \models \varphi(x_1, \ldots, x_\ell) \quad \text{iff} \quad M \models \varphi(\sigma(x_1), \ldots, \sigma(x_\ell)) \tag{1.11.2}$$

for every $\Sigma_1^{(n)}$ formula φ and all $x_1, \ldots, x_\ell \in \bar{M}$, as follows from (1.11.1). Using this we prove that our pseudointerpretation coincides with the ordinary interpretation, i.e.

$$\omega\varrho_i = \omega\varrho_{\bar{M}}^i \quad \text{for} \quad i \leq n \tag{1.11.3}$$

which immediately yields $\sigma : \bar{M} \underset{\Sigma_1^{(n)}}{\longrightarrow} M$.

We proceed by induction on i. The case $i = 0$ is trivial. Hence σ is Σ_1 and $\bar{M}$ is an acceptable structure. So suppose (1.11.3) holds for all $i' < i$. (Hence, σ is $\Sigma_1^{(i-1)}$ in the correct interpretation and $\Sigma_1^{(i)}$ in the pseudointerpretation.) We show first

$$\omega\varrho_{\bar{M}}^i \leq \omega\varrho_i. \tag{1.11.4}$$

To see this, pick $p \in P_M^n \cap X$ and set $\bar{p} = \sigma^{-1}(p)$. Let A be $\Sigma_1^{(i-1)}(M)$ in p such that $A \cap \omega\varrho_M^i \notin M$ and $\bar{A}$ be $\Sigma_1^{(i-1)}(M)$ in $\bar{p}$ by the same definition. Then $\bar{A} \cap \omega\varrho_i \notin \bar{M}$. If it were an element of $\bar{M}$, set $\bar{a} = \bar{A} \cap \omega\varrho_i$. Then

$$(\forall \xi^i)(\xi^i \in \bar{a} \longleftrightarrow \bar{A}(\xi^i))$$

in the pseudointerpretation. This is a $\Pi_1^{(i)}$ statement, hence in M we have

$$(\forall \xi^i)(\xi^i \in \sigma(\bar{a}) \longleftrightarrow A(\xi^i))$$

which means that $A \cap \omega\varrho_M^i = \sigma(\bar{a}) \cap \omega\varrho_M^i \in M$. Contradiction.

As to the converse, let $\bar{A}$ be a $\Sigma_1^{(i-1)}(\bar{M})$ set in $\bar{q}$ and A be a $\Sigma_1^{(i-1)}(M)$ set in $q = \sigma(\bar{q})$ by the same definition. Let $\bar{\alpha} < \omega\varrho_i$ and $\alpha = \sigma(\bar{\alpha})$. Then $A \cap \alpha \in M$ and is in fact an element of H_M^i. Thus

$$(\exists z^i)(\forall \xi^i < \alpha)(\xi^i \in z^i \longleftrightarrow A(\xi^i))$$

holds in M. This is a $\Sigma_1^{(i)}$ statement, hence

$$(\exists z^i)(\forall \xi^i < \bar{\alpha})(\xi^i \in z^i \longleftrightarrow \bar{A}(\xi^i))$$

which means that $\bar{A} \cap \bar{\alpha} \in \bar{M}$. This completes the proof. □(Lemma 1.11.1)

Under a stronger assumption we can get stronger preservation properties of σ.

Lemma 1.11.2. *Let $M = \langle J^B_\beta, D\rangle$ and $\bar{M} = \langle J^{\bar{B}}_{\bar{\beta}}, \bar{D}\rangle$ be acceptable structures. Let $\sigma : \bar{M} \to M$ be $\Sigma^{(n)}_1$-preserving and such that $\sigma \upharpoonright \omega\varrho^{n+1}_M = \mathrm{id}$ and $\mathrm{rng}(\sigma) \cap P^*_M \neq \emptyset$. Then σ is Σ^*-preserving.*

Proof. By induction on m we prove that

a) $\omega\varrho^m_{\bar{M}} = \omega\varrho^m_M$

b) $\sigma : \bar{M} \underset{\Sigma^{(m)}_1}{\longrightarrow} M$

for $m > n$. As is usual for this sort of proof, we shall alternately use a) to prove b) and vice versa in the course of the induction.

Claim 1. Suppose $\sigma : \bar{M} \underset{\Sigma^{(m)}_1}{\longrightarrow} M$, where $m \geq n$. Then $\omega\varrho^{m+1}_{\bar{M}} = \omega\varrho^{m+1}_M$.

Proof. Let $p \in \mathrm{rng}(\sigma) \cap P^*_M$ and $\bar{p} = \sigma^{-1}(p)$. Let Q be $\Sigma^{(m)}_1(M)$ in p such that $Q \cap \omega\varrho^{m+1}_M \notin M$ and $\bar{Q}$ be $\Sigma^{(m)}_1(\bar{M})$ in $\bar{p}$ by the same definition. Then $\bar{Q} \cap \omega\varrho^{m+1}_M \notin \bar{M}$. To see this suppose the contrary. Set $\bar{a} = \bar{Q} \cap \omega\varrho^{m+1}_M$. But then $\sigma(\bar{a}) \cap \omega\varrho^{m+1}_M = \bar{a} = \bar{Q} \cap \omega\varrho^{m+1}_M = Q \cap \omega\varrho^{m+1}_M$ since $\sigma \upharpoonright \omega\varrho^{m+1}_M = \mathrm{id}$. Contradiction. Hence

$$\omega\varrho^{m+1}_{\bar{M}} \leq \omega\varrho^{m+1}_M.$$

To see the converse, let $\bar{q} \in \bar{M}, q = \sigma(q)$, $\bar{Q}$ be $\Sigma^{(m)}_1(\bar{M})$ in $\bar{q}$ and Q be $\Sigma^{(m)}_1(M)$ in q by the same definition. Pick $\bar{\alpha} < \omega\varrho^{m+1}_M$. Then clearly $\bar{Q} \cap \bar{\alpha} = Q \cap \bar{\alpha} \in M$. But $\omega\varrho^{n+1}_M$ is a cardinal in M (otherwise $\omega\varrho^{n+1}_M = \mathbf{On} \cap M$, thus $\sigma = \mathrm{id}$ and the lemma follows trivially) hence $\bar{Q} \cap \bar{\alpha} \in J^B_{\varrho^{n+1}_M} = J^{\bar{B}}_{\varrho^{n+1}_M}$. □(Claim 1)

Claim 2. Suppose $\omega\varrho^m_M = \omega\varrho^m_M$ and $\sigma : \bar{M} \underset{\Sigma^{(m-1)}_1}{\longrightarrow} M$, where $m > n$. Then

$$\sigma : \bar{M} \underset{\Sigma^{(m)}_1}{\longrightarrow} M.$$

Proof. By induction on $\Sigma^{(m)}_1$-formulae. For $\Sigma^{(m-1)}_1$-formulae this holds by our assumption. The induction step for boolean combinations of formulae is trivial. Now suppose $\varphi(z^m, v^0_1, \ldots, v^0_\ell)$ is preserved by σ. We show that $(\exists z^m)\varphi(z^m, v^0_1, \ldots, v^0_\ell)$ is also preserved, which clearly proves the claim. Let $x_1, \ldots, x_\ell \in \bar{M}$ and $M \models (\exists z^m)\varphi(z^m, \sigma(x_1), \ldots, \sigma(x_\ell))$. Pick $z \in H^m_M$ witnessing this. By the induction hypothesis,

$$\bar{M} \models \varphi(z, x_1, \ldots, x_\ell) \quad \text{iff} \quad M \models \varphi(z, \sigma(x_1), \ldots, \sigma(x_\ell))$$

since $\sigma(z) = z$. Hence $\bar{M} \models (\exists z^m)\varphi(z^m, x_1, \ldots, x_\ell)$ since $H^m_{\bar{M}} = H^m_M$.

□(Claim 2, Lemma 1.11.2)

1.12 Witnesses and Solidity

Witnesses are a tool for identifying standard parameters. One of their useful properties is that they are preserved under suitable maps. This means, roughly speaking, that such maps will also preserve standard parameters and other fine structural notions.

Definition. Let M be an acceptable structure and $p \subset M$ be a finite set of ordinals. We say that $\langle Q, r\rangle$ is a *generalized witness* for $\nu \in \mathbf{On} \cap M$ w.r.t. M, p iff

a) Q is acceptable and $\nu \subset Q$;

b) $r \subset Q$ is a finite set of ordinals;

c) for every $\Sigma_1^{(n)}$ formula $\varphi(v_0, \ldots, v_\ell)$ and all $\xi_1, \ldots, \xi_\ell < \nu$

$$M \models \varphi(p - (\nu + 1), \xi_1, \ldots, \xi_\ell) \longrightarrow Q \models \varphi(r, \xi_1, \ldots, \xi_\ell),$$

where n is such that $\omega\varrho_M^{n+1} \leq \nu < \omega\varrho_M^n$.

We could, of course, allow Q to be non-transitive, but in this case we would have to explain in which sense $\Sigma_0^{(n)}$-formulae are satisfied in Q, which would mean a lot of tedious and superfluous work, since the above definition suffices for our applications. In the following we shall often call Q a witness, taking r as a suppressed parameter. Although we allowed p to be an arbitrary finite set of ordinals, we shall be primarily interested in cases where $p \in P_M^*$ (resp. $p \in P_M^n$ for $n \in \omega$).

We begin our study of witnesses with the observation that the clause c) of the above definition can be reformulated as a "uniform" statement.

Lemma 1.12.1. *There is a $\Sigma_1^{(n)}$-formula ψ such that for every M, p, ν, n as above, $\langle Q, r\rangle$ is a generalized witness for ν w.r.t. M, p iff for any $\xi_1, \ldots, \xi_\ell < \nu$,*

$$M \models \psi(p - (\nu + 1), \xi_1, \ldots, \xi_\ell) \longrightarrow Q \models \psi(r, \xi_1, \ldots, \xi_\ell), \tag{1.12.1}$$

Proof. The implication ($\rightarrow$) is clearly trivial. To see the converse, recall that the Σ_1-satisfaction relation for $M^{n,p\restriction n}$ is uniformly $\Sigma_1(M^{n,p\restriction n})$; hence there is a $\Sigma_1^{(n)}$-formula ψ such that

$$\psi(i, \langle p', \xi_1, \ldots, \xi_\ell\rangle) \longleftrightarrow \models^{\Sigma_1}_{M^{n,p\restriction n}} (i, \langle p' \cap \omega\varrho_M^n, \xi_1, \ldots, \xi_\ell, \rangle)$$

where $p' = p - (\nu + 1)$ and $\xi_1, \ldots, \xi_\ell$ are arbitrary less than ν. Then ψ is our desired formula. Clearly, ψ is determined uniformly. The clause c) of the definition of generalized witness follows from (1.12.1) in a straightforward way (use Lemma 1.6.4).

□(Lemma 1.12.1)

The following lemma explains the importance of generalized witnesses.

Lemma 1.12.2. *Let M be an acceptable structure, $n \in \omega$ and $p \in P_M^n$. Suppose that for every $\nu \in p$ there is a generalized witness Q for ν w.r.t. M, p such that $Q \in M$. Then $p = p_{M,n}$.*

Proof. Suppose not. Then $p_{M,n} <^* p$. This means that there is $\nu \in p - p_{M,n}$ such that $p - (\nu + 1) = p_{M,n} - (\nu + 1)$; denote this common value by q. Let $i < n$ be such that $\omega\varrho_M^{i+1} \leq \nu < \omega\varrho_M^i$ and A be a $\Sigma_1^{(i)}(M)$ set in $p_{M,n}$ such that $A \cap \omega\varrho_M^{i+1} \notin M$. Fix a generalized witness $\langle Q, r\rangle$ for ν w.r.t. M, p such that $Q \in M$.

Let X be the closure of $\nu \cup q$ under good $\Sigma_1^{(i)}(M)$ functions. Collapse $M|X$ to some W and let $\sigma : W \to M$ be the inverse of the collapsing map, which is clearly $\Sigma_1^{(i)}$-preserving, as follows from the downward extension of embeddings lemma. Let $\bar{q} = \sigma^{-1}(q)$. Then W is the closure of $\nu \cup \bar{q}$ under good $\Sigma_1^{(i)}(W)$ functions; in particular,

$$W = \tilde{h}_W^{i+1}(\nu \cup \{\bar{q}\}). \tag{1.12.2}$$

Let $\bar{A}$ be the $\Sigma_1^{(i)}(W)$ set in $\bar{p} = \sigma^{-1}(p_{M,n})$ by the same definition as A (notice that $p_{M,n} \in \operatorname{rng}(\sigma)$!). Since $\sigma \restriction \nu = \mathrm{id}$, $\bar{A} \cap \nu = A \cap \nu$. We shall construct a $\Sigma_0^{(i)}$ embedding σ^* of W into Q as follows: if $z \in W$, $z = \bar{f}(\xi_1, \ldots, \xi_\ell, \bar{q})$ for some good $\Sigma_1^{(i)}(W)$ function $\bar{f}$ and some $\xi_1, \ldots, \xi_\ell < \nu$; set

$$\sigma^*(z) = \tilde{f}(\xi_1, \ldots, \xi_\ell, r)$$

where $\tilde{f}$ is the good $\Sigma_1^{(i)}(Q)$ function by the same definition as $\bar{f}$. Then

$$\sigma^* \text{ is correctly defined.} \tag{1.12.3}$$

To see this, note first that if $z = \bar{f}(\xi_1, \ldots, \xi_\ell, \bar{q})$ then there is a $\tilde{z} \in Q$ such that $\tilde{z} = \tilde{f}(\xi_1, \ldots, \xi_\ell, r)$. Clearly there is a $z' = \sigma(z) \in M$ satisfying $z' = f(\xi_1, \ldots, \xi_\ell, q)$, where f is the good $\Sigma_1^{(i)}(M)$ function by the same definition. The statement

$$(\exists z')(z' = f(\xi_1, \ldots, \xi_\ell, q))$$

is $\Sigma_1^{(i)}$; hence by c) of the definition of generalized witness we obtain a $\tilde{z}$ as required above. Thus, all we have to prove is that $\sigma^*(z)$ does not depend on the representation of z. But if $\bar{f}, \bar{g}$ are good $\Sigma_1^{(i)}(W)$ functions such that $\bar{f}(\xi_1, \ldots, \xi_\ell, \bar{q}) = \bar{g}(\eta_1, \ldots, \eta_m, \bar{q})$ for some $\xi_1, \ldots, \xi_\ell, \eta_1, \ldots, \eta_m, < \nu$ then

$$f(\xi_1, \ldots, \xi_\ell, q) = g(\eta_1, \ldots, \eta_m, q) \tag{1.12.4}$$

where f, g are the corresponding good $\Sigma_1^{(i)}(M)$ functions. By the fact that (1.12.4) is a $\Sigma_1^{(i)}$-statement,

$$\tilde{f}(\xi_1, \ldots, \xi_\ell, r) = \tilde{g}(\eta_1, \ldots, \eta_m, r)$$

where $\tilde{f}, \tilde{g}$ are the corresponding good $\Sigma_1^{(i)}(Q)$ functions. □(1.12.3)

The same argument in fact shows that for every $\Sigma_1^{(i)}$ formula $\varphi(v_1, \ldots, v_\ell)$ and every $z_1, \ldots, z_\ell \in W$,

$$W \models \varphi(z_1, \ldots, z_\ell) \longrightarrow Q \models \varphi(\sigma^*(z_1), \ldots, \sigma^*(z_\ell)) \tag{1.12.5}$$

since if z_j is of the form $\bar{f}_j(\xi_1^j, \ldots, \xi_{m_j}^j, \bar{q})$ then

$$\varphi(\bar{f}_1(\xi_1^1, \ldots, \xi_{m_1}^1, \bar{q}), \ldots, f_\ell(\xi_1^\ell, \ldots, \xi_{m_\ell}^\ell, \bar{q}))$$

is again a $\Sigma_1^{(i)}$ statement. But (1.12.5) clearly yields

$$\sigma^* \text{ is } \Sigma_0^{(i)}\text{-preserving.} \tag{1.12.6}$$

Now set $\omega\varrho^* = \sup(\sigma^{*\prime\prime}\omega\varrho_W^i)$, $A^* = A_Q^{i,r\restriction i} \cap J_{\varrho^*}^{B_Q}$ (we suppose $Q = \langle J_{\alpha_Q}^{B_Q}, D_Q\rangle$). Clearly $\varrho^* \leq \varrho_Q^i$. Hence $Q^* = \langle J_{\varrho^*}^{B_Q}, A^*\rangle \in M$. Moreover

$$\sigma^* \restriction H_W^i : W^{i,\bar{q}\restriction i} \underset{\Sigma_0}{\longrightarrow} Q^* \text{ cofinally, } \sigma^* \restriction \nu = \text{id} \text{ and } \sigma^*(\bar{q}) = r. \tag{1.12.7}$$

We now return back to our set $\bar{A}$. Clearly $\bar{A}$ is $\Sigma_1(W^{i,\bar{q}\restriction i})$ in $s = \bar{p} \cap \omega\varrho_W^i$. Let $\tilde{A}$ be $\Sigma_1(Q^*)$ in $\sigma^*(s)$ by the same definition. Then $\tilde{A} \in M$. By (1.12.7),

$$\tilde{A} \cap \omega\varrho_M^{i+1} = \bar{A} \cap \omega\varrho_M^{i+1} = A \cap \omega\varrho_M^{i+1}.$$

Contradiction. □(Lemma 1.12.2)

Both the above lemma and its proof have several important consequences. We shall first focus on the structure W which we constructed above. Note that $\langle W, \bar{q}\rangle$ is a generalized witness for ν w.r.t. M, p.

Definition. Let M be acceptable, $p \subset M$ be a finite set of ordinals and $\nu \in \mathbf{On} \cap M$. Let $W = W_M^{\nu,p}$ be the transitive collapse of $M|X$, where X is the $\Sigma_1^{(n)}$-hull of $\nu \cup p - (\nu+1)$ in M for $n \in \omega$ such that $\omega\varrho_M^{n+1} \leq \nu < \omega\varrho_M^n$. We call W the *standard* witness for ν w.r.t. M, p and the corresponding inverse of the collapsing map $\sigma : W \to M$ the *canonical witness map* of W.

Lemma 1.12.3. *Let M be acceptable and $p \in P_M^n$. Then for every $\nu \in P_M^n$,*

$W_M^{\nu,p} \in M$ iff there is a generalized witness $Q \in M$ for ν w.r.t. M, p.

Proof. The implication ($\leftarrow$) is trivial since $W_M^{\nu,p}$ is a generalized witness. As to the converse, we analyse the proof of Lemma 1.12.2 more closely. Note that by (1.12.2) there is a $\boldsymbol{\Sigma}_1^{(i)}(W)$ set $a \subset \nu$ which codes $W_M^{\nu,p}$. By the same argument as above, a is $\boldsymbol{\Sigma}_1(Q^*)$ and consequently $a \in M$. Now, note that

$$\nu = \text{cr}(\sigma). \tag{1.12.8}$$

If this were not the case, then $\nu \in \text{rng}(\sigma)$. As a consequence, p would be an element of $\text{rng}(\sigma)$, so given any $\Sigma_1^{(i)}(M)$ set A in p satisfying $A \cap \omega\varrho_M^{i+1} \notin M$, the set $\bar{A}$ that is $\Sigma_1^{(i)}(W)$ in $\sigma^{-1}(p)$ by the same definition would agree with A below $\omega\varrho_M^{i+1}$, i.e.

$A \cap \omega\varrho_M^{i+1} = \bar{A} \cap \omega\varrho_M^{i+1}$, which yields a contradiction exactly as before. But (1.12.8) means that ν is regular in W and consequently $\sigma(\nu) > \nu$ is regular in M. Hence $H^M_{\sigma(\nu)}$ is a ZFC$^-$ model containing a as an element, so we can decode a inside $H^M_{\sigma(\nu)}$. Thus $W \in M$. □(Lemma 1.12.3)

Since $p \upharpoonright i \in P^i_M$ for every $p \in P^n_M$ and $i \leq n$, Lemma 1.12.2 gives us the following corollary:

Corollary 1.12.4. *Let M be acceptable, $p \in P^n_M$ and suppose that every $\nu \in p$ has a generalized witness w.r.t. M, p. Then $p \upharpoonright i = p_{M,i}$ for all $i \leq n$. Moreover, if n is such that $\omega\varrho^n_M = \omega\varrho^\omega_M$ (i.e. $p \in P^*_M$) then $p = p_M$.* □(Corollary 1.12.4)

We have thus seen that only elements of standard parameters can have witnesses in M. The next question is whether there are structures containing witnesses. It will later turn out that mice comprise a class of such structures (and it is easy to see that each J_α is such a structure). Hence the following definition does make sense.

Definition. An acceptable structure M is *solid* iff $W^{\nu,p_M}_M \in M$ for all $\nu \in p_M$. M is solid above α iff $W^{\nu,p_M}_M \in M$ whenever $\nu \in p_M - \alpha$.

Notation. Instead of W^{ν,p_M}_M we write briefly W^ν_M.

Lemma 1.12.5. *Let M be solid above α and $p \in M$ be a finite set of ordinals such that*

a) $p \cap \alpha = \emptyset$;

b) $M = \tilde{h}^{n+1}_M(\alpha \cup \{p\})$ *where $n \in \omega$ is such that* $\omega\varrho^{n+1}_M \leq \alpha < \omega\varrho^n_M$;

c) *for every $\nu \in p$ there is a generalized witness $\langle Q, r \rangle$ for ν w.r.t. M, p, such that $Q \in M$.*

Then $p = p_M - \alpha$.

Proof. Note first that b) implies

$$M = \tilde{h}^n_M(\omega\varrho^n_M \cup \{p - \omega\varrho^n_M\})$$

which means that $p - \omega\varrho^n_M \in P^n_M$. By Lemma 1.12.2 and Corollary 1.12.4,

$$p - \omega\varrho^n_M = p_M - \omega\varrho^n_M.$$

Therefore it suffices to prove

$$\nu \in p \longleftrightarrow \nu \in p_M \tag{1.12.9}$$

for ν between α and $\omega\varrho^n_M$. Suppose the contrary. Let ν be minimal such that

$$p - (\nu + 1) = p_M - (\nu + 1).$$

From this we derive a contradiction by showing that (1.12.9) holds for such ν after all.

The implication ($\leftarrow$). Suppose $\nu \notin p$. It follows immediately from the definition of a standard witness and form b) that

$$W_M^{\nu, p_M} = W_M^{\nu, p} = M.$$

Consequently, $\nu \notin p_M$.

The implication ($\rightarrow$). Suppose $\nu \in p$. If $\nu \notin p_M$ then $p_M \subset p - (\nu + 1) \cup \nu$. Hence if A is a $\Sigma_1^{(n)}(M)$ set in p_M such that $A \cap \omega\varrho_M^{n+1} \notin M$ and $\bar{A}$ is $\Sigma_1^{(n)}(W_M^{\nu,p})$ in $\sigma^{-1}(p_M)$ (σ is the canonical witness map), then $A \cap \omega\varrho_M^{n+1} = \bar{A} \cap \omega\varrho_M^{n+1}$ is $\mathbf{\Sigma}_1(Q^*)$, where Q^* is defined exactly as in the proof of Lemma 1.12.2. But $Q^* \in M$ which means $A \cap \omega\varrho_M^{n+1} \in M$. Contradiction. □(Lemma 1.12.5)

We now show that sufficiently preserving maps between solid structures preserve standard parameters.

Lemma 1.12.6. *Let $\bar{M}$, M be acceptable and let $\sigma : \bar{M} \rightarrow M$ be $\Sigma_1^{(n)}$-preserving. Suppose $\bar{p} \in \bar{M}$, $p = \sigma(\bar{p})$, $\bar{\nu} \in \bar{M}$, $\nu = \sigma(\bar{\nu})$ where $\omega\varrho_{\bar{M}}^{n+1} \leq \bar{\nu}$, $\omega\varrho_M^{n+1} \leq \nu$. Let $\langle \bar{Q}, \bar{r} \rangle$ be a generalized witness for $\bar{\nu}$ w.r.t. $\bar{M}$, $\bar{p}$ such that $\bar{Q} \in \bar{M}$ and $\langle Q, r \rangle = \sigma(\langle \bar{Q}, \bar{r} \rangle)$. Then $\langle Q, r \rangle$ is a generalized witness for ν w.r.t. M, p.*

Proof. Notice that there is an $i < n$ such that $\omega\varrho_{\bar{M}}^{i+1} \leq \bar{\nu} < \omega\varrho_{\bar{M}}^{i}$ and $\omega\varrho_M^{i+1} \leq \nu < \omega\varrho_M^{i}$. Let $\varphi(v_0, \ldots, v_\ell)$ be a $\Sigma_1^{(i)}$ formula. Then

$$\bar{M} \models (\forall \xi_1^i, \ldots, \xi_\ell^i < \bar{\nu})[\varphi(\bar{p} - (\bar{\nu} + 1), \xi_1^i, \ldots, \xi_\ell^i) \longrightarrow \bar{Q} \models \varphi(\bar{r}, \xi_1^i, \ldots, \xi_\ell^i)].$$

This statement is clearly $\Pi_1^{(i)}$ and hence preserved upwards. □(Lemma 1.12.6)

Lemma 1.12.7. *Let $\bar{M}$, M, σ, $\bar{p}$, p be as in the previous lemma. Suppose moreover that $\bar{p} \in P_{\bar{M}}^{n+1}$, $p \in P_M^{n+1}$, $\omega\varrho_{\bar{M}}^{n+1} \leq \bar{\alpha}$, $\omega\varrho_M^{n+1} \leq \alpha = \sigma(\bar{\alpha})$ and $\bar{M}$ is solid above $\bar{\alpha}$. Then $\bar{p} - \bar{\alpha} = p_{\bar{M},n+1} - \bar{\alpha} = p_{\bar{M}} - \bar{\alpha}$, $p - \alpha = p_{M,n+1} - \alpha = p_M - \alpha$ and M is solid above α.*

Proof. By a straightforward application of the above lemmata. □(Lemma 1.12.7)

Corollary 1.12.8. *Let $\bar{M}$, M be acceptable, $\sigma : \bar{M} \rightarrow M$ be Σ^*-preserving and $p = \sigma(p_{\bar{M}}) \in P_M^*$. Suppose $\bar{M}$ is solid. Then $p = p_M$ and M is solid.*

□(Corollary 1.12.8)

Notes

Fine structure theory was originally developed in the $\boldsymbol{L}$-context by Jensen in [Je72]. Since then, a much broader class of structures has been studied and a more general theory needed. The $*$-theory was worked out by Jensen and first presented in a compact form in [Je88]. This is a completely general theory and, as presented in this book, can be developed independently of inner model theory. There are other approaches to the general fine structure theory; some of them turn out to be equivalent to the $*$-theory, as far as we deal with the applications to inner models. The notion of a solidity witness was discovered Mitchell and later rediscovered by S. Friedman.

Chapter 2

Extenders and Coherent Structures

In this section, we shall present two representations of extenders and introduce a notion of coherent structure, which is the first step towards the definition of premouse. We assume some familiarity with extenders, although most of their basic properties are stated below. We shall, however, often skip over the proofs or merely sketch them. For an introduction to the theory of extenders we recommend [Ka94] or [MS89]; familiarity of either of these two accounts will be more than sufficient for understanding this section.

An extender is a means of encoding elementary embeddings of models of (fragments of) ZFC in a set-size object. There are various ways to represent extenders. One possible approach is to consider embeddings restricted to the power set of their critical points. The embeddings are then reconstructible from their restrictions.

2.1 Extenders

Definition. $U \subset \mathcal{P}(\kappa)$ is *suitable* for κ iff

a) κ is primitive recursively closed.

b) If $A_1, \dots, A_n \in U$, $\nu_1, \dots, \nu_m \in \kappa$ and B is primitive recursive in the predicates $A_1, \dots, A_n$ and the parameters $\nu_1, \dots, \nu_m$, then $B \cap \kappa \in U$.

Remark. For U as above we have

i) $\kappa \in U$.

ii) $A \cap B, A \cup B, A - B \in U$ for $A, B \in U$.

iii) $\nu < \kappa \longrightarrow \nu, \{\nu\} \in U$.

iv) If we set $U^m = \{A^{(m)};\ A \in U\}$ where

$$A^{(m)} = \{\langle \xi_1, \dots, \xi_m \rangle \in {}^m\kappa;\ \prec \xi_1, \dots, \xi_m \succ \in A\},$$

then for any $B \subset {}^m\mathbf{On}$ we have: if $A_1, \dots, A_n \in {}^mU$ and B is primitive recursive in $A_1, \dots, A_m$ and parameters less than κ, then $B \cap {}^m\kappa \in U$.

Hence, using Gödel's pairing function, we can code tuples of ordinals by ordinals; this is the reason why we require that certain ordinals (like κ in the above definition) are primitive recursively closed. (We, of course, need not do this, but we prefer to, as it simplifies notation.)

Definition (The map representation of extenders). Let $U \subset \mathcal{P}(\kappa)$ be suitable for κ and $\lambda > \kappa$ be primitive recursively closed. A map $F : U \longrightarrow \mathcal{P}(\lambda)$ is an *extender on* U with *length* λ iff whenever $B_{A_1,\ldots,A_n} \subset \mathbf{On}$ is uniformly primitive recursive in $A_1, \ldots, A_n \in U$ and parameters from κ, then

$$F(B_{A_1,\ldots,A_n}) = B_{F(A_1),\ldots,F(A_n)}.$$

The ordinal κ is called the *critical point* of F.

The above definition should be understood in the following way: if $B_{A_1,\ldots,A_n}$ is primitive recursive in $A_1, \ldots, A_n$ and parameters $\nu_1, \ldots, \nu_m < \kappa$, then $F(B_{A_1,\ldots,A_n})$ is primitive recursive in $F(A_1), \ldots, F(A_n), \nu_1, \ldots, \nu_m$ by the same definition.

If U is clear from the context, we often briefly say "extender at κ, λ" instead of "extender on U with length λ". We shall occasionally write length(F) for the length of F.

Let us also mention that V will denote the universe of all sets throughout the whole book.

The following are immediate consequences of the above definition.

i) $F(\nu) = \nu$ and $F(\{\nu\}) = \{\nu\}$ for $\nu < \kappa$;

ii) $F(\kappa) = \lambda$ and $F(\emptyset) = \emptyset$;

iii) $F(A \cap B) = F(A) \cap F(B)$, $F(A \cup B) = F(A) \cup F(B)$ and $F(A - B) = F(A) - F(B)$;

iv) $A \subset B \longleftrightarrow F(A) \subset F(B)$;

v) $\kappa \cap F(A) = A$, as for $\nu < \kappa$ we have

$$\nu \in A \longleftrightarrow \{\nu\} \subset A \longleftrightarrow \{\nu\} \subset F(A) \longleftrightarrow \nu \in F(A).$$

We can extend F to F^m on U^m by setting

$$F^m(A^{(m)}) = (F(A))^{(m)}.$$

To simplify the notation, we write $F(A)$ for $F^m(A)$ for $A \in U^m$. Then

vi) $F({}^m\kappa \cap B_{A_1,\ldots,A_n}) = {}^m\lambda \cap B_{F(A_1),\ldots,F(A_n)}$ whenever $B_{A_1,\ldots,A_n} \subset {}^m\mathbf{On}$ is uniformly primitive recursive in $A_1, \ldots, A_n \in U$ and parameters from κ;

vii) $F(\nu_1 \times \cdots \times \nu_m) = \nu_1 \times \cdots \times \nu_m$ and $F(\langle \nu_1, \ldots, \nu_m \rangle) = \langle \nu_1, \ldots, \nu_m \rangle$ for any $\nu_1, \ldots, \nu_m \in \kappa$.

Definition. $M = \langle |M|, \in, A_1, \ldots, A_n \rangle$ is *suitable for* κ iff

a) M is transitive and rudimentarily closed.

b) If $x \in M$, then there are $\alpha \in \mathbf{On} \cap M$ and $f \in M$ such that $f : \alpha \xrightarrow{\text{onto}} x$.

c) κ is primitive recursively closed.

d) If $a_1, \ldots, a_m \in \mathcal{P}(\kappa) \cap M$, then $J_\kappa^{a_1,\ldots,a_m} \in M$.

It follows from the above definition that $U = \mathcal{P}(\kappa) \cap M$ is suitable.

At the beginning of this section, we noted that extenders are canonical set-size objects coding elementary embeddings of structures. In order to establish the correspondence between extenders and the associated embeddings, we need that the respective structures have sufficient closure properties. Such closure properties are guaranteed by suitability. Property b) will be used in the proof of Łos' Theorem: property d) in reconstructing an extender from its associated embedding. One perhaps needs less for the latter purpose; however, the class of all suitable structures is sufficiently wide for our later applications and its definition is simple enough.

Definition. Let $M = \langle |M|, \in, A_1, \ldots, A_n \rangle$. We write $\pi : M \underset{F}{\longrightarrow} N$ iff

a) N is transitive.

b) $\pi : M \to N$ is Σ_0-preserving and cofinal.

c) M is suitable for $\kappa = \mathrm{cr}(\pi)$.

d) $F = \langle \lambda \cap \pi(x);\ x \in \mathcal{P}(\kappa) \cap M \rangle$ for a primitive recursively closed λ such that $\kappa < \lambda \leq \pi(\kappa)$.

e) $N =$ the $\Sigma_0(N)$-closure of $\mathrm{rng}(\pi) \cup \lambda$.

The structure N is called the *ultrapower* of M by F and π is the associated *ultrapower map*. We also write $N = \mathrm{Ult}(M, F)$.

We shall see below that N really can be obtained by the ultrapower construction. Clause e) can be equivalently written as

$$N = \{\pi(f)(\alpha);\ \alpha < \lambda \ \&\ f \in {}^\kappa M \cap M\}.$$

Lemma 2.1.1. *Let* $\pi : M \underset{F}{\longrightarrow} N$ *and* $\kappa = \mathrm{cr}(\pi)$. *Then F is an extender of length λ on $\mathcal{P}(\kappa) \cap M$ and N, π are uniquely determined.*

Proof. Note first that $F(\kappa) = \lambda$. Hence, once we have proved that F is an extender, we will know that its length is λ. To see that F is an extender, suppose $B_{a_1,\ldots,a_m}$ is uniformly primitive recursive in $a_1, \ldots, a_m \in \mathcal{P}(\kappa) \cap M$. Then

$$\pi(\kappa \cap B_{a_1,\ldots,a_m}) = \pi(\kappa) \cap B_{\pi(a_1),\ldots,\pi(a_m)}$$

since π is Σ_0-preserving: by the fact that κ is primitive recursively closed, $B_{a_1,\ldots,a_m}$ has the same canonical Σ_1-definition over $J_\kappa^{a_1,\ldots,a_m}$ and V. This definition is preserved by π, since the suitability of M guarantees that the map $\pi : J_\kappa^{a_1,\ldots,a_m} \to J_{\pi(\kappa)}^{\pi(a_1),\ldots,\pi(a_m)}$ is elementary. Hence,

$$F(\kappa \cap B_{a_1,\ldots,a_m}) = \lambda \cap B_{\pi(a_1),\ldots,\pi(a_m)} = \lambda \cap B_{\lambda\cap\pi(a_1),\ldots,\lambda\cap\pi(a_m)}.$$

The right equality holds since λ is primitive recursively closed: $B_{\pi(a_1),\dots,\pi(a_m)}$ then has the same canonical Σ_1 definition over $J_{\pi(\kappa)}^{\pi(a_1),\dots,\pi(a_m)}$, $J_{\lambda}^{\pi(a_1),\dots,\pi(a_m)}$ and V. This proves the first part of the lemma.

Now suppose that $\pi' : M \underset{F}{\longrightarrow} N'$. Note first

$$\lambda \cap \pi(x) = F(x) = \lambda \cap \pi'(x) \quad \text{for all } x \in \mathcal{P}(\kappa) \cap M. \tag{2.1.1}$$

Define a map $\sigma : N \to N'$ by

$$\sigma(\pi(f)(\alpha)) = \pi'(f)(\alpha)$$

for $\alpha < \lambda$ and $f \in {}^{\kappa}M \cap M$. The situation can be visualized by the following commutative diagram

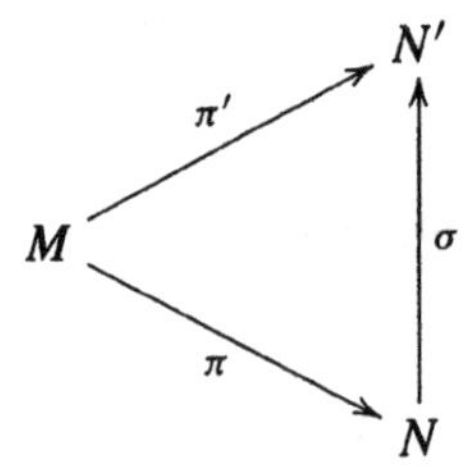

Hence

$$\lambda \cup \mathrm{rng}(\pi') \subset \mathrm{rng}(\sigma). \tag{2.1.2}$$

Notice that it is sufficient to prove

$$\sigma : N \underset{\Sigma_0}{\longrightarrow} N' \quad \text{cofinally,} \tag{2.1.3}$$

as this will imply that σ is Σ_1-preserving, so (2.1.2) will then guarantee that $N' \subset \mathrm{rng}(\sigma)$. By Mostowski's collapsing theorem, $N = N'$ and $\pi = \pi'$.

Proof of (2.1.3). Let $f_1, \dots, f_\ell \in {}^{\kappa}M \cap M$ and $\alpha_1, \dots, \alpha_\ell \in \lambda$. Then for a Σ_0-formula $\varphi(v_1, \dots, v_\ell)$ we have

$$N \models \varphi(\pi(f_1)(\alpha_1), \dots, \pi(f_\ell)(\alpha_\ell)) \longleftrightarrow \langle \alpha_1, \dots, \alpha_\ell \rangle \in \pi(a), \tag{2.1.4}$$

where

$$a = \{\langle \zeta_1, \dots, \zeta_\ell \rangle \in {}^{\ell}\kappa;\ M \models \varphi(f_1(\zeta_1), \dots, f_\ell(\zeta_\ell))\}.$$

We next observe that (2.1.1) implies

$$\langle \alpha_1, \dots, \alpha_\ell \rangle \in \pi(a) \longleftrightarrow \langle \alpha_1, \dots, \alpha_\ell \rangle \in F(a) \longleftrightarrow \langle \alpha_1, \dots, \alpha_\ell \rangle \in \pi'(a)$$

where the right side is equivalent to $N' \models \varphi(\pi'(f_1)(\alpha_1), \dots, \pi'(f_\ell)(\alpha_\ell))$.

□((2.1.3), Lemma 2.1.1)

Definition. Let $M = \langle |M|, \in^M, A^M \rangle$ be a structure. The substructure

$$N = \langle |N|, \in^M \restriction |N|, A^M \restriction |N| \rangle$$

of M is called the *well-founded core of M* iff the following hold:

a) For every $x \in N$ and $y \in M$, $y \in^M x \longrightarrow y \in N$.

b) N is well founded.

c) If N' is a substructure of M satisfying a) and b), and $N \subset N'$, then $N' = N$.

We write $N = \text{wfc}(M)$.

It can be easily seen that wfc(M) always exists and is the largest (in the sense of inclusion) transitive $\in^M$-segment of M.

Definition. Let M be as above. Then $\pi : M \underset{F}{\longrightarrow} N$ *weakly* iff b) – e) of the ultrapower definition hold as above together with

a′) $N = \langle |N|, \in^N, A^N \rangle$ is such that wfc(N) is transitive and $\lambda \subset \text{wfc}(N)$.

Repeating the proof from above we obtain:

Corollary 2.1.2. *Let M be as above, $\pi : M \underset{F}{\longrightarrow} N$ weakly and $\kappa = \text{cr}(\pi)$. Then F is an extender of length $\lambda = F(\kappa)$ on $\mathcal{P}(\kappa) \cap M$ and N, π are uniquely determined up to isomorphism by F.* □(Corollary 2.1.2)

Finally, by an ultrapower construction we obtain:

Lemma 2.1.3. *Let M be suitable for κ and F be an extender on $\mathcal{P}(\kappa) \cap M$ of length λ. Then there are N, π such that $\pi : M \underset{F}{\longrightarrow} N$ weakly.*

Proof. We define a term model $\mathbb{D} = \mathbb{D}(M, F) = \langle D, \cong, \tilde{\in}, \tilde{A} \rangle$ by

$$D = \{\langle \alpha, f \rangle;\ \alpha < \lambda \ \&\ f \in {}^{\kappa}M \cap M\}$$

$$\langle \alpha, f \rangle \cong \langle \beta, g \rangle \longleftrightarrow \langle \alpha, \beta \rangle \in F(\{\langle \eta_1, \eta_2 \rangle;\ f(\eta_1) = g(\eta_2)\})$$

$$\langle \alpha, f \rangle \tilde{\in} \langle \beta, g \rangle \longleftrightarrow \langle \alpha, \beta \rangle \in F(\{\langle \eta_1, \eta_2 \rangle;\ f(\eta_1) \in g(\eta_2)\})$$

$$\tilde{A}(\langle \alpha, f \rangle) \longleftrightarrow \alpha \in F(\{\eta;\ A(f(\eta))\}).$$

By induction on the Σ_0 formulae we get Łos' Theorem

$$\begin{aligned} &\mathbb{D} \models \varphi(\langle \alpha_1, f_1 \rangle, \ldots, \langle \alpha_\ell, f_\ell \rangle) \\ &\longleftrightarrow \langle \alpha_1, \ldots, \alpha_\ell \rangle \in F(\{\langle \eta_1, \ldots, \eta_\ell \rangle;\ M \models \varphi(f_1(\eta_1), \ldots, f_\ell(\eta_\ell))\}). \end{aligned} \tag{2.1.5}$$

Using this it is easy to see that $\cong$ is a congruence relation on $\mathbb{D}$, so there is a natural projection $p : \mathbb{D} \to \mathbb{D}/\cong$. (If the equivalence classes of $\cong$ are proper classes, we replace them by representatives which are sets; this construction is standard). Let $\sigma : (\mathbb{D}/\cong) \to N$ be an isomorphism, where $\text{wfc}(N)$ is transitive. For each $\langle\alpha, f\rangle \in D$ we set

$$[\alpha, f] = \sigma \circ p(\langle\alpha, f\rangle).$$

Then we define $\pi : M \to N$ by

$$\pi(x) = [0, c_x] \quad \text{where } c_x(\eta) = x \quad \text{for all} \quad \eta < \kappa.$$

We shall now verify that in fact $\pi : M \underset{F}{\longrightarrow} N$. First of all, Łos' Theorem gives us immediately

$$\pi : M \underset{\Sigma_0}{\longrightarrow} N \quad \text{cofinally.} \tag{2.1.6}$$

In order to get

$$\lambda \subset \text{wfc}(N) \tag{2.1.7}$$

we first prove

$$\text{if } \alpha < \lambda \text{ and } \langle\beta, f\rangle \,\tilde{\in}\, \langle\alpha, \text{id}\rangle, \text{ then } \langle\beta, f\rangle \cong \langle\gamma, \text{id}\rangle \text{ for some } \gamma < \alpha. \tag{2.1.8}$$

Clearly, $[\alpha, \text{id}]$ represents an ordinal in N. If $[\beta, f]$ represents an ordinal in N smaller than $[\alpha, \text{id}]$, then, by (2.1.8), $[\beta, f] =^N [\gamma, \text{id}]$ for some $\gamma < \alpha$. Hence $[\beta, f] = [\gamma, \text{id}]$, which means that all ordinals below α in N can be represented in the canonical form $[\xi, \text{id}]$. Moreover, $[\xi, \text{id}] \in^N [\zeta, \text{id}] \longleftrightarrow \xi \in \zeta$. Thus, the structure $\langle\{[\xi, \text{id}]; [\xi, \text{id}] \in^N [\alpha, \text{id}]\}, \in^N\rangle$ is isomorphic to $\langle\alpha, \in\rangle$ and, hence, well-founded. Thus $\alpha \subset \text{wfc}(N)$ for all $\alpha < \lambda$, which immediately gives (2.1.7). Moreover, as $\text{wfc}(N)$ is transitive, we also get

$$[\alpha, \text{id}] = \alpha. \tag{2.1.9}$$

Proof of (2.1.8). Let $[\beta, f] \,\tilde{\in}\, [\alpha, \text{id}]$. By Łos' Theorem,

$$\langle\beta, \alpha\rangle \in F(\{\langle\eta_1, \eta_2\rangle;\ f(\eta_1) \in \eta_2\}).$$

Set $A = \{\langle\eta_1, \eta_2\rangle;\ f(\eta_1) = \eta_2\}$. Then

$$\begin{aligned}\langle\beta, \alpha\rangle \;&\in\; F(\{\langle\eta_1, \eta_2\rangle; (\exists\, \eta_3 \in \eta_2)(\langle\eta_1, \eta_3\rangle \in A)\}) \\ &=\; \{\langle\eta_1, \eta_2\rangle;\ (\exists\, \eta_3 \in \eta_2)(\langle\eta_1, \eta_3\rangle \in F(A))\},\end{aligned}$$

since the above set is uniformly primitive recursive in A. Hence there is $\gamma \in \alpha$ s.t. $\langle\beta, \gamma\rangle \in F(A)$, which by Łos' Theorem means that $\langle\beta, f\rangle \cong \langle\gamma, \text{id}\rangle$. □(2.1.8)

To see that N is the $\Sigma_0(N)$-closure of $\lambda \cup \text{rng}(\pi)$, we prove

$$[\alpha, f] = \pi(f)(\alpha) \quad \text{for every} \quad \langle\alpha, f\rangle \in D. \tag{2.1.10}$$

Proof. Note that

$$A = \{\langle\eta_1, \eta_2\rangle;\ f(\eta_1) = c_f(\eta_2)(\mathrm{id}(\eta_1))\} = \kappa \times \kappa.$$

Hence, for every $\alpha \in \lambda$ we have $\langle\alpha, 0\rangle \in F(A)$, which by Łos' Theorem tells us that

$$[\alpha, f] = [0, c_f]([\alpha, \mathrm{id}]) = \pi(f)(\alpha). \qquad \square(2.1.10)$$

To see that

$$\kappa = \mathrm{cr}(\pi) \qquad (2.1.11)$$

we need to verify two things. Firstly, $\pi \upharpoonright \kappa = \mathrm{id}$ and secondly, $\pi(\kappa) > \alpha$ in N for all $\alpha < \lambda$. The former follows from the fact that for $\alpha < \kappa$,

$$A = \{\langle\eta_1, \eta_2\rangle;\ c_\alpha(\eta_1) = \mathrm{id}(\eta_2)\} = \kappa \times \{\alpha\},$$

hence $\langle 0, \alpha\rangle \in F(A) = \lambda \times \{\alpha\}$ and by Łos' Theorem we get $[0, c_\alpha] = [\alpha, \mathrm{id}]$, which means $\pi(\alpha) = \alpha$. The latter follows from the fact that for $\alpha < \lambda$,

$$A = \{\langle\eta_1, \eta_2\rangle; \mathrm{id}(\eta_1) \in c_\kappa(\eta_2)\} = \kappa \times \kappa,$$

hence $\langle\alpha, 0\rangle \in F(A) = \lambda \times \lambda$ and by Łos' Theorem we get $[\alpha, \mathrm{id}] \in^N [0, c_\kappa]$, which means $\alpha \in^N \pi(\kappa)$. $\square$(2.1.11)

It only remains to prove

$$F = \langle\pi(x) \cap \lambda;\ x \in \mathcal{P}(\kappa) \cap M\rangle. \qquad (2.1.12)$$

So pick $x \in \mathcal{P}(\kappa) \cap M$ and $\alpha < \lambda$. Then

$$\begin{aligned}\alpha \in \pi(x) &\longleftrightarrow [\alpha, \mathrm{id}] \in^N [0, c_x]\\ &\longleftrightarrow \langle\alpha, 0\rangle \in F(\{\langle\eta_1, \eta_2\rangle;\ \eta_1 \in x\}) = F(x \times \kappa) = F(x) \times \lambda \longleftrightarrow \alpha \in F(x).\end{aligned}$$

$\square$(Lemma 2.1.3)

If F is an extender on U of length λ and $\pi : M \underset{F}{\longrightarrow} N$ weakly, then the ordinals of N that are less than $\pi(\kappa)$ are exactly those represented by $[\alpha, f]$ for $\alpha < \lambda$ and $f \in U^2$, $f : \kappa \to \kappa$. Hence $\lambda = \pi(\kappa)$ iff $[\alpha, f] \in \lambda$ for all such α, f. This motivates the next definition.

Definition. Let F be an extender on U of length λ. We say that F is *whole* iff whenever $f \in U^2 \cap {}^\kappa\kappa$ and $\alpha < \lambda$, there is $\beta < \lambda$ such that

$$\langle\alpha, \beta\rangle \in F(\{\langle\eta_1, \eta_2\rangle;\ f(\eta_1) = \eta_2\}).$$

Lemma 2.1.4. *Let $\pi : M \underset{F}{\longrightarrow} N$ weakly, where F is at κ, λ. Then F is whole iff $\lambda = \pi(\kappa)$.* $\square$(Lemma 2.1.4)

Corollary 2.1.5. *If F is whole then $\pi(\kappa) \in \mathrm{wfc}(N)$.* $\square$(Corollary 2.1.5)

If U, U' are suitable, $U' \subset U$ and F is an extender on U with length λ, then $F \upharpoonright U'$ is clearly an extender on U'. With this in mind, we shall briefly say that F is an extender *on M at κ, λ* if M is suitable for κ and F is an extender on $\mathcal{P}(\kappa) \cap M$ with length λ.

Definition. Let F be an extender on U of length λ. Then, for a primitive recursively closed $\nu \leq \lambda$, $F|\nu$ is an extender on U such that

$$(F|\nu)(x) = F(x) \cap \nu.$$

That $F|\nu$ satisfies the definition of extender follows from the fact that ν is primitive recursively closed.

Definition. Let F be an extender on U at κ, λ and $A \subset \lambda$ be closed under Gödel's pairing function. We say that A *generates* F iff for every $f \in U^2$ and $\alpha < \lambda$ there is a $g \in U^2$ and a $\beta \in A$ such that

$$\prec\alpha, \beta\succ \in F(\{\prec\eta_0, \eta_1\succ;\ f(\eta_0) = g(\eta_1)\}).$$

A primitive recursively closed ordinal γ is called a *generator* of F iff there is no $f \in U^2$ and $\alpha < \gamma$ such that

$$\prec\alpha, \gamma\succ \in F(\{\prec\eta_0, \eta_1\succ;\ f(\eta_0) = \eta_1\}).$$

To illustrate the meaning of these notions suppose that F is an extender on M and consider the ultrapower $\pi : M \underset{F}{\longrightarrow} M'$. Then β generates F iff $\pi : M \underset{E}{\longrightarrow} M'$ where $E = F|\beta$ and γ is a generator of F iff γ is not a member of the Σ_0-closure of $\mathrm{rng}(\pi) \cup \gamma$ in M. Both can be easily seen using the methods developed above.

2.2 The Hypermeasure Representation of Extenders

Definition. Let F be an extender on U of length λ. $\vec{F} = \langle F_\alpha;\ \alpha < \lambda\rangle$ is the *associated hypermeasure* iff $F_\alpha = \{x \in U;\ \alpha \in F(x)\}$ for all $\alpha < \lambda$.

Definition. $\vec{F} = \langle F_\alpha; \alpha < \lambda\rangle$ is a *hypermeasure on U* iff there is an extender F on U of length λ such that $\vec{F}$ is its associated hypermeasure.

It is this representation of extenders which is usually used in the literature. There, λ is often taken to be arbitrary, so the hypermeasure must then be indexed by elements of $[\lambda]^{<\omega}$ rather than of λ. However, if λ is primitive recursively closed, then finite subsets of λ can be coded by Gödel tuples that are elements of λ, so the two ways of indexing become equivalent. Because of technical reasons, we shall consider just extenders with primitive recursively closed lengths (one reason for this is that we shall use both the map and the hypermeasure representation and the conversion between them is easier if we restrict ourselves to extenders with primitive recursively closed lengths).

In practice we shall often call a hypermeasure an extender, if it is clear from the context what we mean. Hypermeasures can be characterized by a finite list of combinatorial conditions, which is in the literature usually taken as the definition of an extender.

Lemma 2.2.1. *Suppose M is suitable for κ and $\lambda > \kappa$ is primitive recursively closed. The sequence $\vec{F} = \langle F_\alpha; \alpha < \lambda \rangle$ is a hypermeasure iff*

a) *each F_α is a κ-complete ultrafilter on $\mathcal{P}(\kappa) \cap M$.*

b) *$\{\alpha\} \in F_\alpha$ for each $\alpha < \kappa$.*

c) *If $\alpha < \beta < \lambda$ then $\{\prec \eta_1, \eta_2 \succ; \eta_1 \in \eta_2\} \in F_{\prec \alpha, \beta \succ}$.*

d) *Let $\langle \alpha_1, \dots, \alpha_n \rangle \in {}^n\lambda$, and $1 \leq i_1 < \dots, < i_\ell \leq n$. For $x \subset \kappa$ set*

$$x^* = \{\prec \eta_1, \dots, \eta_n \succ; \prec \eta_{i_1}, \dots, \eta_{i_\ell} \succ \in x\}.$$

Then $x^* \in F_{\prec \alpha_1, \dots, \alpha_n \succ} \longleftrightarrow x \in F_{\prec \alpha_{i_1}, \dots, \alpha_{i_\ell} \succ}$.

e) *Let $\langle \alpha_1, \dots, \alpha_n \rangle \in {}^n\lambda$ and $1 \leq i_j \leq n$ for $j = 1, \dots, \ell$. For $x \subset \kappa$ set*

$$\tilde{x} = \{\prec \eta_1, \dots, \eta_n, \prec \eta_{i_1}, \dots, \eta_{i_\ell} \succ \succ; \prec \eta_1, \dots, \eta_n \succ \in x\}.$$

Then $\tilde{x} \in F_{\prec \alpha_1, \dots, \alpha_n, \prec \alpha_{i_1}, \dots, \alpha_{i_\ell} \succ \succ} \longleftrightarrow x \in F_{\prec \alpha_1, \dots, \alpha_n \succ}$.

f) *Let $f \in M$, $f : \kappa \to \kappa$ and $\gamma, \beta < \lambda$ be such that*

$$\{\prec \eta_1, \eta_2 \succ; f(\eta_1) \in \eta_2\} \in F_{\prec \gamma, \beta \succ}.$$

Then there is a $\delta < \beta$ such that $\{\prec \eta_1, \eta_2 \succ; f(\eta_1) = \eta_2\} \in F_{\prec \gamma, \delta \succ}$.

Remark. In a), the κ-completeness w.r.t. M is meant, i.e. $\bigcap \mathrm{rng}(f) \in U$ whenever $f \in M$ and $f : \gamma \to U$ for some $\gamma < \kappa$.

Proof. These conditions are just what we need to carry out the ultrapower construction. Such a construction can be found in [Ka94] and [MS89] and is much like the construction we did above. The conditions a) – e) enable us to prove Łos Theorem for Σ_0-formulae; f) is called normality condition and enables to prove an equivalent of (2.1.8). We define $\mathbb{D}$ similarly as above with the only, rather formal, amendment: instead of $\alpha \in F(x)$ we write $x \in F_\alpha$. The rest of the construction is then almost the same as above. We get a map $\pi : M \to N$ and all of the properties corresponding to (2.1.9) – (2.1.12). In particular, we get: $\alpha \in \pi(x) \leftrightarrow x \in F_\alpha$ for $\alpha < \lambda$. Hence, if we set $F(x) = \pi(x) \cap \lambda$ for $x \in \mathcal{P}(\kappa) \cap M$, it is easy to see that $\pi : M \xrightarrow[F]{} N$ weakly and, consequently, F is an extender by Lemma 2.1.1. Clearly then $\langle F_\alpha; \alpha < \lambda \rangle$ is a hypermeasure associated with F. □(Lemma 2.2.1)

We now adopt the terminology, which we developed for extenders, to hypermeasures. This is useful, since, as already mentioned above, we shall later freely use both the map and the hypermeasure representation, often in the same context, if this simplifies the notation without causing confusion and/or the choice of the representation is inconsequential (e.g. if we form an ultrapower). Hence the following statements have an obvious meaning:

a) $\vec{F}$ is a hypermeasure on M at κ, λ;

b) $\pi : M \underset{\vec{F}}{\longrightarrow} N, \pi : M \underset{\vec{F}}{\longrightarrow} N$ weakly (here we can formulate propositions corresponding to Lemma 2.1.1, 2.1.3 and Corollary 2.1.2);

c) $\vec{F}$ is whole.

2.3 Amenability

In this section we shall prove one useful lemma concerning amenability and introduce two other sorts of amenability, which will play an important role later in connection with the so called Σ^*-ultrapowers.

Lemma 2.3.1. *Let $\pi : M \underset{F}{\longrightarrow} N$, where $\kappa = \mathrm{cr}(\pi)$ is the largest cardinal in M and F is whole. Then $\langle N, F\rangle$ is amenable.*

Proof. Given $a \in N$, we have to show $F \cap a \in N$. We first note the following two facts:

a) $F = \pi \restriction (\mathcal{P}(\kappa) \cap M)$.

b) It is sufficient to prove $F \cap a \in N$ for $a \in \mathrm{rng}(\pi)$, since π is cofinal in N.

Suppose $\langle x, y\rangle \in F \cap a$ and $a = \pi(\bar{a})$. Then $y \in \bigcup^2 a$, hence $x = \pi^{-1}(y) \in \bigcup^2 \bar{a}$. Let $\bar{f} : \kappa \to \bigcup^2 \bar{a}$ be surjective and $\bar{f} \in M$. Such an $\bar{f}$ exists since κ is the largest cardinal in M. Then $f = \pi(\bar{f}) \in N$. If $\xi < \kappa$ is such that $\bar{f}(\xi) = x$, then $y = f(\xi)$ and $x = f(\xi) \cap \kappa$. Hence,

$$F \cap a = \{\langle f(\xi) \cap \kappa; f(\xi)\rangle; \xi < \kappa\} \cap a \in N. \qquad \square(\text{Lemma 2.3.1})$$

Definition. Let F be an extender on U of length λ and $\kappa = \mathrm{cr}(F)$. F is *weakly amenable* iff $\{\xi; X''\{\xi\} \in F_\delta\} \in U$ whenever $X \in U^2$ and $\delta < \lambda$.

The above definition of weak amenability is a generalization of the usual definition for extenders on structures: If F is an extender on M at κ, λ, then F is weakly amenable iff for every $f \in {}^\kappa M \cap M$, and every $\delta < \lambda$, $\{\xi < \kappa; f(\xi)\} \in F_\delta$ and we shall in the following often use this form. Of course, this version can be easily reformulated in the abstract setting – just note the correspondence between f here and X above. If we deal with a weakly amenable extender F on M we shall often say that F is weakly amenable w.r.t. M.

Definition. Let F be an extender on M at κ, λ. F is *Σ_1-amenable w.r.t. M* iff F_δ is $\Sigma_1(M)$ for every $\delta < \lambda$.

Definition. Let F be an extender on M. We say that F is *close* to M iff F is both weakly and Σ_1- amenable w.r.t. M.

Clearly, both definitions have an obvious meaning if we take hypermeasures instead of extenders, and in the following we will usually just state definitions for extenders and leave their conversion to hypermeasures up to the reader.

Let us also mention that either of the properties weakly amenable and Σ_1-amenable can hold without the other. Both can fail.

Lemma 2.3.2. *Let* $\pi : M \underset{F}{\longrightarrow} N$ *weakly and* $\kappa = \mathrm{cr}(F)$. *Then*

a) $\mathcal{P}(\kappa) \cap M \subset N$,

b) $\mathcal{P}(\kappa) \cap N \subset M$ *iff* F *is weakly amenable.*

Proof. a) follows from the fact that $x = \pi(x) \cap \kappa$ for $x \in \mathcal{P}(\kappa) \cap M$. We now prove b). To see the implication $(\rightarrow)$, let $X \in \mathcal{P}({}^2\kappa) \cap M$ and $\delta < \lambda$. Then for $\xi < \kappa$ we have

$$X''\{\xi\} \in F_\delta \longleftrightarrow \delta \in \pi(X''\{\xi\}) = \pi(X)''\{\xi\}.$$

Hence $a = \{\xi < \kappa;\ X''\{\xi\} \in F_\delta\} \in N \cap \mathcal{P}(\kappa)$, since $\pi(X) \in N$. So $a \in M$ by our assumption.

To see $(\leftarrow)$, let $x \in N$, $x = \pi(f)(\alpha)$ for some $f \in {}^\kappa M \cap M$ and $\alpha < \lambda$. For $\xi < \kappa$ we then have

$$\xi \in x \longleftrightarrow \alpha \in F(a_\xi) \longleftrightarrow a_\xi \in F_\alpha,$$

where $a_\xi = \{\eta < \kappa;\ \xi \in f(\eta)\}$. Hence, if we set

$$X = \{\langle \eta_1, \eta_2 \rangle;\ \eta_1 \in f(\eta_2)\},$$

then $X \in M$ and $a_\xi = X''\{\xi\}$; thus $x = \{\xi < \kappa;\ X''\{\xi\} \in F_\alpha\} \in M$ by weak amenability. □(Lemma 2.3.2)

To illustrate a use of Σ_1-amenability we prove the following lemma. The argumentation used here will be of great importance later in connection with Σ^*-ultrapowers.

Lemma 2.3.3. *Let* F *be* Σ_1*-amenable w.r.t.* $\bar{M}$, $\pi : \bar{M} \underset{F}{\longrightarrow} M$ *and* $\kappa = \mathrm{cr}(F)$. *Then*

$$\mathcal{P}(\kappa) \cap \boldsymbol{\Sigma}_1(M) \subset \boldsymbol{\Sigma}_1(\bar{M}).$$

Proof. Let $A \subset \kappa$ be $\Sigma_1(M)$ in $p = \pi(f)(\alpha)$, where $f \in {}^\kappa M \cap M$ and $\alpha < \lambda = \mathrm{length}(F)$. Then there is a $\Sigma_0(M)$-relation $R(v_1, v_2, v_3)$ such that

$$A(x) \longleftrightarrow (\exists z) R(z, x, p).$$

We then have

$$\begin{aligned} A(\xi) &\longleftrightarrow (\exists u)(\exists z \in \pi(u)) R(z, \xi, \pi(f)(\alpha)) \\ &\longleftrightarrow (\exists u)[\alpha \in F(\{\eta < \kappa;\ (\exists z \in u)\bar{R}(z, \xi, f(\eta))\})], \end{aligned}$$

where $\bar{R}$ has the same Σ_0 definition over $\bar{M}$ as R over M. Let a be the set to which we apply F on the rightmost side above. Then $A(\xi) \leftrightarrow (\exists u)(\exists v)(v = a\ \&\ F_\alpha(v))$, where by $v = a$ we mean the Σ_0-definition of a from above. This definition of A is clearly $\mathbf{\Sigma}_1(\bar{M})$. □(Lemma 2.3.3)

The same proof yields the following lemma.

Lemma 2.3.4. *Let $\bar{M}$, M, F be as above. Let $R(x_1, \dots, x_\ell, y_1, \dots, y_m)$ be $\Sigma_1(M)$. Let $f_1, \dots, f_m \in \bar{M}$, where $\kappa = \mathrm{cr}(F)$. Let $\alpha_1, \dots, \alpha_m < \lambda = \mathrm{length}(F)$. Then*

$$\tilde{R} = \{\langle x_1, \dots, x_\ell \rangle, R(\pi(x_1), \dots, \pi(x_\ell), \pi(f_1)(\alpha_1), \dots, \pi(f_m)(\alpha_m))\}$$

is $\Sigma_1(\bar{M})$ in $f_1, \dots, f_m$ and r, where $F_{\prec\alpha_1,\dots,\alpha_m\succ}$ is $\Sigma_1(\bar{M})$ in r (uniformly in the Σ_1-definition of R and the Σ_1 definition of $F_{\prec\alpha_1,\dots,\alpha_m\succ}$ from r). □(Lemma 2.3.4)

2.4 Coherent Structures

Definition. A structure $M = \langle J_\alpha^A, F\rangle$ is *coherent* iff

a) J_α^A is acceptable,

b) F is a whole extender on $J_{\bar{\alpha}}^A$, where $\bar{\alpha} < \alpha$ is such that

- $\kappa = \mathrm{cr}(F)$ is the largest cardinal in $J_{\bar{\alpha}}^A$
- $J_\alpha^A = \mathrm{Ult}(J_{\bar{\alpha}}^A, F)$.

Lemma 2.4.1. *Let $M = \langle J_\alpha^A, F\rangle$ be coherent. Then:*

a) *M is amenable.*

b) *Let $\lambda = F(\kappa)$. Then λ is the largest cardinal in M.*

c) *Let $\bar{\beta} < \bar{\alpha}$ be such that κ is the largest cardinal in $J_{\bar{\beta}}^A$. Let $\omega\beta = \sup \pi''\omega\bar{\beta}$, where $\pi : J_{\bar{\alpha}}^A \xrightarrow[F]{} J_\alpha^A$. Set $\bar{F} = F \restriction J_{\bar{\beta}}^A$. Then $\bar{M} = \langle J_\beta^A, \bar{F}\rangle$ is coherent.*

d) *F is weakly amenable iff $\bar{\alpha} = \kappa^{+M}$.*

Proof. a) follows immediately from Lemma 2.3.1 and both b) and d) are trivial. Hence it is sufficient to prove just c).

To see that $J_{\bar{\beta}}^A$ is suitable for κ it suffices to prove that $J_\kappa^a \in J_{\bar{\beta}}^A$ for each $a \in J_{\bar{\beta}}^A$. This follows from the fact that κ is regular in $J_{\bar{\alpha}}^A$ (since it is a critical point of an embedding). The defining formula for S^a then gives us the same hierarchy $S^a \restriction \kappa$ in V and in J_δ^A, where $\delta < \bar{\beta}$ is such that $a \in J_\delta^A$, since, by regularity of κ, each $S^a \restriction \xi$ is an element of J_κ^A for $\xi < \kappa$. Hence $S^a \restriction \kappa \in J_{\bar{\beta}}^A$ which implies $J_\kappa^a = \bigcup \mathrm{rng}(S^a \restriction \kappa) \in J_{\bar{\beta}}^A$.

Let $\bar{\pi} = \pi \restriction \mathcal{P}(\kappa) \cap J_{\bar{\beta}}^A$. To see that $\bar{\pi} : J_{\bar{\beta}}^A \xrightarrow[\bar{F}]{} J_\beta^A$ note that π is a Σ_0 cofinal map and $\bar{\pi}(x) = \bar{F}(x)$ for all $x \in \mathcal{P}(\kappa) \cap J_{\bar{\beta}}^A$. We now show that J_β^A is the $\Sigma_0(J_\beta^A)$-closure of $\text{rng}(\bar{\pi}) \cup \lambda$. Pick $x \in J^A_\beta$. By cofinality of $\bar{\pi}$, there is $a \in J^A_{\bar{\beta}}$ such that $x \in \pi(a)$. Let $f \in J^A_{\bar{\beta}}$ be a function mapping κ onto a. Then $\bar{\pi}(f) : \lambda \to \pi(a)$ is surjective, so for some $\gamma < \lambda$, we have $x = \bar{\pi}(f)(\gamma)$. □(Lemma 2.4.1)

Note that if $\bar{\beta}$ is as in c), then $\bar{F}$ is not weakly amenable, since M contains more subsets of κ than $\mathcal{P}(\kappa) \cap J^A_{\bar{\beta}}$. This illustrates how easily weak amenability can be violated.

Lemma 2.4.2. *There is a Q-sentence φ such that for every transitive structure M closed under the pairing function,*

$$M \text{ is coherent} \longleftrightarrow M \models \varphi.$$

By Lemma 1.1.3, being an acceptable J-structure is a Q-condition. Hence, Lemma 2.4.2 follows immediately from

Lemma 2.4.3. *There is a Q-sentence φ such that for every acceptable structure $M = \langle J^A_\alpha, F\rangle$,*

$$M \text{ is coherent} \longleftrightarrow M \models \varphi.$$

Proof. We shall first informally state a Q-condition equivalent to coherency and then explain the points which are not immediate. The statement reads

$$(Q\xi)(\exists \kappa, \lambda)\big(\text{i}) \,\&\, \ldots, \&\, \text{vii})\big),$$

where i) – vii) are the following clauses:

i) $F \cap S^A_\xi$ is a map;

ii) $\langle \kappa, \lambda\rangle \in F$ and $\langle \kappa, \lambda\rangle$ is the only pair of ordinals in S^A_ξ with this property;

iii) there are $\bar{\xi} < \lambda$, $\xi' \geq \xi$ such that

$$\text{dom}(F \cap S^A_\xi) \subset \mathcal{P}(\kappa) \cap S^A_{\bar{\xi}} \subsetneqq \text{dom}(F \cap S^A_{\xi'});$$

iv) κ, λ are regular limit cardinals in S^A_ξ;

v) for all $\bar{a}, \bar{b}, a, b \in S^A_\xi$ such that $\langle \bar{a}, \bar{b}\rangle \in F$, if a is $\Sigma_1(J^{\bar{a}}_\kappa)$ in a parameter $\nu < \kappa$ and b is $\Sigma_1(J^{\bar{b}}_\lambda)$ in the same parameter by the same definition, then $\langle a, b\rangle \in F$;

vi) there is $\bar{\xi} < \lambda$ such that $\text{dom}(F \cap S^A_\xi) \subset S^A_{\bar{\xi}}$ and

$$(\forall f \in {}^\kappa\kappa \cap S^A_{\bar{\xi}})(\forall \beta < \lambda)(\exists \gamma < \lambda)(\exists x, y \in S^A_{\xi+1})$$
$$(x = \{\prec\eta_1, \eta_2\succ \in \kappa;\ f(\eta_1) = \eta_2\} \,\&\, \langle x, y\rangle \in F \,\&\, \prec\beta, \gamma\succ \in y);$$

vii) there are $\bar{\xi}$, $\bar{f}$, f such that $\mathrm{dom}(F \cap S^A_\xi) \subset S^A_{\bar{\xi}}$, $\bar{f} : \kappa \xrightarrow{\text{onto}} S^A_{\bar{\xi}}$, $f : \lambda \xrightarrow{\text{onto}} S^A_\xi$ and if we set

$$\begin{aligned}
a &= \{\prec\eta_1, \eta_2\succ;\ f(\eta_1) = f(\eta_2)\} \\
\bar{a} &= \{\prec\eta_1, \eta_2\succ;\ \bar{f}(\eta_1) = \bar{f}(\eta_2)\} \\
b &= \{\prec\eta_1, \eta_2\succ;\ f(\eta_1) \in f(\eta_2)\} \\
\bar{b} &= \{\prec\eta_1, \eta_2\succ;\ \bar{f}(\eta_1) \in \bar{f}(\eta_2)\} \\
c &= \{\eta;\ A(f(\eta))\} \\
\bar{c} &= \{\eta;\ A(\bar{f}(\eta))\}
\end{aligned}$$

then $\langle \bar{a}, a\rangle \in F$, $\langle \bar{b}, b\rangle \in F$ and $\langle \bar{c}, c\rangle \in F$.

Clauses i) – v) express that F is an extender, vi) says that F is whole and vii) that $\pi : J^A_{\bar{\alpha}} \underset{F}{\longrightarrow} J^A_\alpha$ for some $\bar{\alpha} < \lambda$ and π. It is clear that i) – vii) hold in M.

We first note that the above statement is really a Q-condition. For most of the above clauses this is trivial. As to v), note that this condition can be expressed by a formula with quantification restricted to $S^A_{\xi+1}$, since $J^{\bar{b}}_\lambda \in S^A_{\xi+1}$ by the proof of Lemma 2.4.1, hence the statement $J^{\bar{b}}_\lambda \models \varphi(x)$, where φ is Σ_1, is expressible over $S^A_{\xi+1}$. Note also that vii) is a Σ_1-statement.

Now we have to prove that if an acceptable structure satisfies the above statement then it is already coherent. First note that κ, λ are uniquely determined by ii) and both are regular cardinals in J^A_α by iv) and hence primitive recursively closed. By vii), λ is the largest cardinal in J^A_α, hence $J^{\bar{a}}_\kappa, J^{\bar{b}}_\lambda \in J^A_\alpha$ whenever $\bar{a} \in \mathcal{P}(\kappa) \cap J^A_\alpha$ and $\bar{b} \in \mathcal{P}(\lambda) \cap J^A_\alpha$. iii) says that, for some $\bar{\alpha} \leq \lambda$, $\mathrm{dom}(F) = \mathcal{P}(\kappa) \cap J^A_{\bar{\alpha}}$ and, by vii), $\bar{\alpha} \leq \kappa^{+J^A_\alpha}$ and $J^A_{\bar{\alpha}}$ is suitable for κ. Hence, using the fact (see the proof of Lemma 2.1.1) that the canonical Σ_1-definitions of primitive recursive functions in $\bar{b}$ are the same over V as over $J^{\bar{b}}_\lambda$ and similarly for $\bar{a}$ and $J^{\bar{a}}_\kappa$), v) tells us that F is an extender on $J^A_{\bar{\alpha}}$ at κ, λ. It follows immediately that $\bar{\alpha} < \lambda$. To complete the proof, let

$$\pi : J^A_{\bar{\alpha}} \underset{F}{\longrightarrow} N \quad \text{weakly.}$$

We show that $N = J^A_\alpha$. Note that F is whole by vi), hence $\pi(\kappa) = F(\kappa) = \lambda$ and $\pi(\kappa) + 1 \subset \mathrm{wfc}(N)$. Let $\bar{\xi}, \xi, \bar{f}, f$ be as in vii). Then $\bar{f}$ maps κ onto $S^A_{\bar{\xi}}$, so $\pi(\bar{f})$ maps λ onto $\pi(S^A_{\bar{\xi}})$ in N. Define a map $\sigma_\xi : \pi(S^A_{\bar{\xi}}) \to S^A_\xi$ by

$$\sigma_\xi(\pi(\bar{f})(\beta)) = f(\beta) \quad \text{for } \beta < \lambda.$$

For $\beta_1, \beta_2 < \lambda$ we have

$$\pi(\bar{f})(\beta_1) = \pi(\bar{f})(\beta_2) \longleftrightarrow f(\beta_1) = f(\beta_2) \tag{2.4.1}$$

since the left side is equivalent to the fact that $\prec\beta_1, \beta_2\succ$ is an element of

$$F(\{\prec\eta_1, \eta_2\succ;\ \bar{f}(\eta_1) = \bar{f}(\eta_2)\}) = \{\prec\eta_1, \eta_2\succ;\ f(\eta_1) = f(\eta_2)\}.$$

It follows from (2.4.1) that σ_ξ is well defined. The same computation with $\in$ tells us that σ_ξ preserves membership. Hence, $\pi(S^A_{\bar\xi})$ is well-founded. Now note that this holds for cofinally many $\bar\xi$ in $\omega\bar\alpha$ and that $N = \bigcup\{\pi(S^A_{\bar\xi})\,;\ \bar\xi < \bar\alpha\}$. Hence, N is well-founded. From now on take N to be transitive. Notice also that σ_ξ is onto S^A_ξ, hence $\pi(S^A_{\bar\xi}) = S^A_\xi$. This holds for cofinally many $\bar\xi < \omega\bar\alpha$ and cofinally many $\xi < \omega\alpha$. That the predicate $A^N \cap |S^A_\xi|$ is equal to $A \cap |S^A_\xi|$ follows from a computation similar to (2.4.1). Hence, $N = J^A_\alpha$. □(Lemma 2.4.3)

2.5 Extendibility

Definition. Let M be suitable for κ and F be an extender on M with critical point κ. We say that M is *extendible* by F iff $\mathrm{Ult}(M, F)$ exists.

Of course we are just interested in those cases when $\mathrm{Ult}(M, F)$ exists, in other words, when the ultrapower of M by F is well-founded. This is often a difficult question and since we work with arbitrary structures and outer extenders, this clearly cannot be decided inside M. Therefore we will need various criteria which will guarantee extendibility of M. A useful and a well-known one is ω-completeness.

Definition. Let F be an extender on $U \subset \mathcal{P}(\kappa)$ of length λ. F is *ω-complete* iff for every pair of countable sets $X \subset U$, $A \subset \lambda$ satisfying

$$\prec\alpha_1, \alpha_2\succ \in A \longleftrightarrow \alpha_1, \alpha_2 \in A \quad \text{for every} \quad \alpha_1, \alpha_2 < \lambda \tag{2.5.1}$$

there is a map $\mathfrak{k} : A \to \kappa$ preserving the order and $\prec\ \succ$ such that

$$x \in F_\alpha \longrightarrow \mathfrak{k}(\alpha) \in x$$

for all $x \in X$ and $\alpha \in A$.

A map which preserves the ordering of ordinals and Gödel pairs is briefly called *$\prec\ \succ$-respecting*. The condition (2.5.1) is included just for convenience; if A is countable we can easily close it under Gödel pairs. If Y is countable and $\mathfrak{k}$ is as above where $X = Y \cap U$ and $A = Y \cap \lambda$, we say that F *fixes* Y. Hence, F is ω-complete iff F fixes every countable Y.

Definition. Let F be an extender on M at κ, λ and $\bar F$ be an extender on $\bar M$ at $\bar\kappa, \bar\lambda$. We write $\langle\sigma, \mathfrak{f}\rangle : \langle\bar M, \bar F\rangle \longrightarrow \langle M, F\rangle$ iff

a) $\sigma : \bar M \underset{\Sigma_0}{\longrightarrow} M$ and $\sigma(\bar\kappa) = \kappa$,

b) $\mathfrak{f} : \bar\lambda \to \lambda$ is $\prec\ \succ$-respecting,

c) $\alpha \in \bar F(x) \to \mathfrak{f}(\alpha) \in F(\sigma(x))$ for all $\alpha < \bar\lambda$ and $x \in \mathcal{P}(\bar\kappa) \cap \bar M$.

We could now directly prove that every M is extendible by any ω-complete F on M; but we instead state the following characterization of ω-completeness, which immediately implies extendibility. Its proof is a bit more complicated, but the method will be of importance later in connection with iterability criteria.

Lemma 2.5.1. *Let M be suitable for κ and F be an extender on M at κ, λ. Then F is ω-complete iff for every pair $\bar{M}, \bar{F}$ of countable sets and every $\sigma, \mathfrak{h}$ such that*

$$\langle \sigma, \mathfrak{h} \rangle : \langle \bar{M}, \bar{F} \rangle \longrightarrow \langle M, F \rangle,$$

if $\pi : \bar{M} \underset{\bar{F}}{\longrightarrow} N$ weakly, then there is a $\sigma' : N \underset{\Sigma_0}{\longrightarrow} M$ which completes the following commutative diagram.

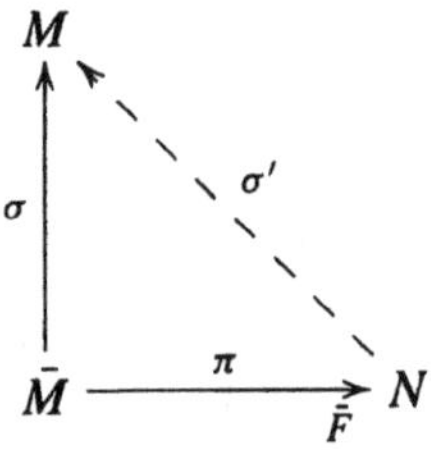

Proof. We prove first ($\leftarrow$). Pick countable X, A such that $A \subset \lambda$ satisfies (2.5.1) and $X \subset \mathcal{P}(\kappa) \cap M$. Let θ be regular large enough such that $X, A, M, F \in H_\theta$. Let Y be the Skolem hull of $\{M, F\} \cup X \cup A$ in H_θ. Transitivize Y, getting thus $\bar{H}$ and an uncollapsing map $\sigma : \bar{H} \to H_\theta$. Let $\langle \bar{M}, \bar{F} \rangle = \sigma^{-1}(M, F)$. Then $\bar{M}$, $\bar{F}$ are countable and clearly $A \subset \sigma''\bar{\lambda}$, where $\bar{\lambda} = \sigma^{-1}{}''\lambda = \sigma^{-1}(\lambda)$. Moreover, $\langle \sigma \restriction \bar{M}, \sigma \restriction \bar{\lambda} \rangle : \langle \bar{M}, \bar{F} \rangle \to \langle M, F \rangle$. Let $\pi : \bar{M} \underset{\bar{F}}{\longrightarrow} N$ weakly and $\sigma' : N \underset{\Sigma_0}{\longrightarrow} M$ be the map completing the above diagram. Then for $x \in X$ and $\alpha \in A$ we have

$$\begin{aligned} \alpha \in F(x) &\longleftrightarrow \sigma^{-1}(\alpha) \in \bar{F}(\sigma^{-1}(x)) \longleftrightarrow \sigma^{-1}(\alpha) \in \pi \circ \sigma^{-1}(x) \\ &\longleftrightarrow \sigma' \circ \sigma^{-1}(\alpha) \in \sigma' \circ \pi \circ \sigma^{-1}(x) = x. \end{aligned}$$

Hence, it suffices to set $\mathfrak{k} = \sigma' \circ \sigma^{-1} \restriction A$, since this map is clearly $\prec\, \succ$-respecting.

To see ($\rightarrow$), suppose we have an uncompleted diagram as above. We want to find the map σ'. Set $A = \mathfrak{h}''\bar{\lambda}$, $X = \sigma''(\mathcal{P}(\bar{\kappa}) \cap \bar{M})$. Clearly both X, A are countable, $X \subset \mathcal{P}(\kappa) \cap M$, $A \subset \lambda$ and A satisfies (2.5.1). Let $\mathfrak{k} : A \to \kappa$ be a map for X, A witnessing ω-completeness of F. Set

$$\sigma'(\pi(f)(\alpha)) = \sigma(f)(\mathfrak{k} \circ \mathfrak{h}(\alpha))$$

for $f \in {}^{\kappa}\bar{M} \cap \bar{M}$ and $\alpha < \bar{\lambda}$. Then

$$\sigma = \sigma' \circ \pi, \tag{2.5.2}$$

since $\sigma' \circ \pi(x) = \sigma'(\pi(c_x)(0)) = \sigma(c_x)(\mathfrak{k} \circ \mathfrak{h}(0)) = c_{\sigma(x)}(\mathfrak{k} \circ \mathfrak{h}(0)) = \sigma(x)$. Moreover,

$$\sigma' : N \underset{\Sigma_0}{\longrightarrow} M. \tag{2.5.3}$$

To see this, suppose $N \models \varphi(\pi(f_1)(\alpha_1), \ldots, \pi(f_\ell)(\alpha_\ell))$, where φ is a Σ_0-formula, $f_1, \ldots, f_\ell \in {}^{\kappa}\bar{M} \cap \bar{M}$ and $\alpha_1, \ldots, \alpha_\ell < \bar{\lambda}$. This means

$$\prec\alpha_1, \ldots, \alpha_\ell\succ \in \bar{F}(u^{\varphi,\bar{M}}_{f_1,\ldots,f_\ell}),$$

where

$$u^{\varphi,\bar{M}}_{f_1,\ldots,f_\ell} = \{\prec\eta_1, \ldots, \eta_\ell\succ \in \bar{\kappa};\ \bar{M} \models \varphi(f_1(\eta_1), \ldots, f_\ell(\eta_\ell))\}.$$

Observing that $u^{\varphi,M}_{f_1,\ldots,f_\ell} = \sigma(u^{\varphi,\bar{M}}_{f_1,\ldots,f_\ell})$ we obtain

$$\prec\mathfrak{h}(\alpha_1), \ldots, \mathfrak{h}(\alpha_\ell)\succ \in F(u^{\varphi,M}_{\sigma(f_1),\ldots,\sigma(f_\ell)}),$$

and, by ω-completeness of F,

$$\prec\mathfrak{k} \circ \mathfrak{h}(\alpha_1), \ldots, \mathfrak{k} \circ \mathfrak{h}(\alpha_\ell)\succ \in u^{\varphi,M}_{\sigma(f_1),\ldots,\sigma(f_\ell)}$$

which means $M \models \varphi(\sigma(f_1)(\mathfrak{k} \circ \mathfrak{h}(\alpha_1), \ldots, \sigma(f_\ell)(\mathfrak{k} \circ \mathfrak{h}(\alpha_\ell)))$. □(Lemma 2.5.1)

Corollary 2.5.2. *Let M, F be as above. If F is ω-complete, then M is extendible by F.*

Proof. Suppose not. This means that there is an infinite decreasing sequence in $\mathbb{D}(M, F)$. In other words, there is a sequence $\langle\langle\alpha_i, f_i\rangle;\ i \in \omega\rangle$ such that

$$\langle\alpha_{i+1}, \alpha_i\rangle \in F(\{\langle\eta_1, \eta_0\rangle;\ f_{i+1}(\eta_1) \in f_i(\eta_0)\}) \tag{2.5.4}$$

for all $i \in \omega$. Let θ be regular and large enough such that $M, F \in H_\theta$. Let X be a countable elementary substructure of H_θ which contains M, F and all α_i, f_i as elements. Let $\bar{H}$ be the transitive collapse of X and $\sigma : \bar{H} \to H_\theta$ the corresponding uncollapsing map. Set $(\bar{M}, \bar{F}, \bar{f}_i, \bar{\alpha}_i) = \sigma^{-1}(M, F, f_i, \alpha_i)$. Then $\bar{M}$ is not extendible by $\bar{F}$, since (2.5.4) holds for $\bar{\alpha}_i$, $\bar{f}_i$ and $\bar{F}$ thus witnessing the non-wellfoundedness of N, where $\pi : \bar{M} \underset{\bar{F}}{\longrightarrow} N$ weakly. On the other hand, $\langle\sigma \restriction \bar{M}, \sigma \restriction \bar{\lambda}\rangle : \langle\bar{M}, \bar{F}\rangle \longrightarrow \langle M, F\rangle$ ($\bar{\lambda} = \text{length}(\bar{F})$) and by the previous lemma, N can be embedded to M. This means that N is well-founded after all, contradiction. □(Corollary 2.5.2)

The previous lemma implies that ω-completeness guarantees extendibility even for class structures: every sequence $\langle\alpha_i, f_i\rangle_i$ witnessing non-wellfoundedness of the ultrapower of M is bounded; hence for sufficiently large θ, $M \cap H_\theta$ cannot be extendible by F.

If we consider extenders whose domain is the entire power set of the critical point, then ω-completeness turns out to be equivalent with extendibility.

Lemma 2.5.3. *Let F be an extender on V at κ, λ. (This means $F \in V$). Then V is extendible by F iff F is ω-complete.*

Proof. ($\leftarrow$) follows from the previous remark. Hence it is sufficient to prove ($\rightarrow$). Pick countable $X \subset \mathcal{P}(\kappa)$ and $A \subset \lambda$ satisfying (2.5.1). Let $\langle x_i;\ i \in \omega\rangle$, $\langle \alpha_i;\ i \in \omega\rangle$ be fixed enumerations of X, A. Let $\pi : V \underset{F}{\longrightarrow} M$. Note first that for every $n \in \omega$ we can find a $\prec\ \succ$-respecting map

$$\bar{k}\colon \{\alpha_0, \ldots, \alpha_{n-1}\} \to \kappa \text{ such that } \alpha_i \in F(x_j) \longrightarrow \bar{k}(\alpha_i) \in x_j. \tag{2.5.5}$$

This is possible since $\alpha_i \in F(x_j)$ implies $\alpha_i \in \pi(x_j)$ and $\langle \pi(x_j);\ j \in n\rangle \in \operatorname{rng}(\pi)$. Hence, the following holds

$$(\exists f)(f : n \to \pi(\kappa)\ \&\ \varphi(f)\ \&\ \psi(f, \langle \pi(x_j)\rangle_{j<n}), \tag{2.5.6}$$

where $\varphi(f)$ is the conjunction of all equalities $f(i) = \prec f(i_1), \ldots, f(i_\ell)\succ$ such that $\alpha_i = \prec\alpha_{i_1}, \ldots, \alpha_{i_\ell}\succ$ and $\psi(f, \langle \pi(x_j)\rangle_{j<n})$ is the conjunction of all equalities $f(i) \in \pi(x_j)$ for $i, j < n$ and such that $\alpha_i \in \pi(x_j)$. The formula (2.5.6) then clearly holds in M, hence by applying π^{-1} we obtain an $\bar{f} : n \to \kappa$ such that $\varphi(\bar{f})\ \&\ \psi(\bar{f}; \langle x_j\rangle_{j<n})$. Then $\bar{k} : \alpha_i \mapsto \bar{f}(i)$ is the desired map.

Now let

$$T = \{\bar{k};\ \bar{k} \text{ satisfies (2.5.5) for some } n \in \omega\}.$$

T is ordered by reverse inclusion and $\langle T, \supset\rangle$ is in fact a tree. Intuitively, T is the set of all approximations of a function $\mathfrak{k} : A \to \kappa$ witnessing the ω-completeness of F. To get such a function $\mathfrak{k}$ we have to show that T has a cofinal branch. In other words, we have to show that T is non-wellfounded.

We first look at $\pi(T)$. It consists of the functions $\tilde{k} : \{\pi(\alpha_i);\ i \in n\} \to \pi(\kappa)$ which are $\prec\ \succ$-respecting and such that $\pi(\alpha_i) \in \pi(F)(\pi(x_j)) \to \tilde{k}(\pi(\alpha_i)) \in \pi(x_j)$ for $i, j \in n$, since T is countable and hence $\pi(T) = \pi''T$. Now note that

$$\pi(\alpha_i) \in \pi(F)(\pi(x_j)) \longrightarrow \alpha_i \in F(x_j) \longrightarrow \alpha_i \in \pi(x_j),$$

hence $\pi^{-1} \restriction \pi(A)$ is a union of a cofinal branch in $\pi(T)$. Hence $\pi(T)$ is not well-founded. By absoluteness of well-foundedness, $M \models$ ($\pi(T)$ is not well-founded). Hence T is not well-founded. □(Lemma 2.5.3)

Note that the above argument goes through not only for $\boldsymbol{V}$ but for any inner model M which is closed under ω-sequences. In fact, after a slight modification it will also go through for a wide class of ω-closed set structures. (One has to get rid of the ordinals α_i in the definition of T, since they need not be elements of the structure; this is left to the reader.)

Lemma 2.5.4. *Let E be an extender on M at κ, λ, where M is suitable for κ. Let $\sigma : M \underset{E}{\longrightarrow} M'$ and $F = \sigma \restriction (\mathcal{P}(\kappa) \cap M)$. If E is ω-complete, then so is F.*

Proof. Let X, A be countable such that $X \subset \mathcal{P}(\kappa) \cap M$ and $A \subset F(\kappa) = \sigma(\kappa)$. Each $\alpha \in A$ is of the form $\sigma(g_\alpha)(\beta_\alpha)$ for some $g_\alpha \in {}^\kappa\kappa \cap M$ and $\beta_\alpha < \lambda$. Thus, if $x \in F_\alpha$

for some $x \in X$ and $\alpha \in A$, then this means $\sigma(g_\alpha)(\beta_\alpha) = \alpha \in F(x) = \sigma(x)$ and by Łos' Theorem:

$$\beta_\alpha \in E(u_{\alpha,x}), \text{ where } u_{\alpha,x} = \{\eta < \kappa;\ g_\alpha(\eta) \in x\}.$$

Let $Y = \{u_{\alpha,x};\ \alpha \in A\ \&\ x \in X\}$ and $B = \{\beta_\alpha;\ \alpha \in A\}$. These two sets are countable, so by ω-completeness of E, there is a $\prec\ \succ$-respecting map $\mathfrak{k} : B \to \kappa$ such that $u \in E_\beta \longrightarrow \mathfrak{k}(\beta) \in u$ whenever $\beta \in B$ and $u \in Y$. Hence,

$$\mathfrak{k}(\beta_\alpha) \in u_{\alpha,x} \text{ for } \alpha, x \text{ as above}$$

and consequently,

$$x \in F_\alpha \longrightarrow g_\alpha \circ \mathfrak{k}(\beta_\alpha) \in x \text{ for } x \in X,\ \alpha \in A.$$

Thus, the map $\mathfrak{k}^* : A \to \kappa$ verifying the ω-completeness of F can be defined by

$$\mathfrak{k}^*(\alpha) = g_\alpha \circ \mathfrak{k}(\beta_\alpha)$$

and it only remains to prove is the $\prec\ \succ$-respectivity of $\mathfrak{k}^*$. In order to do this pick $\alpha, \alpha_1, \ldots, \alpha_\ell \in A$ satisfying $\alpha = \prec\alpha_1, \ldots, \alpha_\ell\succ$. Hence,

$$\sigma(g_\alpha)(\beta_\alpha) \in \prec\sigma(g_{\alpha_1}(\beta_{\alpha_1}), \ldots, \sigma(g_{\alpha_\ell}(\beta_{\alpha_\ell})\succ$$

and by Łos' Theorem, $\prec\beta_\alpha, \beta_{\alpha_1}, \ldots, \beta_{\alpha_\ell}\succ \in E(u)$ where

$$u = \{\prec\eta_0, \ldots, \eta_\ell\succ \in \kappa;\ g_\alpha(\eta_0) = \prec g_{\alpha_1}(\eta_1), \ldots, g_{\alpha_\ell}(\eta_\ell)\succ\}.$$

This implies

$$g_\alpha(\mathfrak{k}(\beta_\alpha)) = \prec g_{\alpha_1}(\mathfrak{k}(\beta_{\alpha_1})), \ldots, g_{\alpha_\ell}(\mathfrak{k}(\beta_{\alpha_\ell}))\succ$$

and we are done. That $\mathfrak{k}^*$ is also order-preserving can be seen in a similar way.

□(Lemma 2.5.4)

Lemma 2.5.5. *Suppose that $\bar{M}$, M are coherent structures with extenders $\bar{F}$, F and that $\sigma : \bar{M} \to M$ is a Σ_0-preserving map such that*

$$M = \bigcup\{\sigma(x);\ x \in \bar{M}\ \&\ \mathrm{card}_M(x) < \bar{\kappa}\}$$

where $\bar{\kappa} = \mathrm{cr}(\bar{F})$. If $\bar{F}$ is ω-complete, then so is F.

Proof. Note that σ is cofinal (and hence Σ_1-preserving). Let $\kappa = \sigma(\bar{\kappa}) = \mathrm{cr}(F)$, and let $\bar{\lambda}$ and λ be the length of $\bar{F}$ and F, respectively. Let $A \subset \lambda$ and $X \subset \mathcal{P}(\kappa) \cap M$ be countable. For every $\alpha \in A$ pick a $g_\alpha \in \bar{M}$ and a β_α such that $g_\alpha : \delta_\alpha \to \bar{\lambda}$ for some $\delta_\alpha < \bar{\kappa}$, $\beta_\alpha < \sigma(\delta_\alpha)$ and $\alpha = \sigma(g_\alpha)(\beta_\alpha)$. Similarly, for every $x \in X$ pick an $f_x \in \bar{M}$ and a γ_x such that $f_x : \varepsilon_x \to \mathcal{P}(\bar{\kappa}) \cap \bar{M}$ for some $\varepsilon_x < \bar{\kappa}$, $\gamma_x < \sigma(\varepsilon_x)$ and $x = \sigma(f_x)(\gamma_x)$. We shall construct a map $\mathfrak{k} : A \to \kappa$ witnessing the ω-completeness of F.

Let $Y \subset \bar{M}$ be countable such that $\bar{M}|Y \prec \bar{M}$ and g_α, $f_x \in Y$ for every $\alpha \in A$ and $x \in X$. Collapse $\bar{M}|Y$ to a transitive structure M^*. Let $\sigma^* : M^* \to M$ be the inverse of the collapsing map. Let $M' = \text{Ult}(M^*, F^*)$, π be the associated ultrapower map and $\tilde{\sigma} : M' \to \bar{M}$ be a Σ_0-preserving map such that $\sigma^* = \tilde{\sigma} \circ \pi$. Such a map exists by ω-completeness of $\bar{F}$.

Pick $\alpha \in A$ and $x \in X$ such that $\alpha \in F(x)$. Then $\sigma(g_\alpha)(\beta_\alpha) \in \sigma(\bar{F}(f_x))(\gamma_x)$ (this involves a slight abuse of notation; but f_x is primitive recursively codable by a subset of $\bar{\kappa}$, so it does not do any harm). Hence

$$\langle \beta_\alpha, \gamma_x \rangle \in \sigma(a) \tag{2.5.7}$$

where

$$a = \{\langle \eta_0, \eta_1 \rangle \in \delta_\alpha \times \varepsilon_x;\ g_\alpha(\eta_0) \in \bar{F}(f_x)(\eta_1)\}.$$

Set $a^* = \sigma^{*-1}(a)$ and similarly for g^*_α, f^*_x. Then

$$\begin{aligned} a^* &= \{\langle \eta_0, \eta_1 \rangle \in \delta^*_\alpha \times \varepsilon^*_x;\ g^*_\alpha(\eta_0) \in F^*(f^*_x)(\eta_1)\} \\ &= \{\langle \eta_0, \eta_1 \rangle \in \delta^*_\alpha \times \varepsilon^*_x;\ g^*_\alpha(\eta_0) \in \pi(f^*_x)(\eta_1)\} \end{aligned} \tag{2.5.8}$$

and

$$a = \sigma^*(a^*) = \tilde{\sigma} \circ \pi(a^*) = \tilde{\sigma}(a^*).$$

Hence, applying $\tilde{\sigma}$ to (2.5.8) yields

$$a = \{\langle \eta_0, \eta_1 \rangle \in \delta_\alpha \times \varepsilon_x;\ \tilde{\sigma}(g^*_\alpha)(\eta_0) \in f_x(\eta_1)\}.$$

This holds since $\tilde{\sigma} \circ \pi(f^*_x) = \sigma^*(f^*_x) = f_x$ and $g^*_\alpha \in M'$, which means that $\tilde{\sigma}(g^*_\alpha)$ is defined (the latter follows from the coherency of M^*). A straightforward computation using (2.5.7) gives us

$$\sigma \circ \tilde{\sigma}(g^*_\alpha)(\beta_\alpha) \in \sigma(f_x)(\varepsilon_x) = x.$$

Hence it suffices to set

$$\mathfrak{k}(\alpha) = \sigma \circ \tilde{\sigma}(g^*_\alpha)(\beta_\alpha);$$

it follows from the above computation that $\mathfrak{k}(\alpha) \in x$ whenever $\alpha \in F(x)$.

To complete the proof it remains to show that $\mathfrak{k}$ is $\prec\ \succ$-respecting. Pick $\alpha, \alpha_1, \dots, \alpha_\ell$ such that $\alpha = \prec\alpha_1, \dots, \alpha_\ell\succ$ and set $d = \delta_\alpha \times \delta_{\alpha_1} \times \dots \times \delta_{\alpha_\ell}$. Then $\langle \beta_\alpha, \beta_{\alpha_1}, \dots, \beta_{\alpha_\ell} \rangle \in \sigma(b)$, where

$$b = \{\langle \eta, \eta_1, \dots, \eta_\ell \rangle \in d;\ g_\alpha(\eta) = \prec g_{\alpha_1}(\eta_1), \dots, g_{\alpha_\ell}(\eta_\ell)\succ\}.$$

We now repeat the argument from the first part of the proof and obtain

$$b = \{\langle \eta, \eta_1, \dots, \eta_\ell \rangle \in d;\ \tilde{\sigma}(g^*_\alpha)(\eta) = \prec\tilde{\sigma}(g^*_{\alpha_1})(\eta_1), \dots, \tilde{\sigma}(g^*_{\alpha_\ell})(\eta_\ell)\succ\}.$$

By applying σ we finally obtain

$$\sigma \circ \tilde{\sigma}(g^*_\alpha)(\beta_\alpha) = \prec\sigma \circ \tilde{\sigma}(g^*_{\alpha_1})(\beta_{\alpha_1}), \dots, \sigma \circ \tilde{\sigma}(g^*_{\alpha_\ell})(\beta_{\alpha_\ell})\succ\}$$

which means

$$\mathfrak{k}(\alpha) = \prec\mathfrak{k}(\alpha_1), \ldots, \mathfrak{k}(\alpha_\ell)\succ.$$

That $\mathfrak{k}$ is order preserving can be seen by the same argument. □(Lemma 2.5.5)

There are many situations when we need to extend a structure using an extender that is not ω-complete. A useful method to ensure extendibility in such cases is to embed the pair consisting of the structure and the extender in question into an extendible pair.

Lemma 2.5.6. *Let $\langle\sigma, \mathfrak{g}\rangle : \langle\bar{M}, \bar{F}\rangle \to \langle M, F\rangle$, where M is extendible by F. Then $\bar{M}$ is extendible by $\bar{F}$. Moreover, there is a unique map $\sigma' : \bar{M}' \underset{\Sigma_0}{\longrightarrow} M'$ completing the following diagram*

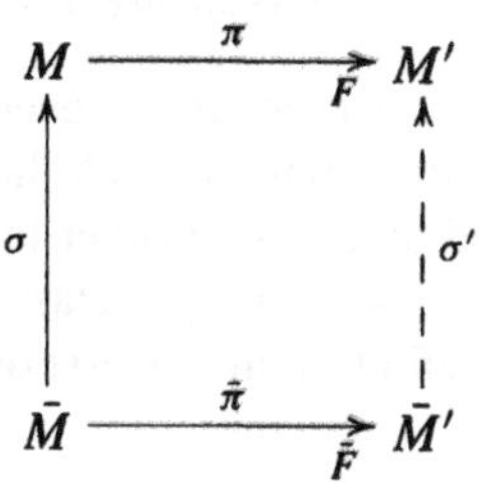

such that $\sigma' \restriction \bar{\lambda} = \mathfrak{g}$, where $\bar{\lambda} = \text{length}(\bar{F})$. The map σ' is defined by

$$\sigma'(\bar{\pi}(f)(\alpha)) = \pi \circ \sigma(f)(\mathfrak{g}(\alpha)).$$

Proof. It is straightforward to verify that σ' must be defined as above. It is also clear that all we have to prove is that σ' is Σ_0-preserving assuming that $\bar{\pi} : \bar{M} \underset{\bar{F}}{\longrightarrow} \bar{M}'$ weakly, since this implies that the structure $\bar{M}'$ is well-founded. Let $\kappa = \text{cr}(F)$ and $\bar{\kappa} = \text{cr}(\bar{F})$. Pick a Σ_0-formula $\varphi(v_1, \ldots, v_\ell)$, functions $f_1, \ldots, f_\ell$ from ${}^{\bar{\kappa}}\bar{M} \cap \bar{M}$ and $\alpha_1, \ldots, \alpha_\ell < \bar{\lambda}$. Then

$$\bar{M}' \models \varphi(\bar{\pi}(f_1)(\alpha_1), \ldots, \bar{\pi}(f_\ell)(\alpha_\ell))$$

is equivalent to

$$\prec\alpha_1, \ldots, \alpha_\ell\succ \in \bar{F}(\{\prec\eta_1, \ldots, \eta_\ell\succ \in \bar{\kappa};\ \bar{M} \models \varphi(f_1(\eta_1), \ldots, f_\ell(\eta_\ell))\}).$$

Using the assumption on $\langle\sigma, \mathfrak{g}\rangle$ (and the fact that $\mathfrak{g}$ is $\prec\ \succ$-respecting) we can rewrite this statement as $\prec\mathfrak{g}(\alpha_1), \ldots, \mathfrak{g}(\alpha_\ell)\succ \in F(u)$ where

$$u = \{\prec\eta_1, \ldots, \eta_\ell\succ \in \kappa;\ M \models \varphi(\sigma(f_1)(\eta_1), \ldots, \sigma(f_\ell)(\eta_\ell))\},$$

which by Łos' theorem is equivalent to

$$M' \models \varphi(\pi \circ \sigma(f_1)(\mathfrak{g}(\alpha_1)), \ldots, \pi \circ \sigma(f_\ell)(\mathfrak{g}(\alpha_\ell)))$$ □(Lemma 2.5.6)

2.6 Strong Cardinals

In this section we shall just state some very basic definitions and properties in connection with strong cardinals. Recall that a cardinal κ is measurable iff there is an elementary embedding $\pi : V \to M$ with critical point κ. It follows then that $V_{\kappa+1} \subset M$ for any such embedding, but the target model M remains in general "thin" above $\kappa + 1$. The idea behind strongness is to impose stronger requirements on the target model in the sense that it should contain larger V_β.

Definition. We say that a cardinal κ is *β-strong* iff there is an elementary embedding $\pi : V \to M$ with critical point κ such that $\beta < \pi(\kappa)$ and $V_\beta \subset M$. We then say that π is a *β-strong embedding*. A cardinal κ is *strong* iff κ is β-strong for every ordinal β.

It can be easily seen that κ is strong iff for every set a there is an elementary embedding $\pi : V \to M$ with critical point κ such that $a \in M$. The above definition has a little drawback: it is stated in a second order language, since we quantify over elementary embeddings of V, which are proper class objects. This drawback can be remedied by reformulating the definition in our first order language with extenders in place of embeddings.

Definition. An extender F on V is called *β-strong* iff V is extendible by F and the corresponding embedding is β-strong.

Lemma 2.6.1. *A cardinal κ is β-strong iff there is a β-strong extender on κ.*

Proof. The direction ($\leftarrow$) is obvious since any β-strong extender produces an embedding $\pi : V \to M$ which is β-strong. To see the converse it suffices to observe that given a β-strong embedding π, the associated extender $F = \pi \restriction \mathcal{P}(\kappa)$ is also β-strong. If $\bar{M} = \mathrm{Ult}(V, F)$ and $\bar{\pi}$ is the associated ultrapower map, we can define an embedding $\sigma : \bar{M} \to M$ by $\sigma(\bar{\pi}(f)(\alpha)) = \pi(f)(\alpha)$ and it is easy to see that $\sigma \restriction \pi(\kappa) = \mathrm{id}$. The cardinal κ, being β-strong, is inaccessible, so there is a surjective map $g : \kappa \to V_\kappa$. Thus, $\pi(g)$ maps $\pi(\kappa)$ surjectively onto $V^M_{\pi(\kappa)}$. Since $\pi(\kappa)$ is inacessible in M, the level $V_\beta = V^M_\beta$ is contained in $\pi(g)''\alpha$ for some $\alpha < \pi(\kappa)$. But $\pi(g)''\alpha = \sigma(\bar{\pi}(g))''\alpha$, so $V_\beta \subset \mathrm{rng}(\sigma)$. It follows that $V_\beta \subset \bar{M}$, which means that F is β-strong. □(Lemma 2.6.1)

If V is of the form J^A_∞, it is easy to see that κ is strong iff for every β there is an embedding $\pi : V \to M$ such that $J^A_\beta \subset M$.

In many situations we consider models with additional predicates. We will therefore need a property which is stronger than strongness in the sense that the corresponding embeddings "stretch" the additional predicates correctly.

Definition. A cardinal κ is *β-strong w.r.t. A* iff there is a β-strong extender $\pi : \langle V, A\rangle \to \langle M, A'\rangle$ such that $A \cap V_\beta = A' \cap V_\beta$. Such an embedding is called *β-strong w.r.t. A*. A cardinal κ is *strong w.r.t. A* iff for every β there is an embedding which is β-strong w.r.t. A.

Definition. An extender on κ is called *β-strong w.r.t. A* iff the associated embedding is.

Exactly as above we can verify that a cardinal is β-strong w.r.t. A iff there is an extender on κ which is β-strong w.r.t. A.

Let us finally mention that the following two notions are much stronger than strongness, resp. strongness with respect to a predicate.

Definition. A cardinal κ is *Woodin* iff for every $A \subset V_\kappa$ there is a $\mu < \kappa$ satisfying

$$\langle V_\kappa, A\rangle \models \mu \text{ is strong w.r.t. } A.$$

Note that κ is Woodin iff for every $A \subset \kappa$ and every $\beta < \kappa$ there is an embedding (in V) which is β-strong w.r.t. A. This follows from the fact that if we have such an embedding in $\langle V_\kappa, A\rangle$ then the corresponding extender is ω-complete in V_κ and therefore in V. Hence, V is extendible by the extender and the associated embedding is clearly β-strong w.r.t. A since it is an extension of the original embedding.

Lemma 2.6.2. *Every Woodin cardinal is inaccessible.*

Proof. We first show that κ is regular. Suppose the contrary. Let $f : \gamma \to \kappa$ be a monotone sequence cofinal in κ, where $\gamma < \kappa$. Suppose moreover that f is chosen such that $\min(\text{rng}(f)) > \gamma$. Let μ be strong w.r.t. f in $\langle V_\kappa, A\rangle$. If $\mu < \gamma$ then we can find a β such that $V_\beta \cap f \neq \emptyset$ and an embedding π which is β-strong w.r.t. f. This yields a contradiction as follows: By our choice of f, $f \cap V_\kappa = \emptyset$; but then $\emptyset \neq f \cap V_\beta = \pi(f \cap V_\kappa) \cap V_\beta = \emptyset$.

Suppose $\mu \geq \gamma$. Pick $\xi < \gamma$. Let β, π be as above and such that $f(\xi) < \beta$. Then $f(\xi) = \pi(f)(\pi(\xi)) \in \text{rng}(\pi)$, but then $f(\xi) < \mu$. Hence f is bounded by μ. Contradiction. This completes the proof of the regularity of κ.

To see that κ is strong limit, suppose for a contradiction that this is not the case, i.e. that we have an injective map $f : \kappa \to \mathcal{P}(\lambda)$ for some $\lambda < \kappa$. Let μ be strong w.r.t. f, A in V_κ, where A is the set of all ordinals greater than $\lambda + 1$. Then $\mu > \lambda + 1$; this follows from the fact that μ is strong w.r.t. A. Pick a $\beta > \mu$ and an embedding π which is β-strong w.r.t. f. Then

$$f \restriction \beta = f \cap V_\beta = \pi(f \cap \mu) \cap V_\beta = \pi(f \restriction \mu) \cap V_\beta$$

and $\text{rng}(f \restriction \beta) = \pi(\text{rng}(f \restriction \mu)) = \text{rng}(f \restriction \mu)$ since $\text{rng}(f \restriction \mu) \subset V_{\lambda+1}$. But this is impossible since f is injective. □(Lemma 2.6.2)

Woodin cardinals are in fact Mahlo (this follows by an argument similar to that above), but need not satisfy large cardinal properties stronger than Mahloness. The following notion is a considerable strengthening of Woodinness.

Definition. A cardinal κ is *superstrong* iff there is an embedding $\pi : V \to M$ which is $\pi(\kappa)$-strong. Such an embedding is called *superstrong*.

It follows then from the above that κ is superstrong iff there is an extender whose associated embedding is superstrong.

Lemma 2.6.3. *Every superstrong cardinal is Woodin.*

Proof. Let $\pi : V \to M$ be a superstrong embedding with critical point κ. Pick an $A \subset \kappa$. We want to show that there is a $\mu < \kappa$ which is strong w.r.t. A in $\langle V_\kappa, A\rangle$. This follows easily from the following statement by the elementarity of π.

$$\langle V_{\pi(\kappa)}, A\rangle \models \kappa \text{ is strong w.r.t. } A \tag{2.6.1}$$

To see (2.6.1), pick a $\beta < \pi(\kappa)$. Let $F' = \pi \restriction \mathcal{P}(\kappa)$ and $F = F'|\beta$. Then F is easily seen to be β-strong w.r.t. A (just imitate the proof of Lemma 2.6.1) and clearly $F \in V_{\pi(\kappa)}$. □(Lemma 2.6.3)

In fact, the above lemma can be considerably strengthened. The cardinal κ is clearly measurable, so it is both measurable and Woodin. If

$$U = \{x \subset \kappa \,;\, \kappa \in \pi(x)\}$$

is the "least" normal measure associated with π then there are clearly U-many Woodin cardinals below κ.

Notes

The notion of coherent sequence of ultrafilters was introduced by Mitchell in [Mi74]. Later, in [Mi1] he introduced the notion of an extender, which he called a "hypermeasure" and extended the notion of coherence to extenders. With later developments, many variants of this notion appeared and coherent extender sequences considered nowadays [MiS94], [Je97] look, of course, different. The notion of extender as defined in this book – the map representation, as well as our notion of coherent structure was introduced by Jensen in [Je97]. The idea of viewing extenders as maps and indexing them at the successor of the image of the critical point in the corresponding ultrapower is due to S. Friedman. The notion of closeness first appeared in [MiS94].

Chapter 3

Fine Ultrapowers

In this chapter we shall develop the theory of fine structural ultrapowers, also called *fine ultrapowers*, which will behave well with respect to fine structure. In particular, we will be able to give a detailed fine structural analysis of structures arising from such ultrapower constructions and the associated ultrapower maps will have strong preservation properties. From now on we restrict ourselves to acceptable J-structures. By the argument from the proof of Lemma 2.4.1, any such structure M is suitable for all κ's which look like uncountable cardinals in M.

3.1 The $*$-Ultrapower Construction

Let F be an extender with critical point κ. Given an acceptable J-structure M and a $p \in R^n_M$, we would like to find a structure $\tilde{M}$ and a $\tilde{p} \in R^n_{\tilde{M}}$ such that $\tilde{M}^{n,\tilde{p}} = \text{Ult}(M^{n,p}, F)$ where κ lies between n-th and $(n+1)$-st projectum of M. One possible approach to this task would be to form $\text{Ult}(M^{n,p}, F)$ and then iteratively apply the upwards extension of embeddings lemma (Lemma 1.4.1) to the associated ultrapower map. It turns out that the outcome of this construction can be equivalently presented as a sort of ultrapower, which differs from the usual ultrapower in that we employ not only the functions $f : \kappa \to M$ which are elements of M, but also those which are *suitably definable* over M. Such an ultrapower is called a $*$-*ultrapower.* The advantage of this approach is that we are able to treat these generalized ultrapowers in the same spirit as the usual ones; moreover, we shall see that the functions which are used in the $*$-ultrapower construction behave well enough so that we can manipulate them as if they were *elements* of M. The construction goes through even if R_M is empty and it does not refer to parameters from M.

Definition. Let $M = \langle J^A_\alpha, B\rangle$ be acceptable. We write $\pi : M \overset{*}{\underset{F}{\longrightarrow}} N$ iff

a) N is transitive;

b) $\pi : M \underset{\Sigma_0^{(n)}}{\longrightarrow} N$ for every n such that $\kappa < \omega\varrho^n_M$, where

c) $\kappa = \text{cr}(\pi)$;

d) $F = \langle \lambda \cap \pi(x);\ x \in \mathcal{P}(\kappa) \cap M\rangle$ for a primitive recursively closed λ such that $\kappa < \lambda \leq \pi(\kappa)$;

e) N is the closure of $\lambda \cup \mathrm{rng}(\pi)$ under functions which are elements of N or good $\Sigma_1^{(n)}(N)$-functions for n with $\omega\varrho_M^{n+1} > \kappa$.

Similarly as for the ordinary ultrapowers we define:

Definition. $N = \mathrm{Ult}^*(M, F)$ iff there is a π such that $\pi : M \xrightarrow[F]{*} N$. If $\mathrm{Ult}^*(M, F)$ exists, we also say that M is $*$-*extendible* by F.

Two immediate consequences of the above definitions are:

Lemma 3.1.1. *If* $\omega\varrho_M^1 \leq \kappa$, *then* $\pi : M \xrightarrow[F]{*} N$ *iff* $\pi : M \xrightarrow[F]{} N$. □(Lemma 3.1.1)

Lemma 3.1.2. *If* $\omega\varrho_M^n > \kappa$ *for all* $n \in \omega$ *and* $\pi : M \xrightarrow[F]{*} N$, *then* $\pi : M \xrightarrow[\Sigma^*]{} N$.
□(Lemma 3.1.2)

In the case of usual ultrapowers $\pi : M \xrightarrow[F]{} N$ (which we shall often call Σ_0-*ultrapowers*) we know that each $x \in N$ can be expressed in the form $x = \pi(f)(\xi)$ for some function $f \in {}^{\kappa}M \cap M$ and a $\xi < \lambda$. We shall now state a corresponding result for Σ^*-ultrapowers. For this purpose we also need to define $\pi(f)$ for the good $\Sigma_1^{(n)}(M)$-functions such that $\kappa < \omega\varrho_M^{n+1}$. Since good functions have functionally absolute definitions, the following suggests itself.

Definition. Let $N = \mathrm{Ult}(M, F)$ and π be the associated ultrapower map. Given a good $\Sigma_1^{(n)}(M)$-function f in p with $\mathrm{dom}(f) = \kappa < \omega\varrho_M^{n+1}$, $\pi(f)$ is the good $\Sigma_1^{(n)}(N)$ function in $\pi(p)$ which has the same functionally absolute $\Sigma_1^{(n)}$-definition and domain $\pi(\kappa)$.

Lemma 3.1.3. *The above definition is correct.*

Proof. Suppose f has two functionally absolute definitions, i.e. there are good $\Sigma_1^{(n)}(M)$ functions $F_M(z^n, w^0)$, $G_M(z^n, w^0)$ and parameters p, q such that

$$F_M(\xi, p) = f(\xi) = G_M(\xi, q)$$

for all $\xi < \kappa$. But then the $\Pi_0^{(n+1)}$-statement

$$(\forall z^{n+1} \in \kappa)(F_M(z^{n+1}, p) = G_M(z^{n+1}, q))$$

is preserved by π since π is $\Sigma_0^{(n+1)}$-preserving. Hence $F_N(\xi, \pi(p)) = G_N(\xi, \pi(q))$ for all $\xi \in \pi(\kappa)$, where F_N is the good $\Sigma_1^{(n)}(N)$ function which has the same functionally absolute $\Sigma_1^{(n)}$ definition as F_M and similarly for G_N. Thus, $\pi(f)$ is independent of its functionally absolute definition. □(Lemma 3.1.3)

Before we prove that each $x \in N$ can be expressed in the form $\pi(f)(\beta)$, let us state one more technical lemma, a sort of "uniformization lemma", which will be useful in many situations. In particular, it will be the key lemma in the proof of Łos' Theorem for Σ^*-ultrapowers.

Lemma 3.1.4. *Let $R(y^m, x_1^{i_1}, \ldots, x_\ell^{i_\ell})$ be a $\Sigma_1^{(m)}(M)$ relation, where $i_j \le m$ for $j = 1, \ldots, \ell$. Let $n \ge m$ be such that $\omega\varrho_M^{n+1} > \kappa$. Suppose that $f_1, \ldots, f_\ell$ are good $\mathbf{\Sigma}_1^{(n)}(M)$ functions such that f_j is to $H_M^{i_j}$ and* $\mathrm{dom}(f_j) = \kappa$ *for $j = 1, \ldots, \ell$. Then there is a good $\mathbf{\Sigma}_1^{(n)}(M)$ function g with* $\mathrm{rng}(g) \subset H_M^m$ *and* $\mathrm{dom}(g) = \kappa$ *satisfying*

$$(\exists z^m)R(z^m, f_1(\xi_1), \ldots, f_\ell(\xi_\ell)) \longrightarrow R(g(\prec\xi_1, \ldots, \xi_\ell\succ), f_1(\xi_1), \ldots, f_\ell(\xi_\ell)) \tag{3.1.1}$$

for all $\xi_1, \ldots, \xi_\ell < \kappa$.

Proof. By Lemma 1.8.3, there is a good $\mathbf{\Sigma}_1^{(m)}(M)$ function $F(x_1^{i_1}, \ldots, x_\ell^{i_\ell})$ uniformizing R. Hence, if we set

$$g^*(\prec\xi_1, \ldots, \xi_\ell\succ) \simeq F(f_1(\xi_1), \ldots, f_\ell(\xi_\ell)),$$

then g^* is a good $\mathbf{\Sigma}_1^{(n)}(M)$ function and (3.1.1) holds with g^* in place of g, however, g^* need not be defined on all of κ and $\mathrm{rng}(g^*)$ need not be a subset of H_M^m. To fix this, we shall redefine g^* in a straightforward way, i.e. we set

$$g(\prec\xi_1, \ldots, \xi_\ell\succ) = \begin{cases} g^*(\prec\xi_1, \ldots, \xi_\ell\succ) & \text{if defined and } \in H_M^m \\ 0 & \text{otherwise.} \end{cases}$$

To see that this definition makes sense, note first that $a \in M$, where

$$a = \{\prec\xi_1, \ldots, \xi_\ell\succ \in \kappa;\ (\exists z^m)(z^m = g^*(\prec\xi_1, \ldots, \xi_\ell\succ))\};$$

since a is a $\mathbf{\Sigma}_1^{(n)}(M)$ subset of κ and $\omega\varrho_M^{n+1} > \kappa$. Define $G(z^n, w^n)$ by

$$y = G(z^0, w^0) \longleftrightarrow (y = z^0 \ \&\ w^0 \in a) \vee (y = 0 \ \&\ w^0 \notin a).$$

Then G is $\mathbf{\Sigma}_0(M)$, hence, it is good and it is easy to verify that

$$g(\prec\xi_1, \ldots, \xi_\ell\succ) = G(g^*(\prec\xi_1, \ldots, \xi_\ell\succ), \prec\xi_1, \ldots, \xi_\ell\succ).$$

It follows then immediately that the function g satisfies all of our requirements, as $g(\prec\xi_1, \ldots, \xi_\ell\succ) = g^*(\prec\xi_1, \ldots, \xi_\ell\succ)$ in all relevant cases. □(Lemma 3.1.4)

Remark. The function g from the above lemma can be regarded as a function to H_M^m, as follows from the remark after the proof of Lemma 1.8.1.

Definition. $\Gamma(\kappa, M)$ is the set of all good $\mathbf{\Sigma}_1^{(n)}(M)$ functions $f : \kappa \to M$ for n such that $\omega\varrho_M^{n+1} > \kappa$. Furthermore, $\Gamma_n(\kappa, M)$ is defined as follows:

$$\Gamma_n(\kappa, M) = \begin{cases} \{f \in \Gamma(\kappa, M);\ \mathrm{rng}(f) \subset H_M^n\} & \text{if } \omega\varrho_M^{n+1} > \kappa \\ \{f \in \Gamma(\kappa, M);\ \mathrm{rng}(f) \in H_M^n\} = {}^\kappa M \cap H_M^n & \text{if } \omega\varrho_M^{n+1} \le \kappa < \omega\varrho_M^n. \end{cases}$$

By good $\mathbf{\Sigma}_1^{(-1)}(M)$ functions $f : \kappa \to M$ we mean those functions which are *elements* of M.

Lemma 3.1.5. *Let $\pi : M \xrightarrow[F]{*} N$. Then each $x \in N$ is of the form $\pi(f)(\beta)$ for some $\beta < \lambda$ and some $f \in \Gamma(\kappa, M)$.*

Proof. We can w.l.o.g. suppose that $\omega\varrho_M^1 > \kappa$. By the definition, there is a good $\Sigma_1^{(n)}(N)$ function $g(z^i, v^j)$ such that $\omega\varrho_N^{n+1} > \kappa$ and $x = g(\beta, p)$ for some $\beta < \lambda$ and $p \in \text{rng}(\pi)$. Let $\bar{g}(z^i, v^j)$ be the good $\Sigma_1^{(n)}(M)$ function by the same functionally absolute definition. By Lemma 3.1.4, there is a good $\mathbf{\Sigma}_1^{(n)}(M)$-function $f : \kappa \to M$ such that for all $\xi \in \kappa$,

$$(\exists y^0)(y^0 = \bar{g}(\xi, \bar{p})) \longrightarrow f(\xi) = \bar{g}(\xi, \bar{p})$$

where $\bar{p} = \pi^{-1}(p)$. Then clearly $f \in \Gamma(\kappa, M)$. Since putting the bounded quantifier $(\forall \xi^{n+1} < \kappa)$ in front of the right side of the above implication yields a $\Pi_0^{(n+1)}$-statement, $\pi(f)(\beta) = g(\beta, p) = x$. □(Lemma 3.1.5)

Łos' Theorem then has the following form:

Lemma 3.1.6. *Let $\varphi(v_1^{i_1}, \ldots, v_\ell^{i_\ell})$ be a $\Sigma_0^{(n)}$-formula for some n satisfying $\omega\varrho_M^n > \kappa$. If $f_1, \ldots, f_\ell \in \Gamma(\kappa, M)$ are such that $f_j \in \Gamma_{i_j}(\kappa, M)$ and $\alpha_1, \ldots, \alpha_\ell < \lambda$, then*

$$N \models \varphi(\pi(f_1)(\alpha_1), \ldots, \pi(f_\ell)(\alpha_\ell))$$

iff

$$\prec\alpha_1, \ldots, \alpha_\ell\succ \in F(\{\prec\eta_1, \ldots, \eta_\ell\succ \in \kappa;\ M \models \varphi(f_1(\eta_1), \ldots, f_\ell(\eta_\ell))\}).$$

Proof. Straightforward. □(Lemma 3.1.6)

Exactly as for Σ_0-ultrapowers, we obtain the following lemma using Łos' Theorem.

Lemma 3.1.7. *Let $\pi : M \xrightarrow[F]{*} N$. Then N, F are uniquely determined.*

□(Lemma 3.1.7)

We now turn to the converse question, namely, whether there is a structure N and a map π such that $\pi : M \xrightarrow[F]{*} N$, provided that M, F are given. As before we shall construct N, π by an ultrapower construction, which, of course, will be slightly modified in order to get ultrapower maps that also preserve $\Sigma_0^{(n+1)}$-formulae. We define a term model $\mathbb{D} = \mathbb{D}^*(M, F)$ as follows:

$$\mathbb{D} = \langle D, \cong, \tilde{\in}, \tilde{A}, \tilde{B}\rangle$$

where

$$D = \{\langle\alpha, f\rangle;\ \alpha < \lambda\ \&\ f \in \Gamma(\kappa, M)\}$$

and

$$\begin{aligned}
\langle\alpha, f\rangle \cong \langle\beta, g\rangle &\longleftrightarrow \langle\alpha, \beta\rangle \in F(\{\langle\eta_1, \eta_2\rangle;\ f(\eta_1) = g(\eta_2)\}) \\
\langle\alpha, f\rangle \tilde{\in} \langle\beta, g\rangle &\longleftrightarrow \langle\alpha, \beta\rangle \in F(\{\langle\eta_1, \eta_2\rangle;\ f(\eta_1) \in g(\eta_2)\}) \\
\tilde{A}(\langle\alpha, f\rangle) &\longleftrightarrow \alpha \in F(\{\eta;\ A(f(\eta))\}) \\
\tilde{B}(\langle\alpha, f\rangle) &\longleftrightarrow \alpha \in F(\{\eta;\ B(f\eta))\}).
\end{aligned}$$

We then prove Łos' Theorem for Σ_0-formulae, which in this case reads:

Lemma 3.1.8. *Let $\langle\alpha_i, f_i\rangle \in D$ for $i = 1, \ldots, \ell$ and $\varphi(v_1, \ldots, v_\ell)$ be a Σ_0-formula. Then*

$$\mathbb{D} \models \varphi(\langle\alpha_1, f_1\rangle, \ldots, \langle\alpha_\ell, f_\ell\rangle)$$

iff

$$\langle\alpha_1, \ldots, \alpha_\ell\rangle \in F(\{\langle\eta_1, \ldots, \eta_\ell\rangle;\ M \models \varphi(f_1(\eta_1), \ldots, f_\ell(\eta_\ell))\}).$$ □(Lemma 3.1.8)

This lemma can be proved by a straightforward induction on Σ_0-formulae using Lemma 3.1.4. Later we shall prove a much stronger version of Łos' Theorem and there we shall give some details of the induction step. Note that if $\pi : M \underset{F}{\overset{*}{\longrightarrow}} N$, then the map

$$\langle\alpha, f\rangle \mapsto \pi(f)(\alpha)$$

is structure preserving, in particular

$$\langle\alpha, f\rangle \tilde{\in} \langle\beta, g\rangle \quad \text{iff} \quad \pi(f)(\alpha) \in \pi(g)(\beta).$$

Hence $\mathbb{D}$ is well-founded. Thus we have

Lemma 3.1.9. *If $\mathbb{D}$ is not well-founded, then there are no π, N such that*

$$\pi : M \underset{F}{\overset{*}{\longrightarrow}} N.$$

□(Lemma 3.1.9)

From now on we shall assume that $\mathbb{D}$ is well-founded. We shall show that under this assumption π and N exist. Using Łos' Theorem we easily obtain

$\mathbb{D}$ satisfies extensionality, *and*
$\cong$ is a congruence relation on $\mathbb{D}$.

Hence, there is a transitive $N = \langle |N|, A', B'\rangle$ and a factorizing map

$$[\ ] : \mathbb{D} \overset{\text{onto}}{\longrightarrow} N,$$

i.e.

$$\begin{aligned}
x \tilde{\in} y &\longleftrightarrow [x] \in [y] \\
x \cong y &\longleftrightarrow [x] = [y] \\
\tilde{A}(x) &\longleftrightarrow A'([x]) \\
\tilde{B}(x) &\longleftrightarrow B'([x])
\end{aligned}$$

for $x, y \in \mathbb{D}$. We define a $\pi : M \to N$ by

$$\pi(x) = [\langle 0, c_x \rangle]$$

where c_x is the constant map with value x. It follows immediately from Łos' Theorem that

$$\pi : M \underset{\Sigma_0}{\longrightarrow} N.$$

Lemma 3.1.10. *Let $n \in \omega$ be such that $\omega\varrho_M^{n+1} \leq \kappa < \omega\varrho_M^n$. Set $\bar{H} = H_M^n$ and $H = \bigcup(\pi''\bar{H})$. Then*

$$\pi \upharpoonright \bar{H} : \bar{H} \underset{F}{\longrightarrow} H.$$

Proof. Pick $x \in H$. We know that $x = [\langle \beta, f \rangle]$ for some $\beta < \lambda$ and $f \in \Gamma(\kappa, M)$. We have to prove that one can pick such an f from $\bar{H}$. Let $a \in \bar{H}$ be such that $x \in \pi(a)$; such an a exists by the definition of H. By Lemma 3.1.4, f can be chosen such that $\text{rng}(f) \subset a$, since

$$\beta \in F(\{\eta < \kappa;\ f(\eta) \in a\}).$$

Hence $f \subset \kappa \times a \in H_M^n$. Since f is at most $\mathbf{\Sigma}_1^{(n-1)}(M)$, $f \in M$ and, by acceptability, $f \in \bar{H}$. □(Lemma 3.1.10)

Notation. Instead of $[\langle \alpha, f \rangle]$ we shall briefly write $[\alpha, f]$.

As an immediate consequence of Lemma 3.1.10 we obtain

$$\kappa = \text{cr}(\pi)$$
$$[\beta, f] = \pi(f)(\beta) \quad \text{for} \quad f \in {}^{\kappa}\bar{H} \cap \bar{H} \quad \text{and} \quad \beta < \lambda.$$

Lemma 3.1.11. *Let M, N, F, π be as above. Then*

a) *N is an acceptable J-structure.*

b) $\pi : M \underset{\Sigma_0^{(n)}}{\longrightarrow} N$ *if $\omega\varrho_M^n > \kappa$.*

c) $\pi : M \underset{\Sigma_2^{(n)}}{\longrightarrow} N$ *if $\omega\varrho_M^{n+1} > \kappa$.*

d) (Łos' Theorem for $\Sigma_0^{(n)}$-formulae). *Let $\varphi(v_1^{i_1}, \dots, v_\ell^{i_\ell})$ be a $\Sigma_0^{(n)}$-formula where $\omega\varrho_M^n > \kappa$. Then for any $f_1, \dots, f_\ell \in \Gamma(\kappa, M)$ such that $f_j \in \Gamma_{i_j}(\kappa, M)$ and any $\alpha_1, \dots, \alpha_\ell < \lambda$, the following holds:*

$$M \models \varphi([\alpha_1, f_1], \dots, [\alpha_\ell, f_\ell])$$

iff

$$\langle \alpha_1, \dots, \alpha_\ell \rangle \in F(\{\langle \eta_1, \dots, \eta_\ell \rangle \in {}^{\ell}\kappa;\ M \models \varphi(f_1(\eta_1), \dots, f_\ell(\eta_\ell))\}).$$

Proof. The proof of Lemma 3.1.11 stretches over several sublemmata. We first introduce a sequence $\langle H_n \rangle$ of transitive subsets of N; these subsets will substitute the H_N^n for a while. We shall interpret variables v^n over H_n and prove b) – d) for this "pseudo-interpretation". Using this partial result we shall inductively prove that in fact $H_n = H_N^n$; the above lemma will then immediately follow. Set

$$\Gamma_n = \Gamma_n(\kappa, M)$$

$$H_n = \{[\beta, f];\ \beta < \lambda \ \&\ f \in \Gamma_n\}.$$

(Note that, for $\omega\varrho_M^{n+1} > \kappa$, the functions from Γ_n can be regarded as functions to H_M^n.) Clearly

$$H_n \quad \text{is transitive.}$$

(Proof: if $[\bar{\beta}, \bar{f}] \in [\beta, f] \in H_n$, then there is an $f^* \in \Gamma_n$ such that $[\bar{\beta}, \bar{f}] = [\beta, f^*]$ by Lemma 3.1.4.)

We say that $N \models \varphi(x_1, \ldots, x_\ell)$ in the *pseudointerpretation* iff $\varphi(x_1, \ldots, x_\ell)$ holds when taking v^i as ranging over H_i for $\omega\varrho_M^i > \kappa$. We then get a version of Łos' Theorem for pseudointerpretations.

Claim 1. (Łos' Theorem for pseudointerpretations). Let $\varphi(v_1^{i_1}, \ldots, v_\ell^{i_\ell})$ be a $\Sigma_0^{(n)}$-formula, where $\omega\varrho_M^n > \kappa$. Let $\alpha_1, \ldots, \alpha_\ell < \lambda$ and $f_1, \ldots, f_\ell \in \Gamma(\kappa, M)$. Then

$$N \models \varphi([\alpha_1, f_1], \ldots, [\alpha_\ell, f_\ell])$$

in the sense of pseudointerpretation iff

$$\langle \alpha_1, \ldots, \alpha_\ell \rangle \in F(\{\langle \eta_1, \ldots, \eta_\ell \rangle \in {}^\ell\kappa;\ M \models \varphi(f_1(\eta_1), \ldots, f_\ell(\eta_\ell))\}).$$

(In particular, this equivalence holds for $\Sigma_1^{(n)}$-formulae if $\omega\varrho_M^{n+1} > \kappa$.)

Proof. We shall proceed by a straightforward induction on $\Sigma_0^{(n)}$-formulae using Lemma 3.1.4. For illustration we show the induction step for formulae of the form

$$(\exists v^n \in z^n)\varphi(v^n, z^n, v_1^{i_1}, \ldots, v_\ell^{i_\ell}).$$

The implication ($\rightarrow$) is trivial, so we shall focus on the converse.

Case 1. $\omega\varrho_M^{n+1} > \kappa$.

Let $f_0 \in \Gamma_n$, $f_j \in \Gamma_{i_j}$ for $j = 1, \ldots, \ell$ and $\alpha_0, \ldots, \alpha_\ell < \lambda$. We suppose that

$$\langle \alpha_0, \ldots, \alpha_\ell \rangle \in F(u) \tag{3.1.2}$$

where

$$u = \{\langle \eta_0, \ldots, \eta_\ell \rangle \in {}^{\ell+1}\kappa;\ M \models (\exists v^n \in f_0(\eta_0))\varphi(v^n, f_0(\eta_0), \ldots, f_\ell(\eta_\ell))\}.$$

Let R be the relation defined by φ over M. By Lemma 3.1.4, there is a function $g \in \Gamma_n$ such that

$$(\exists v^n \in f_0(\eta_0))R(v^n, f_0(\eta_0), \dots, f_\ell(\eta_\ell))$$
$$\longrightarrow g(\prec\eta_0, \dots, \eta_\ell\succ) \in f_0(\eta_0) \;\&\; R(g(\prec\eta_0, \dots, \eta_\ell\succ, f_0(\eta_0), \dots, f_\ell(\eta_\ell)))$$

for all $\eta_0, \dots, \eta_\ell < \kappa$. Let a be the set of all $\prec\eta_0, \dots, \eta_\ell\succ$ satisfying the consequence of the above implication. Then $a \in M$, since it is a $\mathbf{\Sigma}_1^{(m)}(M)$ subset of $\kappa < \omega\varrho_M^{m+1}$ for some $m \leq n$. Moreover, by the above implication, $F(a)$ is a superset of $F(u)$, hence $\prec\alpha_0, \dots, \alpha_\ell\succ \in F(a)$. So, by the induction hypothesis,

$$[\prec\alpha_0, \dots, \alpha_\ell\succ, g] \in [\alpha_0, f_0] \;\&\; N \models \varphi([\prec\alpha_0, \dots, \alpha_\ell\succ, g], [\alpha_0, f_0], \dots, [\alpha_\ell, f_\ell]),$$

which means

$$N \models (\exists v^n \in [\alpha_0, f_0])\varphi(v_n, [\alpha_0, f_0], \dots, [\alpha_\ell, f_\ell]).$$

Case 2. $\omega\varrho_M^{n+1} \leq \kappa < \omega\varrho_M^n$.

We can basically repeat the above proof but we now cannot use Lemma 3.1.4 for the construction of g. In this case, however, we know that $f_0 \in H_M^n$. If we now define g by

$$g(\eta) = \begin{cases} \text{the } <_M\text{-least } z \in f_0(\eta_0) \text{ such that } R(z, f_0(\eta_0), \dots, f_\ell(\eta_\ell)) & \text{if defined} \\ 0 \quad \text{otherwise,} & \end{cases}$$

where $\eta = \prec\eta_0, \dots, \eta_\ell\succ$, then g is a $\mathbf{\Sigma}_0^{(n)}(M)$ subset of $\kappa \times \bigcup \operatorname{rng}(f_0) \in H_M^n$, hence $g \in H_M^n$ by Lemma 1.6.5. The rest of the proof is then the same as above. □(Claim 1)

As an immediate consequence we get:

Claim 2. $\pi : M \underset{\Sigma_0^{(n)}}{\longrightarrow} N$ for $\omega\varrho_M^n > \kappa$ in the pseudointerpretation.

For $\omega\varrho_M^{n+1} > \kappa$ we get a better result.

Claim 3. $\pi : M \underset{\Sigma_2^{(n)}}{\longrightarrow} N$ in the pseudointerpretation for $\omega\varrho_M^{n+1} > \kappa$.

Proof. It follows from the above claim that $\pi : M \underset{\Sigma_1^{(n)}}{\longrightarrow} N$ for such n. Hence, it is sufficient to prove that $\Sigma_2^{(n)}$ formulae are preserved downwards. So suppose

$$N \models (\exists v^n)\varphi(v^n, \pi(x_1), \dots, \pi(x_\ell)),$$

where φ is a $\Pi_1^{(n)}$-formula. Thus, we can pick a $[\beta, g]$ such that $g \in \Gamma_n$ and $N \models \varphi([\beta, g], \pi(x_1), \dots, \pi(x_\ell))$. By Łos' Theorem for pseudointerpretations,

$$\beta \in F(\{\eta;\; M \models \varphi(g(\eta), x_1, \dots, x_\ell)\}),$$

so the set on the right side is nonempty. Since F maps nonempty sets to nonempty sets, there is an $\eta < \kappa$ such that $M \models \varphi(g(\eta), x_1, \ldots, x_\ell)$. Hence,

$$M \models (\exists v^n)\varphi(v^n, x_1, \ldots, x_\ell). \qquad \square\text{(Claim 3)}$$

From this we can already infer clause a) of the lemma.

Claim 4. N is an acceptable J-structure.

Proof. By Lemma 1.1.3, the property of being an acceptable J-structure is a Q-condition. If $\omega\varrho_M^1 \leq \kappa$ then the Σ^*-ultrapower coincides with the Σ_0-ultrapower, hence $\pi : M \underset{\Sigma_0}{\longrightarrow} N$ cofinally and such maps preserve Q-conditions upwards. If $\omega\varrho_M > \kappa$, then $\pi : M \underset{\Sigma_2}{\longrightarrow} N$, since $H_0 = |N|$. Hence π preserves Q-conditions again. $\square$(Claim 4)

Fix the following notation:

$$N = \langle J_{\alpha'}^{A'}, B' \rangle \quad \text{for some } \alpha' \in \mathbf{On}$$
$$\omega\varrho_n = \mathbf{On} \cap H_n.$$

(π is sufficiently preserving to guarantee that $\mathbf{On} \cap H_n$ is of this form.)

Claim 5. $H_n = J_{\varrho_n}^{A'}$ whenever $\omega\varrho_M^n > \kappa$.

Proof. By Claim 4,

$$\pi \restriction H_M^n : J_{\varrho_M^n}^A \underset{\Sigma_2}{\longrightarrow} \langle H_n, A' \cap H_n \rangle \quad \text{if } \omega\varrho_M^{n+1} > \kappa$$

and by Lemma 3.1.10,

$$\pi \restriction H_M^n : J_{\varrho_M^n}^A \underset{\Sigma_0}{\longrightarrow} \langle H_n, A' \cap H_n \rangle \quad \text{cofinally if } \omega\varrho_M^{n+1} \leq \kappa < \omega\varrho_M^n.$$

In both cases $\pi \restriction H_M^n$ is Q-preserving, which tells us that H_n is a J-structure relatively constructed from A'. $\square$(Claim 5)

We are now ready to prove that our pseudointerpretation is in fact the right interpretation.

Claim 6. Let M, N be as above. Then

$$\varrho_n = \varrho_N^n \quad \text{if } \omega\varrho_M^{n+1} > \kappa$$
$$\varrho_n \leq \varrho_N^n \quad \text{if } \omega\varrho_M^{n+1} \leq \kappa < \omega\varrho_M^n.$$

Consequently, $H_n = H_N^n$ if $\omega\varrho_M^{n+1} > \kappa$.

Proof. By induction on n. The case $n = 0$ is trivial, so we go right to the induction step. Suppose $\varrho_i = \varrho_N^i$ for $i < n$; hence $H_i = H_N^i$ for such i. We first prove

$$\varrho_n \leq \varrho_N^n \quad \text{whenever } \omega\varrho_M^n > \kappa. \qquad (3.1.3)$$

Let D be $\mathbf{\Sigma}_1^{(n-1)}(N)$, i.e. $D(z) \leftrightarrow D^*(z, p)$ for all z, where D^* is $\Sigma_1^{(n-1)}(N)$ and $p = [\beta, f]$ for some $\beta < \lambda$ and $f \in \Gamma(\kappa, M)$. Let $x \in H_n$. We show

$$D \cap x \in H_n, \tag{3.1.4}$$

which immediately gives (3.1.3), as follows from Lemma 1.7.4.

Proof of (3.1.4). Let $x = [\gamma, g]$, where $\gamma < \lambda$ and $g \in \Gamma_n$. Set

$$k(\prec\eta_1, \eta_2\succ) = \{z \in g(\eta_1);\ \bar{D}^*(z, f(\eta_2))\}$$

where $\bar{D}^*$ is $\Sigma_1^{(n-1)}(M)$ by the same definition as D^*. Note first that $\mathrm{dom}(k) = \kappa$, since for any fixed $\eta_1, \eta_2 < \kappa$, the set on the right side is a $\mathbf{\Sigma}_1^{(n-1)}(M)$ subset of $f(\eta_1) \in H_M^n$ and therefore is itself an element of H_M^n. Hence $k : \kappa \to H_M^n$. The definition of k reads

$$y = k(\prec\eta_1, \eta_2\succ) \leftrightarrow y \subset g(\eta_1)\ \&\ (\forall z^n \in g(\eta_1))(z^n \in y \leftrightarrow \bar{D}^*(z^n, f(\eta_2))). \tag{3.1.5}$$

Case 1. $\omega\varrho_M^{n+1} > \kappa$. Set

$$y^n = G(w_1^n, w_2^0) \longleftrightarrow y^n \subset w_1^n\ \&\ (\forall z^n \in w_1^n)(z^n \in y \longleftrightarrow \bar{D}^*(z^n, w_2^0)).$$

Then G is a $\Sigma_0^{(n)}$-function to H_M^n and hence is good. G is clearly defined for all w_1^n, w_2^0 and it follows immediately that $k(\xi) = G(g((\xi)_0^2), f((\xi)_1^2))$. Hence $k \in \Gamma(\kappa, M)$. By the fact that (3.1.5) holds for any $\eta_1, \eta_2 < \kappa$,

$$\prec\gamma, \beta\succ \in F(\{\prec\eta_1, \eta_2\succ;\ k(\prec\eta_1, \eta_2\succ) = g(\eta_1) \cap \bar{D}^*(-, f(\eta_2))\}).$$

Using Łos' Theorem for pseudointerpretations we finally obtain

$$[\prec\gamma, \beta\succ, k] = [\gamma, g] \cap D^*(-, [\beta, f]) = X \cap D,$$

since the statement in question is $\Sigma_0^{(n)}$. This proves Case 1.

Case 2. $\omega\varrho_M^{n+1} \leq \kappa < \omega\varrho_M^n$.

The proof is the same but we have to check that $k \in \Gamma(\kappa, M)$, which in this case means that $k \in H_M^n$. But this follows from the fact that $k : \kappa \to H_M^n$ and k is $\mathbf{\Sigma}_0^{(n)}(M)$ (see (3.1.5)), so k is rudimentary over $M^{n,q}$, where q codes the parameters from which f and g are defined. □(3.1.4)

Finally we shall focus on the proof of

$$\varrho_N^n \leq \varrho_n \quad \text{if} \quad \omega\varrho_M^{n+1} > \kappa. \tag{3.1.6}$$

Let $\bar{D}(x_1, x_2)$ and $\bar{p} \in M$ be such that $\bar{D}$ is $\Sigma_1^{(n-1)}(M)$ and

$$\{\xi < \omega\varrho_M^n;\ \bar{D}(\xi, \bar{p})\} \notin M.$$

Let D be $\Sigma_1^{(n-1)}(N)$ by the same definition and $p = \pi(\bar{p})$. We have to show that $x \notin N$, where

$$x = \{\xi < \omega\varrho_n;\ D(\xi, p)\}.$$

Suppose that $x \in N$, i.e. $x = [\beta, f]$ for some $\beta < \lambda$ and $f \in \Gamma(\kappa, M)$. Then

$$(\forall \xi^n)(\xi^n \in [\beta, f] \longleftrightarrow D(\xi^n, p)).$$

This is a $\Pi_1^{(n)}$-statement in the pseudointerpretation, hence $\beta \in F(a)$, where

$$a = \{\eta < \kappa;\ (\forall \xi^n)(\xi^n \in f(\eta) \leftrightarrow \bar{D}(\xi^n, \bar{p}))\}.$$

It follows then immediately that $a \neq \varnothing$, i.e. there is $\eta < \kappa$ such that

$$\{\xi < \omega\varrho_M^n;\ \bar{D}(\xi, \bar{p})\} = f(\eta) \cap \omega\varrho_M^n \in M.$$

Contradiction. □(3.1.6)

By (3.1.6), the clauses b) – d) hold in the ordinary interpretation. □(Lemma 3.1.11)

As a corollary of the proof of (3.1.6) we obtain

Lemma 3.1.12. *$\pi'' P_M^n \subset P_N^n$ if $\omega\varrho_M^{n+1} > \kappa$.* □(Lemma 3.1.12)

Corollary 3.1.13. *If $\omega\varrho_M^{n+1} > \kappa$ and M is n-solid then $\pi(p_M^n) = p_N^n$ and N is n-solid.*

Proof. Apply the previous lemma and Lemma 1.12.6. □(Corollary 3.1.13)

Finally, we are able to infer that N is a $*$-extension of M by F in the sense of our original definition.

Lemma 3.1.14. $\pi : M \overset{*}{\underset{F}{\longrightarrow}} N$.

Proof. a) – d) of the definition of a $*$-ultrapower follow immediately from Lemma 3.1.10 and 3.1.11. We now prove e). In fact, we prove the following:

$$[\beta, f] = \pi(f)(\beta) \quad \text{whenever } \beta < \lambda \text{ and } f \in \Gamma(\kappa, M). \tag{3.1.7}$$

If $f \in M$ this follows exactly as in the case of Σ_0-ultrapowers. Now suppose $f \notin M$. Hence f is a good $\mathbf{\Sigma}_1^{(n)}(M)$ function for some n such that $\omega\varrho_M^{n+1} > \kappa$. Let $R(v_0, v_1, v_2)$ be a $\Sigma_1^{(n)}(M)$ relation such that

$$y = f(x) \longleftrightarrow R(y, x, p)$$

for some parameter $p \in M$. Then clearly

$$\beta \in F(\{\eta \in \kappa;\ R(f(\eta), \eta, p)\}),$$

hence

$$\tilde{R}([\beta, f], \beta, \tilde{p}),$$

where $\tilde{p} = \pi(p)$ and $\tilde{R}$ is $\Sigma_1^{(n)}(N)$ by the same functionally absolute definition (R has such a definition, since R defines f). But this means $[\beta, f] = \pi(f)(\beta)$, since $\tilde{R}$ is a defining relation for $\pi(f)$. □(Lemma 3.1.14)

3.2 Some Special Preservation Properties

The properties of fine ultrapowers developed in the previous section are of a general character. Granting that we have more information about the structures or the extenders in question, we can derive stronger preservation properties of the ultrapower map. Among these belongs the cofinality at the n-th level, where $\omega\varrho_M^n \leq \kappa < \omega\varrho_M^{n+1}$. For instance, this is the case whenever we apply en extender which is either weakly of Σ_1-amenable, or whenever M has a very good parameter. In the following we always assume that $\pi : M \xrightarrow[F]{*} N$.

Lemma 3.2.1. *Let F be weakly amenable and $\omega\varrho_M^{n+1} \leq \kappa < \omega\varrho_M^n$. Then*

$$\pi : M \underset{\Sigma_0^{(n)}}{\longrightarrow} N \text{ cofinally.}$$

(Hence $\varrho_n = \varrho_N^n$.)

Proof. Let $\bar{D}$ be $\Sigma_1^{(n)}(M)$ in $\bar{p}$ such that $\bar{D} \cap \omega\varrho_M^{n+1} \notin M$. Hence $\bar{p} \in P_M^{n+1}$. This means that $\bar{D}$ is $\Sigma_1(M^{n,\bar{p}\restriction n})$ in $\bar{p}(n)$. Let $p = \pi(\bar{p})$. We know that

a) $\pi : M \underset{\Sigma_0^{(n)}}{\longrightarrow} N$;

b) $A^{n,\bar{p}\restriction n}$ is $\Sigma_1^{(n-1)}(M)$ in $\bar{p} \restriction n$.

Let $\tilde{A}$ be $\Sigma_1^{(n-1)}(N)$ in $p \restriction n$ by the same definition as $A^{n,\bar{p}}$. Using the fact that π is $\Sigma_0^{(n)}$-preserving we infer that $\pi(x \cap A^{n,\bar{p}}) = \pi(x) \cap \tilde{A}$ for all $x \in H_M^n$. Hence, if we put $\omega\varrho = \sup(\pi''\omega\varrho_M^n) = \omega\varrho_n$, we obtain

c) $\langle J_\varrho^{A'}, \tilde{A} \cap J_\varrho^{A'} \rangle$ is amenable;

d) $\pi \restriction H_M^n : M^{n,\bar{p}\restriction n} \underset{\Sigma_0}{\longrightarrow} \langle |J_\varrho^{A'}|, \tilde{A} \cap |J_\varrho^{A'}| \rangle$ cofinally.

We first prove

$$\varrho = \varrho_N^n. \tag{3.2.1}$$

Otherwise $\varrho < \varrho_n$. By (3.1.3), $\tilde{A} \cap J_\varrho^{A'} \in N$. Let D be $\Sigma_1(\langle J_\varrho^{A'}, \tilde{A} \cap J_\varrho^{A'} \rangle)$ in $p(n)$ by the same definition as $\bar{D}$. Then clearly $D \cap \omega\varrho_M^{n+1} = \bar{D} \cap \omega\varrho_M^{n+1}$ since $\pi \restriction \omega\varrho_M^{n+1} = \mathrm{id}$. However, since $\tilde{A} \cap J_\varrho^{A'} \in N$, $D \cap \omega\varrho_M^{n+1} \in \mathcal{P}(\kappa) \cap N = \mathcal{P}(\kappa) \cap M$ by weak amenability of F. Contradiction. □(Lemma 3.2.1)

The above proof moreover yields:

Corollary 3.2.2. *Suppose that F is weakly amenable and $\omega\varrho_M^{n+1} \leq \kappa < \omega\varrho_M^n$. Then $\omega\varrho_N^{n+1} \leq \omega\varrho_M^{n+1}$.*

Proof. Once we have the above lemma, we know that the same argument as above will go through for the real interpretations. Hence $\tilde{A} = A_N^{n,p\upharpoonright n}$ and D is $\Sigma_1(N^{n,p\upharpoonright n})$. Since $D \cap \omega\varrho_M^{n+1} \notin N$, the above conclusion holds. □(Corollary 3.2.2)

Note that the proof of Lemma 3.2.1 did not use the full strength of weak amenability but merely the fact that $\mathcal{P}(\omega\varrho_M^{n+1}) \cap M = \mathcal{P}(\omega\varrho_M^{n+1}) \cap N$, where $\omega\varrho_M^{n+1} \le \kappa < \omega\varrho_M^n$. We thus obtain the following generalization:

Corollary 3.2.3. *Let $\varrho = \varrho_M^{n+1}$ be such that $(\omega\varrho)^{+M} < \kappa < \omega\varrho_M^n$. Then*

a) $\pi : M \underset{\Sigma_0^{(n)}}{\longrightarrow} N$ *cofinally.*

b) $\omega\varrho_N^{n+1} \le \omega\varrho_M^{n+1}$.

Proof. All we need to prove is $\mathcal{P}(\omega\varrho) \cap M = \mathcal{P}(\omega\varrho) \cap N$. Since $\pi \upharpoonright \kappa = \text{id}$ and $\pi : M \underset{\Sigma_1}{\longrightarrow} N$, we know $(\omega\varrho)^{+M} = (\omega\varrho)^{+N}$ and the conclusion then follows by acceptability. □(Corollary 3.2.3)

Lemma 3.2.4. *Let $\omega\varrho_M^{n+1} \le \kappa < \omega\varrho_M^n$ and $R_M^n \ne \varnothing$. Then*

a) $\pi : M \underset{\Sigma_0^{(n)}}{\longrightarrow} N$ *cofinally.*

b) $\pi'' R_M^n \subset R_N^n$.

Proof. Pick $x \in N$, say $x = \pi(f)(\beta)$. Let f be $\Sigma_1^{(n-1)}(M)$ in $\bar{p}$. By our assumption, we can fix $\bar{q}_0 \in H_M^n$ such that $\bar{p} = \tilde{h}_M^n(\bar{q}_0, \bar{r})$ for an $\bar{r} \in R_M^n$. Hence, if $\bar{F}$ is a good $\Sigma_1^{(n-1)}(M)$ function defining f, we have $f(\xi) = \bar{G}(\xi, \bar{q}_0, \bar{r})$ for all $\xi < \kappa$, where

$$\bar{G}(z^n, w^n, v^0) \simeq \bar{F}(z^n, \tilde{h}_M^n(w^n, v^0)).$$

$\bar{G}$ is clearly a good $\Sigma_1^{(n-1)}(M)$ function and the right side yields its functionally absolute $\Sigma_1^{(n-1)}$ definition. Let G be the good $\Sigma_1^{(n-1)}(N)$ function by the same functionally absolute definition. Then, putting $q_0 = \pi(\bar{q}_0)$ and $r = \pi(\bar{r})$,

$$\pi(f)(\xi) = G(\xi, q_0, r) \quad \text{for all } \xi < \pi(\kappa)$$

since π is $\Sigma_0^{(n)}$-preserving. Consequently,

$$x = \pi(f)(\beta) = G(\beta, q_0, r).$$

Thus, we have just proved:

$$N \text{ is the closure of } H_n \cup \{r\} \text{ under good } \Sigma_1^{(n-1)}(N) \text{ functions.} \tag{3.2.2}$$

We show that this already implies both a) and b) of the lemma. Set $\sigma = \mathrm{id} \restriction H_n$. Then

$$\sigma : \langle H_n, A_N^{n,r} \cap H_n \rangle \underset{\Sigma_0}{\longrightarrow} N^{n,r}.$$

Using the downward extension of embeddings lemma construct $N^*, r^* \in R^n_{N^*}$ and $\tilde{\sigma} \supset \sigma$ such that $\sigma(r^*) = r$ and

$$\tilde{\sigma} : N^* \underset{\Sigma_0^{(n)}}{\longrightarrow} N.$$

Then $\varrho^n_{N^*} = \varrho_n$. We show that $\tilde{\sigma}$ maps N^* onto N. Pick $x \in N$. By (3.2.2), there is a good $\Sigma_1^{(n-1)}(N)$ function F and $z \in H_n$ such that $x = F(z, r)$. Let F^* be the good $\Sigma_1^{(n-1)}(N^*)$ function by the same functionally absolute definition. By Corollary 1.8.12, $F^*(z, r^*)$ is defined, so let it be x^*. Then $x = \tilde{\sigma}(x^*)$. But then $N^* = N$ and $\tilde{\sigma} = \mathrm{id}$, hence $\varrho_n = \varrho^n_N$ and $r \in R^n_N$. □(Lemma 3.2.4)

Corollary 3.2.5. *Let $\omega\varrho_M^{n+1} \leq \kappa < \omega\varrho_M^n$ and M be sound and solid above κ. Then $\pi(p_M - \kappa) = p_N - \kappa$ and N is sound and solid above κ. More generally, if $\beta \geq \kappa$ and M is sound and solid above β, then $\pi(p_M - \beta) = p_N - \beta'$ where $\beta' = \sup(\pi''\beta)$ and N is sound and solid above β'.*

Proof. Apply the previous lemma, Lemma 1.12.6 and Lemma 1.12.7. □(Corollary 3.2.5)

Lemma 3.2.6. *Suppose F is Σ_1-amenable w.r.t. M and $\omega\varrho_M^{n+1} \leq \kappa < \omega\varrho_M^n$. Then*

$$\pi : M \underset{\Sigma_0^{(n)}}{\longrightarrow} N \text{ cofinally.}$$

Proof. For $n = 0$ this holds trivially. If $n > 0$ we show that F is in fact weakly amenable. Clearly, if $\beta < \lambda$ we have $\langle J^A_{\varrho^1}, F_\beta \cap J^A_{\varrho^1} \rangle$ is amenable, so if $f : \kappa \to \mathcal{P}(\kappa) \cap M$, $f \in M$ then $f^{-1}{}'' F_\beta = f^{-1}{}''(F_\beta \cap J^A_\xi)$ for sufficiently large $\xi < \omega\varrho^1_M$ and the latter is clearly an element of M. □(Lemma 3.2.6)

Lemma 3.2.7. *Let F be Σ_1-amenable w.r.t. M and $\omega\varrho_M^{n+1} \leq \kappa < \omega\varrho_M^n$. Let $D \subset \kappa$ be $\mathbf{\Sigma}_1^{(n)}(N)$. Then D is $\mathbf{\Sigma}_1^{(n)}(M)$.*

Proof. Let D^* be $\Sigma_0^{(n)}(N)$ such that

$$D(\xi) \longleftrightarrow (\exists z^n) D^*(z^n, \xi, p)$$

for all $\xi < \kappa$ and a parameter $p \in N$, say $p = \pi(f)(\beta)$. Let $\bar{D}^*$ be $\Sigma_0^{(n)}(M)$ by the same definition. Then for $\xi < \kappa$,

$$\begin{aligned} D(\xi) &\longleftrightarrow (\exists u \in H^n_M)(\exists z^n \in \pi(u)) D^*(z^n, \xi, \pi(f)(\beta)) \\ &\longleftrightarrow (\exists u \in H^n_M)(a_u \in F_\beta) \end{aligned}$$

where

$$a_u = \{\eta < \kappa; (\exists z^n \in u)\bar{D}^*(z^n, \xi, f(\eta))\}.$$

The first equivalence holds since π is $\Sigma_0^{(n)}$ cofinal (see the previous lemma); the second one follows by Łos' Theorem for $\Sigma_0^{(n)}$ formulae. So all we need to check is that the last relation, $a_u \in F_\beta$, is $\mathbf{\Sigma}_1^{(n)}(M)$. Note that $a_u \in H_M^n$, hence the rightmost side of the above equivalence can be rewritten as

$$(\exists u^n)(\exists a^n)(a^n = a_{u^n} \ \& \ a^n \in F_\beta). \tag{3.2.3}$$

The defining relation $a = a_u$ is easily seen to be $\mathbf{\Sigma}_0^{(n)}(M)$ hence (3.2.3) is $\mathbf{\Sigma}_1^{(n)}(M)$. □(Lemma 3.2.7)

Since κ is primitive recursively closed, we can replace κ by J_κ^A in the above lemma (remember that there is a uniformly $\Sigma_1(M)$ map of κ onto J_κ^A).

Lemma 3.2.8. *Let F, n be as in Lemma* 3.2.6. *Then*

$$\mathcal{P}(J_\kappa^A) \cap \mathbf{\Sigma}_1^{(n)}(M) = \mathcal{P}(J_\kappa^{A'}) \cap \mathbf{\Sigma}_1^{(n)}(N).$$

(Recall that $M = \langle J_\alpha^A, B\rangle$, $N = \langle J_{\alpha'}^{A'}, B'\rangle$.) □(Lemma 3.2.8)

A more careful analysis of the proof of Lemma 3.2.7 yields:

Lemma 3.2.9. *Suppose that F is Σ_1-amenable and $\omega\varrho_M^{n+1} \le \kappa < \omega\varrho_M^n$. Let*

$$R(x_1, \ldots, x_k, y_1, \ldots, y_\ell)$$

be $\Sigma_1^{(n)}(N)$, $f_1, \ldots, f_\ell \in \Gamma(\kappa, M)$ and $\alpha_1, \ldots, \alpha_\ell < \lambda$. Then

$$\bar{R} = \{\langle x_1, \ldots, x_k\rangle;\ R(\pi(x_1), \ldots \pi(x_k), \pi(f_1)(\alpha_1), \ldots, \pi(f_\ell)(\alpha_\ell))\}$$

is $\mathbf{\Sigma}_1^{(n)}(M)$. In fact, if p_i is such that $f_i = p_i$ or f_i is a good $\Sigma_1^{(n-1)}(M)$-function in p_i ($i = 1, \ldots, \ell$) and r is such that $F_{\prec\alpha_1,\ldots,\alpha_\ell\succ}$ is $\Sigma_1(M)$ in r, then $\bar{R}$ is $\Sigma_1^{(n)}(M)$ in $p_1, \ldots, p_\ell, r$ uniformly (i.e. its definition depends only on the $\Sigma_1^{(n)}$-definition of R, the functionally absolute $\Sigma_1^{(n-1)}$-definitions of f_i from p_i ($i = 1, \ldots, \ell$) and the Σ_1-definition of $F_{\prec\alpha_1,\ldots,\alpha_\ell\succ}$ from r). □(Lemma 3.2.9)

3.3 When F Is Close to M

Recall that F is close to M iff F is both weakly and Σ_1-amenable w.r.t. M. For such F, one can establish much closer relationship between the definability over M and that over N. As a consequence we obtain that the $*$-ultrapower embeddings associated with close extenders are fully Σ^*-preserving. In fact, we shall assume a little bit less than closeness, namely

$$\begin{array}{l} \bullet\ F \text{ is } \Sigma_1\text{-amenable w.r.t. } M \\ \bullet\ \text{one of the following holds} \\ \qquad \text{a) } F \text{ is weakly amenable} \\ \qquad \text{b) } (\omega\varrho_M^n)^{+M} < \kappa \text{ whenever } \omega\varrho_M^n \leq \kappa. \end{array} \tag{3.3.1}$$

Lemma 3.3.1. *Assume* (3.3.1). *Let $m \in \omega$ be such that $\omega\varrho_M^m \leq \kappa$. Then*

a) $H_M^m = H_N^m$

b) $\mathbf{\Sigma}_1^{(m)}(M) \cap \mathcal{P}(H_M^m) = \mathbf{\Sigma}_1^{(m)}(N) \cap \mathcal{P}(H_N^m)$

Proof. We proceed by induction on m. Let n be such that $\omega\varrho_M^{n+1} \leq \kappa < \omega\varrho_M^n$. Suppose the lemma holds for all $m' < m$ such that $m' \geq n$. We prove it for m.

To see a), it is sufficient to prove that $\omega\varrho_M^m = \omega\varrho_N^m$, since $J_\kappa^A = J_\kappa^{A'}$. By (3.3.1) we have

$$\mathcal{P}(\omega\varrho_M^m) \cap M = \mathcal{P}(\omega\varrho_M^m) \cap N.$$

By Lemma 3.2.7 if $m = n+1$ or else by the induction hypothesis, we have

$$\mathcal{P}(\omega\varrho_M^m) \cap \mathbf{\Sigma}_1^{(m-1)}(M) = \mathcal{P}(\omega\varrho_M^m) \cap \mathbf{\Sigma}_1^{(m-1)}(N).$$

Hence $a \cap \gamma \in J^A_{\varrho_M^m} \subset N$ for every $\gamma < \omega\varrho_M^m$ and $a \in \mathbf{\Sigma}_1^{(m-1)}(N)$, as follows from the latter of the above equalities. Moreover, if a is $\mathbf{\Sigma}_1^{(m-1)}(M)$ such that $a \cap \omega\varrho_M^m \notin M$ then a is $\mathbf{\Sigma}_1^{(m-1)}(N)$ by the same equality and $a \cap \omega\varrho_M^m \notin N$. Consequently, $\omega\varrho_N^m = \omega\varrho_M^m$.

To see b), recall that $D \subset H_M^m$ is $\mathbf{\Sigma}_1^{(m)}(M)$ iff it is $\mathbf{\Sigma}_1(\langle H_M^m, Q\rangle)$, where $Q \subset H_M^m$ is $\mathbf{\Sigma}_1^{(m-1)}(M)$, and similarly for N (Lemma 1.8.1). Now by Lemma 3.2.7 if $m = n+1$ or else by the induction hypothesis, Q is both $\mathbf{\Sigma}_1^{(m-1)}(M)$ and $\mathbf{\Sigma}_1^{(m-1)}(N)$, hence

$$D \text{ is } \mathbf{\Sigma}_1^{(m)}(M) \quad \text{iff} \quad D \text{ is } \mathbf{\Sigma}_1(\langle H_M^m, Q\rangle) \quad \text{iff} \quad D \text{ is } \mathbf{\Sigma}_1^{(m)}(N)$$

since $H_M^m = H_N^m$. (This holds whenever $D \in \mathbf{\Sigma}_1^{(m)}(M)$ or $\mathbf{\Sigma}_1^{(m)}(N)$.) □(Lemma 3.3.1)

We get the following three lemmata as corollaries of Lemma 3.3.1.

Lemma 3.3.2. $\pi : M \underset{\Sigma^*}{\longrightarrow} N$.

Proof. By the previous lemmata (Lemma 3.2.1, Corollary 3.2.3 or Lemma 3.2.7) we already know that $\pi : M \underset{\Sigma_1^{(n)}}{\longrightarrow} N$ where $\omega\varrho_M^{n+1} \leq \kappa < \omega\varrho_M^n$. For $m > n$ we get the result by induction on m using Lemma 3.3.1. So suppose the lemma holds for all $m' < m$. Let $\bar{D}(v_1, \ldots, v_\ell)$ be $\Sigma_1^{(m)}(M)$ and $D(v_1, \ldots, v_\ell)$ be $\Sigma_1^{(m)}(N)$ by the same definition. Let $z_1, \ldots, z_k \in H_M^m$ and $z_{k+1}, \ldots, z_\ell \in M$. By Lemma 1.6.3, there is a

$\bar{Q}(v_1, \ldots, v_\ell)$ which is $\Sigma_1^{(m-1)}(M)$ and a $Q(v_1, \ldots, v_\ell)$, which is $\Sigma_1^{(m-1)}(N)$ by the same definition and such that

$$\bar{D}_{z_{k+1},\ldots,z_\ell} \quad \text{is} \quad \Sigma_1(\langle H_M^m, \bar{Q}_{z_{k+1},\ldots,z_\ell}\rangle)$$

and

$$D_{\pi(z_{k+1}),\ldots,\pi(z_\ell)} \quad \text{is} \quad \Sigma_1(\langle H_N^m, Q_{\pi(z_{k+1}),\ldots,\pi(z_\ell)}\rangle)$$

by the same definition. Let ψ be the common definition of both. By our assumptions, π is $\Sigma_1^{(m-1)}$-preserving, hence

$$Q_{\pi(z_{k+1}),\ldots,\pi(z_\ell)} = \bar{Q}_{z_{k+1},\ldots,z_\ell}$$

and consequently

$$\begin{aligned} \bar{D}(z_1, \ldots, z_\ell) &\longleftrightarrow \langle H_M^m, \bar{Q}_{z_{k+1},\ldots,z_\ell}\rangle \models \psi(z_1, \ldots, z_k) \\ &\longleftrightarrow \langle H_N^m, Q_{\pi(z_{k+1}),\ldots,\pi(z_\ell)}\rangle \models \psi(z_1, \ldots, z_k) \\ &\longleftrightarrow D(\pi(z_1), \ldots, \pi(z_\ell)). \end{aligned}$$

□(Lemma 3.3.2)

Lemma 3.3.3. *$\pi'' P_M^m \subset P_N^m$ whenever $\omega\varrho_M^m \leq \kappa$.*

Proof. We apply Lemmata 3.3.1 and 3.3.2. Given $p \in P_M^m$, there is an a which is $\Sigma_1^{(m-1)}(M)$ in p such that $a \cap \omega\varrho_M^m \notin M$. Hence $a \cap \omega\varrho_N^m \notin N$ and all we have to show is that a is $\Sigma_1^{(m-1)}(N)$ in $\pi(p)$. Let $\bar{D}$ be $\Sigma_1^{(m-1)}(M)$ such that $\xi \in a \leftrightarrow \bar{D}(\xi, p)$ for all $\xi < \omega\varrho_M^m$, and D a $\Sigma_1^{(m-1)}(N)$ relation by the same definition. Since π is Σ^*-preserving, we have

$$\xi \in a \longleftrightarrow \bar{D}(\xi, p) \longleftrightarrow D(\xi, \pi(p))$$

for all $\xi < \omega\varrho_M^m = \omega\varrho_N^m$. □(Lemma 3.3.3)

Corollary 3.3.4. $\pi'' P_M^* \subset P_N^*$. □(Corollary 3.3.4)

Corollary 3.3.5. *If M is solid above β, then N is solid above $\beta' = \sup(\pi''\beta)$ and $\pi(p_M - \beta) = p_N - \beta'$. Hence, if M is solid, then so is N and $\pi(p_M) = p_N$.*

Proof. Apply the previous corollary and Lemma 1.12.6. □(Corollary 3.3.5)

Lemma 3.3.6. *$\mathcal{P}(\kappa) \cap \Sigma_1^{(m)}(M) = \mathcal{P}(\kappa) \cap \Sigma_1^{(m)}(N)$ whenever $\omega\varrho_M^{m+1} \leq \kappa$. If F is weakly amenable, then this holds for all $m < \omega$.*

Proof. The nontrivial part of the lemma deals with the case $\omega\varrho_M^{m+1} \leq \kappa$. For $m = n$, where $\omega\varrho_M^{n+1} \leq \kappa < \omega\varrho_M^n$, the conclusion follows from Lemma 3.2.7, so we can suppose that $m > n$. The inclusion $\subset$ is an immediate consequence of Lemma 3.3.2, so it is sufficient to focus on the converse.

We proceed by induction. As it is more convenient, we shall work over J_κ^A rather than κ. Hence, the case $m = n$ is precisely Lemma 3.2.8. In order to be able to carry out the induction step, we prove a stronger result, given for $m = n$ by Lemma 3.2.9:

> If $D(x, y)$ is $\Sigma_1^{(m)}(N)$, $f \in \Gamma(\kappa, M)$ is a good $\Sigma_1^{(n-1)}(M)$ function in p and $\beta < \lambda$, then
>
> $$\bar{D} = \{x \in J_\kappa^A;\ D(x, \pi(f)(\beta))\} \tag{3.3.2}$$
>
> is $\Sigma_1^{(m)}(M)$ in p, r uniformly, where r is such that F_β is $\Sigma_1(M)$ in r.

To carry out the induction step, pick an $a \subset J_\kappa^A$ which is $\Sigma_1^{(m)}(N)$ in $p = \pi(f)(\beta)$ and a $D \in \Sigma_1^{(m)}(N)$ such that $x \in a \leftrightarrow D(x, p)$ for all $x \in J_\kappa^A$. Suppose also that f is $\Sigma_1^{(i)}(M)$ in q, where $i < n$. Then for some $D^* \in \Sigma_0^{(m)}(N)$ we can write

$$x \in a \longleftrightarrow (\exists z^m) D^*(z^m, x, p). \tag{3.3.3}$$

Hence there is a $Q \in \Sigma_1^{(m-1)}(N)$ such that $D^*_{p,x}$ is rudimentary over $\langle H_N^m, Q_{p,x}\rangle$ and the definition of Q depends only on the definition of D^*. By the induction hypothesis, there is a $\bar{Q}$ which is $\Sigma_1^{(m-1)}(M)$ whose definition depends just on those of Q, f and F_β and such that

$$Q(z, x, p) \longleftrightarrow \bar{Q}(z, x, q, r) \quad \text{for all } z, x \in J_\kappa^A,$$

where F_β is $\Sigma_1(M)$ in r. Hence $D^*_{p,x}$ is uniformly rudimentary over $\langle H_M^m, \bar{Q}_{q,r,x}\rangle$ which means that there is a $\Sigma_1^{(m)}(M)$ relation $\bar{D}^*$ satisfying

$$D^*(z, x, p) \longleftrightarrow \bar{D}^*(z, x, q, r) \tag{3.3.4}$$

whenever $z \in H_M^m$ and $x \in J_\kappa^A$. Moreover, the transformation from D^* to $\bar{D}^*$ is uniform by Lemma 1.6.3 and the induction hypothesis. Hence, by (3.3.3) and (3.3.4),

$$x \in a \longleftrightarrow (\exists z^m) \bar{D}^*(z^m, x, q, r).$$

This is $\Sigma_1^{(m)}(M)$ in q, r and the definition of this relation depends just on the definitions of D, f and F_β. □(Lemma 3.3.6)

Corollary 3.3.7. $\mathcal{P}(\kappa) \cap \mathbf{\Sigma}^*(M) = \mathcal{P}(\kappa) \cap \mathbf{\Sigma}^*(N)$.

Proof. If $\omega\varrho_M^\omega \leq \kappa$, this follows from Lemma 3.3.6. Otherwise employ Łos' Theorem to express every $\mathbf{\Sigma}^*(N)$-subset of κ by a $\mathbf{\Sigma}^*(M)$-definition. (If F is weakly amenable, Lemma 3.3.6 already yields the full result as $\mathcal{P}(\kappa) \cap \mathbf{\Sigma}^*(N) = \mathcal{P}(\kappa) \cap N$).

□(Corollary 3.3.7)

3.4 Extendibility

We shall now investigate when the extension $\text{Ult}^*(M, F)$ exists, or equivalently, when the ultrapower construction yields a well-founded structure, as we did for the usual ultrapowers. As before, we shall present two methods of verifying *$*$-extendibility* of M by F. We shall again start with ω-completeness.

Lemma 3.4.1. *Let M be an acceptable structure and F be an extender on M at κ, λ. Then F is ω-complete iff for every pair of countable sets $\bar{M}$, $\bar{F}$ and every pair σ, $\mathfrak{h}$ satisfying*

a) $\langle \sigma, \mathfrak{h} \rangle : \langle \bar{M}, \bar{F} \rangle \longrightarrow \langle M, F \rangle$

b) $\sigma : \bar{M} \underset{\Sigma_0^{(n)}}{\longrightarrow} M$, *where* $\omega\varrho_{\bar{M}}^{n+1} \leq \bar{\kappa} < \omega\varrho_{\bar{M}}^{n}$

the following holds: If $\pi : \bar{M} \to \mathbb{D}^*(\bar{M}, \bar{F})$ is the ultrapower map, then there is a map $\sigma' : \mathbb{D}^*(\bar{M}, \bar{F}) \underset{\Sigma_0}{\longrightarrow} M$ such that the following diagram commutes.

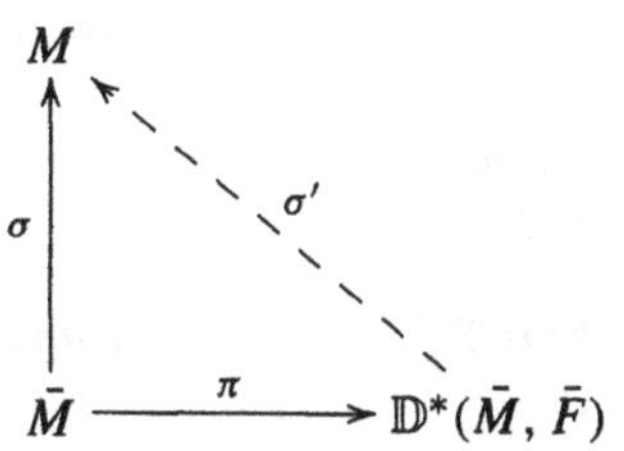

Proof. The proof looks exactly like that for Σ_0-ultrapowers, the only difference being, that now we have to consider functions from $\Gamma(\bar{\kappa}, \bar{M})$ instead of just those from ${}^{\bar{\kappa}}\bar{M} \cap \bar{M}$.

□(Lemma 3.4.1)

Exactly as before we get:

Lemma 3.4.2. *Let M be an acceptable structure and F be an extender on M. If F is ω-complete, then* $\text{Ult}^*(M, F)$ *exists.* □(Lemma 3.4.2)

Lemma 3.4.1 is formulated for the structure $\mathbb{D}^*(\bar{M}, \bar{F})$ instead of a structure with transitive well-founded core (cf. Lemma 2.5.1) since we did not introduce the concept of $\pi : M \underset{F}{\overset{*}{\longrightarrow}} N$ weakly. We could do this, but we will not need such a concept in the sequel.

Note also that once we know that the map σ' in Lemma 3.4.1 exists, it follows that we can transitivise $\text{Ult}^*(\bar{M}, \bar{F})$ to a structure $\bar{N}$. Hence (with a slight abuse of notation),

$$\pi : \bar{M} \underset{\bar{F}}{\overset{*}{\longrightarrow}} \bar{N} \quad \text{and} \quad \sigma' : \bar{N} \underset{\Sigma_0}{\longrightarrow} M.$$

It is then easy to verify that σ' is in fact $\Sigma_0^{(n)}$-preserving. One just has to modify the proof of Lemma 2.5.1 in a straightforward way. We thus obtain:

Corollary 3.4.3. *Let M be an acceptable structure and F be an ω-complete extender on M at κ, λ. Let $\bar{M}, \bar{F}$ be countable and $\sigma, \mathfrak{h}$ be such that* a), b) *of Lemma* 3.4.1 *hold. Then:*

a) *There are $\bar{N}, \pi$ such that $\pi : \bar{M} \overset{*}{\underset{\bar{F}}{\longrightarrow}} \bar{N}$.*

b) *There is a Σ_0-preserving map $\sigma' : \bar{N} \to M$ such that the following diagram commutes:*

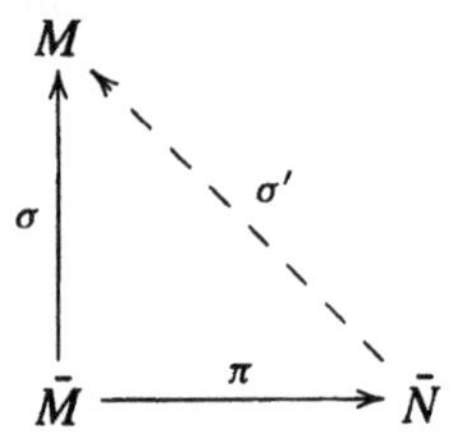

The map is defined by

$$\sigma'(\pi(f)(\beta)) = \sigma(f)(\mathfrak{k} \circ \mathfrak{h}(\beta))$$

where $\mathfrak{k} : \mathfrak{h}''\bar{\lambda} \to \kappa$ is a map witnessing the ω-completeness of F w.r.t. $X = \mathcal{P}(\kappa) \cap \sigma''\bar{M}$ and $A = \mathfrak{h}''\bar{\lambda}$.

c) *The map σ' defined in* b) *is $\Sigma_1^{(n-1)}$-preserving and, if π is $\Sigma_0^{(n)}$ cofinal, then σ' is in fact $\Sigma_0^{(n)}$-preserving.* □(Corollary 3.4.3)

And, exactly as for coarse ultrapowers (with the obvious modifications in the proof), we get:

Lemma 3.4.4. *Let $\langle \sigma, \mathfrak{g} \rangle : \langle \bar{M}, \bar{F} \rangle \to \langle M, F \rangle$, where F is at κ, λ and $\bar{F}$ at $\bar{\kappa}, \bar{\lambda}$, and let $\sigma : \bar{M} \to M$ be $\Sigma_0^{(n)}$-preserving for n such that $\omega\varrho_{\bar{M}}^{n+1} \leq \bar{\kappa} < \omega\varrho_{\bar{M}}^{n}$. Suppose that M is $*$-extendible by F. Then $\bar{M}$ is $*$-extendible by $\bar{F}$. Moreover, if*

$$\pi : M \overset{*}{\underset{F}{\longrightarrow}} N \quad \text{and} \quad \bar{\pi} : \bar{M} \overset{*}{\underset{\bar{F}}{\longrightarrow}} \bar{N},$$

then there is a unique Σ_0-preserving map $\sigma' : \bar{N} \to N$ such that $\sigma' \upharpoonright \bar{\lambda} = \mathfrak{g}$ and the following diagram commutes:

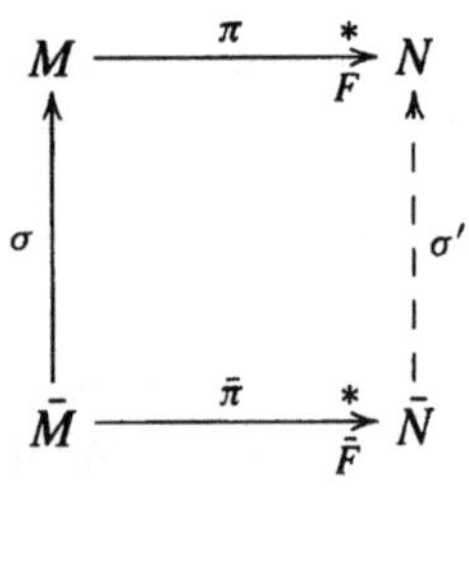

The map σ' is defined by

$$\sigma'(\bar{\pi}(f)(\beta)) = \pi \circ \sigma(f)(\mathfrak{g}(\beta)),$$

is $\Sigma_1^{(n-1)}$-preserving and, if $\bar{\pi}$ is $\Sigma_0^{(n)}$ cofinal, then σ' is actually $\Sigma_0^{(n)}$-preserving.
□(Lemma 3.4.4)

The previous lemma can be considerably strengthened. Of course, first we have to strengthen the assumption.

Definition. Let $\bar{M}$, M be acceptable, F be an extender on M at κ, λ and $\bar{F}$ be an extender on $\bar{M}$ at $\bar{\kappa}$, $\bar{\lambda}$. We write

$$\langle\sigma, \mathfrak{f}\rangle : \langle\bar{M}, \bar{F}\rangle \overset{*}{\longrightarrow} \langle M, F\rangle$$

iff

a) $\langle\sigma, \mathfrak{f}\rangle : \langle\bar{M}, \bar{F}\rangle \to \langle M, F\rangle$.

b) Let $\bar{\alpha} < \bar{\lambda}$, $\alpha = \mathfrak{f}(\bar{\alpha})$. Then $\bar{F}_{\bar{\alpha}}$ is $\Sigma_1(\bar{M})$ in a parameter $\bar{p}$ and F_α is $\Sigma_1(M)$ in $p = \sigma(\bar{p})$ by the same definition.

Notice that the extender $\bar{F}$ from the above definition is automatically Σ_1-amenable w.r.t. $\bar{M}$.

Lemma 3.4.5. *Let $\bar{M}$, M be acceptable, F be an extender on M at κ, λ and $\bar{F}$ be an extender on $\bar{M}$ at $\bar{\kappa}$, $\bar{\lambda}$. Assume that*

a) $\langle\sigma, \mathfrak{g}\rangle : \langle\bar{M}, \bar{F}\rangle \overset{*}{\longrightarrow} \langle M, F\rangle$;

b) $\sigma : \bar{M} \underset{\Sigma^*}{\longrightarrow} M$;

c) *$\bar{F}$, F are weakly amenable;*

d) *F is Σ_1-amenable w.r.t. M.*

Then the map $\sigma' : \bar{N} \to N$, defined as in Lemma 3.4.4, is Σ^*-preserving.

Proof. Let $n \in \omega$ be such that $\omega\varrho_{\bar{M}}^{n+1} \leq \bar{\kappa} < \omega\varrho_{\bar{M}}^{n}$. Since σ is Σ^*-preserving, $\omega\varrho_M^{n+1} \leq \kappa < \omega\varrho_M^n$. We now prove that σ' is $\Sigma_1^{(m)}$-preserving for $m \geq n$.

Let $B_N(x_1, \ldots, x_\ell)$ be $\Sigma_1^{(m)}(N)$ and let $B_{\bar{N}}(x_1, \ldots, x_\ell)$ be $\Sigma_1^{(m)}(\bar{N})$ by the same definition. Let furthermore $\bar{f}_1, \ldots, \bar{f}_\ell \in \Gamma(\bar{\kappa}, \bar{M})$ be such that each $\bar{f}_i$ is a good $\Sigma_1^{(n-1)}(\bar{M})$ function in a parameter $\bar{p}_i$ for $i = 1, \ldots, \ell$ and let $\bar{\alpha}_1, \ldots, \bar{\alpha}_\ell < \bar{\lambda}$. Setting $f_i = \sigma(\bar{f}_i)$, each f_i is in $\Gamma(\kappa, M)$ and is a good $\Sigma_1^{(n-1)}(M)$ function in $p_i = \sigma(\bar{p}_i)$. Finally set $\alpha_i = \mathfrak{g}(\bar{\alpha}_i)$. We want to prove:

$$B_{\bar{N}}(\bar{\pi}(\bar{f}_1)(\bar{\alpha}_1), \ldots, \bar{\pi}(\bar{f}_\ell)(\bar{\alpha}_\ell)) \longleftrightarrow B_N(\pi(f_1)(\alpha_1), \ldots, \pi(f_\ell)(\alpha_\ell)).$$

There is a $\Sigma_1^{(m)}(M)$-relation $B_M^*(x_1, \ldots, x_\ell, y)$, whose definition depends only on the definition of B_M, the functionally absolute $\Sigma_1^{(n-1)}(M)$ definitions of f_i from p_i ($i = 1, \ldots, \ell$) and the $\Sigma_1(M)$ definition of $F_{\prec\alpha_1,\ldots,\alpha_\ell\succ}$ from a parameter r, such that

$$B_N(\pi(f_1)(\alpha_1), \ldots, \pi(f_\ell)(\alpha_\ell)) \longleftrightarrow B_M^*(p_1, \ldots, p_\ell, r). \tag{3.4.1}$$

This follows from Lemma 3.2.9 if $m = n$ and from (3.3.2) if $m > n$. Similarly we obtain a $\Sigma_1^{(m)}(\bar{M})$-relation $B_{\bar{M}}^*(x_1, \ldots, x_\ell, y)$ such that

$$B_{\bar{N}}(\bar{\pi}(f_1)(\alpha_1), \ldots, \bar{\pi}(f_\ell)(\alpha_\ell)) \longleftrightarrow B_{\bar{M}}^*(\bar{p}_1, \ldots, \bar{p}_\ell, \bar{r}), \tag{3.4.2}$$

where $\sigma(\bar{r}) = r$ and $\bar{F}_{\prec\bar{\alpha}_1,\ldots,\bar{\alpha}_\ell\succ}$ is $\Sigma_1(M)$ definable from $\bar{r}$ by the same definition as $F_{\prec\alpha_1,\ldots,\alpha_\ell\succ}$ from r. Hence, B_M^* and $B_{\bar{M}}^*$ have the same $\Sigma_1^{(m)}$ definitions, so by the preservation properties of σ we have

$$B_M^*(p_1, \ldots, p_\ell, r) \longleftrightarrow B_{\bar{M}}^*(\bar{p}_1, \ldots, \bar{p}_\ell, \bar{r}). \tag{3.4.3}$$

The result then follows immediately from (3.4.1), (3.4.2) and (3.4.3).

□(Case 2, Lemma 3.4.5)

A closer look at the proof of the above lemma yields:

Corollary 3.4.6. *Suppose* $\langle\sigma, \mathfrak{g}\rangle : \langle\bar{M}, \bar{F}\rangle \overset{*}{\longrightarrow} \langle M, F\rangle$, *where* $\bar{M}$, M, $\bar{F}$, F *are as before and* $\sigma : \bar{M} \underset{\Sigma_1^{(n)}}{\longrightarrow} M$, *where*

$$\omega\varrho_{\bar{M}}^{n+1} \le \bar{\kappa} < \omega\varrho_{\bar{M}}^{n} \quad \textit{and} \quad \omega\varrho_M^{n+1} \le \kappa < \omega\varrho_M^n.$$

If $\sigma : \bar{M} \underset{\Sigma_1^{(n)}}{\longrightarrow} M$, *then* $\sigma' : \bar{N} \underset{\Sigma_1^{(n)}}{\longrightarrow} N$. □(Corollary 3.4.6)

Corollary 3.4.7. *Suppose the assumptions of Corollary* 3.4.6 *hold and* $m > n$. *If* $\sigma : \bar{M} \underset{\Sigma_h^{(m)}}{\longrightarrow} M$, *then* $\sigma' : \bar{N} \underset{\Sigma_h^{(m)}}{\longrightarrow} N$ *whenever* $h \in \omega$. □(Corollary 3.4.7)

Definition. A map $\sigma : \bar{M} \to M$ is *strongly* $\Sigma_\ell^{(m)}$-preserving iff σ is $\Sigma_\ell^{(m)}$-preserving and $\sigma^{-1}{}'' H_M^{m+1} \subset H_{\bar{M}}^{m+1}$ (or equivalently $\sigma^{-1}{}'' \omega\varrho_M^{m+1} \subset \omega\varrho_{\bar{M}}^{m+1}$).

If $\omega\varrho_{\bar{M}}^{n+1} \le \bar{\kappa} < \omega\varrho_{\bar{M}}^{n}$, $\pi(\bar{\kappa}) = \kappa$ and σ is strongly $\Sigma_0^{(n)}$-preserving, then $\omega\varrho_M^{n+1} \le \kappa < \omega\varrho_M^n$. Hence, in Corollary 3.4.7, if we make the additional assumption that σ is strongly $\Sigma_h^{(m)}$-preserving for $h > 0$, then σ' will be strongly $\Sigma_\ell^{(m)}$-preserving as well, since $H_M^{m+1} = H_N^{m+1}$ and $\sigma' \restriction H_{\bar{N}}^{m+1} = \sigma \restriction H_{\bar{M}}^{m+1}$.

Lemma 3.4.8. *Suppose* $\bar{M}$, M, $\bar{F}$, F *are as above and*

a) $\langle\sigma, \mathfrak{g}\rangle : \langle\bar{M}, \bar{F}\rangle \overset{*}{\longrightarrow} \langle M, F\rangle$;

b) $\bar{F}, F$ *are weakly amenable;*

c) F *is* Σ_1*-amenable w.r.t.* M*;*

d) $\sigma : \bar{M} \underset{\Sigma_h^{(m)}}{\longrightarrow} M$ *strongly, where* $\omega\varrho_{\bar{M}}^{m+1} \leq \bar{\kappa}$.

Then $\sigma' : \bar{N} \underset{\Sigma_h^{(m)}}{\longrightarrow} N$ *strongly.*

□(Lemma 3.4.8)

3.5 k-Ultrapowers

We now modify the fine ultrapower construction by restricting the set of functions $\Gamma(\kappa, M)$. This will lead to ultrapower maps with weaker preservation properties on the one hand, but we obtain ultrapowers with simpler inner structure while still retaining a useful amount of fine structural information on the other hand. In many situations it proves to be useful to compromise and replace $*$-ultrapowers by these restricted ones.

Definition. Let $k \leq \omega$. $\Gamma^k(\kappa, M)$ is the set of all good $\mathbf{\Sigma}_1^{(n)}(M)$ functions $f : \kappa \to M$ for $n < k$ satisfying $\omega\varrho_M^{n+1} > \kappa$.

Obviously, $\Gamma^0(\kappa, M)$ is simply the set of all $f : \kappa \to M$ which are elements of M and $\Gamma^k(\kappa, M) = \Gamma(\kappa, M)$ whenever $\omega\varrho_M^{k+1} \leq \kappa$.

Definition. Let $M = \langle J_\alpha^A, B\rangle$ be acceptable and F be an extender at κ, λ.

$$\mathbb{D}^k(M, F) = \langle D, \tilde{=}, \tilde{\in}, \tilde{A}, \tilde{B}\rangle,$$

where

$$D = \{\langle \alpha, f\rangle;\ \alpha < \lambda \ \&\ f \in \Gamma^k(\kappa, M)\}$$

and $\tilde{=}, \tilde{\in}, \tilde{A}, \tilde{B}$ are defined exactly as in the case of $*$-ultrapowers.

If $\mathbb{D}^k(M, F)$ is well-founded, we identify it with its transitive isomorph N, which is called the *k-ultrapower* of M by F and denoted by $\mathrm{Ult}^k(M, F)$. If π is the associated *k-ultrapower map,* we write

$$\pi : M \underset{F}{\overset{k}{\longrightarrow}} N.$$

By the above remarks, ω-ultrapowers are just $*$-ultrapowers, k-ultrapowers agree with $*$-ultrapowers whenever $\omega\varrho_M^{k+1} \leq \kappa$ and 0-ultrapowers agree with the usual, also called *coarse,* ultrapowers. If $\omega\varrho_M^k > \kappa$, the theory of k-ultrapowers can be developed in the same way as that of $*$-ultrapowers; the arguments of the previous sections go through.

Lemma 3.5.1. *Let* $\omega\varrho_M^k > \kappa$ *and* $\pi : M \underset{F}{\overset{k}{\longrightarrow}} N$*, where* F *is a* (κ, λ) *extender.*

a) $\pi : M \underset{Q}{\longrightarrow} N$.

b) $\pi : M \underset{\Sigma_0^{(k)}}{\longrightarrow} N$.

c) $\pi : M \underset{\Sigma_2^{(n)}}{\longrightarrow} N$ *if* $n < k$.

d) (Łos' Theorem for $\Sigma_0^{(k)}$-formulae). *Let* $\varphi(v_1^{i_1}, \dots, v_\ell^{i_\ell})$ *be a* $\Sigma_0^{(k)}$*-formula, let* $f_j \in \Gamma_{i_j}^k(\kappa, M)$ *for* $j = 1, \dots, \ell$ *and let* $\alpha_1, \dots, \alpha_\ell < \lambda$. *Then*

$$N \models \varphi(\pi(f_1)(\alpha_1), \dots, \pi(f_\ell)(\alpha_\ell)) \quad \textit{iff} \quad \langle \alpha_1, \dots, \alpha_\ell \rangle \in F(u_{f_1, \dots, f_\ell}^{\varphi, M})$$

where $u_{f_1,\dots,f_\ell}^{\varphi,M} = \{\langle \eta_1, \dots, \eta_\ell \rangle;\ M \models \varphi(f_1(\eta_1), \dots, f_\ell(\eta_\ell))\}$.

e) $N = \{\pi(f)(\alpha);\ \alpha < \lambda\ \&\ f \in \Gamma^k(\kappa, M)\}$.

f) $H_N^n = \{\pi(f)(\alpha);\ \alpha < \lambda\ \&\ f \in \Gamma_n^k(\kappa, M)\}$, *where*

$$\Gamma_n^k(\kappa, M) = \begin{cases} \{f \in \Gamma^k(\kappa, M);\ \operatorname{rng}(f) \subset H_M^n\} & \textit{if } n < k \\ \{f \in \Gamma^k(\kappa, M);\ f \in H_M^n\} = {}^\kappa(H_M^n) \cap H_M^n & \textit{if } n = k. \end{cases}$$

g) $\pi : M \underset{\Sigma_0^{(k)}}{\longrightarrow} N$ *cofinally if* $R_M^k \neq \varnothing$; *in this case* $\pi'' R_M^k \subset R_N^k$.

Proof. a) follows from b) and the fact that π is cofinal if $k = 0$. b) – d) as well as e) and f) follow by the proof of Lemma 3.1.11 and g) follows from the proof of Lemma 3.2.4. □(Lemma 3.5.1)

The lemmata on extendibility can be rephrased in the obvious way.

Lemma 3.5.2. *Let M be acceptable and F an* ω*-complete extender on M at* κ, λ. *Let* $\bar{M}, \bar{F}$ *be countable where* $\bar{F}$ *is an extender at* $\bar{\kappa}, \bar{\lambda}$. *Suppose* $\omega\varrho_{\bar{M}}^k > \bar{\kappa}$ *and* $\sigma, \mathfrak{h}$ *is a pair of maps satisfying:*

a) $\langle \sigma, \mathfrak{h} \rangle : \langle \bar{M}, \bar{F} \rangle \longrightarrow \langle M, F \rangle$

b) $\sigma : \bar{M} \underset{\Sigma_0^{(k)}}{\longrightarrow} M$.

Then $\bar{M}$ *is k-extendible by* $\bar{F}$. *Furthermore, if* $\bar{\pi}$ *is the associated k-ultrapower embedding and* $\bar{N} = \operatorname{Ult}^k(\bar{M}, \bar{F})$, *then there is a* $\Sigma_1^{(k-1)}$*-preserving* $\sigma' : \bar{N} \to M$ *such that* $\sigma = \sigma' \circ \bar{\pi}$; *the map* σ' *is defined by*

$$\sigma'(\bar{\pi}(f)(\alpha)) = \sigma(f)(\mathfrak{k} \circ \sigma(\alpha)) \quad \textit{for } f \in \Gamma^k(\bar{\kappa}, \bar{M}) \textit{ and } \alpha < \bar{\lambda},$$

where $\mathfrak{k} : \mathfrak{h}''\bar{\lambda} \to \kappa$ *is a map witnessing the* ω*-completeness of F w.r.t.* $X = \mathcal{P}(\kappa) \cap \sigma''\bar{M}$ *and* $A = \mathfrak{h}''\bar{\lambda}$. *Also, if* $\bar{\pi}$ *is cofinal (i.e. cofinal at the k-th level), then* σ' *is* $\Sigma_0^{(k)}$*-preserving.* □(Lemma 3.5.2)

Lemma 3.5.3. *Let $\langle \sigma, \mathfrak{g}\rangle : \langle \bar{M}, \bar{F}\rangle \longrightarrow \langle M, F\rangle$, where F is at κ, λ and $\bar{F}$ is at $\bar{\kappa}, \bar{\lambda}$. Let $\sigma : \bar{M} \to M$ be $\Sigma_0^{(k)}$-preserving and $\omega\varrho^k_{\bar{M}} > \bar{\kappa}$. Suppose that M is m-extendible by F, where $m \geq k$. Then $\bar{M}$ is k-extendible by $\bar{F}$. Moreover, if*

$$\bar{\pi} : \bar{M} \xrightarrow[\bar{F}]{k} \bar{N} \quad \text{and} \quad \pi : M \xrightarrow[F]{m} N$$

then there is a unique Σ_0-preserving map $\sigma' : \bar{N} \to N$ such that $\sigma' \circ \bar{\pi} = \pi \circ \sigma$ and $\sigma' \restriction \bar{\lambda} = \mathfrak{g}$. The map is defined by

$$\sigma'(\bar{\pi}(f)(\alpha)) = \pi \circ \sigma(f)(\mathfrak{g}(\alpha)) \quad \text{for } f \in \Gamma^k(\bar{\kappa}, \bar{M}) \text{ and } \alpha < \bar{\lambda}$$

and is in fact $\Sigma_1^{(k-1)}$-preserving. Also, if $\bar{\pi}$ is cofinal, then σ' is $\Sigma_0^{(k)}$-preserving. □(Lemma 3.5.3)

Note that if σ is $\Sigma_0^{(h)}$-preserving for some $h < k$ and $\omega\varrho^{h+1}_{\bar{M}} \leq \bar{\kappa} < \omega\varrho^h_{\bar{M}}$, then σ' is $\Sigma_0^{(k-1)}$-preserving and if $\bar{\pi}$ is cofinal, then σ' is $\Sigma_0^{(k)}$-preserving; this follows from the fact that in this case $\mathrm{Ult}^h(\bar{M}, \bar{F}) = \mathrm{Ult}^k(\bar{M}, \bar{F})$.

Lemma 3.5.4. *Let $\langle \sigma, \mathfrak{g}\rangle : \langle \bar{M}, \bar{F}\rangle \stackrel{*}{\longrightarrow} \langle M, F\rangle$, where F is at κ, λ and $\bar{F}$ is at $\bar{\kappa}, \bar{\lambda}$. Let $\sigma : \bar{M} \to M$ be $\Sigma_1^{(k)}$-preserving and $m, \ell \geq k$. Suppose one of the following holds:*

- *$m = k = \ell$;*
- *there is an $h \leq k$ such that $\omega\varrho^{h+1}_{\bar{M}} \leq \bar{\kappa} < \omega\varrho^h_{\bar{M}}$ and $\omega\varrho^{h+1}_M \leq \kappa < \omega\varrho^h_M$.*

Suppose further that the ultrapower maps

$$\pi : M \xrightarrow[F]{m} N \quad \text{and} \quad \bar{\pi} : \bar{M} \xrightarrow[\bar{F}]{\ell} \bar{N}$$

are cofinal at the h-th level. Then the map $\sigma' : \bar{N} \to N$ defined by $\sigma'(\bar{\pi}(f)(\alpha)) = \pi \circ \sigma(f)(\mathfrak{g}(\alpha))$ for $f \in \Gamma^k(\bar{\kappa}, \bar{M})$ and $\alpha < \bar{\lambda}$ is $\Sigma_1^{(k)}$-preserving.

Proof. That $\mathrm{Ult}^k(\bar{M}, \bar{F})$ exists and σ' is $\Sigma_0^{(k)}$-preserving follows by standard arguments. Note that both ultrapowers are in fact h-ultrapowers. That σ' is $\Sigma_1^{(k)}$-preserving follows by the proof of Lemma 3.4.5. □(Lemma 3.5.4)

Lemma 3.5.2 implies that M is k-extendible by any ω-complete extender on M. We have seen that the k-ultrapower embedding $\pi : \bar{M} \to \bar{N}$ is cofinal whenever $R^k_{\bar{M}} \neq \varnothing$ and $\omega\varrho^k_{\bar{M}} > \kappa$; this means that the maps σ' from Lemmata 3.5.2 and 3.5.3 will be $\Sigma_0^{(k)}$-preserving in this case. Finally notice that given $k < m$, any M which is m-extendible by F is also k-extendible by F since the pair $\langle \mathrm{id}, \mathrm{id}\rangle : \langle M, F\rangle \longrightarrow \langle M, F\rangle$ satisfies the assumptions of Lemma 3.5.3.

3.6 Pseudoultrapowers

Another useful modification of the ultrapower construction originates in the problem of extending embeddings to larger domains. We have already discussed a problem of this sort in the section on upwards extensions of embeddings in the fine structure chapter. In fact, an ultrapower is nothing other than such an extension if we consider the respective extender to be the original embedding. In this section we shall allow a wider class of embeddings than those representable by extenders. The ultrapower construction will use these embeddings in the same way we have used extenders; to distinguish between ultrapowers by extenders and those by arbitrary embeddings we shall call the latter *pseudoultrapowers*. Most of the requisite lemmata will go through with minor amendments; however, since we won't make any special assumptions about the embeddings we are extending, we will be able to establish only very general properties of the extensions, basically just properties of the sort discussed in the previous section on k-ultrapowers.

Throughout the section we shall consider acceptable structures Q, Q' and M along with a map σ satisfying:

- $Q = J_\tau^A$, $Q' = J_{\tau'}^{A'}$;
- $\sigma : Q \underset{\Sigma_0}{\longrightarrow} Q'$ cofinally;
- $M = \langle J_\alpha^A, B\rangle$ and τ is a cardinal in M.

Definition. Given $k \le \omega$, $\Gamma^k(\sigma, M)$ is the set (class) of all good $\mathbf{\Sigma}_1^{(n)}(M)$ functions $f : \gamma \to M$ such that $\gamma < \tau, n < k$ and $\omega\varrho_M^{n+1} \ge \tau$. We write $\Gamma(\sigma, M)$ instead of $\Gamma^\omega(\sigma, M)$.

As for ultrapowers by extenders, the case $k = 0$ corresponds to the coarse pseudoultrapower construction, i.e. the construction where we consider just those functions which are *elements* of M. In the above definition we mentioned that the collection of functions used in the pseudoultrapower construction can be a proper class; this will happen if M is a proper class J-structure, i.e. a proper class ZFC model. In this case, the fine pseudoultrapower will again be nothing more than the corresponding coarse pseudoultrapower.

Definition. The term model $\mathbb{D}^k = \mathbb{D}^k(M, \sigma)$ is defined as follows. (If $k = \omega$, we write $\mathbb{D}^*(M, \sigma)$.)

$$\mathbb{D}^k = \langle D, \tilde{=}, \tilde{\in}, \tilde{A}, \tilde{B}\rangle$$

where

$$D = D^k(M, \sigma) = \{\langle \alpha, f\rangle;\ \alpha \in \sigma(\mathrm{dom}(f)) \ \&\ f \in \Gamma^k(\sigma, M)\}$$

and

$$\langle\alpha, f\rangle \tilde{=} \langle\beta, g\rangle \longleftrightarrow \prec\alpha, \beta\succ \in \sigma(u^{=,M}_{f,g})$$
$$\langle\alpha, f\rangle \tilde{\in} \langle\beta, g\rangle \longleftrightarrow \prec\alpha, \beta\succ \in \sigma(u^{\in,M}_{f,g})$$
$$\tilde{A}(\langle\alpha, f\rangle) \longleftrightarrow \alpha \in \sigma(u^A_f)$$
$$\tilde{B}(\langle\alpha, f\rangle) \longleftrightarrow \alpha \in \sigma(u^B_f).$$

Here,

$$u^{=,M}_{f,g} = \{\prec\eta_1, \eta_2\succ;\ \eta_1 \in \mathrm{dom}(f)\ \&\ \eta_2 \in \mathrm{dom}(g)\ \&\ f(\eta_1) = g(\eta_2)\}$$

and $u^{\in,M}_{f,g}$, u^A_f and u^B_f are defined similarly.

These definitions are easily seen to be correct. By induction on Σ_0 formulae we prove the following version of Łos' Theorem.

Lemma 3.6.1. *Let $\alpha_1, \ldots, \alpha_\ell < \tau'$, $f_1, \ldots, f_\ell \in \Gamma^k(\sigma, M)$ and $\varphi(v_1, \ldots, v_\ell)$ be a Σ_0 formula. Then*

$$\mathbb{D}^k \models \varphi(\langle\alpha_1, f_1\rangle, \ldots, \langle\alpha_\ell, f_\ell\rangle) \quad \textit{iff} \quad \prec\alpha_1, \ldots, \alpha_\ell\succ \in \sigma(u^{\varphi,M}_{f_1,\ldots,f_\ell})$$

where

$$u^{\varphi,M}_{f_1,\ldots,f_\ell} = \{\prec\eta_1, \ldots, \eta_\ell\succ;\ \Big(\mathop{\&}_{i=1}^{\ell} \eta_i \in \mathrm{dom}(f_i)\Big)\ \&\ M \models \varphi(f_1(\eta_1), \ldots, f_\ell(\eta_\ell))\}.$$

Proof. We proceed almost literally as in the case of standard $*$-ultrapowers. The only nontrivial step in the induction is the implication ($\longleftarrow$) for formulae of the form $(\exists z \in u)\psi$; here we make use of Lemma 3.1.4 (see the proof of Claim 1 of Lemma 3.1.11 for more details). □(Lemma 3.6.1)

By Łos' theorem, $\tilde{=}$ is a congruence relation on $\mathbb{D}^k$ and $\tilde{\in}$ is extensional. For the rest of the construction we shall assume that $\mathbb{D}$ is well-founded and identify it with its transitive isomorph $N = \langle |N|, A_N, B_N\rangle$ via the map

$$[\quad] : \mathbb{D}^k \longrightarrow N.$$

Then

$$x \tilde{=} y \longleftrightarrow [x] = [y]$$
$$x \tilde{\in} y \longleftrightarrow [x] \in [y]$$
$$\tilde{A}(x) \longleftrightarrow A_N([x])$$
$$\tilde{B}(x) \longleftrightarrow B_N([x])$$

for every $x, y \in D$. We shall again briefly write $[\alpha, f]$ instead of $[\langle\alpha, f\rangle]$

Definition. $\tilde{\sigma} : M \to N$ is defined by $\tilde{\sigma}(x) = [0, c_x]$.

Lemma 3.6.2. $\tilde{\sigma} \upharpoonright Q = \sigma$.

Proof. Set

$$\tilde{Q} = \bigcup \tilde{\sigma}''Q.$$

Then $\tilde{Q}$ is transitive. We shall verify that $\tilde{Q}$ is the collection of all $x \in N$ of the form $[\alpha, f]$ for some $\alpha \in \sigma(\mathrm{dom}(f))$ and $f \in Q$. To see the inclusion $\supset$, pick α, f as above and set $a = \mathrm{rng}(f)$. That $[\alpha, f] \in [0, c_a] = \tilde{\sigma}(a)$ follows easily from Łos' Theorem.

To see the converse, let $x = [\alpha, f]$ for some $\alpha \in \sigma(\gamma)$ where $\gamma = \mathrm{dom}(f)$ and $f \in \Gamma^k(\sigma, M)$. If $a \in Q$ is such that $x \in \tilde{\sigma}(a)$, then $\alpha \in \sigma(\{\eta \in \gamma;\ f(\eta) \in a\})$ and $f \cap (\gamma \times a) \in Q$ since τ is a cardinal in M (this intersection is in M by the fact that f does not have sufficiently large definability degree necessary for being a new bounded subset of Q). Using this we define a function $f' : \gamma \to Q$, setting $f'(\eta) = f(\eta)$, if $f(\eta) \in a$, and $f'(\eta) = 0$, otherwise; then f' is clearly an element of Q. Since

$$\{\eta \in \gamma;\ f(\eta) \in a\} \subset \{\eta \in \gamma;\ f(\eta) = f'(\eta)\},$$

$\alpha \in \sigma(\{\eta \in \gamma;\ f(\eta) = f'(\eta)\})$; Łos' Theorem then yields $[\alpha, f] = [\alpha, f']$. This completes the proof of the inclusion $\subset$.

Now define a map $k^* : \tilde{Q} \to Q$ by

$$k^*([\alpha, f]) = \sigma(f)(\alpha)$$

for α, f as above. Then $\mathrm{dom}(k^*) = \tilde{Q}$ and k^* is well defined, since given any $\langle \alpha, f\rangle, \langle \beta, g\rangle \in D$ with $\mathrm{dom}(f) = \gamma = \mathrm{dom}(g)$ (note that we can assume this w.l.o.g.),

$$\begin{aligned}[\alpha, f] = [\beta, g] &\longleftrightarrow \prec\alpha, \beta\succ \in \sigma(\{\prec\eta_1, \eta_2\succ \in \gamma;\ f(\eta_1) = g(\eta_2)\}) = \\ &= \{\prec\eta_1, \eta_2\succ \in \sigma(\gamma);\ \sigma(f)(\eta_1) = \sigma(g)(\eta_2)\} \longleftrightarrow \sigma(f)(\alpha) = \sigma(g)(\beta).\end{aligned}$$

Similarly we obtain $[\alpha, f] \in [\beta, g] \longleftrightarrow \sigma(f)(\alpha) \in \sigma(g)(\beta)$, which means that k^* is structure-preserving. Moreover, k^* is onto Q since every element of Q is obviously of the form $\sigma(f)(\alpha)$ for some α, f as above. It follows that $\tilde{Q} = Q$ and $k^* = \mathrm{id}$. As an immediate consequence we obtain

$$\tilde{\sigma}(x) = [0, c_x] = k^*([0, c_x]) = \sigma(c_x)(0) = c_{\sigma(x)}(0) = \sigma(x)$$

for every $x \in Q$. □(Lemma 3.6.2)

Definition. The map $\tilde{\sigma}$ is called the *canonical k-extension* of σ to M. The structure N is called the *k-pseudoultrapower* of M by σ.

Our treatment of pseudoultrapowers is fairly general, and, consequently, the statement of the next theorem is somewhat complicated. We therefore recommend the reader to first focus on the special case where Q has a largest cardinal, say κ. It is easy to see that any $\langle \alpha, f\rangle \in D$ is $\tilde{=}$-equivalent to some $\langle \alpha^*, f^*\rangle \in D$ where $f^* : \kappa \to M$

and $\alpha^* < \kappa' = \sigma(\kappa)$. In this case it is convenient to modify the definition of D by restricting ourselves only to pairs of this form, that is,

$$D = \{\langle\alpha, f\rangle;\ \alpha < \sigma(\kappa)\ \&\ f \in \Gamma^k(\sigma, M)\ \&\ \mathrm{dom}(f) = \kappa\}. \tag{3.6.1}$$

In the following we shall consider the above to be our silent convention and use the modified version of D automatically.

Lemma 3.6.3. *Let $\tilde{\sigma} : M \to N$ be the canonical k-extension of σ.*

a) $\tilde{\sigma} : M \underset{Q}{\longrightarrow} N$.

b) *N is an acceptable J-structure, say $N = \langle J^{A'}_{\alpha'}, B'\rangle$. By Lemma 3.6.2, N is an end-extension of Q'.*

c) $\tilde{\sigma} : M \underset{\Sigma_0^{(n)}}{\longrightarrow} N$ *whenever $n \leq k$ and $\omega\varrho^n_M \geq \tau$.*

d) $\tilde{\sigma} : M \underset{\Sigma_2^{(n)}}{\longrightarrow} M$ *whenever $n < k$ and $\omega\varrho^{n+1}_M \geq \tau$.*

e) $N = \{\tilde{\sigma}(f)(\alpha);\ \alpha \in \sigma(\mathrm{dom}(f))\ \&\ f \in \Gamma^k(\sigma, M)\}$.

f) $H^n_N = \{\tilde{\sigma}(f)(\alpha);\ \alpha \in \sigma(\mathrm{dom}(f))\ \&\ f \in \Gamma^k_n(\sigma, M)\}$, *where*

$$\Gamma^k_n(\sigma, M) = \begin{cases} \{f \in \Gamma^k(\sigma, M);\ \mathrm{rng}(f) \subset H^n_M\} & \text{if } n < k \\ \{f \in \Gamma^k(\sigma, M);\ f \in H^n_M\} & \text{if } n = k. \end{cases}$$

g) (Łos' Theorem for $\Sigma_0^{(n)}$-formulae). *Let $\varphi(v_1^{i_1}, \ldots, v_\ell^{i_\ell})$ be a $\Sigma_0^{(n)}$-formula where $\omega\varrho^n_M \geq \tau$ and $n \leq k$. Then for any $f_1, \ldots, f_\ell \in \Gamma^k(\sigma, M)$ such that $f_j \in \Gamma^k_{i_j}(\sigma, M)$ and any $\alpha_1, \ldots, \alpha_\ell < \tau'$,*

$$N \models \varphi(\tilde{\sigma}(f_1)(\alpha_1), \ldots, \tilde{\sigma}(f_\ell)(\alpha_\ell)) \text{ iff } \prec\alpha_1, \ldots, \alpha_\ell\succ\ \in \sigma(u^{\varphi,M}_{f_1,\ldots,f_\ell}).$$

h) *Let $\omega\varrho^n_M \geq \tau$ and either $n = k$ or else $n \leq k$ and $\omega\varrho^{n+1}_M < \tau$. If $R^n_M \neq \varnothing$, then $\tilde{\sigma}$ is $\Sigma_0^{(n)}$ cofinal and $\tilde{\sigma}'' R^n_M \subset R^n_N$.*

Proof. The proof closely follows that of Lemma 3.1.11. The same argument shows that each H^k_n is transitive, where H^k_n is the obvious analogue of H_n from the proof of Lemma 3.1.11. We then prove Łos' Theorem for pseudointerpretations which in this case reads:

Claim 1. Let $\varphi(v_1^{i_1}, \ldots, v_\ell^{i_\ell})$ be a $\Sigma_0^{(n)}$- formula where n is as in g). Let $\alpha_1, \ldots, \alpha_\ell < \tau'$ and $f_j \in \Gamma^k_{i_j}(\sigma, M)$ for $j = 1, \ldots, \ell$. Then

$$N \models \varphi([\alpha_1, f_1], \ldots, [\alpha_\ell, f_\ell])$$

in the sense of pseudointerpretation iff $\prec\alpha_1, \ldots, \alpha_\ell\succ \in \sigma(u^{\varphi,M}_{f_1,\ldots,f_\ell})$.

As an immediate consequence we obtain c) and d) for pseudointerpretations. Setting $n = 0$, we get a); this follows from the fact that either $\tilde{\sigma}$ is Σ_0-preserving and cofinal (if $\Gamma^k(\sigma, M) \subset M$) or $\tilde{\sigma}$ is Σ_2-preserving. From a) we immediately obtain b). For n as above we set

$$\omega\varrho_n = \mathbf{On} \cap H^k_n$$

and as before verify that $H^k_n = J^{A'}_{\varrho_n}$. Finally, we show that the ϱ_n correspond with the real projecta of N.

Claim 2. If n is as in d), then $\omega\varrho_n = \omega\varrho^n_N$ and $H^k_n = H^n_N$. If $n = k$ or else $n \leq k$ and $\omega\varrho^{n+1}_M < \tau$ then $\omega\varrho_n \leq \omega\varrho^n_N$ and $H^k_n \subset H^n_N$.

It follows from Claim 2 that c), d) and Łos' Theorem hold for ordinary interpretations. f) and the version of Łos' Theorem stated in g) are consequences of e). h) follows by the proof of Lemma 3.2.4. To see e), pick $\langle\alpha, f\rangle \in D$. Let $R(y, \xi, p)$ be a defining relation for f with a functionally absolute $\Sigma^{(n)}_1(M)$ definition in p and $R'(y, \xi, p')$ a defining relation for $\tilde{\sigma}(f)$ with the same functionally absolute $\Sigma^{(n)}_1(N)$ definition in $p' = \tilde{\sigma}(p)$. Then $\alpha \in \sigma(\{\eta;\ R(\eta, f(\eta), p)\})$, so by Łos' Theorem, $R'([\alpha, \mathrm{id}], [\alpha, f], p')$. By Lemma 3.6.2, $[\alpha, \mathrm{id}] = \sigma(\mathrm{id})(\alpha) = \alpha$. Since R' is a defining relation for $\tilde{\sigma}(f)$, we have $[\alpha, f] = \tilde{\sigma}(f)([\alpha, \mathrm{id}]) = \tilde{\sigma}(f)(\alpha)$.

Proof of Claim 2. Although the proof is much like the corresponding part of the proof of Lemma 3.1.11, we give the details. We proceed by induction on n. The case $n = 0$ is trivial, so let us focus on the induction step. Suppose the claim holds for $n-1$. First, we verify that $\omega\varrho_n \leq \omega\varrho^n_N$ for n as in c). Let $a \in H_n$. Then $a = [\alpha, f]$ for some $\langle\alpha, f\rangle \in D$ such that $f \in \Gamma^k_n(\sigma, M)$. Let E be $\Sigma^{(n-1)}_1(N)$ in a parameter $p = [\beta, g]$ where $\langle\beta, g\rangle \in D$ and E^* be $\Sigma^{(n-1)}_1(N)$ such that $E(z) \leftrightarrow E^*(z, p)$. We have to prove that $E \cap a \in H_n$. As before, this will follow from the fact that $E \cap a = [\prec\alpha, \beta\succ, k^*]$, where k^* is defined by $k^*(\prec\eta_1, \eta_2\succ) = g(\eta_1) \cap \bar{E}^*(-, f(\eta_2))$; here, $\bar{E}^*$ is $\Sigma^{(n-1)}_1(M)$ with the same definition as E^*. To see this, we apply Łos' theorem for pseudointerpretations to the defining relation of k^*, which is easily seen to be $\Sigma^{(n)}_0$; granted that $k^* \in \Gamma^k(\sigma, M)$, we obtain $[\prec\alpha, \beta\succ, k^*] = [\alpha, f] \cap E^*(-, [\beta, g]) = a \cap E$. To see that $k^* \in \Gamma^k_n(\sigma, M)$, we observe that each $k^*(\prec\eta_1, \eta_1\succ)$, being a bounded $\Sigma^{(n-1)}_1$ subset of H^n_M, must itself be in H^n_M. The map k^* is $\Sigma^{(n)}_0(M)$ definable from parameters. If n is as in d), then k^* is obviously in $\Gamma^k(\sigma, M)$; otherwise $g \in H^n_M$, so the same must hold for k^*.

It remains to prove that $\omega\varrho^n_N \leq \omega\varrho_n$ whenever n is as in d). Pick $E(-,-)$ which is $\Sigma^{(n-1)}_1(M)$ and $p \in M$ such that $E(-, p) \cap \omega\varrho^n_M \notin M$. Let E' be $\Sigma^{(n-1)}_1(N)$ by the same definition and $p' = \tilde{\sigma}(p)$. We show that $E'(-, p') \cap \omega\varrho_n \notin N$. Suppose the contrary. Then $E'(-, p') \cap \omega\varrho_n$ is of the form $[\alpha, f]$ for some $\langle\alpha, f\rangle \in D$. In N we have $(\forall\xi^n)(\xi^n \in [\alpha, f] \longleftrightarrow E'(\xi^n, p'))$ with respect to the pseudointerpretation. By Łos' Theorem, $\alpha \in \sigma(u)$ where

$$u = \{\eta;\ \bar{M} \models (\forall\xi^n)(\xi^n \in f(\eta) \leftrightarrow E(\xi^n, p))\}.$$

Hence $u \neq \varnothing$, which means that there is an $\eta < \tau$ such that $E(-, p) \cap \omega\varrho_M^n = f(\eta) \in M$. Contradiction. □(Claim 2, Lemma 3.6.3)

Lemma 3.6.4. *Let $\omega\varrho_M^n \geq \tau$ and either $n = k$ or else $n \leq k$ and $\omega\varrho_M^{n+1} < \tau$. Let γ be regular in M and $\tau \leq \gamma < \omega\varrho_M^n$. Then $\tilde{\sigma}(\gamma) = \sup(\sigma''\gamma)$. If $\omega\varrho_M^n$ is a successor cardinal in M, then the same equality also holds of $\gamma = \omega\varrho_M^n$.*

Proof. Let $\beta < \tilde{\sigma}(\gamma)$, say $\beta = \tilde{\sigma}(f)(\delta)$ where $\delta < \tau'$ and $f \in \Gamma_n^k(\sigma, M)$. We shall find an $\varepsilon < \gamma$ such that $\beta < \tilde{\sigma}(\varepsilon)$. Since n is as above, $f \in M$. Assume w.l.o.g. that $\operatorname{rng}(f) \subset \gamma$. Set $\varepsilon = \sup(\operatorname{rng} f) + 1$; then $\varepsilon < \gamma$ as γ is regular in M. The rest follows from the fact that $\operatorname{rng}(\tilde{\sigma}(f)) \subset \tilde{\sigma}(\varepsilon)$.

To see the second part of the lemma, it suffices to show that any $f \in \Gamma^k(\sigma, M)$ is bounded in $\gamma = \omega\varrho_M^n$ (and therefore an element of M). This is clear unless $k \leq n$ and $\omega\varrho_M^{n+1} < \tau$. In this case, assuming that $f \in \Gamma^k(\sigma, M)$ is cofinal in $\omega\varrho_M^n$, we obtain a contradiction as follows. We shall construct a $\Sigma_1^{(n-1)}(M)$ surjection of β to γ, where β is the cardinal predecessor of γ in M. Using such a surjection, one can produce a $\mathbf{\Sigma}_1^{(n-1)}(M)$ subset of β coding the ordinal type γ. Such a set cannot be an element of M, which means that $\omega\varrho_M^n \leq \beta$, contradicting our assumptions on n. Let g_ξ be the $\leq_M$-least surjection from β to ξ in the canonical well ordering $\leq_M$. By acceptability, the sequence $\langle g_\xi;\ \xi < \gamma\rangle$ is lightface Δ_1 definable over J_γ^A. It is then easy to verify that the map g defined by

$$g(\eta, \xi) = g_{f(\eta)}(\xi)$$

is a $\mathbf{\Sigma}_1^{(n-1)}(M)$ definable surjection from $\beta \times \beta$ to γ. □(Lemma 3.6.4)

Corollary 3.6.5. *Suppose Q has a largest cardinal κ and $\kappa' = \sigma(\kappa)$. Then $\tilde{\sigma}(\tau) = \tau' = (\kappa')^{+N}$.*

Corollary 3.6.6. *Suppose n is as in Lemma 3.6.4 and there is no $\mathbf{\Sigma}_1^{(n-1)}(M)$ partial map singularizing τ, i.e. a map whose domain is a bounded subset of τ and whose range is cofinal in τ. Then $\tilde{\sigma}(\tau) = \tau'$. Thus, τ' is regular in N.*

Proof. It suffices to observe that there is no $f \in \Gamma^k(\sigma, M)$ singularizing τ. This is a consequence of the fact that any partial map f as above can be turned into a $\mathbf{\Sigma}_1^{(n-1)}(M)$ map whose domain is some $\gamma < \tau$ and whose range is a cofinal subset of τ (the set a of all those ξ for which $f(\xi) < \tau$ is an element of M, so set $f'(\xi) = f(\xi)$ if $\xi \in a$ and $f'(\xi) = 0$ otherwise). Such a map is good by definition and, hence, an element of $\Gamma^k(\sigma, M)$. The rest follows by the proof of Lemma 3.6.3. □(Corollary 3.6.6)

If τ is regular in M, then, considering k to be the least number such that there is a (good) $\mathbf{\Sigma}_1^{(k)}(M)$ function singularizing τ (if such a function exists), we obtain "maximal" k-extensions such that the associated map $\tilde{\sigma}$ sends τ cofinally into its image. Denote this k by $k(M)$. By the proof of Lemma 3.6.3, if Q has a largest cardinal then such k satisfies $\omega\varrho_M^{k+1} < \tau \leq \omega\varrho_M^k$; so the k-extension agrees with the ω-extension in this case. The following lemma gives another criterion for cofinality of extensions (which might be of interest, although we won't make any substantial use of it below).

Definition. Let τ be regular in M, $n = k(M)$ if $k(M)$ exists and $n = \omega$ otherwise. The canonical $k(M)$-extension of σ to M is briefly called the *canonical extension* of σ to M.

Lemma 3.6.7. *Let τ be regular in M and let $\tilde{\sigma}$ be the canonical extension of σ. Set $n = k(M)$. Then $\tilde{\sigma}$ is $\Sigma_0^{(n)}$-preserving and cofinal. Moreover, τ' is regular in N and there is a good $\mathbf{\Sigma}_1^{(n)}(N)$ function singularizing τ'. It follows that $\omega\varrho_N^{n+1} \leq \tau'$; if κ is the a largest cardinal in Q, then $\omega\varrho_N^{n+1} \leq \kappa' = \sigma(\kappa)$.*

Proof. That $\tilde{\sigma}$ is $\Sigma_0^{(n)}$-preserving follows from Lemma 3.6.3. Let

$$\omega\varrho = \sup(\tilde{\sigma}''\omega\varrho_M^n).$$

Pick a p such that there is a function f singularizing τ which is $\Sigma_1^{(n)}(M)$ in p and set $p' = \tilde{\sigma}(p)$. Then $\tilde{\sigma} \upharpoonright M^{n,p} : M^{n,p} \to \tilde{M}$ is Σ_0-preserving and cofinal where $\tilde{M} = \langle J_\varrho^{A'}, A_{M'}^{n,p'} \cap J_\varrho^{A'} \rangle$. We can w.l.o.g. suppose that $\mathrm{rng}(f) \subset \tau$, hence f is rudimentary over $M^{n,p}$. Let f' be rudimentary over $\tilde{M}$ by the same definition; then $\mathrm{dom}(f')$ is bounded in τ' and $\mathrm{rng}(f') \supset \tilde{\sigma}''(\mathrm{rng}(f))$. Since $\tilde{\sigma}$ maps τ cofinally into τ' (see Corollary 3.6.6), $\mathrm{rng}(f')$ is cofinal in τ', thus f' singularizes τ'. If $\tilde{\sigma}$ were non-cofinal, $\tilde{M}$ would be an element of N. Then $f' \in N$, which would contradict the fact that τ' is regular in N. So $\tilde{\sigma}$ is cofinal and therefore $\Sigma_1^{(n)}$-preserving. It follows that $\tilde{\sigma}(f)$ is well-defined and is a good $\mathbf{\Sigma}_1^{(n)}(N)$ function singularizing τ'. The statements concerning $\omega\varrho_N^{n+1}$ are immediate consequences of the above and the proof of Lemma 3.6.4. □(Lemma 3.6.7)

Lemma 3.6.8. *Let $\tilde{\sigma}$ be the canonical k-extension of σ to M, $\omega\varrho_M^n \geq \tau$ and $n \leq k$.*

a) *If M is n-solid, then so is N and $\tilde{\sigma}(p_M^n) = p_N^n$. Moreover, if M is n-sound then so is N.*

b) *Suppose $\omega\varrho_M^{n+1} \leq \beta$ and M is sound and solid above β. Then N is sound and solid above $\beta' = \sup(\tilde{\sigma}''\beta)$ and $\tilde{\sigma}(p_M - \beta) = p_N - \beta' = p_N - \tilde{\sigma}(\beta)$.*

Note that b) does not give more information than a) unless $\beta < \omega\varrho_M^n$. We have stated a) just for the sake of completeness and since the assumptions in this case are weaker – we do not need soundness in order to prove facts about solidity.

Proof. To see a) we apply Lemma 1.12.2. Hence it is sufficient to observe that $\tilde{\sigma}(p_M^n) \in P_N^n$, which follows from the proof of Lemma 3.6.3, and that $\tilde{\sigma}$ preserves generalized witnesses upwards (see Lemma 1.12.6). Soundness of N then follows from h) of Lemma 3.6.3. b) can be shown in a similar manner using Lemma 1.12.7.

□(Lemma 3.6.8)

The same proof yields:

Lemma 3.6.9. *Suppose $k(M)$ exists and $\tilde{\sigma}$ is the canonical extension of σ to M. If M is solid above τ, then N is solid above τ' and $\tilde{\sigma}(p_M - \tau) = p_N - \tau'$. Moreover, if M is sound above τ, then N is sound above τ'. The same statement holds with κ, κ' in place of τ, τ' if κ is the cardinal predecessor of τ in M and $\kappa' = \sigma(\kappa)$. (In fact, we could formulate an analogous result for any $\beta \geq \tau$ as in the previous lemma; the point is that in the current case results on solidity and soundness can be separated.)*

In the rest of this section we shall present two simple extendibility criteria for pseudoultrapowers. The first one is also known as the "Interpolation Lemma".

Lemma 3.6.10 (Interpolation Lemma). *Let $\omega\varrho_M^n \geq \tau$ and either $n = k$ or else $n \leq k$ and $\omega\varrho_M^{n+1} < \tau$. Suppose*

$$\sigma : M \underset{\Sigma_0^{(n)}}{\longrightarrow} M^*,$$

where $M^* = \langle J_{\alpha^*}^{A^*}, B^* \rangle$. Let $\tau^* = \sigma(\tau)$, $\tau' = \sup(\sigma''\tau)$ and $A' = A^* \cap J_{\tau'}^{A^*}$. Let $Q = J_\tau^A$ and $Q' = J_{\tau'}^{A'}$.

a) *The canonical k-extension $\tilde{\sigma} : M \to N$ of $\sigma \restriction Q : Q \to Q'$ exists.*

b) *There is a unique $\Sigma_0^{(n-1)}$-preserving map $\sigma' : N \to M^*$ such that $\sigma = \sigma' \circ \tilde{\sigma}$ and $\sigma' \restriction \tau' = \mathrm{id}$; the map σ' is defined by*

$$\sigma'(\tilde{\sigma}(f)(\alpha)) = \sigma(f)(\alpha)$$

for $\langle \alpha, f \rangle \in D^k(M, \sigma)$ and is in fact $\Sigma_1^{(n-1)}$-preserving (here $0 - 1 = 0$).

c) *If $\tilde{\sigma}$ is cofinal (i.e. cofinal at the n-th level), then σ' is $\Sigma_0^{(n)}$-preserving.*

The situation is illustrated by the following diagram.

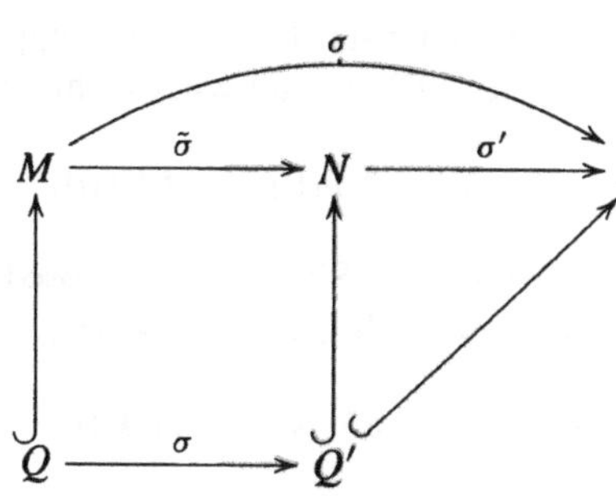

Proof. Define a map $\sigma_{\mathbb{D}} : \mathbb{D} \to M^*$ (here $\mathbb{D} = \mathbb{D}^k(M, \sigma)$) by: $\sigma_{\mathbb{D}}([\alpha, f]) = \sigma(f)(\alpha)$. Łos' Theorem for Σ_0-formulae guarantees that $\sigma_{\mathbb{D}}$ is well defined and Σ_0-preserving, so $\mathbb{D}$ is well-founded. This proves a). Let $\tilde{\sigma} : M \to N$ be the corresponding canonical extension.

That σ' is determined by the properties stated in b) *uniquely* follows from this simple computation: $\sigma'(\tilde{\sigma}(f)(\alpha)) = \sigma' \circ \tilde{\sigma}(f)(\sigma'(\alpha)) = \sigma(f)(\alpha)$. To see that σ' is $\Sigma_1^{(n-1)}$-preserving we apply Łos' Theorem for pseudoultrapowers and the fact that σ is $\Sigma_1^{(n-1)}$-preserving: Given any $\langle\alpha_1, f_1\rangle, \ldots, \langle\alpha_\ell, f_\ell\rangle \in D^k(M,\sigma)$ and a $\Sigma_0^{(n)}$ formula $\varphi(v_1, \ldots, v_\ell)$,

$$N \models \varphi(\tilde{\sigma}(f_1)(\alpha_1), \ldots, \tilde{\sigma}(f_\ell)(\alpha_\ell)) \longleftrightarrow \prec\alpha_1, \ldots, \alpha_\ell\succ \in \sigma(u^{\varphi,M}_{f_1,\ldots,f_\ell})$$
$$\longleftrightarrow M^* \models \varphi(\sigma(f_1)(\alpha_1), \ldots, \sigma(f_\ell)(\alpha_\ell))$$

since $\sigma(u^{\varphi,M}_{f_1,\ldots,f_\ell}) = \{\prec\eta_1, \ldots, \eta_\ell\succ;\ M^* \models \varphi(\sigma(f_1)(\eta_1), \ldots, \sigma(f_\ell)(\eta_\ell))\}$.

Finally, if $\tilde{\sigma}$ is $\Sigma_0^{(n)}$ and cofinal, then every element of H^n_N can be represented as $\tilde{\sigma}(f)(\alpha)$ where f is in H^n_M. As $\sigma(f) \in H^n_{M^*}$, the same applies also to $\sigma'(\sigma(f)(\alpha)) = \sigma(f)(\alpha)$. This shows that σ' maps H^n_N into $H^n_{M^*}$. By the above computation, σ' preserves $\Sigma_0^{(n)}$-formulae. □(Lemma 3.6.10)

Definition. A set X is *ω-closed* iff ${}^\omega X \subset X$. A structure is ω-closed iff its support is.

Lemma 3.6.11. *Let $\omega\varrho^n_M \geq \tau$ and either $n = k$ or else $n \leq k$ and $\omega\varrho^{n+1}_M < \tau$. Let H be transitive and ω-closed and $\sigma : H \to H_\theta$ be fully elementary, where θ is regular. Suppose $Q \in H$ and $Q' = \bigcup\sigma''Q$.*

a) *The canonical k-extension $\tilde{\sigma} : M \to N$ of $\sigma \restriction Q : Q \to Q'$ exists (note that $\sigma \restriction Q$ is Σ_0-preserving and cofinal).*

b) *Suppose $\tilde{\sigma}$ is cofinal. If $\bar{M}$ is countable, $\omega\varrho^n_{\bar{M}} \geq \bar{\tau}$, $\bar{\sigma} : \bar{M} \to N$ is $\Sigma_0^{(n)}$-preserving and $\bar{\sigma}(\bar{\tau}) = \tau'$, then there is a $\Sigma_0^{(n)}$-preserving embedding $\sigma^* : \bar{M} \to M$ such that $\sigma^*(\bar{\tau}) = \tau$.*

Proof. The ω-closedness of H plays a similar role as the ω-completeness in case of extenders; the arguments are of the same spirit. To see a), suppose for a contradiction that $\mathbb{D}(M,\sigma)$ is ill-founded. Let $\langle[\alpha_i, f_i];\ i \in \omega\rangle$ witness this. Set

$$u_i = \{\langle\eta_1, \eta_0\rangle;\ f_{i+1}(\eta_1) \in f_i(\eta_0)\}.$$

Then $\langle\alpha_{i+1}, \alpha_i\rangle \in \sigma(u_i)$ for every $i \in \omega$. Since H is ω-closed, it contains the whole sequence $u = \langle u_i;\ i \in \omega\rangle$ as an element. In H_θ we thus have

$$(\exists k')[k' : \omega \to \tau' \ \&\ (\forall i \in \omega)(\langle k'(i+1), k'(i)\rangle \in \sigma(u)_i)],$$

as is witnessed by the sequence $\langle\alpha_i;\ i \in \omega\rangle$. By the elementarity of σ, there is $k : \omega \to \tau$ in H such that $\langle k(i+1), k(i)\rangle \in u_i$ for all $i \in \omega$. This means that $\langle f_i(k(i));\ i \in \omega\rangle$ is an infinite $\in$-decreasing sequence, which is impossible.

The proof of b) is similar, but longer. Fix an enumeration $\langle\bar{x}_i;\ i \in \omega\rangle$ of $\bar{M}$. Let $x_i = \bar{\sigma}(\bar{x}_i)$. For every $i \in \omega$ pick an $f_i \in \Gamma^k(\sigma, M)$ and an $\alpha_i \in \sigma(\mathrm{dom}(f_i))$ such that $x_i = \tilde{\sigma}(f_i)(\alpha_i)$. Let $\langle\varphi_j;\ j \in \omega\rangle$ be a recursive enumeration of all formulae which are

$\Sigma_0^{(n)}$ or $\Pi_0^{(n)}$, ℓ_j be the number of free variables in φ_j and m_i^j be the type of the i-th free variable in φ_j ($i < \ell_j$). We define a set a coding the truth in $\bar{M}$ as follows: a consists of all pairs $\langle j, g\rangle$ satisfying

- $g : \ell_j \longrightarrow \omega$;
- $\bar{x}_{g(i)} \in H_{\bar{M}}^{m_i^j}$ for every $i < \ell_j$;
- $\bar{M} \models \varphi_j(\bar{x}_{g(0)}, \ldots, \bar{x}_{g(\ell_j-1)})$.

Since a is essentially a subset of ω, a is an element of H and $\sigma(a) = a$. Furthermore, the following statements are equivalent

- $\bar{M} \models \varphi_j(\bar{x}_{g(0)}, \ldots, \bar{x}_{g(\ell_j-1)})$
- $N \models \varphi_j(\tilde{\sigma}(f_{g(0)})(\alpha_{g(0)}), \ldots, \tilde{\sigma}(f_{g(\ell_j-1)})(\alpha_{g(\ell_j-1)}))$
- $\langle \alpha_{g(0)}, \ldots, \alpha_{g(\ell_j-1)}\rangle \in \sigma(u_g^{j,M}) \stackrel{\text{def}}{=} u^{\varphi_j,M}_{f_{g(0)},\ldots,f_{g(\ell_j-1)}}$

Then $\langle u_g^{j,M};\ \langle j, g\rangle \in a\rangle$ is an element of H by the ω-closedness and the sequence $\langle \alpha_i;\ i \in \omega\rangle$ witnesses the statement

$$(\exists k : \omega \to \tau')(\forall \langle j, g\rangle \in a)(\langle k(g(0)), \ldots, k(g(\ell_j - 1))\rangle \in \sigma(u_g^{j,M}))$$

in H_θ. Hence there is a $k : \omega \to \tau$ in H such that $\langle k(g(0)), \ldots, k(g(\ell_j - 1))\rangle \in u_g^{j,M}$ or, equivalently,

$$M \models \varphi_j(f_{g(0)}(k(g(0)), \ldots, f_{g(\ell_j-1)}(k(g(\ell_j - 1)))) \tag{3.6.2}$$

for every $\langle j, g\rangle$. Define $\sigma^* : \bar{M} \to M$ by

$$\sigma^*(\bar{x}_i) = f_i(k(i)).$$

Then $\sigma^*(\bar{\tau}) = \tau$; to see this it suffices to set $f_i = c_\tau$ where $\bar{\tau} = \bar{x}_i$. To see that σ^* is sufficiently elementary, we first observe that (3.6.2) yields

$$\bar{M} \models \varphi(z_1, \ldots, z_\ell) \longleftrightarrow M \models \varphi(\sigma^*(z_1), \ldots, \sigma^*(z_\ell))$$

for every $\ell \in \omega$, $z_1, \ldots, z_\ell \in \bar{M}$ and a $\Sigma_0^{(n)}$-formula φ. Since this also guarantees the inclusion $(\sigma^*)''H_{\bar{M}}^m \subset H_M^m$ for $m < n$, the embedding σ^* is $\Sigma_1^{(n-1)}$-preserving. To complete the proof it remains to prove the above inclusion for $m = n$. This follows from the fact that $\tilde{\sigma}$ is cofinal at the n-th level: if $\bar{x}_i \in H_{\bar{M}}^n$, then $x_i \in H_N^n$, hence $x_i = \tilde{\sigma}(f_i)(\alpha_i)$ where $f_i \in H_M^n$. Consequently, $\sigma^*(\bar{x}_i) = f_i(k(i)) \in H_M^n$.

□(Lemma 3.6.11)

The above lemma motivates the following notion, which generalizes the concept of ω-completeness from extenders to arbitrary maps.

Definition. Given transitive structures M, N and a Σ_0-preserving embedding $\sigma : M \to N$, we say that σ is κ-complete iff for every embedding $\tau : \bar{M} \to N$ of size at most κ there is a Σ_0-preserving $\pi : \bar{M} \to M$ such that $\tau = \sigma \circ \pi$.

If σ, Q, Q' are as in the above lemma, then $\sigma \restriction Q$ is ω-complete. This follows by a minor modification of the above proof. We fix an enumeration of Σ_0-formulae $\langle \varphi_j;\ j \in \omega \rangle$. Given a countable Σ_0-preserving map $\tau : \bar{Q} \to Q'$, let $\langle x_i;\ i \in \omega \rangle$ be an enumeration of $\mathrm{rng}(\tau)$, $b = \{i \in \omega;\ x_i \in \mathrm{rng}(\sigma)\}$ and for $i \in \omega$ set $y_i = x_i$ if $i \in b$ and $y_i = \varnothing$ otherwise. By the closure properties of σ, both $\langle y_i;\ i \in \omega \rangle$ and b are in the range of σ and $\sigma(b) = b$; let $\langle \bar{y}_i;\ i \in \omega \rangle$ be the σ-preimage of $\langle y_i;\ i \in \omega \rangle$. Define a to be the set of all pairs $\langle j, g \rangle$ where $j \in \omega$, ℓ_j is the number of free variables of φ_j and $g : \ell_j \to \omega$ is such that $Q' \models \varphi_j(x_1, \ldots, x_\ell)$; we again have $\sigma(a) = a$. Then, in H_θ,

$$(\exists z \in {}^\omega Q')[(\forall \langle j, g \rangle \in a)(Q' \models \varphi_j(z_1, \ldots, z_\ell)) \ \&\ (\forall i \in b)(z_i = y_i)].$$

Using the elementarity of σ we infer that the same statement holds in H with Q and $\langle \bar{y}_i;\ i \in \omega \rangle$ in place of Q' and $\langle y_i;\ i \in \omega \rangle$. Pick a sequence $\langle \bar{z}_i;\ i \in \omega \rangle$ witnessing this and define $\pi : \bar{M} \to M$ by

$$\pi(\tau^{-1}(x_i)) = \bar{z}_i;$$

then $\sigma(\bar{z}_i) = \sigma(\bar{y}_i) = y_i$ for all $i \in b$, so $\tau = \sigma \circ \pi$ as desired, and it is easy to see that π is Σ_0-preserving.

A minor modification of the proof of Lemma 3.6.11 yields a generalization of this lemma to ω-complete embeddings. We leave the proof to the reader. It follows directly from the formulation of the next lemma that the canonical extension of an ω-complete embedding is itself ω-complete and the same obviously applies to κ-complete embeddings.

Lemma 3.6.12. *Let $\omega\varrho_M^n \geq \tau$ and either $n = k$ or else $n \leq k$ and $\omega\varrho_M^{n+1} < \tau$. Suppose $\sigma : Q \to Q'$ is ω-complete and cofinal.*

a) *The canonical k-extension $\tilde{\sigma} : M \to N$ of σ exists.*

b) *Suppose $\tilde{\sigma}$ is $\Sigma_0^{(n)}$ cofinal. If $\bar{M}$ is countable, $\omega\varrho_{\bar{M}}^n \geq \bar{\tau}$, $\tau : \bar{M} \to N$ is $\Sigma_0^{(n)}$-preserving and $\bar{\sigma}(\bar{\tau}) = \tau'$, then there is a $\Sigma_0^{(n)}$-preserving embedding $\pi : \bar{M} \to M$ such that $\sigma \circ \pi = \tau$.*

□(Lemma 3.6.12)

Notice that an extender F on M is ω-complete iff the ultrapower map $\sigma : M \to \mathrm{Ult}(M, F)$ is ω-complete in the sense of the above definition, so the above definition generalizes the usual notion of ω-completeness. The verification of this fact is standard and again left to the reader.

Let us finally state one more criterion on the existence of fine structural canonical extensions, which uses considerably weaker assumptions. We stress that in this case $k(M)$ exists, i.e. τ is regular in M and definably singularized over M.

Lemma 3.6.13. *Suppose* $\mathrm{cf}(\tau) > \omega$ *and* $\tilde{h}^{n+1}_M(\tau \cup \{r\}) = M$ *for some* $r \in M$, *where* $n = k(M)$. *Then the canonical extension* $\tilde{\sigma} : M \to N$ *exists.*

Proof. Suppose the contrary. Let $\langle [\alpha_i, f_i];\ i \in \omega \rangle$ be an infinite descending chain in $\mathbb{D}(M, \sigma)$. Pick a sufficiently large regular θ so that $M \in H_\theta$. Form a countable elementary substructure X of H_θ such that $Q, M, r, f_i \in X$ for all i, the transitive collapse $\bar{H}$ of X and the uncollapsing map $\bar{\sigma} : \bar{H} \to H_\theta$. Let $\bar{\sigma}(\bar{M}, \bar{Q}, \bar{\tau}, \bar{r}) = (M, Q, \tau, r)$.

Since τ is not ω-cofinal, $\tilde{\tau} = \sup(\bar{\sigma}''\bar{\tau}) < \tau$. The map $\bar{\sigma} \restriction \bar{Q} : \bar{Q} \to \tilde{Q}$, where $\tilde{Q} = J^A_{\tilde{\tau}}$, is Σ_0-preserving and cofinal. By applying the interpolation lemma to the triple $\bar{M}, M$ and $\bar{\sigma}$ we obtain the canonical extension $\bar{\sigma}_0 : \bar{M} \to \tilde{M}$ of $\bar{\sigma} \restriction \bar{Q}$ and a $\Sigma_0^{(n)}$-preserving map $\bar{\sigma}_1 : \tilde{M} \to M$ with critical point $\tilde{\tau}$ such that $\bar{\sigma}_1(\tilde{\tau}) = \tau$. The canonical extension $\bar{\sigma}_0$ is $\Sigma_0^{(n)}$-preserving and cofinal by Lemma 3.6.7. The elementarity of $\bar{\sigma}$ also guarantees that $\tilde{h}^{n+1}_{\bar{M}}(\bar{\tau} \cup \{\bar{r}\}) = \bar{M}$ and the proof of h) in Lemma 3.6.3 yields the corresponding statement for $\tilde{M}$, that is, $\tilde{h}^{n+1}_{\tilde{M}}(\tilde{\tau} \cup \{\tilde{r}\}) = \tilde{M}$ where $\tilde{r} = \bar{\sigma}(\bar{r})$. It follows that $\tilde{M}$ can be encoded in a subset of ${}^5\tau$ which is $\Sigma_1^{(n)}(\tilde{M})$ in r as a structure $\langle D, E, I \rangle$, where $E, I \subset D$ and

- $D = \mathrm{dom}(\tilde{h}^{n+1}_{\tilde{M}}(-, \tilde{r})) \cap \tilde{\tau}$;
- $\xi_0\, E\, \xi_1 \longleftrightarrow \tilde{h}^{n+1}_{\tilde{M}}(\xi_0, \tilde{r}) \in \tilde{h}^{n+1}_{\tilde{M}}(\xi_1, \tilde{r})$;
- $\xi_0\, I\, \xi_1 \longleftrightarrow \tilde{h}^{n+1}_{\tilde{M}}(\xi_0, \tilde{r}) = \tilde{h}^{n+1}_{\tilde{M}}(\xi_1, \tilde{r})$.

I is a congruence relation on $\langle D, E \rangle$ and $\tilde{M}$ can be obviously obtained from $\langle D, E, I \rangle$ by collapsing $\langle D, E \rangle / I$. It follows that the sets D, E, I are $\Sigma_1(\tilde{M}^{n,\tilde{r} \restriction n})$ in $\tilde{r}_n = \tilde{r} \cap \omega\varrho^n_{\tilde{M}}$.

We show that $\tilde{M} \in Q$. Notice that the least $\gamma < \tau$ which can be mapped cofinally into τ by a good $\Sigma_1^{(n)}(M)$ function in some q is definable over M and the same applies to the least corresponding parameter q. Thus, $\gamma < \tilde{\tau}$ and $q \in \mathrm{rng}(\bar{\sigma}) \subset \mathrm{rng}(\bar{\sigma}_1)$, i.e. there is a $\tilde{q} \in \tilde{M}$ such that $\bar{\sigma}_1(\tilde{q}) = q$. It follows that the map $\bar{\sigma}_1$ is non-cofinal: Otherwise $\bar{\sigma}_1$ would be $\Sigma_1^{(n)}$-preserving, so given any good $\Sigma_1^{(n)}(M)$ function g in q mapping γ cofinally into τ, if $\tilde{g}$ were a good $\Sigma_0^{(n)}(\tilde{M})$ function in $\tilde{q}$ by the same functionally absolute definition, then $\tilde{g}(\xi)$ would be defined whenever $g(\xi)$ and $\bar{\sigma}_1(\tilde{g}(\xi)) = g(\xi)$. Consequently, $\mathrm{rng}(\bar{\sigma}_1)$ would be cofinal in τ, contradicting the fact that $\bar{\sigma}_1(\tilde{\tau}) = \tau > \tilde{\tau}$. Set

- $\omega\tilde{\varrho} = \sup(\bar{\sigma}_1''(\omega\varrho^n_{\tilde{M}}))$;
- $A^* = A^{n,\tilde{r} \restriction n}_{\tilde{M}} \cap J^{A^M}_{\tilde{\varrho}}$ and $M^* = \langle J^{A^M}_{\tilde{\varrho}}, A^* \rangle$.

Then A^*, and therefore also M^*, is an element of M. Since

$$\bar{\sigma}_1 \restriction \tilde{M}^{n,r \restriction n} : \tilde{M}^{n,\tilde{r} \restriction n} \to M^*$$

is Σ_0-preserving and cofinal, the same list of Σ_1-formulae which determines D, E, I over $\tilde{M}^{n,\tilde{r}\upharpoonright n}$ from $\tilde{r}_n$, determines D, E, I over M' from $\bar{\sigma}_1(\tilde{r}_n)$. It follows that $D, E, I \in M$ and since these objects are essentially subsets of $\tilde{\tau}$, they must be elements of Q by the fact that τ is an M-cardinal. Then $\tilde{M}$ can be reconstructed inside Q by collapsing.

Now return to the sequence $[\alpha_i, f_i]$ representing an infinite descending chain of ordinals in the putative pseudoultrapower. Let $\bar{f}_i$ be the $\bar{\sigma}$-preimage of f_i and $\tilde{f}_i = \bar{\sigma}_0(\bar{f}_i)$ for all $i \in \omega$. Each $\bar{f}_i$ is a good $\Sigma_1^{(n-1)}(\bar{M})$ function in some $\bar{p}_i \in \bar{M}$, where f_i is a good $\Sigma_1^{(n-1)}(M)$ function in $p_i = \bar{\sigma}(\bar{p}_i)$ and, similarly, each $\tilde{f}_i$ is a good $\Sigma_1^{(n-1)}(\tilde{M})$ function in $\tilde{p}_i = \bar{\sigma}_0(\bar{p}_i) \in \tilde{M}$ (recall $n = k(\bar{M})$). It follows that each $\tilde{f}_i$ is an element of Q. Furthermore, $f_i = \bar{\sigma}_1(\tilde{f}_i)$ and $\mathrm{dom}(f_i) = \mathrm{dom}(\tilde{f}_i) < \tilde{\tau}$. Given any pair of indices i, j we have

$$\{\prec\eta_0, \eta_1\succ;\ f_i(\eta_0) \in f_j(\eta_1)\} = \{\prec\eta_0, \eta_1\succ;\ \tilde{f}_i(\eta_0) \in \tilde{f}_j(\eta_1)\} \stackrel{\text{def}}{=} a_{ij}$$

by the preservation degree of $\bar{\sigma}_1$. By applying the fact that $[\alpha_i, f_i]$ represents an infinite descending chain of ordinals in $\mathbb{D}(M, \sigma)$ we obtain

$$\prec\alpha_{i+1}, \alpha_i\succ \in \sigma(a_{i+1,i}) = \{\prec\eta_0, \eta_1\succ;\ \sigma(\tilde{f}_i)(\eta_0) \in \sigma(\tilde{f}_j)(\eta_1)\}$$

for all i, which means that

$$\sigma(\tilde{f}_0)(\alpha_0) \ni \sigma(\tilde{f}_1)(\alpha_1) \ni \sigma(\tilde{f}_2)(\alpha_2) \ni \cdots \ni \sigma(\tilde{f}_n)(\alpha_n) \ni \cdots$$

constitutes an infinite descending chain of ordinals. Contradiction, so $\mathbb{D}(M, \sigma)$ is well-founded after all and the canonical extension of σ to M exists. □(Lemma 3.6.13)

Notes

Fine ultrapowers arise naturally in the fine structure theory. The present approach to the theory of such ultrapowers using good functions is due to Jensen, and appeared in a compact form together with the $*$-theory in [Je88]. Although the use of $*$-theory is not necessary in order to develop the theory of fine ultrapowers, it makes the resemblance between fine ultrapowers and usual ultrapowers more visible. Also, the $*$-theory enables us to formulate and prove useful results about fine ultrapowers (e.g. the fact that iteration maps are Σ^*-preserving), whose formulation without $*$-theory is not easily conceivable. Pseudoultrapowers arise naturally in covering arguments, and were first time used in Jensen's proof of the covering lemma for L. Although we use the term "pseudoultrapower", the construction is nothing other than the usual ultrapower construction with one exception, that instead of an extender we use a map of a more general nature. Many people call such maps *long* extenders, as they can be represented as "extenders" whose generators exceed the image of the critical point.

Chapter 4

Mice and Iterability

In this chapter we shall introduce structures called *premice*, which are the building stones of the core model we are aiming to construct. In general, premice are J-structures constructed relative to coherent extender sequences; since we stay at the level where no extenders with more than one generator exist, it will be more convenient to consider sequences of measures instead. We shall develop a standard technique for comparing premice; this amounts to iterating out the least disagreements in the corresponding measure sequences. It might happen that partial measures are used in the course of a comparison; in this case we truncate the corresponding structures in order to remove those sets which are not measurable by the applied measures. This comparison process requires sufficient *iterability* of the premice involved, i.e. the possibility of forming iterated ultrapowers which arise in the comparison process. We shall see that, as long as we deal with models below one strong cardinal, we can successfully carry out the comparison process using *linear* iterations. Iterable premice are colloquially called *mice*. In order for the comparison process to yield a meaningful outcome, we have to bring the fine structure into play: The application of fine ultrapowers and the fact that sufficiently iterable premice are solid (in the sense of the section on solidity in Chapter 1) will guarantee that at least one side of an arbitrary coiteration does not contain any truncations. As a consequence of iterability and solidity we obtain a condensation lemma, which can be considered a generalization of the condensation lemma for $\boldsymbol{L}$.

4.1 Premice

Definition. A *premouse* is an acceptable J-structure $M = \langle J_\alpha^E, E_{\omega\alpha}\rangle$ satisfying:

a) $E \subset \{\langle \nu, x\rangle;\ \nu < \omega\alpha\ \&\ x \subset \nu\}$. Set $E_\nu = \{x;\ \langle \nu, x\rangle \in E\}$.

b) Given any $\nu \leq \omega\alpha$, either $E_\nu = \varnothing$ or else ν is a limit ordinal, S_ν^E has a largest cardinal κ and E_ν is a normal measure over S_ν^E with critical point κ and $M\|\nu \stackrel{\text{def}}{=} \langle S_\nu^E, E_{\omega\nu}\rangle$ is an amenable structure.

c) (Coherency) Let $\nu \leq \omega\alpha$ and

$$\pi : S_\nu^E \underset{E_\nu}{\longrightarrow} N \quad \text{weakly,}$$

where $N = \langle |N|, E'\rangle$. Then $\nu + 1 \subset \text{wfcore}(N)$, $E' \upharpoonright \nu = E \upharpoonright \nu$ and $E'_\nu = \varnothing$.

d) $M\|\nu$ is sound for every $\nu < \alpha$.

Although all ultrapowers we are interested in will be well-founded, we stated the coherency condition in the above definition more generally. The only reason we did so is the following: In the next lemma we prove that being a premouse is a Q-condition which requires us to avoid any reference to well-foundedness. For well-founded ultrapowers we can reformulate the coherency condition in the obvious way.

Since E_ν is an amenable measure, $\nu = \kappa^{+N}$ where $N = S^{E'}_{\nu'}$ (otherwise there is a surjection $f : \kappa \to \mathcal{P}(\kappa) \cap S^{E'}_{\nu}$ in N' and a diagonalization argument would yield a set $a \subset \kappa$ which is not a member of $S^{E'}_{\nu} = S^{E}_{\nu}$ – contradiction to the amenability of E_ν). It follows that S^E_ν is a ZFC$^-$ model, $\omega\nu = \nu$ and $S^E_\nu = J^E_\nu$ whenever $E_\nu \neq \varnothing$. By this fact, the coherency condition can be stated in the following form: If $J^{E'}_{\nu'} = \mathrm{Ult}(J^E_\nu, E_\nu)$ then $E' \restriction \nu = E \restriction \nu$ and $E'_\nu = \varnothing$.

We stress that E_ν is a normal measure over J^E_ν, but it need not be a measure over M. This means that E_ν need not measure all subsets of its critical point that are elements of M; in other words, we allow new subsets of $\mathrm{cr}(E_\nu)$ to occur in M at some level $\nu' \geq \nu$. This is an important point since, as we shall see later in connection with the construction of the $\boldsymbol{K}^c$ model, we cannot a priori guarantee that no measures will be collapsed at later stages of the construction.

Definition. If $M = \langle J^E_\alpha, E_{\omega\alpha}\rangle$ is a premouse then α is called the *height* of M and is denoted by $\mathrm{ht}(M)$.

Definition. If $M = \langle J^E_\alpha, E_{\omega\alpha}\rangle$ is a premouse and $E_{\omega\alpha} \neq \varnothing$, then $E_{\omega\alpha} = E_\alpha$ is called the *top measure* of M and is denoted by E^M_{top}. Furthermore, E^M denotes the predicate E.

Definition. A premouse M without a top measure, i.e. such that $E^M_{\mathrm{top}} = \varnothing$, is called *passive*. If M has a top measure, i.e. if $E^M_{\mathrm{top}} \neq \varnothing$, then M is called *active*.

Lemma 4.1.1. *If M is a premouse and $\nu < \mathrm{ht}(M)$, then both $M||\nu$ and $J^{E^M}_\nu$ are premice.*

Proof. This follows by a straightforward verification of a) – d) in $M||\nu$. □(Lemma 4.1.1)

Lemma 4.1.2. *There is a Q-sentence Ψ such that for any transitive structure M,*

$$M \text{ is a premouse} \longleftrightarrow M \models \Psi.$$

Proof. We already know that being an acceptable structure is a Q-condition. Thus, it suffices to verify that the clauses a) – d) from the above definition can be expressed in a Q-fashion. Suppose first $M = J^E_\alpha$. The verification of a) is trivial. b) can be expressed as the conjunction of the following statements:

- $(\forall\nu)[(\forall z)(\langle\nu, z\rangle \notin E) \vee (\nu \text{ is a limit ordinal } \& \ (\exists\kappa < \nu)(\langle\nu,\kappa\rangle \in E)]$
- $(\forall\nu,\kappa)(\forall y)[(\langle\nu,\kappa\rangle \in E \ \&\ y = \langle S^E_\nu, E_\nu\rangle) \longrightarrow$
  - κ is the largest cardinal in y

- E_ν is a normal measure over y
- y is an amenable structure]

Both statements are Π_1. For the latter, note that the consequent of the above implication is Σ_0 in y and ν since $E_\nu, y \in M$.

To see c), we first observe that $\nu + 1$ (and thus also $J^E_{\nu+1}$) is always contained in the well-founded core of the ultrapower by E_ν. This follows from the fact that any $\gamma < \nu$ is in the ultrapower represented by the function f^*_γ, where

$$f^*_\gamma(\eta) = \mathrm{otp}(f''_\gamma \eta)$$

for a surjection $f_\gamma : \kappa \to \gamma$. Such a representation does not depend on the choice of the surjection f: if g is another surjection of κ onto γ, then $f''\eta = g''\eta$ on a closed unbounded set (this follows by the standard closure argument), so $[f^*] = [g^*]$ in the ultrapower. Thus, we can take the surjection f_γ to be the $<_E$-least one. The same closure argument yields:

$$\begin{array}{lll} \beta < \gamma \longrightarrow & f^*_\beta(\eta) < f^*_\gamma(\eta) & \text{for c.u.b. many } \eta; \\ \beta = \gamma \longrightarrow & f^*_\beta(\eta) = f^*_\gamma(\eta) & \text{for c.u.b. many } \eta. \end{array}$$

It follows that $\beta < \gamma$ iff $[f^*_\beta] < [f^*_\gamma]$ and $\beta = \gamma$ iff $[f^*_\beta] = [f^*_\gamma]$ in the ultrapower.

We now prove that any $[f] < [f^*_\gamma]$ in the ultrapower can be represented by some $[f^*_\beta]$. Since $f(\eta) < \mathrm{otp}(f''_\gamma \eta)$ on a set $x \in E_\nu$, we can let $\bar\eta < \eta$ be such that $f(\eta) = \mathrm{otp}[f_\gamma(\bar\eta) \cap f''_\gamma \eta]$. We have obtained a regressive function on x; by the normality of E_ν, there is an $y \in E_\nu$, $y \subset x$ such that $\bar\eta$ has a constant value, say η_0, for every $\eta \in y$. Let $\beta = f_\gamma(\eta_0)$. Then $f(\eta) = \mathrm{otp}(\beta \cap f''_\gamma \eta)$ for every $\eta \in y$ and it follows easily that $f(\eta) = \tilde f^*(\eta)$ on a set from E_ν, where $\tilde f : \kappa \to \beta$ is a surjection defined by

$$f(\eta) = \begin{cases} f_\gamma(\eta) & \text{if } f_\gamma(\eta) < \beta \\ \min \mathrm{rng}(f_\gamma) & \text{otherwise.} \end{cases}$$

Thus, $[f] = [f^*_\beta]$ by the above remarks.

This shows that $\nu \subset \mathrm{wfcore}(N)$. To see that ν is also an element of the well-founded core, it suffices to observe that $\nu = \kappa^{+N}$ or, more precisely, if $[f] < [\mathrm{id}^+]$ then $[f] = [f_\gamma]$ in the ultrapower, where $\gamma < \nu$ (here $\mathrm{id}^+ : \kappa \to \kappa$ is defined by $\mathrm{id}^+(\eta) = \eta^+$). But if this failed, N would contain a surjection of κ onto $\mathcal{P}(\kappa) \cap S^{E'}_\nu = \mathcal{P}(\kappa) \cap S^E_\nu$ and using a diagonalization argument we would get a contradiction with the weak amenability of E_ν exactly as above.

By the above observation, it suffices to express the statements $E' \restriction \nu = E \restriction \nu$ and $E'_\nu = \varnothing$ in a Q-fashion over M. This can be done using the Łos theorem. Thus, $x \in E'_\xi$ iff $\langle f^*_\xi, g_x \rangle \in^N E'$, where g_x represents x in the ultrapower, i.e. $g_x(\eta) = x \cap \eta$. The former equality is equivalent to

$$\{\eta;\ \langle f^*_\xi(\eta), x \cap \eta \rangle \in E\} \in E_\nu \longleftrightarrow \langle \xi, x \rangle \in E$$

and the left side of this equivalence can be expressed in a Π_1 fashion as

$$(\forall \xi, \kappa < \nu, x, f, z,)\psi(\xi, \kappa, \nu, x, f, z),$$

where $\psi(\xi, \kappa, \nu, x, f, z)$ is the conjunction of the following formulae:

- $\langle \nu, \kappa \rangle \in E \;\&\; f = f_\xi^*$;
- $z = \{\eta < \kappa;\ \langle f(\eta), x \cap \eta \rangle \in E\} \longrightarrow (\langle \nu, z \rangle \in E \longleftrightarrow \langle \xi, x \rangle \in E)$.

Similarly, $E'_\nu = \varnothing$ is equivalent to $E'_{\mathrm{id}^+} = \varnothing$, which can be rewritten (using the Łos theorem) as

$$(\forall \kappa < \nu, z)(\langle \nu, \kappa \rangle \in E \;\&\; z = \{\eta < \kappa;\ \langle \eta^+, \eta \rangle \notin E\} \longrightarrow \langle \nu, z \rangle \in E).$$

Clearly any premouse M satisfies the above statements; to make sure that the clauses a) – c) follow from these statements, we add an additional technical statement (which holds automatically in any premouse) and which guarantees that $f_\xi^* \in S_\nu^E$ whenever $\xi < \nu$ and $E_\nu \neq \varnothing$:

$$E_\nu \neq \varnothing \longrightarrow [(\forall \xi < \nu)(\exists \zeta < \nu)(\xi \leq \zeta \;\&\; S_\zeta^E \models \mathrm{ZFC}^-)].$$

Note that all the above statements are in fact Π_1. For the last statement, to express that S_ζ^E is a ZFC^- model we need an arithmetization of the language of ZFC^-, which can be done uniformly in a first order (in fact Π_1) fashion over $S_{\zeta+1}^E$ (we shall present an argument of the sort below). This guarantees that the statement in question is Π_1.

Finally we have to express clause d) as a Q-statement over M. There is a recursive list of first order sentences $\langle \varphi_n;\ n \in \omega \rangle$ such that given an acceptable structure $P = \langle J_\beta^A, B \rangle$, P is sound iff $(\forall n \in \omega)(P \models \varphi_n)$. This sequence of sentences is an element of J_2. We only consider formulae consisting of variables $\langle v_i;\ i \in \omega \rangle$, symbols $\in, =, \neg$, $\&, \exists$, and symbols for A, B. By a slight abuse of notation, we shall simply write A, B (thus, we won't formally distinguish between relations and the corresponding symbols). This enables us to talk about acceptable J-structures. For $x : \omega \longrightarrow J_\beta^A$, $x \in J_\beta^A$ and a formula $\varphi(v_{i_1}, \dots, v_{i_n})$ we say that $\langle J_\beta^A, B \rangle \models \varphi[x]$ iff $\langle J_\beta^A, B \rangle \models \varphi(x(i_1), \dots, x(i_n))$. Recall that the statement "φ is a formula" is $\Delta_1(J_1)$. Granted that $P \in M$, we define a universal satisfaction relation R as follows

$R(\varphi, a, A, B, x, h) \longleftrightarrow$
- φ is a formula
- $A \subset a$, $B \subset a$, $x \in a \cap {}^\omega a$ and $h \in \{0, 1\}$
- $\langle a, A, B \rangle$ is the J-structure $\langle J_\beta^A, B \rangle$
- $h = 0 \longleftrightarrow \langle J_\beta^A, B \rangle \models \varphi[x]$.

All of these clauses except the last one can be expressed in a Σ_1-fashion over J_α^E in a straightforward way; the last clause can be, too, by formalizing the induction on the complexity of φ:

$(\exists g)(\exists f)(\exists n \in \omega)$

- $f : n \longrightarrow$ Formulae
- $g : n \times (a \cap {}^{\omega}a) \longrightarrow \{0, 1\}$
- $f(n-1) = \varphi$ and $g(n-1) = h$

<u>and</u> $(\forall i < n)$

- $f(i)$ is atomic <u>or</u>
- $(\exists i_1, i_2 < i)(f(i) = \text{"}f(i_1)\ \&\ f(i_2)\text{"})$ <u>or</u>
- $(\exists i_1 < i)(f(i) = \text{"}\neg f(i_1)\text{"})$ <u>or</u>
- $(\exists i_1 < i)(\exists j \in \omega)(f(i) = \text{"}(\exists v_j) f(i_1)\text{"})$

<u>and</u> $(\forall i < n)(\forall i_1, i_2 < i)(\forall j_1, j_2 \in \omega)$

1. $f(i) = \text{"}v_{j_1} \in v_{j_2}\text{"}$
$$\longrightarrow [(g(i,x) = 0\ \&\ x(j_1) \in x(j_2)) \vee (g(i,x) = 1\ \&\ x(j_1) \notin x(j_2))]$$
2. $f(i) = \text{"}v_{j_1} = v_{j_2}\text{"}$
$$\longrightarrow [(g(i,x) = 0\ \&\ x(j_1) = x(j_2))(g(i,x) = 1\ \&\ x(j_1) \neq x(j_2))]$$
3. $f(i) = \text{"}A(v_{j_1})\text{"} \longrightarrow (g(i,x) = 0 \longleftrightarrow x(j_1) \in A)$
4. $f(i) = \text{"}B(v_{j_1})\text{"} \longrightarrow (g(i,x) = 0 \longleftrightarrow x(j_1) \in B)$
5. $f(i) = \text{"}f(i_1)\ \&\ f(i_2)\text{"} \longrightarrow g(i,x) = g(i_1,x) \cup g(i_2,x)$
6. $f(i) = \text{"}\neg f(i_1)\text{"} \longrightarrow g(i,x) = 1 - g(i_1,x))$
7. $f(i) = \text{"}(\exists v_{j_1}) f(i_1)\text{"} \longrightarrow g(i,x) = \bigcap_{z \in a} g(i_1, s(x, j_1, z))$

where s is defined by

$$s(x, j, z) = x - \{\langle j, x(j)\rangle\} \cup \{\langle j, z\rangle\}.$$

This long statement is Σ_1. Moreover, since any finite sequence of formulae is an element of J_1, we can restrict the quantifier for f to J_1. Hence, if we denote the above formula by Φ, then $\Phi(\varphi, |J^A_\beta|, A, B, x, h)$ is of the form

$$(\exists g)\Phi_0(g, \varphi, |J^A_\beta|, A, B, x, h)$$

where Φ_0 is a Σ_0-formula. Let $\langle J^A_\beta, B\rangle \in J^E_\alpha$, $x \in J^A_\beta$, $x : \omega \to J^A_\beta$, $h \in \{0, 1\}$ and φ be a first order formula. Then:

- $\langle J^A_\beta, B\rangle \models \varphi[x]$ iff $J^E_\alpha \models \Phi(\varphi, |J^A_\beta|, A, B, x, 0)$
- $\langle J^A_\beta, B\rangle \models \neg\varphi[x]$ iff $J^E_\alpha \models \Phi(\varphi, |J^A_\beta|, A, B, x, 1)$
- If $\langle J^A_\beta, B\rangle \models \varphi[x]$ and $\gamma < \omega\alpha$ is such that $\langle J^A_\beta, B\rangle \in S^E_\gamma$, then $S^E_\gamma \models \Phi(\varphi, |J^A_\beta|, A, B, x, 0)$

- If $S^E_{\gamma_1} \models \Phi(\varphi, |J^A_\beta|, A, B, x, h)$ and $\gamma_1 < \gamma_2 \leq \omega\alpha$, then $S^E_{\gamma_2} \models \Phi(\varphi, |J^A_\beta|, A, B, x, h)$.

This means that $\langle J^A_\beta, B\rangle \models \varphi[x]$ iff for γ such that $\langle J^A_\beta, B\rangle \in S^E_\gamma$,

$$S^E_\gamma \models (\forall z)[\Phi(\varphi, |J^A_\beta|, A, B, x, z) \longrightarrow z = 0].$$

Consequently, we can express the statement "$(\forall\nu)(\forall n \in \omega)(\langle J^E_\nu, E_\nu\rangle \models \varphi_n)$" as follows

$$(\forall\gamma)\Big(S^E_\gamma \models (\forall\nu)(\forall n \in \omega)(\forall z)(\Phi(\varphi_n, |J^E_\nu|, E \restriction \nu, E_\nu, c_0, z) \to z = 0)\Big)$$

where $c_0 : \omega \to \mathbf{On}$, $c_0(n) = 0$ for all n. This is a Π_1-statement over J^E_α.

The above calculations yield a finite list of sentences such that if Ψ_0 is their conjunction, the statement of the lemma holds for any transitive structure M of the form $\langle |M|, E\rangle$, i.e. M is a premouse iff $M \models \Psi_0$. This takes care of passive premice. Otherwise we have to add a finite list of sentences expressing c) and d) for the top measure. But these sentences are all Π_1 (see the above calculations) except for the technical statement requiring that S^E_ζ is a ZFC$^-$ model for cofinally many ζ's, which is expressible in a Q-fashion. □(Lemma 4.1.2)

As an immediate corollary we obtain:

Lemma 4.1.3. *Let $\bar{M}, M$ be transitive and $\sigma : \bar{M} \to M$.*

a) *If M is a premouse and σ is Σ_1, then $\bar{M}$ is a premouse.*

b) *If $\bar{M}$ is a premouse and σ is Σ_2 or Σ_0 cofinal, then M is a premouse.*

c) *If M is a premouse, σ is Σ_0 and $\bar{M}$ is a J-structure without the top predicate, then $\bar{M}$ is a premouse.*

d) *If M is a premouse and F is a measure over M, then* $\mathrm{Ult}^k(M, F)$ *is a premouse for every $k \in \omega$.*

e) *If M is a premouse and F is a measure over M, then* $\mathrm{Ult}^*(M, F)$ *is a premouse.*

Proof. The only nontrivial clause is c). Let $\bar{\nu} = \mathrm{ht}(\bar{M})$ and $\omega\tilde{\nu} = \sup(\sigma''\omega\bar{\nu})$. Set $E = E^M$. Then $\sigma : \bar{M} \to J^E_{\tilde{\nu}}$ cofinally. By Lemma 4.1.1, $J^E_{\tilde{\nu}}$ is a premouse. By a), $\bar{M}$ is a premouse. □(Lemma 4.1.3)

4.2 Iterations

Given two premice M, N, we would like to compare them, i.e. find a sensible way of deciding which of them is "larger". In general, the sequences E^M, E^N are completely different, so neither M nor N has to be a lengthening of the other. If ν is the least point of disagreement between E^M, E^N, form the ultrapowers $\mathrm{Ult}(M, E^M_\nu)$ and $\mathrm{Ult}(N, E^N_\nu)$ (of course, we form the ultrapower only if ν indexes a measure in the respective premouse; if say E^M_ν is partial, we form the ultrapower of the maximal initial segment over which is E^M_ν total). After forming the ultrapowers, the measures E^M_ν, E^N_ν vanish, so by the coherency condition, the resulting premice agree up to $\nu + 1$. One might attempt to repeat forming ultrapowers in this way in order to get rid of all disagreements between the two measure sequences. This is really possible and the crucial point here is the ability to iterate the process of forming the ultrapowers.

Definition. Let M be a premouse. An *iteration* of M of *length* θ with *indices* $\langle\langle\nu_i, \alpha_i\rangle;\ i + 1 < \theta\rangle$ is a sequence $\langle M_i;\ i < \theta\rangle$ of premice together with a sequence of *iteration maps* $\langle\pi_{ij};\ i \leq j < \theta\rangle$ satisfying:

a) $M_0 = M$;

b) the maps π_{ij} commute;

c) $\nu_i \leq \alpha_i \leq \mathrm{ht}(M_i)$;

d) if $E^{M_i}_{\omega\nu_i} = \varnothing$, then $M_{i+1} = M_i||\alpha_i$ and $\pi_{i,i+1} = \mathrm{id} \restriction (M_i||\alpha_i)$;

e) if $E^{M_i}_{\omega\nu_i} \neq \varnothing$, then $E^{M_i}_{\nu_i}$ is a measure *on* $M_i||\alpha_i$ (i.e. $E^{M_i}_{\nu_i}$ measures all subsets of its critical point which are elements of $M_i||\alpha$) and

$$\pi_{i,i+1} : M_i||\alpha_i \underset{E^{M_i}_{\nu_i}}{\overset{*}{\longrightarrow}} M_{i+1};$$

f) $\{i < \theta;\ \omega\alpha_i \in M_i\}$ is finite;

g) if λ is a limit ordinal, then

$$\langle M_\lambda, \langle\pi_{i\lambda};\ i < \lambda\rangle\rangle = \lim_{i \leq j < \lambda} (M_i, \pi_{ij}).$$

Given a $k \in \omega$, a k-iteration is defined similarly except that in clause e), we form k-ultrapowers instead of $*$-ultrapowers if $\alpha_j = \mathrm{ht}(M_j)$ for all $j \leq i$, and $*$-iterations otherwise, i.e. if $\alpha_j < \mathrm{ht}(M_j)$ for some $j \leq i$.

Note that iteration maps are in general partial since π_{ij} cannot be defined everywhere on M_i if $\alpha_k < \mathrm{ht}(M_k)$ for some k such that $i \leq k < j$. If $\alpha_k = \mathrm{ht}(M_k)$ for all such k, then π_{ij} is total. For $i < \theta$, the premouse M_i is called the *i-th iterate* and we set

- $\kappa_i \simeq \mathrm{cr}(E^{M_i}_{\nu_i}) \simeq \mathrm{cr}(\pi_{i,i+1})$.

The indices i for which $\omega\alpha_i \in M_i$ are called *truncation points* and the process of replacing M_i by $M_i\|\alpha_i$ is called a *truncation* of M_i. Thus, in a k-iteration we form k-ultrapowers up to the first truncation point and $*$-ultrapowers from the first truncation point on.

Since the measures used in the iterations are internal (i.e. each $E_{\nu_i}^{M_i}$ is rudimentary over M_i), each $E_{\nu_i}^{M_i}$ is close to M_i and the corresponding ultrapower maps are Σ^*-preserving. It follows that the iteration maps are Σ^*-preserving (provided that they are total).

Definition. An iteration $\Im$ which does not contain any truncation (i.e. $\alpha_i = \mathrm{ht}(M_i)$ whenever $i+1 < \mathrm{length}(\Im)$) is called *simple*.

Definition. An iteration of M is called *standard* iff

a) if $E_{\omega\nu_i}^{M_i} = \varnothing$, then $\alpha_i = \mathrm{ht}(M_i)$;

b) if $E_{\omega\nu_i}^{M_i} \neq \varnothing$, then $\alpha_i =$ the maximal α such that $E_{\nu_i}^{M_i}$ is an measure *on* $M_i||\alpha$

for every $i < \theta$. For a standard iteration we consider $\langle \nu_i;\ i+1 < \theta \rangle$ to be the sequence of iteration indices.

Thus, if i is a truncation point in a standard iteration and α_i is the height of the corresponding truncate, then there is a subset of κ_i in $M_i||(\alpha_i+1)$ which is not measured by $E_{\nu_i}^{M_i}$, i.e. which does not belong to $M_i||\alpha_i$. Such a set is definable over $M_i||\alpha_i$ and, since $M_i||\alpha_i$ is sound, also $*$-definable. It follows that $\omega\varrho_{M_i||\alpha_i}^{\omega} \leq \kappa_i$.

Definition. An iteration $\Im$ of M is *normal* iff $\Im$ is standard and $\nu_i < \nu_j$ whenever $i < j$.

Normal iterations arise in the comparison process. Standard iterations can be considered to be concatenations of normal iterations. Although normal iterations will be our main object of interest, in order to develop most of their useful properties we have to consider the class of all standard iterations.

Definition. A *degenerate iteration* of a premouse M is a sequence $\Im$ of premice $\langle M_i \rangle_i$ of length θ such that

a) each proper initial segment of $\Im$ is an iteration;

b) $\{i < \theta;\ \alpha_i < \mathrm{ht}(M_i)\}$ is infinite (and hence cofinal in θ).

Definition. A premouse M is *iterable* iff there are no degenerate iterations of M and for every iteration $\Im$ of M can be *continued* , i.e. setting $\theta = \mathrm{length}(\Im)$, the following holds:

a) if $\theta = i+1$ for some i, then for every $\nu \leq \mathrm{ht}(M_i)$ and every α such that $E_{\nu}^{M_i}$ is a measure on $M_i||\alpha$, the ultrapower $\mathrm{Ult}^*(M_i||\alpha, E_{\nu}^{M_i})$ exists;

b) if θ is a limit, then the corresponding direct limit is well-founded.

The notions *normally iterable* and *k-iterable* are defined in the obvious way.

Definition. A premouse M is called *mouse* iff M is iterable.

It is often useful to consider a weaker notion of iterability, where one is only allowed to apply measures with sufficiently large critical points, resp. indices.

Definition. Let α be an ordinal. We say that an iteration $\mathfrak{I}$ of a premouse M is

a) *above* α iff $\kappa_i \geq \alpha$ for all $i < \text{length}(\mathfrak{I})$

b) *beyond* α iff $\nu_i \geq \alpha$ for all $i < \text{length}(\mathfrak{I})$.

The notions of iterability, normal iterability and k-iterability above (beyond) α as well as further variations are defined in the obvious way. In the rest of this section we state some simple facts about iterations.

Lemma 4.2.1. *Let $\mathfrak{I} = \langle M_i, \pi_{ij};\ i \leq j < \theta \rangle$ be a normal (k-)iteration. Then*

a) $M_j || \nu_i = \langle J_{\nu_i}^{E^{M_j}}, \varnothing \rangle$ *whenever* $i < j$;

b) $\nu_i < \kappa_j$ *whenever* $i < j$ *and* $E_{\nu_i}^{M_i} \neq \varnothing \neq E_{\nu_j}^{M_j}$;

c) κ_i *is inaccessible in* $M_i || \alpha_i$ *and in* M_j *whenever* $i < j$;

d) $\mathcal{P}(\kappa_i) \cap M_j = \mathcal{P}(\kappa_i) \cap (M_i || \alpha_i)$ *whenever* $i < j$;

e) $\kappa_i^{+M_j} = \kappa_i^{+M_i || \alpha_i} = \nu_i$ *whenever* $i < j$;

f) $\text{cr}(E_{\text{top}}^{M_j}) > \kappa_i$ *whenever* $i < j$.

Proof. We proceed by induction. Suppose $\mu < \theta$ and the lemma holds for all i, j such that $i < j < \mu$. Suppose first that μ is a successor; say $\mu = h + 1$. If $E_{\nu_h}^{M_h} = \varnothing$, then $M_\mu = M_h$ and all clauses except b) follow trivially. To see b), suppose $E_{\nu_\mu}^{M_\mu} \neq \varnothing \neq E_{\nu_i}^{M_i}$ for an $i < \mu$. Then both κ_μ, κ_i are inaccessible in $M_\mu || \nu_\mu$ and $\nu_i = \kappa_i^{+M_\mu || \nu_\mu}$. This is automatic for κ_μ; for κ_i it follows from the induction hypothesis. It follows that $\nu_i \leq \kappa_\mu$ since κ_μ is the largest cardinal in $M_\mu || \nu_\mu$. But ν_μ is a successor cardinal in $M_\mu || \nu_\mu$, so the strict inequality must hold. Now let $E_{\nu_h}^{M_h} \neq \varnothing$. Then

$$\pi_{h\mu} : M_h || \alpha_h \xrightarrow[E_{\nu_h}^{M_h}]{*} M_\mu.$$

Clause a) follows immediately from the definition of premouse, c) – e) follow from the weak amenability of $E_{\nu_h}^{M_h}$ and f) follows from the fact that $\text{cr}(E_{\alpha_h}^{M_h}) \geq \kappa_h$ and $\pi_{h\mu}(\kappa_h) > \nu_h$. Finally, b) follows by an argument similar to that above.

Now suppose that μ is a limit. Let $i_0 < \mu$ be such that there are no truncation points between i_0 and μ. Suppose w.l.o.g. that $E_{\nu_i}^{M_i} \neq \varnothing$ for cofinally many $i < \mu$. Pick such an i. By the induction hypothesis, $\text{cr}(\pi_{i+1,\mu}) > \nu_i$. Hence $\pi_{i+1,\mu} \upharpoonright (\nu_i + 1) = \text{id}$ and a) follows immediately. As $\pi_{i+1,\mu}$ is Σ_1-preserving, we have c) – f). Finally, b) follows by the same argument as before. □(Lemma 4.2.1)

Lemma 4.2.2. *Let $\mathfrak{I}$ be a simple normal iteration of M and $\kappa = \kappa_0$. Then for every $i < \text{length}(\mathfrak{I})$,*

a) $\mathcal{P}(\kappa) \cap M = \mathcal{P}(\kappa) \cap M_i$;

b) $\omega\varrho^n_{M_i} = \omega\varrho^n_M$ *whenever* $\omega\varrho^n_M \leq \kappa$;

c) $\mathcal{P}(\kappa) \cap \mathbf{\Sigma}^{(n)}_1(M) = \mathcal{P}(\kappa) \cap \mathbf{\Sigma}^{(n)}_1(M_i)$ *for every* $n \in \omega$;

d) $\mathcal{P}(\kappa) \cap \mathbf{\Sigma}^*(M) = \mathcal{P}(\kappa) \cap \mathbf{\Sigma}^*(M_i)$.

Proof. By induction on i. Assume w.l.o.g. that $E^{M_h}_{\nu_h} \neq \varnothing$ for all h. For successor i, the induction step follows immediately from Lemmata 3.3.1 and 3.3.6. From now on suppose that i is a limit and that the lemma holds for $\iota < i$. In particular, clauses a) – d) hold for $n = 1$.

To see a), it suffices to show that $\mathcal{P}(\kappa) \cap M_i = \mathcal{P}(\kappa) \cap M_1$. Given any $x \in M_i$ such that $x \subset \kappa$, there is an $\iota < i$ such that $x = \pi_{\iota,i}(x_\iota)$ for some $x_\iota \in M_\iota$. But since $\text{cr}(\pi_{\iota,i}) > \kappa$, we conclude that $\pi_{\iota,i}(x_\iota) = x_\iota$, so $x = x_\iota \in M_\iota$. By the induction hypothesis, $x \in M$. This proves the inclusion $\supset$, the converse is obvious.

To see b), it suffices to prove that $\omega\varrho^n_{M_i} = \omega\varrho^n_{M_1}$ for n as above. We use the fact that $\pi_{1,i}$ is Σ^*-preserving and that its critical point is larger than κ.

Now notice that, given a $\xi < \kappa_1$, we have $\xi < \omega\varrho^n_{M_1}$ iff $M_1 \models (\exists\xi^n)(\xi = \xi^n)$ which, by the above properties of $\pi_{1,i}$, is equivalent to $M_i \models (\exists\xi^n)(\xi = \xi^n)$ and thus to $\xi < \omega\varrho^n_{M_i}$. It follows easily that $\omega\varrho^n_{M_i} = \omega\varrho^n_{M_1}$, and so $\omega\varrho^n_{M_i} = \omega\varrho^n_M$.

Finally we prove c); d) is an obvious consequence of c). Again, the inclusion $\subset$ is obvious. To see the converse, let A_i be $\Sigma^{(n)}_1(M_i)$ in some p_i, let $\iota < i$ and $p_\iota \in M_\iota$ be such that $\pi_{\iota,i}(p_\iota) = p_i$ and let A_ι be $\Sigma^{(n)}_1(M_\iota)$ in p_ι by the same definition as A_i. Then $A_i \cap \kappa = A_\iota \cap \kappa \in \mathcal{P}(\kappa) \cap \mathbf{\Sigma}^{(n)}_1(M_\iota) = \mathcal{P}(\kappa) \cap \mathbf{\Sigma}^{(n)}_1(M)$. □(Lemma 4.2.2)

Note that the conclusions of the above lemma also hold for any simple iteration of M above κ and a straightforward reformulation of the lemma also holds for k-iterations. Furthermore, if $\mathfrak{I}$ is above $\omega\varrho^m_M$, then a) – d) hold with $\omega\varrho^m_M$ in place of κ.

Lemma 4.2.3. *Let $\mathfrak{I}$ be a normal (k-)iteration of M of length θ, $i < \theta$ and $E^{M_i}_{\nu_i} \neq \emptyset$. Then*

$$x \in E^{M_i}_{\nu_i} \longleftrightarrow \kappa_i \in \pi_{i,\theta}(x)$$

for $x \in \mathcal{P}(\kappa_i) \cap M_i$.

Proof. By Lemma 4.2.1 clause b), $\text{cr}(\pi_{i+1,\theta}) > \nu_i$. Hence, given $x \subset \kappa_i$,

$$x \in E^{M_i}_{\nu_i} \longleftrightarrow \kappa_i \in \pi_{i,i+1}(x) \longleftrightarrow \kappa_i = \pi_{i+1,\theta}(\kappa_i) \in \pi_{i+1,\theta} \circ \pi_{i,i+1}(x) = \pi_{i,\theta}(x).$$

□(Lemma 4.2.3)

Lemma 4.2.4. *Let $\langle M_i, \pi_{ij};\ i \leq j < \theta + 1\rangle$ be a simple iteration of M and $x \in M_\theta$. Then:*

a) $x = \pi_{0,\theta}(f)(\prec\kappa_{i_1},\ldots,\kappa_{i_\ell}\succ)$ *for some* $f \in \Gamma(\kappa_0, M_0)$, $\ell \in \omega$ *and ordinals* $i_1,\ldots,i_\ell < \theta$.

b) *If the iteration is normal,* $x = \pi_{0,\theta}(f)(\alpha)$ *for some* $f \in \Gamma(\kappa_0, M_0)$ *and an ordinal* $\alpha < \nu = \sup\{\nu_i;\ i < \theta\}$.

A straightforward reformulation also holds for k-iterations.

Proof. Clause a) follows by a straightforward induction on i using the fact that elements of an ultrapower of any N are of the form $\pi(f)(\kappa)$, where π is the corresponding ultrapower map, $f \in \Gamma(\kappa, N)$ and $\kappa = \mathrm{cr}(\pi)$. Clause b) follows from a) and the facts that $\pi_{i+1,\theta}(\kappa_i) = \kappa_i$ and ν (as defined above) is primitive recursively closed, so every Gödel tuple of ordinals can be replaced by a single ordinal. □(Lemma 4.2.4)

Lemma 4.2.5. *Let $\Im$ be a simple iteration of M resulting in M′ and π be the associated iteration map. Then* $\pi'' P^*_M \subset P^*_{M'}$. *An analogous conclusion holds for* P^n_M *in place of* P^*_M.

Proof. We proceed by induction on the length of $\Im$ using Corollary 3.3.4 at successor steps. □(Lemma 4.2.5

4.3 Copying and the Dodd–Jensen Lemma

Given a sufficiently preserving map $\sigma : \bar{M} \to M$, any iteration of $\bar{M}$ can be turned into an iteration of M; this process is called the *copying process.* An important consequence of this fact is the Dodd–Jensen Lemma, which asserts that mice cannot have iterates which are both simple and nonsimple and that iteration maps are "minimal" in the obvious sense.

Lemma 4.3.1. *Let $\bar{M}$ be a premouse, M be a mouse and* $\sigma : \bar{M} \to M$ *be* Σ^**-preserving. Let* $\bar{\Im} = \langle \bar{M}_i, \bar{\pi}_{ij};\ i \le j < \theta\rangle$ *be an iteration of $\bar{M}$ with indices* $\langle\langle \bar{\nu}_i, \bar{\alpha}_i\rangle;\ i < \theta\rangle$. *Then there is a unique iteration* $\Im = \langle M_i, \pi_{ij};\ i \le j < \theta\rangle$ *of M and a unique sequence of maps* $\langle \sigma_i;\ i < \theta\rangle$ *such that*

a) $\sigma_0 = \sigma$;

b) $\sigma_i : \bar{M}_i \underset{\Sigma^*}{\longrightarrow} M_i$;

c) $\sigma_j \restriction \bar{\nu}_i = \sigma_i \restriction \bar{\nu}_i$ *for all* $i \le j < \theta$;

d) *for every $i \leq j < \theta$, the following diagram commutes:*

$$\begin{array}{ccc} M_i & \xrightarrow{\pi_{ij}} & M_j \\ \uparrow{\scriptstyle \sigma_i} & & \uparrow{\scriptstyle \sigma_j} \\ \bar{M}_i & \xrightarrow{\bar{\pi}_{ij}} & \bar{M}_j \end{array}$$

The iteration $\Im$ is called the copy *of $\bar{\Im}$ via σ and the maps σ_i are called* copy maps.

Proof. We shall iteratively apply Lemma 3.4.5. Set $\sigma_0 = \sigma$. Suppose that $\langle M_i, \pi_{ij};\ i \leq j < \lambda\rangle$, $\langle \sigma_i;\ i < \lambda\rangle$ are defined and that $\langle\langle \bar{\nu}_i, \bar{\alpha}_i\rangle;\ i < \theta\rangle$ are the iteration indices of $\bar{\Im}$.
Case 1. $\lambda = j + 1$ for some j.

Set $\nu_j = \sigma_j(\bar{\nu}_j)$ and

$$\alpha_j = \begin{cases} \sigma_j(\bar{\alpha}_j) & \text{if } j \text{ is a truncation point} \\ \mathrm{ht}(M_j) & \text{otherwise.} \end{cases}$$

If j is a truncation point, set $\sigma_j^* = \sigma_j \upharpoonright (\bar{M}_j||\bar{\alpha}_j)$. Then $\sigma_j^* : \bar{M}_j||\bar{\alpha}_j \to M_j||\alpha_j$ is fully elementary and therefore Σ^*-preserving and, trivially, $\sigma_j^* \upharpoonright \omega\bar{\alpha}_j = \sigma_j \upharpoonright \omega\bar{\alpha}_j$. Hence $\langle \sigma_j^*, \sigma_j \upharpoonright \bar{\nu}_j\rangle : \bar{M}_j||\bar{\alpha}_j \stackrel{*}{\longrightarrow} M_j||\alpha_j$ and the rest of the assumptions of Lemma 3.4.5 is also satisfied. Consequently, there is a unique Σ^*-preserving map $\sigma_\lambda : \bar{M}_\lambda \to M_\lambda = \mathrm{Ult}^*(M_j||\alpha_j, E_{\nu_j}^{M_j})$ making the above diagram commute. If j is not a truncation point, we do the same construction but with σ_j in place of σ_j^*.
Case 2. λ is a limit.

There is a $j < \lambda$ such that both the iterations from $\bar{M}_j$ to $\bar{M}_\lambda$ and that from M_j to M_λ are simple, so we can define a Σ^*-preserving map $\sigma_\lambda : \bar{M}_\lambda \to M_\lambda$ by $\sigma_\lambda(\bar{\pi}_{i\lambda}(x)) = \pi_{i\lambda} \circ \sigma_i(x)$ for $j < i < \lambda$ and $x \in \bar{M}_i$. □(Lemma 4.3.1)

Lemma 4.3.2. *Let $\sigma : \bar{M} \to M$ be Σ^*-preserving, where $\bar{M}$ is transitive and M is a mouse. Then $\bar{M}$ is a mouse.*

Proof. $\bar{M}$ is a premouse by Lemma 4.1.3. $\bar{M}$ cannot have a degenerate iteration, since the copying process would turn any such iteration into a degenerate iteration of M, which is impossible. Let $\bar{\Im}$ be an iteration of $\bar{M}$ of length θ and $\Im$ be an iteration of M which is the copy of $\bar{\Im}$ via σ. Then $\bar{\Im}$ can be continued in the same way as in the proof of the previous lemma. □(Lemma 4.3.2)

Lemma 4.3.3. *Let M be a countably iterable premouse (i.e. there are no countable degenerate iterations and each countable iteration can be continued). Then M is a mouse.*

Proof. Suppose not. Let $\mathfrak{I} = \langle M_i, \pi_{ij};\ i \leq j < \theta \rangle$ be an iteration of M witnessing that M is not iterable, i.e. an iteration which is either degenerate or else cannot be continued. By the absoluteness of well-foundedness, H_τ thinks that $\mathfrak{I}$ is a witness to the non-iterability of M. Let τ be a regular cardinal large enough that $\mathfrak{I} \in H_\tau$. Let $X \prec H_\tau$ be a countable elementary substructure such that $\mathfrak{I} \in X$ and $\bar{H}$ be the transitive isomorph of X. The *uncollapsing map* (i.e. the inverse to the corresponding collapsing map) $\sigma : \bar{H} \to H$ is fully elementary. Let $\bar{\mathfrak{I}} = \sigma^{-1}(\mathfrak{I})$ and $\bar{M} = \sigma^{-1}(M)$. Then in $\bar{H}$, $\bar{\mathfrak{I}}$ is an iteration of $\bar{M}$ which is either degenerate or else cannot be continued. Again by the absoluteness of well-foundedness, this holds not only in $\bar{H}$, but also in the real world. Since $\bar{\mathfrak{I}}$ is countable, it witnesses that $\bar{M}$ is not countably iterable. On the other hand,

$$\sigma \restriction \bar{M} : \bar{M} \underset{\Sigma^*}{\longrightarrow} M,$$

hence by Lemma 4.3.1 and the fact that M is countably iterable, $\bar{\mathfrak{I}}$ cannot be degenerate and can be continued after all. Contradiction. □(Lemma 4.3.3)

The above proof thus shows that if M is not iterable then we can construct a countable premouse $\bar{M}$ elementarily embeddable into M which is not countably iterable. Hence we have:

Corollary 4.3.4. *Suppose every countable premouse elementarily embeddable into M is countably iterable. Then M is iterable.* □(Corollary 4.3.4)

This means that countable premice provide us with complete information about iterability. Also, iterability of countable premice can be decided inside H_{ω_1} since only countable iterations are to be checked and these are all members of H_{ω_1}.

Lemma 4.3.5. *Let M be a countable premouse. There is a Π_1 formula φ such that*

$$M \text{ is iterable} \longleftrightarrow H_{\omega_1} \models \varphi(M).$$

Proof. It suffices to verify that M is countably iterable, which is equivalent to saying that M is iterable in the sense of H_{ω_1}. If M is iterable, then it is also iterable in the sense of any ZF^- model $H \in H_{\omega_1}$ containing M by the absoluteness of well foundedness (any iteration in the sense of H is really an iteration). On the other hand, if $\mathfrak{I}$ is a countable iteration of M, take the least elementary substructure of H_{ω_1} containing the transitive closure of $\{\mathfrak{I}\}$; the transitive collapse H of this structure is a ZF^- model containing $\mathfrak{I}$, so the possibility of extending $\mathfrak{I}$ can be decided inside H. Clearly $H \in H_{\omega_1}$, so the iterability of M turns out to be equivalent to the statement

$$(\forall H)[(H \text{ transitive } \& \ H \models \mathrm{ZF}^- \ \& \ M \in H) \longrightarrow H \models \text{“}M \text{ is iterable”}].$$

This statement is Π_1; “$H \models \mathrm{ZF}^-$” can be expressed as a Σ_0 statement with the parameter M and the same applies to “($H \models$ “M is iterable”)”. □(Lemma 4.3.5)

The above lemma enables us to compute the complexity of iterability in terms of descriptive set theory. A countable mouse can be encoded into a subset of ω via a bijection $h : \omega \to M$. More precisely, using h we turn M into a structure $\langle D, E, I\rangle$ on ω by setting

$$D = \mathrm{dom}(h) \cap \omega$$
$$E = \{\langle m, n\rangle \in D;\ h(m) \in h(n)\}$$
$$I = \{\langle m, n\rangle \in D;\ h(m) = h(n)\}.$$

This structure can be recursively encoded in a real. Of course, this real depends on the recursive isomorphism between ω and $\omega \times (\omega \times \omega) \times (\omega \times \omega)$ used for the encoding of $\langle D, E, I\rangle$, so from now on we shall keep this isomorphism, say r, fixed.

If a is a real encoding M, we can reconstruct M from a inside any ZF$^-$ model H containing a by collapsing the above structure, which is equal to $r''a$. (r, being recursive, is automatically in H.) So M is iterable iff

$$(\forall H)(\psi_1(H, a) \longrightarrow \psi_2(H, a),$$

where

- $\psi_1(H, a) \longleftrightarrow$ (H transitive & $H \models$ ZF$^-$ & $a \in H$) and
- $\psi_2(H, a) \longleftrightarrow H \models (\exists M)$("$a$ codes M and M is iterable").

holds in H_{ω_1}. This is a Π_1-statement about the real a (a subset of ω) over H_{ω_1}, so it is Π^1_2. By Shoenfield's absoluteness lemma, such a statement is absolute relative to models containing a, so we infer that M is iterable iff M is iterable in any transitive ZF$^-$ model containing ω_1 (as a subset) and a (as an element). We can do even better:

Theorem 4.3.6. *Let H be a transitive ZF$^-$ model containing ω_1 (as a subset) and $M \in H$ be countable in the sense of V. Then M is iterable iff*

$$H \models \text{"}M \text{ is iterable"}.$$

Proof. We show that the iterability of M in the sense of H implies its iterability in the sense of V; the converse is trivial. Suppose that M is iterable in H but not in V. Given a $\mathrm{Col}(\omega, M)$-generic filter G over V, $\omega_1^V = \omega_1^{V[G]}$ as we force with a countable poset. In particular, $V[G] \models (\omega_1 \subset H[G])$. By the remarks preceding this theorem, M is not iterable in $V[G]$ and therefore not in $H[G]$. So for some $p \in \mathrm{Col}(\omega, M)$ we have $H \models (p \Vdash$ "$\check{M}$ is not iterable"); then $H \models$ ("$\check{M}$ is not iterable") by the homogeneity of $\mathrm{Col}(\omega, M)$. Since G was chosen arbitrarily, this statement really holds in V. Suppose n is such that the statement '($\Vdash$ "$\check{M}$ is not iterable")' can be expressed by a Σ_n formula; let X be the Σ_n Skolem hull of $\{M\}$ in H, $\bar{H}$ the transitive collapse of X and σ the associated uncollapsing map. Both $\bar{H}, \sigma$ are elements of H. Let $\bar{M}$ be the preimage of M under σ. The map σ is Σ_n-preserving, so $\bar{H} \models$ ($\Vdash$ "$(\bar{M})\check{}$ is not iterable"). Since $\bar{H}$ is countable in H, there is a $\mathrm{Col}(\omega, \bar{M})$-generic filter $\bar{G}$ over $\bar{H}$ in H. Then $\bar{H}[\bar{G}] \models$ "$\bar{M}$ is not iterable", so there is an iteration of $\bar{M}$ in $\bar{H}[\bar{G}] \subset H$ which cannot be continued. Hence $\bar{M}$ is not iterable in H. But $\bar{M}$ must be iterable in H, as $\sigma \restriction$

$\bar{M} : \bar{M} \to M$ is a fully elementary embedding which is an element of H and, by our assumption, M is iterable in H. Contradiction. □(Theorem 4.3.6)

This proof easily generalizes for mice with overlapping extenders up to one Woodin cardinal, that is, for non-linear iteration trees which pick unique branches. Later we shall give an alternative proof of this theorem based on the so-called "universal" iterations, which goes through even for uncountable M, but whose generalization to non-linear iterations is not straightforward.

The proof of the previous theorem yields a somewhat weaker result for uncountable premice: If $M \in H$ is a premouse, H is a transitive ZFC$^-$-model and $(\mathrm{card}(M)^+)^V \subset H$, then the iterability of M in H already implies the iterability of M in V.

Lemmata on iterability similar to those above also hold for k-iterations, resp. iterations above a certain point; we state two useful results along these lines.

Lemma 4.3.7. *Let $\bar{M}, M$ be premice.*

a) *If M is iterable above $\sigma(\omega\varrho_M^{n+1})$ where*

$$\sigma : \bar{M} \underset{\Sigma_1^{(n)}}{\longrightarrow} M \text{ strongly,}$$

then $\bar{M}$ is iterable above $\omega\varrho_{\bar{M}}^{n+1}$.

b) *If M is normally iterable above $\sigma(\omega\varrho_M^{n+1})$ where*

$$\sigma : \bar{M} \underset{\Sigma_0^{(n)}}{\longrightarrow} M,$$

then $\bar{M}$ is normally iterable above $\omega\varrho_{\bar{M}}^{n+1}$.

Moreover, if $\bar{M}'$ is an iterate, resp. (simple) normal iterate of $\bar{M}$ above $\omega\varrho_{\bar{M}}^{n+1}$, there is an iterate, resp. (simple) normal iterate M' of M above $\omega\varrho_M^{n+1}$ and a map σ' such that

$$\sigma' : \bar{M}' \underset{\Sigma_1^{(n)}}{\longrightarrow} M' \text{ strongly, resp. } \sigma' : \bar{M}' \underset{\Sigma_0^{(m)}}{\longrightarrow} M'$$

where $m \leq n$ is maximal such that all critical points in the iteration from $\bar{M}$ to $\bar{M}'$ are strictly less than the m-th projecta, and $\sigma' \circ \pi_{\bar{M}\bar{M}'} = \pi_{MM'} \circ \sigma$ where $\pi_{\bar{M}\bar{M}'}, \pi_{MM'}$ are the corresponding iteration maps. (In other words, the diagram from the proof of Lemma 4.3.1 commutes).

Proof. It suffices to show that any iteration $\bar{\mathfrak{I}}$ of $\bar{M}$ above $\omega\varrho_{\bar{M}}^{n+1}$ can be successfully copied into an iteration of M above $\sigma(\omega\varrho_{\bar{M}}^{n+1})$ via σ. Notice that $\mathfrak{I}$ is above $\sigma(\omega\varrho_M^{n+1})$, so the copying construction does not break down because of the lack of iterability of M. The only issue is to verify that the copying construction can be carried out. This requires an argument since we now work with weaker maps than before. Note that it is sufficient to consider only simple iterations of $\bar{M}$, since after any truncation, all copy maps are Σ^*-preserving by Lemma 4.3.1. We shall proceed by induction on the length of $\Im$.

At successor steps, define σ_{i+1} exactly as in the proof of Lemma 4.3.1. This is possible, since the induction hypothesis (or the initial assumption on σ, if $i = 0$) guarantees that σ_i is strongly $\Sigma_1^{(n)}$-preserving and $\kappa_i \geq \omega\varrho_{\bar{M}}^{n+1} = \omega\varrho_{\bar{M}_i}^{n+1}$ (see the remark after Lemma 4.2.2) in case a) and σ_i is $\Sigma_0^{(n_i)}$-preserving, where $\omega\varrho_{\bar{M}_i}^{n_i+1} \leq \kappa_i < \omega\varrho_{\bar{M}_i}^{n_i}$ in case b). Then σ_{i+1} is correctly defined and strongly $\Sigma_1^{(n)}$-preserving by Lemma 3.4.8 in the former case and $\Sigma_0^{(n_i)}$-preserving by Lemmata 3.3.1 and 3.4.4 in the latter case.

At limit steps μ define $\sigma_\mu : \bar{M}_\mu \to M_\mu$ in the obvious way. First consider the case a). That σ_μ is $\Sigma_1^{(n)}$-preserving follows by a straightforward computation, we verify the strongness. If $\xi = \sigma_\mu(\bar{\xi}) < \omega\varrho_{M_\mu}^{n+1}$, pick an $i < \mu$ such that $\bar{\xi} \in \operatorname{rng}(\bar{\pi}_{i,\mu})$ and $\xi \in \operatorname{rng}(\pi_{i,\mu})$. Set $\bar{\xi}_i = \bar{\pi}_{i,\mu}^{-1}(\bar{\xi})$, $\xi_i = \pi_{i,\mu}^{-1}(\xi)$. Then $\xi_i = \sigma_i(\bar{\xi}_i)$ and $\xi_i < \omega\varrho_{M_i}^{n+1}$ since $\pi_{i,\mu}$ is Σ^*-preserving and $\bar{\xi}_i < \omega\varrho_{\bar{M}_i}^{n+1}$ since σ_i is strongly $\Sigma_1^{(n)}$-preserving (here we use the induction hypothesis). It follows that $\bar{\xi} < \omega\varrho_{\bar{M}_\mu}^{n+1}$, which completes the proof. In case b), there is a $j < \mu$ and an $m \in \omega$ such that $n_i = m$ for all $i \geq j$. By Lemma 4.2.2, $\omega\varrho_{\bar{M}_j}^{m+1} = \omega\varrho_{\bar{M}_i}^{m+1} = \omega\varrho_{\bar{M}_\mu}^{m+1}$ for all such i. Hence σ_μ is $\Sigma_0^{(m)}$-preserving.

□(Lemma 4.3.7)

Lemma 4.3.8. *Suppose $\bar{M}$, M are premice and $\sigma : \bar{M} \to M$.*

a) *If σ is $\Sigma_1^{(k)}$-preserving and M is m-iterable for some $m \geq k$, then $\bar{M}$ is normally k-iterable. In particular, if M is m-iterable, then M is normally k-iterable.*

b) *If $R_M^k \neq \emptyset \neq R_{\bar{M}}^k$, σ is $\Sigma_1^{(k)}$-preserving and M is k-iterable, then $\bar{M}$ is k-iterable.*

c) *If σ is Σ_0-preserving and M is k-iterable for some $k \leq \omega$, then $\bar{M}$ is 0-iterable. In particular, every mouse is a 0-mouse.*

Proof. For the proof of a), resp. b), we iteratively apply Lemma 3.5.3, resp. Lemma 3.5.4, in the copying construction. Clause c) follows from the fact that the ultrapowers used in 0-iterations are 0-ultrapowers, i.e. one does not need to know anything about the projecta of $\bar{M}$ in order to continue such iterations. For this reason, the copying construction does not break down. □(Lemma 4.3.8)

We are now ready to prove the most important result of this section, the Dodd–Jensen lemma. In practice, one often uses various formulations of the lemma, we state two useful ones below.

Theorem 4.3.9 (Dodd–Jensen Lemma). *Let M be a mouse, $\Im$ be an iteration of M resulting in M' and $\pi : M \to M'$ be the corresponding iteration map. Suppose that there is a Σ^*-preserving map $\sigma : M \to M'$. Then $\Im$ is simple and $\pi(\xi) \leq \sigma(\xi)$ for all $\xi \in M$.*

Proof. We first prove that $\mathfrak{I}$ is simple. Suppose this fails. For $i \in \omega$ we shall construct mice M_i, iterations $\mathfrak{I}_i$ from M_i to M_{i+1} with iteration maps $\pi_{i,i+1}$ and Σ^*-preserving maps $\sigma_i : M_i \to M_{i+1}$ as follows:

- $M_0 = M, M_1 = M', \mathfrak{I}_0 = \mathfrak{I}, \pi_{01} = \pi$ and $\sigma_0 = \sigma$;
- given $M_i, M_{i+1}, \mathfrak{I}_i, \pi_{i,i+1}, \sigma_i$ we set:
  - $\mathfrak{I}_{i+1}$ is an iteration of M_{i+1} which is a copy of $\mathfrak{I}_i$ via σ_i,
  - M_{i+2} is the resulting mouse of $\mathfrak{I}_{i+1}$,
  - $\pi_{i+1,i+2}$ is the iteration map corresponding to $\mathfrak{I}_{i+1}$,
  - $\sigma_{i+1} : M_{i+1} \xrightarrow[\Sigma^*]{} M_{i+2}$ is the unique map determined by the copying construction (as defined in Lemma 4.3.1)

We obtain the following diagram

$$\begin{array}{ccccccccc}
M_1 & \xrightarrow{\pi_{12}} & M_2 & \xrightarrow{\pi_{23}} & M_3 & \xrightarrow{\pi_{34}} & M_4 & \xrightarrow{\pi_{45}} & \cdots \\
\uparrow \sigma_1 & & \uparrow \sigma_2 & & \uparrow \sigma_3 & & \uparrow \sigma_4 & & \\
M_0 & \xrightarrow{\pi_{01}} & M_1 & \xrightarrow{\pi_{12}} & M_2 & \xrightarrow{\pi_{23}} & M_3 & \xrightarrow{\pi_{34}} & \cdots
\end{array}$$

where all iterations $\mathfrak{I}_i$ are non-simple, so the maps π_{ij} are partial whenever $i < j \in \omega$. Let $\mathfrak{I}^*$ be the concatenation of all $\mathfrak{I}_i$, $i \in \omega$. Then $\mathfrak{I}^*$ is a degenerate iteration of $M_0 = M$. Contradiction.

Hence $\mathfrak{I}$ must be simple. Then π is then total and so are the maps π_{ij}. Towards the second part of the theorem, suppose for a contradiction that there is a $\xi \in M_0$ such that $\sigma_0(\xi) < \pi_{01}(\xi)$. By the commutativity of the above diagram,

$$\sigma_i \circ \pi_{i-1,i} = \pi_{i,i+1} \circ \sigma_{i-1} \tag{4.3.1}$$

for all $i > 0$. Furthermore, for $k \in \omega$ we can form the compositions $\sigma_k \circ \sigma_{k-1} \circ \cdots \circ \sigma_0$ since the target structure of σ_i is a domain of σ_{i+1}. An iterative application of (4.3.1) yields

$$\sigma_k \circ \sigma_{k-1} \circ \cdots \circ \sigma_0(\xi) < \sigma_k \circ \sigma_{k-1} \circ \cdots \circ \sigma_1 \circ \pi_{0,1}(\xi) = \pi_{k,k+1} \circ \sigma_{k-1} \circ \cdots \circ \sigma_0(\xi) \tag{4.3.2}$$

since

$$\begin{aligned}
\sigma_k \circ \sigma_{k-1} \circ \cdots \circ \sigma_1 \circ \pi_{0,1}(\xi) &= \sigma_k \circ \sigma_{k-1} \circ \cdots \circ \sigma_2 \circ \pi_{1,2} \circ \sigma_0(\xi) = \\
&= \sigma_k \circ \sigma_{k-1} \circ \cdots \circ \sigma_3 \circ \pi_{2,3} \circ \sigma_1 \circ \sigma_0(\xi) = \cdots = \\
&= \sigma_k \circ \pi_{k-1,k} \circ \sigma_{k-2} \circ \cdots \circ \sigma_0(\xi) = \pi_{k,k+1} \circ \sigma_{k-1} \circ \cdots \circ \sigma_0(\xi).
\end{aligned}$$

Let $\langle M^*, \pi_{i\omega};\ i \in \omega\rangle$ be the direct limit of $\langle M_i, \pi_{ij};\ i \le j < \omega\rangle$. (Clearly M^* exists since M^* is a simple iterate of M and the latter is a mouse). Now apply $\pi_{k+1,\omega}$ to both the leftmost and the rightmost side of (4.3.2); we obtain

$$\pi_{k+1,\omega} \circ \sigma_k \circ \sigma_{k-1} \circ \cdots \circ \sigma_0(\xi) < \pi_{k,\omega} \circ \sigma_{k-1} \circ \cdots \sigma_0(\xi).$$

Thus, $\langle \pi_{k+1,\omega} \circ \sigma_k \circ \cdots \circ \sigma_0(\xi);\ k \in \omega\rangle$ constitutes an infinite decreasing sequence of ordinals. Contradiction. □(Lemma 4.3.9)

Corollary 4.3.10. *Let M be a mouse and M' be an iterate of M. Then M cannot be both a simple and a non-simple iterate of M.* □(Corollary 4.3.10)

Corollary 4.3.11. *Let M' be a simple iterate of a mouse M. Then the iteration map $\pi : M \to M'$ is uniquely determined, i.e. does not depend on the actual iteration.* □(Corollary 4.3.11)

Definition. If M' is a simple iterate of a mouse M, we denote the unique iteration map from M to M' by $\pi_{MM'}$.

Lemma 4.3.12. *Let M, M' be premice and $\sigma : M \to M'$.*

a) *If M is iterable above $\omega\varrho_M^{n+1}$, $\Im$ is an iteration of M above $\omega\varrho_M^{n+1}$ resulting in M' and σ is strongly $\Sigma_1^{(n)}$-preserving, then $\Im$ is simple and $\pi_{MM'}(\xi) \le \sigma(\xi)$ for $\xi \in M$.*

b) *If M' is iterable (above $\omega\varrho_M^{n+1}$), $\Im$ is a normal iteration of M above $\omega\varrho_M^{n+1}$ resulting in M' and σ is $\Sigma_0^{(n)}$-preserving, then $\Im$ is simple.*

Proof. It suffices to verify that we can build the diagram as in the proof of Theorem 4.3.9. Suppose first that the iteration $\Im$ from M to M' is non-simple. By Lemma 4.3.7, the copying process can be carried out up to the first truncation point (for case a) notice that M', being an iterate of M above $\omega\varrho_M^{n+1}$ is iterable above this point). From this point on all of the copying maps are Σ^*-preserving so the copying process does not break down.

Now consider case a) where $\Im$ is simple. Set $M_0 = M$, $M_1 = M'$, $\Im_0 = \Im$ and $\Im_1 =$ the copy of $\Im$ via σ. Then $\Im_1$ an iteration above $\omega\varrho_{M'}^{n+1} = \omega\varrho_M^{n+1}$, hence $\Im_1$ can be copied onto an $\Im_2$ via σ_1 which is the copying map obtained at the end of the copying of $\Im_0$ onto $\Im_1$; this is possible since $\sigma_1 : M_1 \to M_2$ and is strongly $\Sigma_1^{(n)}$-preserving, as follows from Lemma 4.3.7. Hence the requisite diagram can be built in this case as well. □(Lemma 4.3.12)

Corollary 4.3.13. *If M' is iterable (above $\omega\varrho_M^{n+1}$), M' is a normal iterate of M above $\omega\varrho_M^{n+1}$ and σ is $\Sigma_0^{(n)}$-preserving, then M is iterable (above $\omega\varrho_M^{n+1}$).*

Proof. By the previous lemma, clause b), M' is a simple iterate of M. The iteration map $\pi : M \to M'$ is Σ^*-preserving, so the conclusion follows from Lemma 4.3.2 □(Corollary 4.3.13)

Finally we mention a version of these results for k-iterations. However, for our immediate purposes, that is, developing the core model theory up to one strong cardinal, only the instance for $k = 0$ will be of use and there is no need for introducing other sorts of iterations. Only later, when we reach the level of superstrong extenders, will general k-iterations prove to be a useful tool for dealing with various technical obstacles.

Lemma 4.3.14. *Let M, M' be premice and $\sigma : M \to M'$.*

a) *Suppose $R^k_M \neq \varnothing \neq R^k_{M}$ if $k > 0$, σ is $\Sigma_1^{(k)}$-preserving and M is k-iterable. If $\Im$ is a k-iteration of M resulting in M', then $\Im$ is simple. Furthermore, if π is the associated iteration map, then $\pi(\xi) \leq \sigma(\xi)$ for all ordinals $\xi \in M$.*

b) *Suppose σ is $\Sigma_0^{(k)}$-preserving and M' is k-iterable. If $\Im$ is a normal k-iteration of M resulting in M', then $\Im$ is simple.*

c) *Let σ be cardinal preserving and Σ_0-preserving, and M be 0-iterable. If $\Im$ is a 0-iteration of M resulting in M' then $\Im$ is simple. Furthermore, if π is the associated iteration map, then $\pi(\xi) \leq \sigma(\xi)$ for all ordinals $\xi \in M$.*

Proof. We imitate the proof of the Dodd–Jensen Lemma; the main point is to verify that the copying construction can be carried out and that a copy of an iteration $\Im$ is simple iff $\Im$ is simple. To guarantee the latter we pose the extra requirement on cardinal preservation in c). The former then follows from the proof of Lemma 4.3.8. □(Lemma 4.3.14)

Corollary 4.3.15. *Let M be k-iterable and $R^k_M \neq \varnothing$ if $k > 0$.*

a) *If M' is a k-iterate of M, then M' cannot be both a simple and a nonsimple k-iterate of M.*

b) *If M' is a simple k-iterate of M, then the k-iteration map from M to M' is uniquely determined, i.e. does not depend on the actual iteration.*

□Corollary 4.3.15)

4.4 Comparison Process

We now describe a method for comparing premice. When dealing with levels of the constructible universe, we know that they are compatible in the sense that one of them is an initial segment of the other. This is, of course, false when we allow many measures to be on the sequence. Despite this, we still attempt to obtain a pair of premice compatible in the above sense by iterating away the least disagreement in the respective measure sequences. It turns out that this process terminates after sufficiently many steps.

Definition. Let M^0, M^1 be premice. A pair of iterations

$$\begin{aligned}\mathfrak{I}^0 &= \langle M_i^0, \pi_{ij}^0;\ i \leq j < \theta + 1\rangle,\\ \mathfrak{I}^1 &= \langle M_i^1, \pi_{i,j}^1;\ i \leq j < \theta + 1\rangle\end{aligned}$$

is a *coiteration* of M^0, M^1 of length $\theta + 1$ iff

a) $M_0^k = M^k$ for $k = 0, 1$ (i.e. $\mathfrak{I}^k$ is an iteration of M^k).

b) Both iterations $\mathfrak{I}^0$, $\mathfrak{I}^1$ are standard.

c) If $i < \theta + 1$, then ν_i is the least ν satisfying $E_\nu^{M_i^0} \neq E_\nu^{M_i^1}$, provided that such a ν exists.

d) ν_i is defined for every $i < \theta$.

Given $\ell_0, \ell_1 \in \omega$, an (ℓ_0, ℓ_1)-coiteration of M^0, M^1 is defined similarly except that we require $\mathfrak{I}^k$ to be an ℓ_k-iteration for $k = 0, 1$.

The iterations of any coiteration are normal. This can be verified by a simple inductive argument. At successor steps we observe that if the M_i^k agree up to ν_i, then the M_{i+1}^k agree at least up to $\nu_i + 1$ due to the coherency condition. At limit steps μ we use the induction hypothesis: the fact that $\mathfrak{I}^k \restriction \mu$ is normal means that $\kappa_{i+1} > \nu_i$ for all $i < \mu$, so the direct limits M_μ^k agree with M_i^k up to ν_i. It follows that the M_μ^k agree at least up to the supremum of all ν_i, i.e. $\nu_\mu \geq \nu_i$ for all $i < \mu$.

Also note that if ν_i indexes a measure on both sides of the coiteration, then

$$\operatorname{cr}\big(E_{\nu_i}^{M_i^0}\big) = \operatorname{cr}\big(E_{\nu_i}^{M_i^1}\big)$$

since these critical points are the largest cardinals in the respective structures $M^0||\nu_i$, $M^1||\nu_i$. Denote this common critical point by κ_i. It follows that $\nu_i < \kappa_j$ whenever $i < j$.

Given a pair of premice, we can attempt to coiterate them even if neither of them is fully iterable; the crucial point is that they are sufficiently iterable to carry out the coiteration.

Definition. A pair of premice is *coiterable* iff their coiteration does not break down because of lack of iterability on either side.

Definition. Let $\langle \mathfrak{I}^0, \mathfrak{I}^1\rangle$ be a coiteration of M^0, M^1 of length $\theta + 1$. We say that $\langle \mathfrak{I}^0, \mathfrak{I}^1\rangle$ *terminates* iff ν_θ is undefined.

That $\langle \mathfrak{I}^0, \mathfrak{I}^1\rangle$ as above terminates can be equivalently expressed as follows: $M_\theta^0 = M_\theta^1||\alpha$ for some $\alpha \leq \operatorname{ht}(M_\theta^1)$ or vice versa.

Lemma 4.4.1 (Comparison lemma). *Let M^0, M^1 be coiterable premice and θ be a regular cardinal larger than the size of each of them. Then the coiteration of M^0, M^1 terminates below θ (i.e. its length is strictly less than θ).*

Proof. Suppose the contrary, i.e. that the iterations $\langle \Im^0, \Im^1 \rangle$ of M^0, M^1 are of length at least θ. Pick a regular $\tau > \theta$. Let X be an elementary substructure of H_τ such that

- $\mathrm{card}(X) < \theta$;
- $\Im^0, \Im^1 \in X$;
- $X \cap H_\theta$ is transitive.

This can be easily arranged since θ is regular. Let $\bar{H}$ be the transitive isomorph of X and let $\sigma : \bar{H} \to H$ be the uncollapsing map. Set $\bar{\theta} = \mathrm{cr}(\sigma) = X \cap \theta$ and $\bar{\Im}^k = \sigma^{-1}(\Im^k)$ for $k = 0, 1$. Then $\sigma(\bar{\theta}) = \theta$. For $i < \bar{\theta}$ we have $M_i^k \in X$ since M_i^k is definable from $\Im^k$ and i, which are elements of X. The same also applies to π_{ij}^k, where $\pi_{ij}^k : M_i^k \to M_j^k$ is the iteration map from M_i^k to M_j^k $(k = 0, 1)$ and $i \le j < \bar{\theta}$. Hence we can set $\bar{M}_i^k = \sigma^{-1}(M_i^k)$, $\bar{\pi}_{ij}^k = \sigma^{-1}(\pi_{ij}^k)$. Assume w.l.o.g. that $k = 0$. The situation is described by the following diagram:

$$
\begin{array}{ccccccc}
M_0^0 & \xrightarrow{\pi_{0,i}^0} & M_i^0 & \xrightarrow{\pi_{i,\bar{\theta}}^0} & M_{\bar{\theta}}^0 & \xrightarrow{\pi_{\bar{\theta},\theta}^0} & M_\theta^0 \\
\uparrow \sigma & & \uparrow \sigma & & & \nearrow \sigma & \\
\bar{M}_0^0 & \xrightarrow{\bar{\pi}_{0,i}^0} & \bar{M}_i^0 & \xrightarrow{\bar{\pi}_{i,\bar{\theta}}^0} & \bar{M}_{\bar{\theta}}^0 & &
\end{array}
$$

Note that the size of M_i^0 is less than θ whenever $i < \bar{\theta}$. This follows by a straightforward induction on i using the assumption that M_0^0 has cardinality less than θ, the fact that $*$-ultrapowers do not increase cardinality and that the size of the direct limit of a diagram shorter than θ remains below θ if it consists of structures of size less than θ.

It follows that $M_i^0 \in X \cap H_\tau$, hence $\bar{M}_i^0 = M_i^0$ and $\sigma \restriction M_i^0 = \mathrm{id} \restriction M_i^0$ for all $i < \bar{\theta}$. Consequently, $\pi_{ij}^0 \in X \cap H_\tau$, which implies $\bar{\pi}_{ij}^0 = \pi_{ij}^0$ whenever $i \le j < \bar{\theta}$. But then $\bar{M}_{\bar{\theta}}^0 = M_{\bar{\theta}}^0$, since they are both direct limits of the same diagram. Finally, $\bar{\pi}_{i,\bar{\theta}}^0 = \pi_{i,\bar{\theta}}^0$ for $i < \bar{\theta}$.

We now determine the relationship between the maps $\pi_{\bar{\theta},\theta}^0$ and σ. Note that if i is a truncation point then $i < \bar{\theta}$, hence the iteration $\Im^0$ is simple from $\bar{\theta}$ on. Let $\xi \in M_{\bar{\theta}}^0$. Then $\xi = \pi_{i,\bar{\theta}}^0(\xi_i)$ for some $\xi_i \in M_i^0$, where $i < \bar{\theta}$. Clearly $\xi_i < \bar{\theta}$ and $\sigma(\xi_i) = \xi_i$. Note also that, by our arrangements, $\pi_{i\theta}^0 \in \mathrm{rng}(\sigma)$. Then

$$\sigma(\xi) = \sigma(\pi_{i,\bar{\theta}}^0(\xi_i)) = \sigma(\pi_{i,\bar{\theta}}^0)(\sigma(\xi_i)) = \pi_{i,\theta}^0(\xi_i) = \pi_{\bar{\theta},\theta}^0 \circ \pi_{i,\bar{\theta}}^0(\xi_i) = \pi_{\bar{\theta},\theta}^0(\xi),$$

hence

$$\pi_{\bar{\theta},\theta}^0 = \sigma \restriction M_{\bar{\theta}}^0.$$

The same equality holds for the other side of the coiteration, i.e.

$$\pi^1_{\bar\theta,\theta} = \sigma \upharpoonright M^1_{\bar\theta}.$$

This in particular means that $\kappa_{\bar\theta} = \bar\theta$ and that we form an ultrapower on both sides at the $\bar\theta$-th step. Equivalently,

$$E^{M^0_{\bar\theta}}_{\nu_{\bar\theta}} \neq \emptyset \neq E^{M^1_{\bar\theta}}_{\nu_{\bar\theta}}.$$

By the normality of $\mathfrak{I}^0, \mathfrak{I}^1$, we can compute these measures from the maps $\pi^k_{\bar\theta,\theta}$ ($k = 0, 1$). Thus,

$$x \in E^{M^0_{\bar\theta}}_{\nu_{\bar\theta}} \longleftrightarrow \kappa_{\bar\theta} \in \pi^0_{\bar\theta,\theta}(x) \longleftrightarrow \kappa_{\bar\theta} \in \sigma(x) \longleftrightarrow \kappa_{\bar\theta} \in \pi^1_{\bar\theta,\theta} \longleftrightarrow x \in E^{M^1_{\bar\theta}}_{\nu_{\bar\theta}}$$

for $x \in \mathcal{P}(\kappa_{\bar\theta}) \cap M^0_{\bar\theta} = \mathcal{P}(\kappa_{\bar\theta}) \cap M^1_{\bar\theta}$. In other words, $E^{M^0_{\bar\theta}}_{\nu_{\bar\theta}} = E^{M^1_{\bar\theta}}_{\nu_{\bar\theta}}$. This contradicts the fact that $\mathfrak{I}^0, \mathfrak{I}^1$ constitute a coiteration. □(Lemma 4.4.1)

If the coiteration of premice M, N terminates and M', N' are the resulting premice, we briefly say that M, N *coiterate* to M', N'.

Lemma 4.4.2. *Coiterate premice M, N to M', N'. If the iteration on the M-side is non-simple, then N' is an initial segment of M', i.e. $N' = M'||\alpha$ for some $\alpha \leq \mathrm{ht}(M')$.*

Proof. Suppose the contrary. Then M' is a proper initial segment of N', i.e. $M' = N'||\alpha$ for some $\alpha < \mathrm{ht}(N')$. Hence M' is sound. Let i be the last truncation point on the M-side and $\pi : M_i||\alpha_i \to M'$ be the corresponding iteration map. Then $\omega\varrho^\omega_{M_i||\alpha_i} \leq \kappa_i$ since we have truncated, i.e. $M_i||(\alpha_i + 1)$ contains a subset of κ_i which is not measured by $E^{M_i}_{\nu_i}$. Let n be such that $\omega\varrho^{n+1}_{M_i||\alpha_i} \leq \kappa_i < \omega\varrho^n_{M_i||\alpha_i}$. Pick a $\bar p \in P^{n+1}_{M_i||\alpha_i}$. By Lemma 4.2.5, $p = \pi(\bar p) \in P^{n+1}_{M'}$. Let $x = \tilde h^n_{M'}(\xi, p)$ for some $\xi < \omega\varrho^{n+1}_{M'}$. Then $x = \pi(\bar x)$, where $\bar x = \tilde h^n_{M_i||\alpha_i}(\xi, \bar p)$. Hence $\tilde h^n_{M'}(\omega\varrho^{n+1}_{M'} \cup \{p\}) \subset \mathrm{rng}(\pi)$, but the latter is a proper subset of M' since it does not contain $\kappa_i = \mathrm{cr}(\pi)$. It follows that $p \notin R^{n+1}_{M'}$, which implies that M' is not $(n + 1)$-sound and therefore not sound. Contradiction. □(Lemma 4.4.2)

If we had truncated on both sides of a coiteration, we might have thrown out too much information for the comparison process to be meaningful. In the next chapter we shall show that once we deal with sufficiently iterable premice, one side of a coiteration is always simple. For the moment we content ourselves with the following.

Corollary 4.4.3. *Coiterate premice M, N to M', N'. If neither side of the coiteration is simple, then $M' = N'$.* □(Corollary 4.4.3)

Finally we note that all results of this section also go through for (k, ℓ)-coiterations.

4.5 Some Iterability Criteria

So far we have developed the theory of mice without actually constructing any. In this section we present several criteria which guarantee various degrees of iterability and allow inductive constructions of sufficiently iterable premice.

Theorem 4.5.1. *Let M be a premouse such that*

a) *$M\|\beta$ is a mouse for all $\beta < \mathrm{ht}(M)$;*

b) *all total measures in M are ω-complete.*

Then M is a mouse.

The strategy in proving Theorem 4.5.1 is to establish first a weaker result, namely the normal iterability of M, and then perform a very special simple normal iteration of M which will turn all total measures into measures satisfying a strong form of ω-completeness. We shall see that this form of ω-completeness allows a direct proof of full iterability; in our case, we obtain full iterability of the simple iterate M' arising from the above iteration. Since the associated iteration map $\pi_{MM'}$ is Σ^*-preserving, M turns out to be fully iterable. The proof of the above theorem thus factorizes into the proofs of the following lemmata.

Lemma 4.5.2. *Let M be a premouse satisfying the assumptions of Theorem 4.5.1. Then M is normally iterable and every non-simple normal iterate of M is a mouse.*

In order to state the second lemma we introduce a very technical notion of *strong ω-completeness.* This notion looks rather ad hoc and should be considered to be a part of the proof of Theorem 4.5.1.

Definition. Let M be a premouse and E_ν^M be a total measure in M. E_ν^M is *strongly ω-complete* iff E_ν^M is ω-complete and for every countable $A \subset M$ there is an $H \subset M$ such that

i) $\omega\varrho_M^{n+1} \cup A \subset H$ and H is closed under good $\Sigma_1^{(n)}(M)$ functions;

ii) $\omega\varrho_M^n \cap H$ is unbounded in $\omega\varrho_M^n$;

iii) $\bigcap(E_\nu^M \cap H) \neq \varnothing$,

where $\kappa = \mathrm{cr}(E_\nu^M)$ and $n \in \omega$ is such that $\omega\varrho_M^{n+1} \leq \kappa < \omega\varrho_M^n$.

Lemma 4.5.3. *Let M be a premouse such that*

a) *$M\|\beta$ is a mouse for all $\beta < \mathrm{ht}(M)$;*

b) *all total measures in M are strongly ω-complete.*

Then M is a mouse.

Proof of Lemma 4.5.2. Suppose the contrary. Let $\mathfrak{I} = \langle M_i, \pi_{ij};\ i \le j < \theta\rangle$ be a normal iteration of M witnessing that M is not normally iterable. Pick a regular cardinal τ large enough that $\mathfrak{I} \in H_\tau$. Let X be a countable elementary substructure of H_τ such that $\mathfrak{I} \in X$, $\bar{H}$ the transitive collapse of X, and σ the uncollapsing map. Set $\bar{\mathfrak{I}} = \sigma^{-1}(\mathfrak{I})$, $\bar{M} = \sigma^{-1}(M)$. Then $\bar{\mathfrak{I}}$ is a normal iteration of $\bar{M}$ witnessing that $\bar{M}$ is not normally iterable and $\sigma : \bar{M} \to M$ is Σ^*-preserving. Let $\bar{\theta} = \text{length}(\bar{\mathfrak{I}})$, and let $\bar{M}_i$ and $\bar{\pi}_{ij}$ be the premice and iteration maps of $\bar{\mathfrak{I}}$, respectively. We obtain a contradiction by showing that the longest simple initial segment of $\bar{\mathfrak{I}}$ can be "embedded" into M in the way described by the following diagram, which implies that $\bar{\mathfrak{I}}$ cannot constitute a counterexample to the normal iterability of $\bar{M}$ after all.

(4.5.1)

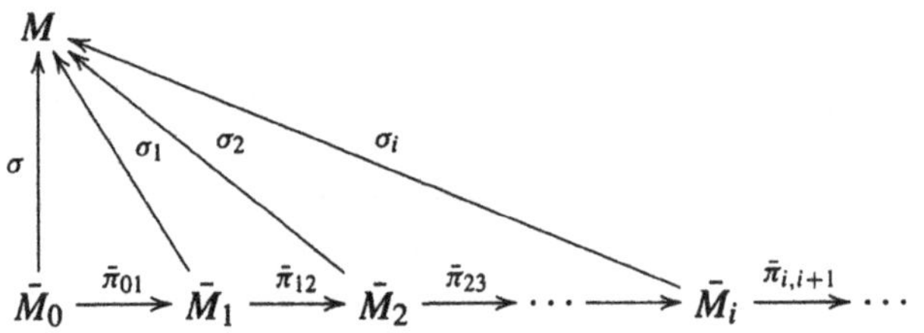

Here,

$$\sigma_i \text{ is } \Sigma_0^{(n_i)}\text{-preserving for all } i < \bar{\theta}, \tag{4.5.2}$$

where

$$n_i = \begin{cases} \omega & \text{if } \omega\varrho^{\omega}_{\bar{M}_i} > \bar{\kappa}_i \\ \text{the unique } n \in \omega \text{ such that } \omega\varrho^{n_i+1}_{\bar{M}_i} \le \bar{\kappa}_i < \omega\varrho^{n_i}_{\bar{M}_i} & \text{otherwise.} \end{cases}$$

If the iteration $\bar{\mathfrak{I}}$ is non-simple, we get the contradiction as follows. Setting ι to be the first truncation point, we observe that $\sigma_\iota : \bar{M}_\iota||\alpha_\iota \to M||\sigma_\iota(\alpha_\iota)$ is Σ^*-preserving, which guarantees the full iterability of $\bar{M}_\iota||\alpha_\iota$ on the one hand (by our assumptions, $M||\sigma_\iota(\alpha_\iota)$ is a mouse), and that $\bar{\mathfrak{I}} \restriction [\iota, \bar{\theta})$ witnesses the failure of iterability of $\bar{M}_\iota||\alpha_\iota$ on the other hand.

From now on suppose that $\mathfrak{I}$ is simple. Hence $\bar{\mathfrak{I}}$ is simple. Set $\bar{M}_0 = \bar{M}$, $\sigma_0 = \sigma \restriction \bar{M}_0$. We now iteratively apply the argument used in the proof of Corollary 3.4.3. As usually, we assume w.l.o.g. that $\mathfrak{I}$ is direct, i.e. $E^{\bar{M}_i}_{\bar{\nu}_i} \neq \varnothing$ whenever $i < \bar{\theta}$. Suppose that σ_i have already been constructed for all $i < \mu$.

If μ is limit, we observe that there is an $i_0 < \mu$ such that n_i has a constant value, say n, for all i which are between i_0 and μ. For $x \in \bar{M}_\mu$ pick $i > i_0$ and $x_i \in \bar{M}_i$ such that $x = \pi_{i,\mu}(x_i)$ and set $\sigma_\mu(x) = \sigma_i(x_i)$. Then σ_μ is well defined since the diagram constructed up to μ commutes and a straightforward argument yields that σ_μ is $\Sigma_0^{(n)}$-preserving. By the remarks below Lemma 4.2.2, $\omega\varrho^{n+1}_{\bar{M}_\mu} \le \bar{\kappa}_i$ for each such i. By the normality of $\bar{\mathfrak{I}}$, $\bar{\kappa}_\mu \ge \sup\{\bar{\kappa}_i;\ i < \mu\}$ and therefore (4.5.2) also holds for $i = \mu$.

Now suppose that $\mu = i + 1$ and

$$\text{if } E^{\bar{M}_i}_{\bar{\nu}_i} \text{ is total in } \bar{M}_i, \text{ then } E^{M}_{\tilde{\nu}_i} \text{ is total in } M, \text{ where } \tilde{\nu}_i = \sigma_i(\bar{\nu}_i). \tag{4.5.3}$$

Set $\tilde{\nu}_i = \sigma_i(\bar{\nu}_i)$. By (4.5.3), $E^M_{\tilde{\nu}_i}$ is a total measure in M, hence $E^M_{\tilde{\nu}_i}$ is ω-complete. Clearly

$$\langle \sigma_i, \sigma_i \upharpoonright \bar{\nu}_i \rangle : \langle \bar{M}_i, E^{\bar{M}_i}_{\bar{\nu}_i} \rangle \longrightarrow \langle M, E^M_{\tilde{\nu}_i} \rangle.$$

This together with (4.5.2) and the fact that $\bar{M}_i$ is countable enables us to define σ_μ by

$$\sigma_\mu(\pi_{i\mu}(f)(\bar{\kappa}_i)) = \sigma_i(f)(\kappa^*_i),$$

for $f \in \Gamma(\bar{\kappa}_i, \bar{M}_i)$ and a fixed $\kappa^*_i \in \bigcap(E^M_{\tilde{\nu}_i} \cap \mathrm{rng}(\sigma_i))$. Then σ_μ is $\Sigma_0^{(n_i)}$-preserving and hence $\Sigma_0^{(n_\mu)}$-preserving, since $n_\mu \le n_i$. It is also clear that the extended diagram commutes.

Granted that (4.5.3) does not fail in the course of the construction, we can "embed" the longest simple initial segment of $\bar{\mathfrak{I}}$ into M as desired and draw the contradiction. By induction on i we now verify that all σ_i are cardinal preserving, which will guarantee (4.5.3). The case $i = 0$ is trivial since σ_0 is fully elementary. Note also that if $\bar{M}_0$ has a largest cardinal, say $\bar{\lambda}_0$, then $\bar{\lambda}_0$ is mapped to the largest cardinal in M. At the induction step $i \mapsto i+1$ we consider two cases. Either $\bar{\kappa}_i < \omega\varrho^1_{\bar{M}_i}$, in which case σ_{i+1} is Σ_1-preserving, or else $\omega\varrho^1_{\bar{M}_i} \le \bar{\kappa}_i$ and the ultrapower map $\bar{\pi}_{i,i+1}$ is Σ_0-preserving and cofinal. In the former case, σ_{i+1} is automatically cardinal preserving. In the latter case there are two possibilities. If $\bar{M}_i$ has a largest cardinal $\bar{\lambda}_i$, then this cardinal is mapped by $\bar{\pi}_{i,i+1}$ to the largest cardinal $\bar{\lambda}_{i+1}$ of $\bar{M}_{i+1}$, so $\sigma_{i+1}(\bar{\lambda}_{i+1}) = \sigma_i(\bar{\lambda}_i)$ is, by the induction hypotheses, the largest cardinal of M. If $\bar{M}_i$ does not have a largest cardinal, then the same applies to $\bar{M}_{i+1}$ and $\bar{\pi}_{i,i+1}$ maps cardinals of $\bar{M}_i$ cofinally to $\bar{M}_{i+1}$. By the commutativity of the diagram (4.5.1), σ_{i+1} maps cofinally many cardinals of $\bar{M}_{i+1}$ to cardinals of M. In both cases the respective conclusion guarantees that σ_{i+1} is cardinal preserving. The verification at limit steps is straightforward.

To see that every non-simple normal iterate N of M is a mouse it suffices to verify this for proper segments of *simple* normal iterates. Here we imitate the above proof; the conclusion follows from the fact that $\bar{N} = \sigma^{-1}(N)$ is Σ^*-embeddable into a proper initial segment of M or into an iterate of such a segment. □(Lemma 4.5.2)

In the above construction, we keep losing the preservation degree of σ_i as the critical points of $\bar{\mathfrak{I}}$ increase. It follows that if $\bar{\mathfrak{I}}$ were not normal, the construction would break down at the first index i such that $\bar{\kappa}_i < \omega\varrho^{n_i+1}_{\bar{M}_i}$. The strategy of the proof of Lemma 4.5.3 is to define the maps σ_i more carefully to guarantee that they are Σ^*-preserving. Since this will automatically hold at limit stages in $\mathfrak{I}$, we shall focus on verifying that

$$\sigma_{i+1} \text{ is } \Sigma^*\text{-preserving provided that } \sigma_i \text{ is.} \tag{4.5.4}$$

Assuming that σ_i is Σ^*-preserving, σ_i maps total measures to total measures, so that no extra argument like that at the end of the proof of Lemma 4.5.2 is needed.

Proof of Lemma 4.5.3. Suppose that the lemma fails, pick a simple iteration $\mathfrak{I} = \langle M_i, \pi_{ij}; i \le j < \theta \rangle$ of M witnessing this fact and construct a countable $\bar{\mathfrak{I}} =$

$\langle \bar{M}_i, \bar{\pi}_{ij};\ i \le j < \bar{\theta} \rangle$ as before. Build the diagram (4.5.1); to prove (4.5.4) we use the fact that all total measures E^M_ν are strongly ω-complete. As before set $\tilde{\nu}_i = \sigma_i(\bar{\nu}_i)$ and $\tilde{\kappa}_i = \sigma_i(\bar{\kappa}_i)$. If $\omega\varrho^\omega_M > \tilde{\kappa}_i$, define σ_{i+1} as in the proof of Lemma 4.5.2; assuming that $\sigma_i : \bar{M}_i \to M$ is Σ^*-preserving, the same will also hold for σ_{i+1}. From now on suppose that $\omega\varrho^\omega_M \le \tilde{\kappa}_i$; let n_i be such that $\omega\varrho^{n_i+1}_M \le \tilde{\kappa}_i < \omega\varrho^{n_i}_M$. We shall construct the following diagram, where all maps are Σ^*-preserving:

(4.5.5)

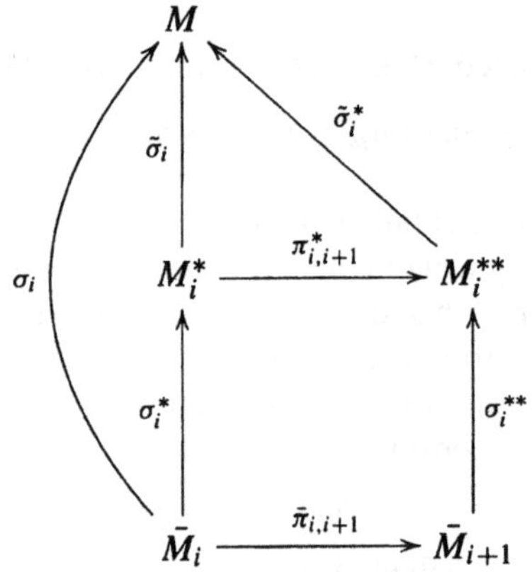

Then $\sigma_{i+1} = \tilde{\sigma}_i^* \circ \sigma_i^{**}$ will be Σ^*-preserving, which completes the proof.

The heart of the construction is the upper triangle of the above diagram, more precisely, the premouse M_i^* and the maps $\tilde{\sigma}_i$ and $\tilde{\sigma}_i^*$. We let $\sigma_i^* = \tilde{\sigma}_i^{-1} \circ \sigma_i$. The rectangle is obtained by copying the i-th step of the iteration $\mathfrak{I}$ via σ_i^*. It follows immediately that all maps in the rectangle are Σ^*-preserving.

The construction of M_i^*, $\tilde{\sigma}_i$ and $\tilde{\sigma}_i^*$. Pick a $p \in P^*_M$ and set $A_i = \operatorname{rng}(\sigma_i) \cup \{p\}$. Since $E^M_{\tilde{\nu}_i}$ is a total measure in M, $E^M_{\tilde{\nu}_i}$ is strongly ω-complete. Let $H_i \subset M$ witness this. Set

$$\begin{aligned} M_i^* &= \text{the transitive collapse of } H_i, \\ \tilde{\sigma}_i &= \text{the associated uncollapsing map.} \end{aligned}$$

Then M_i^* is a premouse and

$$\tilde{\sigma}_i : M_i^* \underset{\Sigma_1^{(n_i)}}{\longrightarrow} M$$

by Lemma 1.11.1. Since $\omega\varrho^{n_i+1}_M \subset \operatorname{rng}(\tilde{\sigma}_i)$ and $\operatorname{rng}(\tilde{\sigma}_i) \cap P^*_M \ne \varnothing$, $\tilde{\sigma}_i$ is Σ^*-preserving by Lemma 1.11.2. Furthermore, $\tilde{\sigma}_i$ is cofinal at the n_i-th level. Pick a $\hat{\kappa}_i \in \bigcap(E^M_{\tilde{\nu}_i} \cap H_i)$ and, letting $\kappa_i^* = \operatorname{cr}(\pi_{i,i+1})$, define $\tilde{\sigma}_i^*$ by

$$\tilde{\sigma}_i^*(\pi^*_{i,i+1}(f)(\kappa_i^*)) = \tilde{\sigma}_i(f)(\hat{\kappa}_i)$$

for $f \in \Gamma(\kappa_i^*, M_i^*)$. Then clearly $\tilde{\sigma}_i^* \circ \pi^*_{i,i+1} = \tilde{\sigma}_i$.

Claim. $\tilde{\sigma}_i^*$ is $\Sigma_0^{(n_i)}$-preserving.

Proof. Let $\varphi(v_1, \ldots, v_\ell)$ be a $\Sigma_0^{(n_i)}$ formula and $f_j \in \Gamma(\kappa_i^*, M_i^*)$ for $j = 1, \ldots, \ell$, where $\kappa_i^* = \sigma_i^*(\bar{\kappa}_i)$ and $\nu_i^* = \sigma_i^*(\bar{\nu}_i)$. Then

$$M_i^{**} \models \varphi(\pi_{i,i+1}^*(f_1)(\kappa_i^*), \ldots, \pi_{i,i+1}^*(f_\ell)(\kappa_i^*))$$
$$\longleftrightarrow u_{f_1,\ldots,f_\ell}^{M_i^*,\varphi} \in E_{\nu_i^*}^{M_i^*} \longleftrightarrow u_{\tilde{\sigma}_i(f_1),\ldots,\tilde{\sigma}_i(f_\ell)}^{M,\varphi} \in E_{\bar{\nu}_i}^{M}$$

where

$$u_{f_1,\ldots,f_\ell}^{M_i^*,\varphi} = \{\eta \in \kappa_i^*;\ M_i^* \models \varphi(f_1(\eta), \ldots, f_\ell(\eta))\}$$

and $u_{\tilde{\sigma}_i(f_1),\ldots,\tilde{\sigma}_i(f_\ell)}^{M,\varphi} = \tilde{\sigma}_i(u_{f_1,\ldots,f_\ell}^{M_i^*,\varphi})$, i.e.

$$u_{\tilde{\sigma}_i(f_1),\ldots,\tilde{\sigma}_i(f_\ell)}^{M,\varphi} = \{\eta \in \bar{\kappa}_i;\ M \models \varphi(\tilde{\sigma}_i(f_1)(\eta), \ldots, \tilde{\sigma}_i(f_\ell)(\eta))\}.$$

Since

$$\hat{\kappa}_i \in u_{\tilde{\sigma}_i(f_1),\ldots,\tilde{\sigma}_i(f_\ell)}^{M,\varphi},$$

we immediately obtain

$$M \models \varphi(\tilde{\sigma}_i(f_1)(\hat{\kappa}_i), \ldots, \tilde{\sigma}_i(f_\ell)(\hat{\kappa}_i)).$$ □(Claim)

Note that $\tilde{\sigma}_i^*$ is cofinal at the n_i-th level by the fact that the same holds for $\tilde{\sigma}_i$. Hence $\tilde{\sigma}_i^*$ is $\Sigma_1^{(n_i)}$-preserving. Also, $\omega\varrho_{M_i^{**}}^{n_i+1} = \omega\varrho_{M_i^*}^{n_i+1} = \omega\varrho_M^{n_i+1}$. It follows that $\omega\varrho_M^{n_i+1} \subset \mathrm{rng}(\tilde{\sigma}_i^*)$, since $\tilde{\sigma}_i^* \upharpoonright \kappa_i^* = \tilde{\sigma}_i \upharpoonright \kappa_i^* = \mathrm{id}$. Since $p \in \mathrm{rng}(\tilde{\sigma}_i^*)$, Lemma 1.11.2 guarantees that $\tilde{\sigma}_i^*$ is Σ^*-preserving. □(Lemma 4.5.3)

Proof of Theorem 4.5.1. Let M satisfy the assumptions of the theorem. By Lemma 4.5.2, M is normally iterable. Define an iteration $\mathfrak{I} = \langle M_i, \pi_{ij};\ i \leq j\rangle$ of M as follows: $\nu_i \simeq$ the least $\nu \leq \mathrm{ht}(M_i)$ satisfying

i) $E_\nu^{M_i}$ is a total measure;

ii) $\nu > \sup\{\nu_{i'};\ i' \leq i\}$

iii) at least one of the following holds
   a) i is a successor ordinal
   b) $\mathrm{cf}(\mathrm{cr}(E_\nu^{M_i})) = \omega$ or $\{i' < i;\ \pi_{i',i}(\nu_{i'}) = \nu\}$ is bounded in i.

$\mathfrak{I}$ is clearly a simple normal iteration of M. Let θ be regular and greater than $\max\{\omega_1, \mathrm{card}(M)\}$.

Claim 1. The iteration $\mathfrak{I}$ terminates below θ.

Proof. Suppose the contrary. Let S_0 be the set of all ordinals less than θ with uncountable cofinality and

$$f(i) = \text{the least } i' < i \text{ such that } \nu_i \in \text{rng}(\pi_{i',i})$$

for $i \in S_0$. Since f is regressive, f has a constant value, say i^*, on a stationary $S_1 \subset S_0$. Since $\text{card}(M_{i^*}) < \theta$ (this can be verified by a straightforward induction), there is a $\nu^* \in M_{i^*}$ and a stationary $S \subset S_1$ such that $\nu_j = \pi_{i^*,j}(\nu^*)$ for all $j \in S$ (we suppose w.l.o.g. that $i^* + 1 \cap S = \varnothing$). Hence $\pi_{ij}(\nu_i) = \nu_j$ whenever $i, j \in S$ and $i < j$. Fix a $j \in S \cap \lim(S)$. Then clearly the latter statement in b) of the clause iii) above fails for j. Furthermore, $\kappa_j = \sup\{\kappa_i;\ i < j\}$: the inequality $\geq$ follows from the normality of $\Im$; to see the converse pick a $\xi < \kappa_j$ and observe that if $i \in S \cap j$ is such that $\xi = \pi_{i,j}(\xi_i)$ for some $\xi_i \in M_i$, then $\xi_i < \kappa_i$ and, consequently, $\xi = \xi_i < \kappa_i$. It follows that $\text{cf}(\kappa_j) > \omega$. This means that both statements in b) of the above condition iii) fail. But then ν_j cannot be an iteration index. Contradiction. □(Claim 1)

Let $\tilde{M}$ be the last premouse in the iteration $\Im$. Then

$$\text{every total measure in } \tilde{M} \text{ is } \omega\text{-complete.} \tag{4.5.6}$$

To see this, suppose that $E_\nu^{\tilde{M}}$ is a total measure and $\kappa = \text{cr}(E_\nu^{\tilde{M}})$. Let

$$j = \begin{cases} \text{the least } h < \text{length}(\Im) \text{ such that } \kappa_h \geq \kappa \text{ if such an } h \text{ exists} \\ \text{length}(\Im) - 1 \quad \text{otherwise} \end{cases}$$

Then j is a limit and $\nu_i > \nu$ if $j < i$. The latter is trivial; the former follows from the fact that if $j = i + 1$, then $\kappa_i < \kappa$ (otherwise $\nu_i = \nu$, hence $E_\nu^{\tilde{M}} = E_\nu^{M_j} = \varnothing$) and $E_\nu^{M_j} = E_\nu^{\tilde{M}}$ would be a total measure, hence $j < \text{length}(\Im)$ and $\nu_j \leq \nu$, which contradicts the definition of j. Since $E_\nu^{M_j}$ is not applied at the j-th step of $\Im$, we have:

- $\text{cf}(\kappa) > \omega$;
- $\{i < j;\ \pi_{ij}(\nu_i) = \nu\}$ is cofinal in j;
- $\kappa = \sup\{\kappa_i;\ i < j\}$ and $\text{cf}(j) > \omega$.

The last clause easily follows from the two preceding ones. Recall again that $E_\nu^{\tilde{M}} = E_\nu^{M_j}$. Towards the ω-completeness of $E_\nu^{\tilde{M}}$, pick a countable $X \subset E_\nu^{\tilde{M}}$. Given $x \in X$, there is an $i < j$ such that $x \in \text{rng}(\pi_{ij})$; since j is not ω-cofinal, we can fix an $\iota < j$ such that $X \subset \text{rng}(\pi_{\iota,j})$ and $\pi_{\iota,j}(\nu_\iota) = \nu$. For $x \in X$ set $x_\iota = \pi_{\iota,j}^{-1}(x)$. Then $x_\iota \in E_{\nu_\iota}^{M_\iota}$, hence $\kappa_\iota \in \pi_{\iota,\iota+1}(x_\iota)$, and since $\kappa_\iota < \kappa_{\iota+1}$, $\kappa_\iota \in \pi_{\iota+1,j} \circ \pi_{\iota,\iota+1}(x_\iota) = \pi_{\iota,j}(x_\iota) = x$. Thus, $\kappa_\iota \in \bigcap(E_\nu^{\tilde{M}} \cap X)$. This completes the proof of (4.5.6).

To prove the strong ω-completeness of the total measures of $\tilde{M}$, we have to analyze how new elements arise in the course of the iteration $\Im$. The following two lemmata summarize how the definability over a simple normal iterate of M can be reduced to the definability over M.

Lemma 4.5.4. *Let $\Im = \langle N_i, \pi_{ij};\ i \le j < \theta + 1\rangle$ be a simple normal iterate of N strictly above κ (i.e. $\kappa_0 > \kappa$). Suppose $\omega\varrho_N^{n+1} \le \kappa < \omega\varrho_N^n$ for an $n \in \omega$. Set $\omega\varrho^{n+1} = \omega\varrho_{N_i}^{n+1}$ ($i \le \theta$). Given a finite $b \subset N_\theta$, there is a finite $b^\natural \subset N_0 = N$ such that every $x \in H_{\kappa_0}^N = H_{\kappa_0}^{N_\theta}$, which is $\Sigma_1^{(n)}(N_\theta)$-definable in parameters from $b \cup \omega\varrho^{n+1}$, is $\Sigma_1^{(n)}(N)$-definable in parameters from $b^\natural \cup \omega\varrho^{n+1}$.*

Proof. We proceed by induction on θ. If $\theta = 1$, the iteration consists of a single ultrapower by $E_{\nu_0}^N$ with the ultrapower map π. For every $z \in b - \text{rng}(\pi)$ pick a $q_z \in N$ such that $z = \pi(f_z)(\kappa_0)$ for some $f_z \in \Gamma(\kappa, N)$ where f_z is a good $\Sigma_1^{(i)}(N)$ function in q_z for some $i \in \omega$. For $z \in \text{rng}(\pi)$ set $q_z = (\pi^{-1})(z)$. Let

$$b^\natural = \{q_z;\ z \in b\} \cup \{\kappa_0\}.$$

Then $b^\natural$ is finite since b is. If $x \in H_{\kappa_0}^N$, $\pi(x) = x$. If moreover x is $\Sigma_1^{(n)}(N_1)$-definable in parameters from $\omega\varrho^{n+1} \cup b$, there is a $\Sigma_1^{(n)}(N_1)$ relation R, $\xi < \omega\varrho^{n+1}$ and $x_1, \ldots, x_\ell \in b$ such that x is the unique object satisfying $R(x, \xi, x_1, \ldots, x_\ell)$. Let R' be the $\Sigma_1^{(n)}(N)$-relation given by Lemma 3.2.9; the $\Sigma_1^{(n)}$-definition of R' is canonically determined by the $\Sigma_1^{(n)}(N_1)$-definition of R and the objects $x_1, \ldots, x_\ell$ (strictly speaking, by the functions f_z for $z = x_1, \ldots, x_\ell$). Then

$$R(\pi(z), \pi(\eta), x_1, \ldots, x_\ell) \longleftrightarrow R'(z, \eta, q_{x_1}, \ldots, q_{x_\ell}, \kappa_0).$$

Consequently, x is the unique element z of N satisfying $R'(z, \xi, q_{x_1}, \ldots, q_{x_\ell}, \kappa_0)$.

Suppose $\Im$ has a successor length greater than 1, i.e. $\theta = \bar{\theta} + 1$ for some $\bar{\theta} > 0$. First apply the induction hypothesis to the iteration $\langle N_{\bar{\theta}}, N_\theta, \pi_{\bar{\theta},\theta}\rangle$, whose length is 1, and b. We obtain a finite set $b' \subset N_{\bar{\theta}}$ such that every $x \in H_{\kappa_0}^N$, which is $\Sigma_1^{(n)}(N_\theta)$-definable in parameters from $\omega\varrho^{n+1} \cup b$, is $\Sigma_1^{(n)}(N_{\bar{\theta}})$-definable in parameters from $\omega\varrho^{n+1} \cup b'$. Then apply the induction hypothesis to the iteration $\Im \restriction \theta$ and b'. We obtain a finite set $b^\natural \subset N$, which clearly satisfies the statement of the lemma.

Finally suppose that θ is a limit; since b is finite, there is an $i < \theta$ such that $b \subset \text{rng}(\pi_{i,\theta})$. Set $b' = (\pi^{-1})''b$ and apply the induction hypothesis to $\Im \restriction (i+1)$ and b'. We obtain a set $b^\natural$ which again clearly satisfies all desired requirements.

□(Lemma 4.5.4)

Proof of Theorem 4.5.1 – *continuation.* Suppose $E_\nu^{\tilde{M}}$ is a total measure in $\tilde{M}$. We show:

$$E_\nu^{\tilde{M}} \text{ is strongly } \omega\text{-complete.}$$

Let n be such that $\omega\varrho_{\tilde{M}}^{n+1} \le \kappa < \omega\varrho_{\tilde{M}}^n$, where $\kappa = \text{cr}(E_\nu^{\tilde{M}})$. Pick a countable $A \subset \mathcal{P}(\kappa) \cap E_\nu^{\tilde{M}}$; suppose w.l.o.g. that $p \in A$ for some $p \in P_{\tilde{M}}^{n+1}$. Set

$$H = \text{the closure of } A \cup \omega\varrho^{n+1} \text{ under good } \Sigma_1^{(n)}(\tilde{M}) \text{ functions.}$$

Claim 2. H witnesses the strong ω-completeness of $E_\nu^{\tilde{M}}$.

Proof. We verify all clauses of the definition of strong ω-completeness. Clause i) follows immediately from the definition of H. To see that H is cofinal in $H^n_{\tilde{M}}$, let M^* be the transitive collapse of $\tilde{M}|H$ and $\sigma^* : M^* \to \tilde{M}$ be the uncollapsing map. Since $p \in H = \operatorname{rng}(\sigma^*)$, σ^* is $\Sigma_1^{(n)}$ preserving by Lemma 1.11.1. Let $p^* = (\sigma^*)^{-1}(p)$. Suppose σ^* is not cofinal at the n-th level. Let $A^* = A^{n,p^*\restriction n}_{M^*}$, $\tilde{A} = A^{n,p\restriction n}_{\tilde{M}}$, $\omega\tilde{\varrho} = \sup((\sigma^*)''\omega\varrho^n_{M^*})$ and $N = \langle |J^{E^{\tilde{M}}}_{\tilde{\varrho}}|, \tilde{A} \cap |J^{E^{\tilde{M}}}_{\tilde{\varrho}}| \rangle$. Then

$$\sigma^* \restriction H^n_{M^*} : (M^*)^{n,p^*\restriction n} \underset{\Sigma_0}{\longrightarrow} N \quad \text{cofinally.}$$

Let $\tilde{B}$ be a $\Sigma_1^{(n)}(\tilde{M})$ set in p such that $\omega\varrho^{n+1} \cap \tilde{B} \notin \tilde{M}$ and B^* be the $\Sigma_1^{(n)}(M^*)$ set in p^* by the same definition. Then

- $B^* \cap \omega\varrho^{n+1} = \tilde{B} \cap \omega\varrho^{n+1}$ since $\sigma^* \restriction \omega\varrho^{n+1} = \mathrm{id}$;
- $B^* \cap H^n_{M^*}$ is $\Sigma_1((M^*)^{n,p^*\restriction n})$ in $p^*(n)$.

Let B_N have the same $\Sigma_1(N)$ definition in $p(n) = \sigma^*(p^*(n))$. Note that $N \in \tilde{M}$. As an immediate consequence we obtain $B \cap \omega\varrho^{n+1} = B_N \cap \omega\varrho^{n+1} \in \tilde{M}$; the equality follows from the preservation properties of $\sigma^* \restriction H^n_{M^*}$ and the fact that $\sigma^* \restriction \omega\varrho^{n+1} = \mathrm{id}$. This contradicts our choice of B, so σ^*, and thus H must be cofinal after all.

Finally we prove that $\bigcap(E_\nu^{\tilde{M}} \cap H) \neq \varnothing$. Define j exactly as in the proof of (4.5.6). Recall that j is a limit, has uncountable cofinality and for unboundedly many $i < j$ we have $\pi_{ij}(\nu_i) = \nu$. Also note that $\kappa_j > \kappa$ (provided that $j \neq \operatorname{length}(\Im) - 1$), since otherwise $E_\nu^{\tilde{M}} = \varnothing$, which contradicts our assumption. It follows that every $x \in \mathcal{P}(\kappa) \cap \tilde{M}$ is an element of $H^{\tilde{M}}_{\kappa_j} = H^{M_j}_{\kappa_j}$ and $\pi_j(x) = x$, where π_j is the iteration map from M_j to $\tilde{M}$. If n is such that $\omega\varrho^{n+1}_{\tilde{M}} \leq \kappa < \omega\varrho^n_{\tilde{M}}$, then the same holds for $M_{j'}$ in place of $\tilde{M}$, where $j' \geq j$, and all $\omega\varrho^{n+1}_{M_{j'}}$ have a common value, say $\omega\varrho^{n+1}$. Since j is limit, there is an $\iota_0 < j$ such that $\omega\varrho^{n+1} < \kappa < \omega\varrho^n_{M_i}$ for $i \geq \iota_0$. Now apply Lemma 4.5.4 to A and the iteration $\Im \restriction [j, \operatorname{length}(\Im))$. To every finite $a \subset A$ there is a finite $a^\natural$ such that any $z \in H^{\tilde{M}}_{\kappa_j}$, which is $\Sigma_1^{(n)}(\tilde{M})$-definable in parameters from $a \cup \omega\varrho^{n+1}$, is $\Sigma_1^{(n)}(M_j)$-definable in parameters from $a^\natural \cup \omega\varrho^{n+1}$. Set

$$A^\natural = \bigcup\{a^\natural;\ a \in [A]^{<\omega}\}.$$

Clearly, $A^\natural$ is countable. Let $\iota \geq \iota_0$ be such that $A^\natural \subset \operatorname{rng}(\pi_{\iota,j})$ and $\pi_{\iota,j}(\nu_\iota) = \nu$. It follows from the above that

$$H \cap \mathcal{P}(\kappa) \subset \operatorname{rng}(\pi_{\iota,j}).$$

Now we use the same argument as in the proof of (4.5.6). The above inclusion guarantees that any $x \in E_\nu^{\tilde{M}} \cap H$ is of the form $\pi_{\iota,j}(x_\iota)$ for some $x_\iota \in E_{\nu_\iota}^{M_\iota}$); using the fact that $\pi_{\iota,j}$

is a normal iteration map starting with an ultrapower by $E_{\nu_\iota}^{M_\iota}$ we infer that $\kappa_\iota \in x$. Thus, $\kappa_\iota \in \bigcap(E_\nu^{\tilde{M}} \cap H)$, which proves the claim and also the theorem. □(Theorem 4.5.1)

We have seen above that iterability is equivalent to countable iterability, which is a sort of "local" property. The next theorem establishes the equivalence of iterability to another property with "local" flavour – iterability in the sense of small models. We stress that we deal with *linear* iterations and that no generalizations of these results would hold for iteration trees which arise in the comparison of mice with many overlapping extenders (these notions will be introduced in the third part of this book).

Theorem 4.5.5. *Let H be a* ZF$^-$ *model such that $\omega_1 \subset H$. Suppose $M \in H$ and* H $\models$ *(M is iterable). Then M is iterable.*

Proof. By Lemma 4.3.3, it suffices to show that M is countably iterable. We first construct a "universal" simple iteration $\Im^U(M)$ of M such that every simple iteration of M in V can be embedded into $\Im^U(M)$. The definition of $\Im^U(M)$ will be sufficiently absolute so that $\Im^U(M) \in H$ for every $M \in H$.

We start with a "one-step" iteration $\Im^*(M)$. Let $\langle \vartheta_i;\ i < \delta(M)\rangle$ be the increasing enumeration of all ordinals indexing total measures in M (we also include $\mathrm{ht}(M)$ if $E_{\mathrm{top}}^M \neq \varnothing$). $\Im^*(M)$ is the simple iteration $\langle M_i, \pi_{ij};\ i \leq j,\ j < \delta(M)+1\rangle$ obtained by "iterating every total measure once", i.e.

$$\begin{aligned} M_0 &= M \\ M_{i+1} &= \mathrm{Ult}^*(M_i, E_{\pi_{0,i}(\vartheta_i)}^{M_i}) \\ M_\mu &= \lim_{i \leq j < \mu} (M_i, \pi_{ij}). \end{aligned}$$

$\Im_\theta^U(M)$ is a concatenation of θ many one-step iterations. More precisely,

$$\Im_\theta^U(M) = \langle M_i^k, \pi_{ij}^{k\ell};\ \langle k, i\rangle \leq \langle \ell, j\rangle\rangle$$

where $k, \ell < \theta, i < \delta(M_0^k)+1, j < \delta(M_0^\ell)+1$ and $\leq$ is the lexicographical ordering of ordinals; here $M_0^0 = M, M_{\delta_k}^k = M_0^{k+1}$ and $\langle M_i^k, \pi_{ij}^{kk};\ i \leq j < \delta(M_0^k)+1\rangle = \Im^*(M_0^k)$. If $M, \theta \in H$ then this construction can be carried out inside H, so $\Im_\theta^U(M)$ also belongs to H. Informally, we call $\Im_\theta^U(M)$ the *universal iteration of M with length θ*.

Let $\theta \in \mathbf{On}$ and $\Im = \langle \bar{M}_k, \bar{\pi}_{k\ell};\ k \leq \ell < \theta\rangle$ be a simple iteration of $\bar{M}$. Given a Σ^*-preserving map $\sigma : \bar{M} \to M$, we define an "embedding" $\langle \sigma_k;\ k < \theta\rangle$ of $\Im$ into $\Im_\theta^U(M)$ with Σ^*-preserving "copy" maps $\sigma_k : \bar{M}_k \to M_0^k$ so that the diagram

$$\begin{array}{ccc} M_0^k & \xrightarrow{\ \pi_{0,0}^{k,\ell}\ } & M_0^\ell \\ \uparrow{\scriptstyle \sigma_k} & & \uparrow{\scriptstyle \sigma_\ell} \\ \bar{M}_k & \xrightarrow{\ \bar{\pi}_{k\ell}\ } & \bar{M}_\ell \end{array}$$

commutes. Set $\sigma_0 = \sigma$. At limit stages μ we embed $\bar{M}_\mu$ into the direct limit M_0^μ of all M_i^k $(k < \mu)$ in the usual way. At the successor step $k \mapsto k+1$ define σ_{k+1} to be the unique map completing the following commutative diagram.

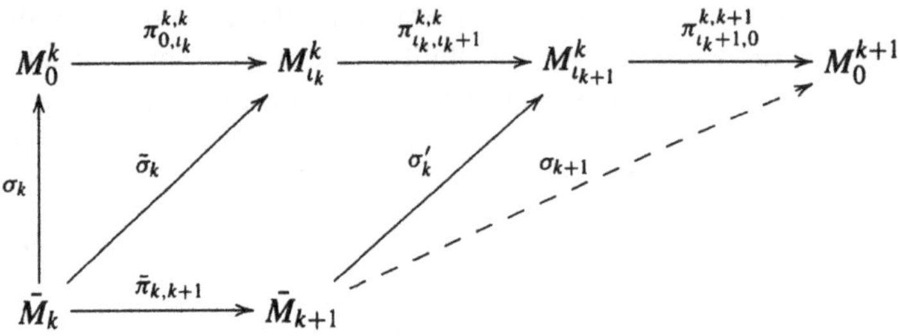

The maps $\bar{\pi}_{k,k+1}$ and $\pi^{k,k}_{\iota_k,\iota_k+1}$ are the ultrapower maps associated with

$$\mathrm{Ult}(\bar{M}_k, E^{\bar{M}_k}_{\nu_k}) \quad \text{and} \quad \mathrm{Ult}(M^k_{\iota_k}, E^{M^k_{\iota_k}}_{\tilde{\sigma}_k(\nu_k)}),$$

respectively. Here, $\sigma_k(\nu_k)$ is the ι_k-th ordinal indexing a total measure in M_0^k. The parallelpiped in the middle represents a copy of a single step $k \mapsto k+1$ in $\Im$ via $\tilde{\sigma}_k = \pi^{k,k}_{0,\iota_k} \circ \sigma_k$. Then

$$\sigma_{k+1} = \pi^{k,k+1}_{\iota_k+1,0} \circ \sigma'_k,$$

so σ_{k+1} is Σ^*-preserving. This completes the "copying" construction for simple iterations.

Now if $\Im$ is an arbitrary iteration of M of countable length θ with truncation points $\varepsilon_1, \ldots, \varepsilon_m$, we construct an embedding of $\Im$ into an iteration $\Im^*(M, \Im) \in H$ by concatenating the embeddings of the simple sections of $\Im$ into the corresponding universal simple iterations. Let $\Im^U_{\alpha,\beta}(M)$ denote the universal iteration of M of length $\mathrm{otp}(\beta - \alpha)$ with superscripts of the premice (and maps) ranging from α (inclusive) to β (exclusive). Here is a formal definition of $\Im^*(M, \Im)$. Set:

- $\varepsilon_0 = 0$, $\sigma_0^0 = \mathrm{id} \restriction M$;
- $\sigma^*_{\varepsilon_s} = \sigma_{\varepsilon_s} \restriction (M_{\varepsilon_s} || \alpha_{\varepsilon_s}$ for $s \leq m)$;
- $\langle \sigma_i^s;\ \varepsilon_s \leq i \leq \varepsilon_{s+1} \rangle$ is the embedding of the iteration $\Im \restriction [\varepsilon_s, \varepsilon_{s+1}]$ (i.e. the section of $\Im$ starting with $M_{\varepsilon_s} || \alpha_{\varepsilon_s}$ and ending with $M_{\varepsilon_{s+1}}$) into $\Im^U_{\varepsilon_s, \varepsilon_{s+1}+1}(M_{\varepsilon_s} || \alpha_{\varepsilon_s})$ via $\sigma^*_{\varepsilon_s}$ for $s < m$;
- $\langle \sigma_i^m;\ \varepsilon_m \leq i < \theta \rangle$ is the embedding of $\Im \restriction [\varepsilon_m, \theta)$ into $\Im^U_{\varepsilon_m,\theta}(M_{\varepsilon_m} || \alpha_{\varepsilon_m})$ via $\sigma^*_{\varepsilon_m}$.

Then $\langle \sigma_i^s \rangle_{s,i}$ is an embedding of $\Im$ into $\Im^*(M, \Im)$, where

$$\Im^*(M, \Im) = \Im^U_{0,\varepsilon_1}(M)^\frown \Im^U_{\varepsilon_1,\varepsilon_2}(M_{\varepsilon_1} || \alpha_{\varepsilon_1})^\frown \ldots {}^\frown \Im^U_{\varepsilon_m,\theta}(M_{\varepsilon_m} || \alpha_{\varepsilon_m}).$$

If θ is countable, then $\Im^*(M, \Im) \in H$ since M is a mouse in the sense of H and $\Im^*(M, \Im)$ is constructed from $M, \langle \varepsilon_1, \alpha_{\varepsilon_1}, \ldots, \varepsilon_m, \alpha_{\varepsilon_m} \rangle \in H$ inside H. Since $\Im$ can be embedded into $\Im^*(M, \Im)$, it follows that $\Im$ can be continued. We show that there are no degenerate countable iterations of M either.

Suppose the contrary; fix the length θ of some countable degenerate iteration. Let R be the set of all pairs $\langle n, \Im \rangle$ such that $\Im$ is a universal iteration of M of length at most θ with exactly n truncations. Define a relation $\mathcal{R} = \langle R, <_{\mathcal{R}} \rangle$ by

$$\langle n', \Im' \rangle <_{\mathcal{R}} \langle n, \Im \rangle \text{ iff } n < n' \text{ and } \Im' \text{ is a lengthening of } \Im.$$

Then $\mathcal{R}$ is a strict partial ordering and $\mathcal{R} \in H$. Furthermore, $\mathcal{R}$ is well-founded in H: Otherwise any infinite decreasing sequence $\langle \langle n_i, \Im_i \rangle; i \in \omega \rangle \in H$ in $\mathcal{R}$ would determine a degenerate iteration of M, namely $\bigcup \Im_i$. By the absoluteness of well-foundedness, $\mathcal{R}$ is well-founded in V. Let $\Im$ be a degenerate iteration of M with length θ and $\langle \varepsilon_i; i < \omega \rangle$ be the sequence of its truncation points. Then

$$\langle \langle n, \Im^*(M, \Im \restriction \varepsilon_n) \rangle; n \in \omega \rangle$$

is an infinite decreasing chain in $\mathcal{R}$, contradicting the well-foundedness of $\mathcal{R}$.

□(Theorem 4.5.5)

Notice that any premouse satisfying the assumptions of Theorem 4.5.1 is 0-iterable. In fact, 0-iterability follows from much weaker assumptions.

Theorem 4.5.6. *Let M be a premouse satisfying:*

a) *$M||\gamma$ is a mouse for every $\gamma < \mathrm{ht}(M)$;*

b) *E^M_{top} is ω-complete.*

Then M is 0-iterable.

Proof. Set $\kappa = \mathrm{cr}(E^M_{\mathrm{top}})$ and $\tau = \mathrm{ht}(M)$. We first observe that J^E_τ is iterable. By assumption a) of the theorem, $J^E_\tau \models (\forall \zeta)(\langle J^E_\zeta, E_{\omega\zeta} \rangle$ is iterable) (iterability is easily reflected downwards). By the ω-completeness of E^M_{top}, M is extendible (and in fact iterable) by E^M_{top}. Let $N = \mathrm{Ult}(M, E^M_{\mathrm{top}})$ and $\tau' = \mathrm{ht}(N)$. Then $J^E_{\tau'}$, being elementarily equivalent to J^E_τ, also thinks that each of its proper initial segment is iterable (the ultrapower map $\pi : M \to N$ is fully elementary since J^E_τ is a ZFC$^-$-model). In particular, J^E_τ, being a proper initial segment of $J^E_{\tau'}$ (this follows from the coherency condition), is iterable in the sense of $J^E_{\tau'}$. But $\tau' > \kappa \geq \omega_1$; the latter inequality follows from the ω-completeness of E^M_{top}. By Theorem 4.5.5, J^E_τ is iterable in the real world.

We now construct a simple 0-iterate M^* of M which is 0-iterable. That M^* is 0-iterable will follow from the fact that the 0-iteration map from M to M^* is Σ_0-preserving. Perform the same iteration as in the proof of Theorem 4.5.1 with the following amendments:

- Use Σ_0-ultrapowers instead of $*$-ultrapowers;
- never use any top measure.

Since iterating M without using any top measure is equivalent to iterating J_τ^E, this iteration does not break down, and as before we can prove that the it terminates. Let $M^* = \langle J_{\tau^*}^{E^*}, F^* \rangle$ be the resulting premouse, σ be the associated iteration map and $\kappa^* = \mathrm{cr}(F^*)$. Then:

- F^* is ω-complete;
- $E_\gamma^{M^*}$ is ω-complete whenever it is a total measure in M^* and $\gamma < \kappa^*$.

The latter follows exactly as in the proof of Theorem 4.5.1. The former is a consequence of Lemma 2.5.5; it suffices to observe that every $x \in M^*$ is an element of $\sigma(\mathrm{rng}(f))$ for some $f \in {}^{\kappa_0}M \cap M$, where $\kappa_0 < \kappa$ is the first critical point in the above iteration (see Lemma 4.2.4). Furthermore, $\sigma : J_\tau^E \to J_{\tau^*}^{E^*}$ is fully elementary; the argument we used to show that J_τ^E is iterable then yields that every proper initial segment of M^* is iterable. By Theorem 4.5.1, M^* is a mouse, which completes the proof. (The use of Theorem 4.5.1 is superfluous here; as we have already observed above, a slight modification of the argument from the proof of Lemma 4.5.2 already yields full 0-iterability of M^*.) □(Theorem 4.5.6)

4.6 Bicephali

In this section we introduce a tool for proving that two measures are equal, the so-called bicephali. More precisely, given a premouse J_ν^E, we would like to have a condition guaranteeing the uniqueness of the measure which can be put on top of J_ν^E. In order to prove the uniqueness, we consider two possible candidates for the top measure and use a comparison argument to show that they agree. The condition we need would thus ensure that this comparison can be carried out.

Definition. A *prebicephalus* is a structure $M = \langle J_\nu^E, F^0, F^1 \rangle$ such that both F^0, F^1 are measures (i.e. they are nonempty) and each $M^h = \langle J_\alpha^E, F^h \rangle$ is a premouse ($h = 0, 1$).

Note that $\mathrm{cr}(F^0) = \mathrm{cr}(F^1)$. Furthermore, being a prebicephalus is a Q-property, as follows from the proof of Lemma 4.1.2. The notions of iteration and iterability can be easily modified for the prebicephali context. It turns out that 0-iterability will suffice for our current purposes. For a prebicephalus $M = \langle J_\alpha^E, F^0, F^1 \rangle$, let $E_\alpha^{M,i} = F^i$ for $i = 1, 2$ and $E_\nu^{M,i} = E_\nu$ for $\nu < \alpha$.

Definition. Let $M = \langle J_\alpha^E, E_\alpha^{M,0}, E_\alpha^{M,1} \rangle$ be a prebicephalus. A *bicephalus iteration* (briefly *iteration*) $\Im$ of M of length θ is a sequence $\langle M_i;\ i < \theta \rangle$ together with sequences of *iteration indices* $\langle \nu_i, h_i, \eta_i;\ i < \theta \rangle$ and *iteration maps* $\langle \pi_{ij};\ i \le j < \theta \rangle$ such that

a) M_i is a prebicephalus and $M_0 = M$;

b) π_{ij} is a partial map from M_i to M_j for $i \leq j$ and $\pi_{ii} = \text{id} \upharpoonright M_i$;

c) $\nu_i \leq \eta_i \leq \alpha_i = \text{ht}(M_i)$ and $h_i \in \{0, 1\}$;

d) there are only finitely many $i < \theta$ such that $\eta_i < \alpha_i$;

e) the maps π_{ij} commute;

f) if $E_{\nu_i}^{M_i,h_i} = \varnothing$, then $M_{i+1} = M_i||\eta_i$ and $\pi_{i,i+1} = \text{id} \upharpoonright (M_i||\eta_i)$;

g) if $F = E_{\nu_i}^{M_i,h_i} \neq \varnothing$ and $\eta_{i'} = \alpha'_i$ for all $i' \leq i$, then $\pi_{i,i+1} : M_i \underset{F}{\longrightarrow} M_{i+1}$;

h) if $F = E_{\nu_i}^{M_i,h_i} \neq \varnothing$ and $\eta_{i'} < \alpha'_i$ for some $i' \leq i$, then

$$\pi_{i,i+1} : M_i||\eta_i \overset{*}{\underset{F}{\longrightarrow}} M_{i+1};$$

i) if $j < \theta$ is a limit, then $\langle M_j, \pi_{ij};\ i < j\rangle = \lim\langle M_i, \pi_{i',i};\ i' \leq i < j\rangle$.

We also set $\kappa_i = \text{cr}(E_{\nu_i}^{M_i,h_i})$.

As before, an iteration $\mathfrak{I}$ of a prebicephalus with limit length is *degenerate* iff the clause d) fails; in this case there are cofinally many such i's (in θ). As before, any i with $\eta_i < \alpha_i$ is called a *truncation point.*

Definition. A prebicephalus M is a *bicephalus* iff M is *iterable,* i.e. iff M does not have any degenerate iterations and every iteration of M can be continued.

Note that any non-simple iterate of a bicephalus is a mouse. For this reason, all lemmata concerning the copying construction and comparison process transfer to the bicephalus context.

Lemma 4.6.1. *Let M be a bicephalus, $\bar{M}$ a prebicephalus and $\sigma^h : \bar{M} \to M$ a Σ_0-preserving map. Then $\bar{M}$ is a bicephalus.* □(Lemma 4.6.1)

Definition. Let M and N be prebicephali. A *coiteration* of M, N of length $\theta + 1$ is a pair of bicephali iterations $\langle \mathfrak{I}^M, \mathfrak{I}^N\rangle$ of length $\theta + 1$ satisfying:

a) $M_0 = M$ and $N_0 = N$.

b) Both $\mathfrak{I}^M, \mathfrak{I}^N$ are *standard,* i.e. η_i^M is the maximal $\eta \leq \text{ht}(M_i)$ such that $E_{\nu_i}^{M_i,h_i}$ measures all subsets of κ_i^M in $M_i||\eta$ (and similarly for N).

c) If $i < \theta + 1$ and there is a ν such that $E_\nu^{M_i,h} \neq E_\nu^{N_i,k}$ for some $h, k \in \{0, 1\}$, then $\nu_i = \nu_i^N = \nu_i^M$ is the least such ν.

d) $\langle h_i^M, h_i^N \rangle$ is the lexicographically least pair $\langle h, k \rangle$ such that

$$E_{\nu_i}^{M_i,h} \neq E_{\nu_i}^{M_i,k}.$$

e) If $i < \theta$ then ν_i (and thus also h_i^M and h_i^N) is defined.

A bicephalus coiteration of length $\theta + 1$ *terminates* iff ν_θ is undefined.

Lemma 4.6.2. *Let M, N be bicephali and $\theta > \max(\mathrm{card}(M), \mathrm{card}(N))$ be regular. Then:*

a) *The coiteration of M, N terminates at some $\bar{\theta} < \theta$.*

b) *One side of the coiteration is simple.*

c) *If there is a truncation on the N-side then $M_{\bar{\theta}}$ is a segment of $N_{\bar{\theta}}$.*

Proof. We simply repeat the proofs of the corresponding lemmata for premice; for a) it is Lemma 4.4.1 and for c) Lemma 4.4.2. Clause b) will follow from the results of the next chapter, namely from Lemma 5.3.1. □(Lemma 4.6.2)

Lemma 4.6.3 (Bicephalus lemma). *Let $M = \langle J_\alpha^E, F^0, F^1 \rangle$ be a bicephalus. Then $F^0 = F^1$.*

Proof. Coiterate M against itself. More formally, coiterate M with N, where $M = N$. Let M_θ, N_θ be the result of the coiteration. By the previous lemma, one side of the coiteration is simple; let it be w.l.o.g. the M-side. We can also assume w.l.o.g. that M_θ is a segment of N_θ. Let G^0, G^1 be the top measures of M_θ. We first observe that $G^0 = G^1$. Then we use the iteration maps to pull this equality backwards. Hence, if $x \in \mathcal{P}(\kappa) \cap M$ then

$$x \in F^0 \longleftrightarrow \pi_{0,\theta}(x) \in G^0 \longleftrightarrow \pi_{0,\theta}(x) \in G^1 \longleftrightarrow x \in F^1$$

where $\kappa = \mathrm{cr}(F^0) = \mathrm{cr}(F^1)$ and $\pi_{0,\theta}$ is the iteration map on the M-side of the coiteration.

To see that the top measures of M_θ agree, note that either M_θ is a non-simple iterate of N, in which case the conclusion follows immediately since N_θ is a mouse; or else M_θ is a simple iterate of N, in which case $M_\theta = N_\theta$ and the conclusion follows from the fact that the iteration terminates at θ, i.e. that $E_{\mathrm{top}}^{M_\theta,h} = E_{\mathrm{top}}^{N_\theta,k}$ for any choice of $h, k \in \{0, 1\}$. □(Lemma 4.6.3)

The uniqueness of the top measure is thus a consequence of iterability. The following lemma provides an iterability criterion in terms of ω-completeness.

Lemma 4.6.4. *Let $M = \langle J_\alpha^E, F^0, F^1 \rangle$ be a prebicephalus such that*

a) *$M \| \gamma$ is iterable whenever $\gamma < \alpha$;*

b) F^0, F^1 *are* ω*-complete.*

Then M is a bicephalus.

Proof. Notice the resemblance of the assumptions of Lemma 4.6.4 to those of Theorem 4.5.6. We shall therefore imitate the proof of that theorem in the bicephali context. First we observe that J_α^E is 0-iterable, which guarantees that M is 0-iterable using indices $\nu_i < \mathrm{ht}(M_i)$. We can therefore perform the same iteration of M as before to obtain a prebicephalus M' all of whose total measures are ω-complete, and the iteration map $\pi : M \to M'$. Using the argument from the proof of Lemma 4.5.2 we infer that M' is 0-iterable (also see the remark at the end the proof of Theorem 4.5.6). By Lemma 4.6.1, M is a bicephalus. □(Lemma 4.6.4)

Corollary 4.6.5. *Let M be a prebicephalus satisfying the assumptions of the previous lemma. Then the top measures of M agree.* □(Corollary 4.3.13)

Notes

The notions of mouse and iterability were introduced by Jensen in connection with core models below one measurable cardinal; these models are also called *Dodd–Jensen* core models and their theory was developed by Dodd and Jensen in [DJ81], [DJ81] and [DJ81a]. The complete treatment of their original approach can be also found in [D82]. Their approach differs substantially from that in our book, as they did not use partial extenders on the measure sequences and did not allow truncations in the course of iterations. The idea of using partial extenders together with truncations is due to Baldwin and Mitchell and the theory based on these two ingredients was fully developed by Mitchell and Steel [MiS94]. The Dodd–Jensen lemma also originates in the above mentioned work of Dodd and Jensen, although its formulation, of course, differs from the present one. The proof of iterability by realizing the premice of countable iterations (i.e. by embedding them into the original premouse) is originally due to Jensen. In the case of non-linear iteration, the main issue becomes the existence of cofinal well-founded branches. The basic result here is due to Martin and Steel [MS94] and the adaptation of the method to fine structural iteration is due to Mitchell–Steel [MiS94] and Steel [St96].

Chapter 5
Solidity and Condensation

The key result of this chapter is the solidity theorem asserting that iterability guarantees solidity. This fact will lead to a canonical well-ordering of mice and the condensation lemma, and will play a crucial role in the $\boldsymbol{K}^c$-construction as well as in our analysis of extender models.

5.1 Cores and Coiterations

We shall first focus on coiterations. Given a "sufficient amount" of solidity, we show that one side of any coiteration must be simple.

Definition. A mouse M is *presolid* iff $M||\beta$ is solid for every $\beta < \mathrm{ht}(M)$.

Lemma 5.1.1. *There is a Π_1-sentence φ such that for every premouse M,*

$$M \text{ is presolid} \quad \text{iff} \quad M \models \varphi.$$

Proof. Given a premouse N, the statements "$p = p_N$" and "$W_N^{\nu,p} \in N$ for every $\nu \in p$" can be uniformly expressed by an effective sequence of formulae. In other words, there is a fixed (i.e. not depending on N) effective sequence of formulae $\langle \varphi_n;\ n \in \omega \rangle$ such that the above two statements hold iff $N \models \varphi_n(p, \nu)$ for all n. Clearly, $\langle \varphi_n \rangle_n$ is an element of $J_2 \subset N$. That $M||\beta$ is solid amounts to saying that $(\forall n)(M||\beta \models \varphi_n)$, which is expressible over M using the satisfaction predicate from the proof of Lemma 4.1.2. The conclusion then follows from the fact that the satisfaction predicate is Π_1 over M. □(Lemma 5.1.1)

Lemma 5.1.2. *Suppose M is a solid premouse. Then every simple iterate of M is solid. The same also holds of n-solidity ($n \in \omega$).*

Proof. If M' is a simple iterate of M and $\pi : M \to M'$ is the corresponding iteration map, then $\pi(p_M) \in P^*_{M'}$ by Lemma 4.2.5. The lemma then follows immediately from Corollary 1.12.8. □(Lemma 5.1.2)

Corollary 5.1.3. *Let M be a presolid premouse. Then:*

a) *Every simple iterate of M is presolid.*

b) *Every non-simple iterate of M is solid.*

The same conclusions also hold for Σ_0-iterations. □(Corollary 5.1.3)

Lemma 5.1.4. *Let M and N be presolid coiterable premice. One side of the coiteration of M with N is simple.*

Proof. Suppose we truncate on both sides. By Corollary 4.4.3, M coiterates with N to a common premouse Q. Let i and j be the last truncation point on the M- and the N-side, respectively. Set $M^* = M_i||\alpha_i$ and $N^* = N_j||\alpha_j$. Both M^* and N^* are sound, $\omega\varrho^\omega_{M^*} \leq \kappa_i$ and $\omega\varrho^\omega_{N^*} \leq \kappa_j$. Since Q is a simple iterate of M^* above its ultimate projectum, $\omega\varrho^\omega_{M^*} = \omega\varrho^\omega_Q$. Since M^* is a non-simple iterate of M, M^* must be solid. This means that $\pi_{M^*Q}(p_{M^*}) = p_Q$ where π_{M^*Q} is the associated iteration map. Let X be the closure of $\omega\varrho^\omega_Q \cup \{p_Q\}$ under good $\Sigma^*(Q)$-functions. Since M^* is sound, M^* is the transitive collapse of $Q|X$ and π_{M^*Q} is the uncollapsing map. The same argument applied to N^* yields an analogous conclusion for N^* and π_{N^*Q}. It follows that $M^* = N^*$ and $\pi_{M^*Q} = \pi_{N^*Q}$.

Suppose that $i \neq j$ and w.l.o.g. $i < j$. By the normality of both iterations, $E^{N_j}_{\nu_i} = E^{M_j}_{\nu_i} = \emptyset$, since we have already used the measure with index ν_i at the i-th stage of the coiteration on M-side. On the other hand, $E^{N_j}_{\nu_i} = E^{M^*}_{\nu_i} \neq \emptyset$ since $N^* = M^*$. But this is a contradiction, so $i = j$. Since $\pi_{M^*Q} = \pi_{N^*Q}$, for each $x \in \mathcal{P}(\kappa_i) \cap M^*$ we have:

$$x \in E^{M^*}_{\nu_i} \longleftrightarrow \kappa_i \in \pi_{M^*Q}(x) \longleftrightarrow \kappa_i \in \pi_{N^*Q}(x) \longleftrightarrow x \in E^{N^*}_{\nu_i}.$$

Then $E^{M_i}_{\nu_i} = E^{N_i}_{\nu_i}$, which is impossible, as ν_i is a coiteration index. Contradiction.

□(Lemma 5.1.4)

Corollary 5.1.5. *Let M, N be presolid 0-coiterable premice. Then one side of their 0-coiteration is simple.* □(Corollary 5.1.5)

The above results can be improved for a pair of premice such that one of them is embeddable into the another. In this case we can specify which side of the coiteration is simple. This fact has several applications, e.g. it yields a sort of iterability criterion, it enables us to analyze the relationship between M and transitivized hulls taken over M and, as we shall see later, it plays a crucial role in the proof of the solidity theorem and condensation lemma.

Lemma 5.1.6. *Let $\bar{M}$ and M be premice, and let M be iterable and presolid. Suppose $\sigma : \bar{M} \to M$ is $\Sigma^{(n)}_0$-preserving and the critical point of the first measure applied on the $\bar{M}$-side in the coiteration of $\bar{M}$ with M is at least $\omega\varrho^{n+1}_{\bar{M}}$. Let $\bar{M}'$, M' be the result of this coiteration. Then:*

a) *$\bar{M}$ is coiterable with M.*

b) *The iteration from $\bar{M}$ to $\bar{M}'$ is simple and $\bar{M}'$ is an initial segment of M'.*

c) *$\bar{M}$ is iterable.*

Proof. By Lemma 4.2.4, $\bar{M}$ is normally iterable above the $(n+1)$-st projectum. Since the first measure applied on the $\bar{M}$-side is at least $\omega\varrho_{\bar{M}}^{n+1}$, the whole iteration on this side is above $\omega\varrho_{\bar{M}}^{n+1}$, which yields a).

By Lemma 5.1.1, $\bar{M}$ is presolid. Suppose b) fails. Then the iteration from M to M' would be simple and M' would be a proper initial segment of $\bar{M}'$, as follows from Lemma 5.1.4 and a simple discussion. Let $\pi : M \to M'$ be the iteration map. Then $\pi \circ \sigma : \bar{M} \to M'$ would be $\Sigma_0^{(n)}$-preserving and M' would be a non-simple normal iterate of $\bar{M}$. This contradicts Lemma 4.3.12.

Finally c) follows from b) since the iteration map $\bar{\pi} : \bar{M} \to \bar{M}'$ is Σ^*-preserving and $\bar{M}'$, being an iterate of M, is itself iterable. □(Lemma 5.1.6)

Lemma 5.1.7. *Let M be iterable and presolid, $\omega\varrho_M^{n+1} \leq \alpha < \omega\varrho_M^n$ and $p \in P_M^{n+1}$. Let $\bar{M}$ be the transitive collapse of the $\Sigma_1^{(n)}$-hull of $\alpha \cup \{p\}$, let $\sigma : \bar{M} \to M$ be the uncollapsing map and let $\sigma(\bar{p}) = p$. Then $\omega\varrho_{\bar{M}}^{n+1} = \omega\varrho_M^{n+1} \stackrel{\text{def}}{=} \omega\varrho^{n+1}$ and*

a) *$\bar{M}$ coiterates with M simply above $\omega\varrho^{n+1}$ to a common mouse;*

b) *$\bar{M}$ is a mouse;*

c) $\mathcal{P}(\omega\varrho^{n+1}) \cap \bar{M} = \mathcal{P}(\omega\varrho^{n+1}) \cap M$;

d) $\mathcal{P}(\omega\varrho^{n+1}) \cap \Sigma_1^{(m)}(\bar{M}) = \mathcal{P}(\omega\varrho^{n+1}) \cap \Sigma_1^{(m)}(M)$ *for all $m \in \omega$;*

e) $\omega\varrho_{\bar{M}}^m = \omega\varrho_M^m \stackrel{\text{def}}{=} \omega\varrho^m$ *for every $m > n$;*

f) *σ is Σ^*-preserving and* $\sup(\sigma''\omega\varrho_{\bar{M}}^n) = \omega\varrho_M^n$;

g) $\sigma'' P_{\bar{M}}^m \subset P_M^m$ *and* $(\sigma^{-1})'' P_M^m \subset P_{\bar{M}}^m$;

h) *p can be* lengthened *to a parameter $q \in P_M^m$ for every $m > n$;*

i) *if $\sigma(\bar{p}) = p_M$ then $\bar{p} = p_{\bar{M}}$. In particular, if $p = p_M$ then $\bar{p} = p_{\bar{M}}$. The analogous conclusion holds for p_M^m in place of p_M for every $m > n$.*

Proof. By the downward extension of embeddings lemma, σ is $\Sigma_1^{(n)}$-preserving. The argument from the proof of Lemma 1.11.2 shows that the $(n+1)$-st projecta $\omega\varrho^{n+1}$ of M and $\bar{M}$ are equal and that $\bar{p} \in P_{\bar{M}}^{n+1}$. We show that the critical point of the first disagreement between M and $\bar{M}$ is at least $\omega\varrho^{n+1}$, which will enable us to apply the previous lemma. By our assumptions on α, $\sigma \restriction \omega\varrho^{n+1} = \mathrm{id}$, so $E^{\bar{M}} \restriction \omega\varrho^{n+1} = E^M \restriction \omega\varrho^{n+1}$. This means that the first index used in the coiteration ν is at least $\omega\varrho^{n+1}$. Now, if $\nu > \omega\varrho^{n+1}$, then the corresponding critical point is greater or equal to $\omega\varrho^{n+1}$, so the only potential problem occurs when $\nu = \omega\varrho^{n+1}$ and $E_\nu^{\bar{M}} \neq E_\nu^M$. But then $\nu = \mathrm{cr}(\sigma)$ and ν is an extender index in M or $\bar{M}$. Setting κ to be the corresponding critical point, ν is the successor of κ both in M and $\bar{M}$. It follows that $\kappa^{+M} = \sigma(\nu) > \omega\varrho^{n+1} = \kappa^{+M}$, which is a contradiction. This shows that both sides of the coiteration of M with $\bar{M}$ are above $\omega\varrho^{n+1}$. The assumptions of the previous lemma are thus satisfied and $\bar{M}$ is coiterable with M. Furthermore, the $\bar{M}$-side of the coiteration is simple and above $\omega\varrho^{n+1}$, and $\bar{M}'$ is an initial segment of M' for the pair of premice at the end of the coiteration. $\bar{M}'$ can be thus considered an iterate of M and the corresponding iteration is just the iteration from M to M' succeeded by a truncation, if necessary. We show that the iteration from M to $\bar{M}'$ is simple, which will complete the proof of a). Suppose the contrary. Let $A(\eta, \xi)$ be a $\Sigma_1^{(n)}(M)$ relation such that $a = A(-, p) \cap \omega\varrho^{n+1} \notin M$. Let $\bar{A}$ be the $\Sigma_1^{(n)}(\bar{M})$ relation by the same definition. Since σ is $\Sigma_1^{(n)}$-preserving, $\bar{A}(-, \bar{p}) \cap \omega\varrho^{n+1} = a$. The iteration on the $\bar{M}$-side is above $\omega\varrho^{n+1}$, so, by Lemma 4.2.2, $a \in \Sigma_1^{(n)}(\bar{M}')$. Setting ι to be the last truncation point in the iteration from M to $\bar{M}'$ (ι equal to the length of the coiteration is possible), the iteration from $M_\iota||\alpha_\iota$ to $\bar{M}'$ is simple and above $\omega\varrho^{n+1}$. As a result, we have that $\omega\varrho^{n+1}$ is the $(n+1)$-st projectum of $M_\iota||\alpha_\iota$ and that $a \in \Sigma_1^{(n)}(M_\iota||\alpha_\iota)$. Hence $a \in M_\iota$. An iterative (if necessary) application of Lemma 4.2.1 yields that $a \in M$, which contradicts our choice of a and completes the proof of a). Clause b) follows immediately from the previous lemma.

Denote the common simple iterate of M with $\bar{M}$ by N. (Thus $M = N = \bar{M}'$.) Clauses c) – e) are consequences of Lemma 4.2.2; it suffices to observe that the corresponding quantities computed for M (resp. $\bar{M}$) agree with those computed for N. The first part of f) follows from e) by the inductive argument from the proof of Lemma 1.11.2. We prove the second part. Pick $\bar{A}(\eta, \xi)$ as above; then there is a $\Sigma_1(\bar{M}^{n,\bar{p}\restriction n})$ relation $\bar{B}(\eta, \xi)$ such that $\bar{B}(\eta, \bar{p}^n) \longleftrightarrow \bar{A}(\eta, \bar{p})$ for all $\eta < \omega\varrho^{n+1}$. If $\omega\varrho = \sup(\sigma''\omega\varrho^n_{\bar{M}}) < \omega\varrho^n_M$, then $a = \bar{B}(-, \bar{p}^n) \cap \omega\varrho^{n+1} = B(-, p^n) \cap \omega\varrho^{n+1}$ where $B(\eta, \xi)$ is $\Sigma_1(\tilde{M})$ by the same definition as $\bar{B}$ and $\tilde{M} = \langle |J^E_\varrho|, A^{n,p\restriction n} \cap |J^E_\varrho| \rangle$ (here $E = E^M$). Since $\tilde{M} \in M$, a must be an element of M as well, but this contradicts c).

To see g), pick $\bar{q}, q$ such that $\sigma(\bar{q}) = q$ and relations $\bar{A}$ (resp. A) which are $\Sigma_1^{(m-1)}(\bar{M})$ (resp. $\Sigma_1^{(m-1)}(M)$) by the same definition. By f), $\omega\varrho^m_{\bar{M}} \cap \bar{A}(-, \bar{q}) = \omega\varrho^m_M \cap A(-, q)$. Setting a to be this common value and applying c) we obtain $\bar{q} \in P^m_{\bar{M}}$ iff $a \notin \bar{M}$ iff $a \notin M$ iff $q \in P^m_M$. Clause i) is an immediate consequence: if $\bar{p} \neq p_{\bar{M}}$, then $p_{\bar{M}} <^* \bar{p}$ and $\sigma(p_{\bar{M}}) <^* p_M$. But $p_{\bar{M}} \in P^m_M$ by g). Contradiction.

It remains to prove h). Let $\sigma(\bar{p}) = p$. Note that $\bar{p} \in R^{n+1}_{\bar{M}}$. Let $\bar{q} \in P^m_{\bar{M}}$. To each $\alpha \in \bar{q} - \omega\varrho^{n+1}$ there is a $\xi_\alpha < \omega\varrho^{n+1}$ such that $\alpha = \tilde{h}^{n+1}_{\bar{M}}(\xi_\alpha, \bar{p})$. Using this fact it is easy to see that any relation which is $\Sigma_1^{(m-1)}(\bar{M})$ in $\bar{q}$ is $\Sigma_1^{(m-1)}(\bar{M})$ in $\bar{p}^* = \bar{p} \cup (\bar{q} \cap \omega\varrho^{n+1}) \cup \{\xi_\alpha;\ \alpha \in q - \omega\varrho^{n+1}\}$. It follows that $\bar{p}^* \in P^m_{\bar{M}}$. By g), $p^* = \sigma(\bar{p}^*) \in P^m_M$. Since $\bar{p}^*$ is a lengthening of $\bar{p}$, p^* is a lengthening of p.

□(Lemma 5.1.7)

Corollary 5.1.8. *Let M be a presolid mouse. Then $p^n_M = p^m_M - \omega\varrho^n_M$ whenever $n < m$. The same conclusion holds with p_M in place of p^m_M.*

Proof. This is an immediate consequence of clause h) of the previous lemma and Lemma 1.9.7. □(Corollary 5.1.8)

The above lemma shows, among other things, that transitivized hulls of projecta containing good parameters keep a considerable amount of information about the original structure and are moreover sound. For this reason it is crucial to consider these hulls in various arguments and constructions (see e.g. the proof of Lemma 5.1.4) since the original structure might contain a lot of "random" information, which can accumulate, for example, in the course of some iteration. Such hulls are called *cores*, as their transitive collapses constitute the "minimal iterable premice" possessing the relevant fine structural characteristics of the original structure. The property described by clause c) is also called the *universality* of good parameters. Note also that the $\Sigma_1^{(n)}$-hull of $\alpha \cup \{p\}$ from Lemma 5.1.7 is equal to the Σ^*-hull of $\alpha \cup \{p\}$.

Definition. Let M be a mouse and $\omega\varrho_M^\omega \leq \alpha \in \mathbf{On} \cap M$. We say that $\bar{M}$ is the *core* of *M above* α and write $\bar{M} = \mathrm{core}_\alpha(M)$, iff there is a Σ^*-preserving map $\sigma : \bar{M} \to M$ such that

a) $\sigma \restriction \alpha = \mathrm{id}$;

b) $\sigma(p_{\bar{M}}) = p_M$;

c) $\bar{M}$ is the closure of $\alpha \cup p_{\bar{M}}$ under good $\Sigma^*(\bar{M})$ functions.

The map σ is called the *core map above* α. If $\alpha = \omega\varrho_M^n$, we call $\bar{M}$ the *n-th core* of M and write $\mathrm{core}^n(M)$; if $\alpha = \omega\varrho_M^\omega$, we call $\bar{M}$ briefly the *core* of M and write $\mathrm{core}(M)$.

Corollary 5.1.9. *Given a presolid iterable premouse M, $\mathrm{core}^n(M)$ is n-sound for every $n \in \omega$. In particular, $\mathrm{core}(M)$ is sound.* □(Corollary 5.1.9)

5.2 The Solidity Theorem

We are now ready to prove the central result of this chapter.

Theorem 5.2.1. *Every iterable premouse is solid.*

Proof. We proceed by the induction on the length of mice. In order to carry out the induction step, we have to prove that every presolid mouse is solid. Fix such a mouse M; for $\nu \in p_M$ we show that the standard witness W_M^ν is an element of M. Let n be such that $\omega\varrho_M^{n+1} \leq \nu < \omega\varrho_M^n$. The canonical witness map $\sigma : W \to M$ is $\Sigma_1^{(n)}$-preserving. Since W is the transitive collapse of the $\Sigma_1^{(n)}(M)$-hull of $\nu \cup \{p_M - (\nu+1)\}$, $W = \tilde{h}_W^{n+1}(\nu \cup \{q\})$ where q is the σ-preimage of $\nu \cup \{p_M - (\nu + 1)\}$; as a consequence of this we have that $\omega\varrho_W^{n+1} \leq \nu$. Recall also that $\nu = \mathrm{cr}(\sigma)$ (see the proof of Lemma 1.12.3). Set $\tau = \sigma(\nu)$.

Let $D = \mathrm{dom}(\tilde{h}_W^{n+1}) \cap \nu$ and E, I be relations on $D \times D$ defined as follows:

$$\xi \, E \, \zeta \longleftrightarrow \tilde{h}_W^{n+1}(\xi, q) \in \tilde{h}_W^{n+1}(\zeta, q);$$
$$\xi \, I \, \zeta \longleftrightarrow \tilde{h}_W^{n+1}(\xi, q) = \tilde{h}_W^{n+1}(\zeta, q).$$

Then I is a congruence relation on $\mathbb{D} = \langle D, E \rangle$ and $\mathbb{D}/I$ is isomorphic to W. It follows that W can be obtained from $\mathbb{D}/I$ by collapsing. Note also that the sets D, E, I are $\Sigma_1^{(n)}(W)$ in q and the whole triple can be recursively encoded in a single $a \subset \nu$, which is also $\Sigma_1^{(n)}(W)$ in q. (E.g. set $\prec \xi_1, \xi_2, \xi_3, \xi_4, \xi_5 \succ \in a$ iff $\xi_1 \in D$, $\xi_2 \, E \, \xi_3$ and $\xi_4 \, I \, \xi_5$.) Our strategy is to show that $a \in M$. Using acceptability we infer that in fact $a \in J_\tau^E$, where $E = E^M$, and, since J_τ^E is a ZFC$^-$ model, we can decode the triple D, E, I from a, construct $\mathbb{D}/I$ and collapse it *inside* J_τ^E. This shows that W is in fact an element of $M\|\tau$.

To see that $a \in M$ we shall utilize a comparison argument. We coiterate W with M and, setting Q to be the last model on the W-side, we prove:

Claim 1. The iteration from W to Q is simple and above ν.

Claim 2. The iteration from M to Q is non-simple and beyond ν.

By Claim 1 and Lemma 4.2.2, a is $\mathbf{\Sigma}_1^{(n)}(Q)$. By Claim 2, there is a last truncation point on the M-side, call it ι. Suppose first that $\kappa_\iota \geq \nu$. Then a is $\mathbf{\Sigma}_1^{(n)}(M_\iota\|\alpha_\iota)$ since the iteration from $M_\iota\|\alpha_\iota$ to Q is simple and above ν. Hence $a \in M_\iota$. If the whole M-side of the coiteration is above ν, we infer that $a \in M$ using Lemma 4.2.1. Otherwise the first iteration index is ν and the above argument yields only that $a \in M_1$, where $M_1 = \mathrm{Ult}^*(M\|\alpha_0, E_\nu^M)$. Let $\pi_{0,1}^M$ be the associated ultrapower map. Then $a = \pi_{0,1}^M(f)(\kappa_0)$ for some $f \in \Gamma(\kappa_0, M)$. Let $\langle f_\xi^*;\ \xi < \nu \rangle$ be the sequence of functions from the proof of Lemma 4.1.2 such that $\pi_{0,1}^M(f_\xi^*)(\kappa_0) = \xi$ for all $\xi < \nu$. This sequence is Σ_1-definable over J_ν^E in κ_0 (and therefore an element of $M\|\tau$). Given that $\xi < \nu$, we have

$$\xi \in a \longleftrightarrow \pi_{0,1}^M(f_\xi^*)(\kappa_0) \in \pi_{0,1}^M(f)(\kappa_0) \longleftrightarrow \{\eta;\ f_\xi^*(\eta) \in f(\eta)\} \in E_\nu^M.$$

It follows that the relation $\xi \in a$ is definable over $M\|\alpha_0$. But notice that $\alpha_0 < \mathrm{ht}(M)$ since E_ν^M is a partial measure. Consequently, $a \in M$.

It remains to discuss the case where $\kappa_\iota < \nu$. This is possible only if $\iota = 0$ and $\nu_0 = \nu$. The above argument yields that a is $\mathbf{\Sigma}_1^{(n)}(M_1)$; say $\xi \in a$ iff $A_1(\xi, r)$ where A_1 is a $\Sigma_1^{(n)}(M_1)$ relation and $r = \pi_{0,1}^M(g)(\kappa_0)$ for some $g \in \Gamma(\kappa_0, M)$. Let B_1 be a $\Sigma_0^{(n)}(M_1)$ relation such that $A_1(\xi, \zeta) \longleftrightarrow (\exists z^n) B_1(z^n, \xi, \zeta)$ and B be $\Sigma_0^{(n)}(M)$ by the same definition as B_1. Then

$$\xi \in a \longleftrightarrow (\exists \vartheta < \omega \varrho_{M\|\alpha_0}^n)(\exists z^n \in \pi_{0,1}^M(S_\vartheta^E)) B_1(z^n, \pi_{0,1}^M(f_\xi^*)(\kappa_0), \pi_{0,1}^M(g)(\kappa_0))$$

where, using Łos' theorem, the right side can be equivalently expressed as

$$(\exists\vartheta < \omega\varrho^n_{M||\alpha_0})[\{\eta < \kappa_0;\ (\exists z^n \in S^E_\vartheta)B(z^n, f^*_\xi(\eta), g(\eta))\} \in E^M_\nu],$$

which is again definable over $M||\alpha_0$. So, as before, we infer that $a \in M$.

We shall now return back to the coiteration of W with M and the above claims. First of all we have to show that these premice are coiterable, which is not obvious. Here we intend to apply Lemma 5.1.6; for this purpose it suffices to verify that the critical point of the first measure applied on the W-side of the coiteration of M with W is at least ν. This follows easily once we have shown that M and W agree below $\nu + 1$ whenever ν indexes a measure in W.

Claim 3. Suppose $E^W_\nu \neq \varnothing$. Then $E^M_\nu = E^W_\nu$.

Notice that Claim 3 together with Lemma 5.1.6 immediately imply not only the coiterability of W with M, but also Claim 1. Furthermore, the fact that the M-side of the coiteration is beyond ν is trivial. Thus, it suffices to focus on the proofs of Claim 2 and Claim 3.

Proof of Claim 3. Let $\kappa = \mathrm{cr}(E^W_\nu)$. Then $\nu = \kappa^{+W}$ and, since ν is the critical point of σ, $\nu < \tau$ where $\tau = \sigma(\nu) = \kappa^{+M}$. Let δ be maximal such that $\nu = \kappa^{+M||\delta}$ and $M^* = M||\delta$. Then:

- $\omega\varrho^\omega_{M^*} \leq \kappa$.
- $\mathcal{P}(\kappa) \cap W = \mathcal{P}(\kappa) \cap M^*$.
- M^* is extendible by E^W_μ and $\tilde{M} = \mathrm{Ult}^*(M^*, E^W_\mu)$ is iterable.

The first two clauses are trivial; the last one requires an argument. Since ν indexes a measure in W, τ indexes a measure in M with the same critical point κ. Furthermore, for every $x \in \mathcal{P}(\kappa) \cap W$ we have $x \in E^W_\nu$ iff $x = \sigma(x) \in E^M_\tau$, i.e. $E^W_\nu = E^M_\tau \cap J^E_\nu$ (here $E = E^W \restriction \nu = E^M \restriction \nu$). Let $M' = \mathrm{Ult}^*(M, E^M_\tau)$ and π' be the associated ultrapower map. Then $\pi'(M^*)$ is an initial segment of M' and is therefore iterable. We now show that M^* is Σ^*-extendible by E^W_ν by constructing a Σ_0-preserving embedding $\sigma^* : \mathbb{D}^*(M^*, E^W_\nu) \to \pi'(M^*)$. The embedding σ^* is defined in the obvious way, i.e. $\sigma^*([f]) = \pi'(f)(\kappa)$ for every $f \in \Gamma(\kappa, M^*)$. For a Σ_0 formula φ and $f_1, \ldots, f_\ell \in \Gamma(\kappa, M^*)$ we obtain

$$\mathbb{D}^*(M^*, E^W_\nu) \models \varphi([f_1], \ldots, [f_\ell]) \longleftrightarrow u^{M^*,\varphi}_{f_1,\ldots,f_\ell} \in E^W_\nu \longleftrightarrow u^{M^*,\varphi}_{f_1,\ldots,f_\ell} \in E^M_\tau.$$

The latter equivalence follows from the equality $E^W_\nu = E^M_\tau \cap J^E_\nu$. We recall that

$$u^{M^*,\varphi}_{f_1,\ldots,f_\ell} = \{\eta \in \kappa;\ M^* \models \varphi(f_1,(\eta), \ldots, f_\ell(\eta))\}.$$

Note also that $f_1, \ldots, f_\ell \in M$, so applying Łos' Theorem to the rightmost side of the above equivalence yields

$$M' \models [\pi'(M^*) \models \varphi(\pi'(f_1)(\kappa), \ldots, \pi'(f_\ell)(\kappa))]$$

which can be equivalently expressed by

$$\pi'(M^*) \models \varphi(\pi'(f_1)(\kappa), \ldots, \pi'(f_\ell)(\kappa)).$$

This clearly guarantees that σ^* is both correctly defined and Σ_0-preserving. It follows that $\tilde{M} = \mathrm{Ult}^*(M^*, E_\nu^W)$ exists and is a premouse. By a slight abuse of notation we can write that $\sigma^* : \tilde{M} \to \pi'(M)$. An argument similar to that above shows that σ^* is in fact $\Sigma_0^{(m)}$-preserving, where m is such that $\omega\varrho_{M^*}^{m+1} \leq \kappa < \omega\varrho_{M^*}^m$; furthermore, it is easy to see that $\sigma^* \restriction \kappa = \mathrm{id}$. The weak amenability of E_ν^W w.r.t. M^* guarantees that $\omega\varrho_{\tilde{M}}^{m+1} \leq \omega\varrho_{M^*}^{m+1}$. That M^* is a mouse will follow from Lemma 5.1.6 once we have proved that the critical point of $E_\vartheta^{\tilde{M}}$ is at least $\omega\varrho_{\tilde{M}}^{m+1}$, provided that the least index of disagreement ϑ between $\tilde{M}$ and $\pi'(M^*)$ indexes an extender in $\tilde{M}$. But this follows immediately from the fact that κ is a limit cardinal in $\tilde{M}$.

In the following we shall analyze the coiteration of M^* with $\tilde{M}$. M^* agrees with W below ν and ν is a cardinal in M^*, therefore $\tilde{W} = \mathrm{Ult}(J_\nu^E, E_\nu^W)$ is an initial segment of $\tilde{M}$. From the coherency condition we infer that $E^{\tilde{M}} \restriction \nu = E^W \restriction \nu = E^{M^*} \restriction \nu$ and $E_\nu^{\tilde{M}} = E_\nu^{\tilde{W}} = \varnothing$. Furthermore, $\kappa^{+\tilde{M}} = \nu = \kappa^{+M^*}$. We thus conclude:

- M^* coiterates with $\tilde{M}$ above κ;
- the $\tilde{M}$-side of the coiteration is above ν.

Let $\pi : M^* \to \tilde{M}$ be the ultrapower map. Since E_ν^W is weakly amenable w.r.t. M^*, π is $\Sigma_0^{(m)}$-preserving and cofinal, and the power set of κ in the sense of M^* agrees with that in the sense of $\tilde{M}$. From the latter we infer that if A is $\Sigma_1^{(m)}(M^*)$ in p_{M^*} such that $a = A \cap \kappa \notin M^*$ (note that such A exist) then $a \notin \tilde{M}$. By the proof of clause a) of Lemma 5.1.7, M^* coiterates with $\tilde{M}$ simply to a common mouse, say Q. Let π^* and $\tilde{\pi}$ denote the iteration map from M^* and $\tilde{M}$ to Q, respectively. Using further arguments from the proof of Lemma 5.1.7 and the above established properties of the coiteration and the map π we show:

- $\omega\varrho_{M^*}^{m+1} = \omega\varrho_Q^{m+1} = \omega\varrho_{\tilde{M}}^{m+1}$; denote this common value by $\omega\varrho^{m+1}$;
- $\tilde{\pi} \circ \pi(p) \in P_Q^{m+1}$, where $p = p_{M^*} - \omega\varrho^{m+1}$;
- π is Σ^*-preserving.

The premouse M^*, being a proper initial segment of M, is both sound and, by our induction hypothesis, solid. Hence $W_{M^*}^{\alpha,p} \in M^*$ for every $\alpha \in p$. Since $\tilde{\pi} \circ \pi$

is $\Sigma_1^{(m)}$-preserving, $\tilde{\pi} \circ \pi(W_{M^*}^{\alpha,p})$ (notice the slight abuse of notation) is a generalized witness for $\tilde{\pi} \circ \pi(\alpha)$ for all $\alpha \in p$. By Lemma 1.12.2, $\tilde{\pi} \circ \pi(p) = p_{Q,m+1}$. Since π^* is an iteration map, $\pi^*(p) \in P_Q^{m+1}$; the Dodd–Jensen lemma then guarantees that $\pi^*(p) \leq^* \tilde{\pi} \circ \pi(p)$. But no parameter from P_Q^{m+1} can be strictly less than $p_{Q,m+1}$, so we conclude

$$\pi^*(p) = p_{Q,m+1} = \tilde{\pi} \circ \pi(p). \tag{5.2.1}$$

Since the coiteration is above κ, we have

$$(\tilde{\pi} \circ \pi) \restriction \kappa = \mathrm{id} = \pi^* \restriction \kappa. \tag{5.2.2}$$

By the soundness of M^*, every $x \in M^*$ is of the form $\tilde{h}_{M^*}^{m+1}(\xi, p)$ for some $\xi < \kappa$. From this, (5.2.1), (5.2.2) and the preservation properties of the maps π^* and $\tilde{\pi} \circ \pi$ we infer

$$\pi^*(x) = \tilde{h}_Q^{m+1}(\xi, \pi^*(p)) = \tilde{h}_Q^{m+1}(\xi, \tilde{\pi} \circ \pi(p)) = \tilde{\pi} \circ \pi(x),$$

which means that $\tilde{\pi} \circ \pi = \pi^*$. In particular, $\mathrm{cr}(\pi^*) = \kappa$. It follows that the first measure applied on the M^*-side of the coiteration must be $E_\nu^{M^*}$, since ν is the only ordinal which can index a total measure with critical point κ on the M^*-sequence. This shows that $E_\nu^{M^*} \neq \varnothing$. Moreover, given $a \in \mathcal{P}(\kappa) \cap M^*$,

$$a \in E_\nu^{M^*} \longleftrightarrow \kappa \in \pi^*(a) = \tilde{\pi} \circ \pi(a) \longleftrightarrow \kappa \in \pi(a) \longleftrightarrow a \in E_\nu^W.$$

The middle equivalence follows from the fact that $\pi(a)$ agrees with $\tilde{\pi} \circ \pi(a)$ below ν, which in turn is a consequence of our earlier observation that the $\tilde{M}$-side of the coiteration is above ν. From the above computation we immediately obtain $E_\nu^W = E_\nu^{M^*} = E_\nu^M$ (where the latter equality is trivial). □(Claim 3)

Proof of Claim 2. As we have already seen above, it suffices to prove the first part of the claim. Suppose for a contradiction that the iteration from M to Q is simple. We show:

- $\omega\varrho_W^{n+1} = \omega\varrho_Q^{n+1} = \omega\varrho_M^{n+1} \stackrel{\text{def}}{=} \omega\varrho^{n+1}$;
- $\mathcal{P}(\omega\varrho^{n+1}) \cap W = \mathcal{P}(\omega\varrho^{n+1}) \cap Q = \mathcal{P}(\omega\varrho^{n+1}) \cap M$.

The equalities on the left hand sides follow from Claim 1 and the fact that $\omega\varrho_W^{n+1} \leq \nu$. To see the equalities on the right hand sides we show that the iteration from M to Q is above $\omega\varrho_M^{n+1}$. This is immediate if the iteration is above ν. But if the critical point κ_i of the first ultrapower is less than ν (it follows that $i = 0$, but this is not relevant for the argument), then κ_i cannot be smaller than $\omega\varrho_M^{n+1}$, otherwise $\omega\varrho_Q^{n+1} \geq \omega\varrho_{M_{i+1}}^{n+1} > \pi_{i,i+1}^M(\kappa) > \nu$ (here $\pi_{i,i+1}^M$ is the associated ultrapower map), which contradicts our earlier computations.

Using the above equalities we derive a contradiction as follows. We first recall that, by Corollary 5.1.8, $p_M - \omega\varrho^{n+1} = p_{M,n+1}$. Pick $r \in P_W^{n+1}$. Then $r = r_1 \cup r_2$ where $r_1 = r - \nu$ and $r_2 = r \cap \nu$. Let $A(\xi, \zeta_1, \zeta_2)$ be a $\Sigma_1^{(n)}(W)$ relation such that

$a = A(-, r_1, r_2) \cap \omega\varrho^{n+1} \notin W$. By the construction of W, $r_1 = \tilde{h}_W^{n+1}(\xi, q)$ for some $\xi < \nu$ (recall that $\sigma(q) = p_M - (\nu + 1)$). This means that for $\eta < \omega\varrho^{n+1}$,

$$A(\eta, r_1, r_2) \longleftrightarrow A(\eta, \tilde{h}_W^{n+1}(\xi, q), r_2) \overset{\text{def}}{\longleftrightarrow} B(\eta, q, \xi, r_2).$$

Then B is also $\Sigma_1^{(n)}(W)$. Let B' be $\Sigma_1^{(n)}(M)$ by the same definition. Since σ is $\Sigma_1^{(n)}$-preserving, for every $\eta < \omega\varrho^{n+1}$ we have

$$B(\eta, q, \xi, r_2) \longleftrightarrow B'(\eta, p_M - (\nu + 1), \xi, r_2),$$

i.e. $a = B'(-, p_M - (\nu+1), \xi, r_2) \cap \omega\varrho^{n+1}$. Since M, W have the same power sets of $\omega\varrho^{n+1}$, $a \notin M$. This means that we can define a new $\Sigma_1^{(n)}(M)$ subset of $\omega\varrho_M^{n+1}$ from the parameter $p_M - (\nu + 1) \cup \{\xi\} \cup r_2 <^* p_{M,n+1}$. Contradiction.

□(Claim 2, Theorem 5.2.1)

5.3 Consequences of Solidity

In this section we state most of the more or less immediate consequences of solidity and of the above proof. We begin by recapitulating the results of the previous two sections, which were stated under the superfluous assumption of presolidity. Notice that, given an iterable premouse M, each of its initial segments is itself iterable and therefore solid. (Thus, M is both solid and presolid).

Lemma 5.3.1. *Let M, N be a pair of coiterable premice. Suppose that M is iterable. Then at least one side of their coiteration is simple. The same conclusion also holds for k-coiterations.*

Proof. This follows by the proof of Lemma 5.1.4. Define M^*, N^* as in that proof; then M^* and Q, being non-simple iterates of M, are iterable. The iteration map $\pi_{N^*Q} : N^* \to Q$ is Σ^*-preserving, which also implies that N^* is iterable. By Theorem 5.2.1, both M^* and N^* are solid. This suffices to carry out the rest of the proof.

□(Lemma 5.3.1)

Lemma 5.3.2. *Let $M = \langle J_\nu^E, F \rangle$ be 0-iterable and $\bar{\nu} < \nu$ be such that*

$$M^* = \langle J_{\bar{\nu}}^E, F \cap J_{\bar{\nu}}^E \rangle$$

is a premouse. Then $F \cap J_{\bar{\nu}}^E = E_{\bar{\nu}}$.

Proof. This follows by the proof of Claim 3 from the proof of Theorem 5.2.1. Let M^* be as in that proof, $M' = \mathrm{Ult}(M, F)$ and let π' be the associated ultrapower map. Then $\pi'(M^*)$, being a proper initial segment of the 0-iterable premouse M', is $*$-iterable. The rest of the proof can be repeated literally. □(Lemma 5.3.2)

Lemma 5.3.3. *The lemmata and corollaries of Section* 5.1 *hold for any mice (resp. for* 0*-mice, if applicable).* □(Lemma 5.3.3)

Definition. Let M be a mouse, $n \in \omega$ and $\omega\varrho_M^{n+1} \leq \alpha < \omega\varrho_M^n$. We say that M is *sound above* α iff $M = \tilde{h}_M^{n+1}(\alpha \cup \{p_M\})$.

This definition generalizes the definition of n-soundness; indeed, M is n-sound iff M is sound above $\omega\varrho_M^n$ and M is sound iff M is sound above $\omega\varrho_M^\omega$. We stress that the main reason why this definition makes sense is that p_M is a lengthening of each $p_{M,n}$ (this is a consequence of solidity, see Corollary 5.1.8).

Lemma 5.3.4. *Let M be a mouse, $\omega\varrho_M^{n+1} \leq \alpha < \omega\varrho_M^n$ and $\bar{M} = \mathrm{core}_\alpha(M)$.*

a) *$\bar{M}$ is sound above α. In particular, if $\alpha = \omega\varrho_M^{n+1}$ then $\bar{M}$ is $(n+1)$-sound.*

b) *$\bar{M}$, M coiterate simply to a common mouse Q above $\omega\varrho^{n+1}$, where $\omega\varrho^{n+1} = \omega\varrho_{\bar{M}}^{n+1} = \omega\varrho_Q^{n+1} = \omega\varrho_M^{n+1}$.*

c) *If $\alpha = \omega\varrho^{n+1}$, the following diagram commutes:*

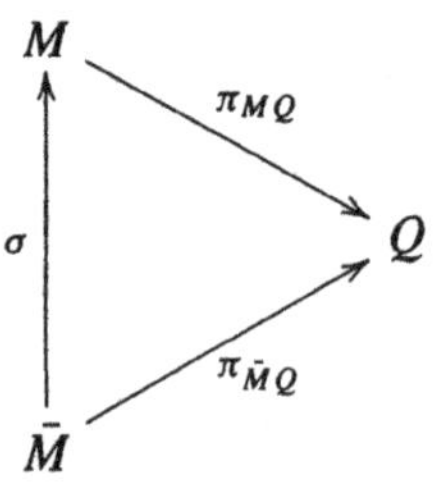

Here σ is the core map. In particular,

d) $\mathrm{rng}(\pi_{\bar{M}Q}) \subset \mathrm{rng}(\pi_{MQ})$ *and*

e) $\mathcal{P}(\omega\varrho^{n+1}) \cap \bar{M} = \mathcal{P}(\omega\varrho^{n+1}) \cap M$ *and* $(\omega\varrho^{n+1})^{+\bar{M}} = (\omega\varrho^{n+1})^{+M}$.

Proof. Clause a) only quotes our previous results and d) is an immediate consequence of c). For b) we use Lemma 5.3.2 to see that the $\bar{M}$-side of the coiteration is above α; the rest follows by the proof of Lemma 5.1.7. Finally we verify c). Here we shall proceed as in the proof of Claim 3 of Theorem 5.2.1. We observe:

- $(\pi_{MQ} \circ \sigma) \restriction \omega\varrho^{n+1} = \mathrm{id} \restriction \omega\varrho^{n+1} = \pi_{\bar{M}Q} \restriction \omega\varrho^{n+1}$.
- $\pi_{MQ} \circ \sigma(p_{\bar{M}}) = p_Q = \pi_{\bar{M}Q}(p_{\bar{M}})$.

The former is a consequence of b); the latter can be seen as follows. Using the facts that $\sigma(p_{\bar{M}}) = p_M$ and $\pi_{MQ}, \pi_{\bar{M}Q}$ are iteration maps we infer that $\pi_{MQ} \circ \sigma(p_{\bar{M}}), \pi_{\bar{M}Q}(p_{\bar{M}}) \in P_Q^*$. The premouse $\bar{M}$, being iterable, is solid and therefore $W_M^\nu \in M$ for every $\nu \in p_M$. Since all maps in the above diagram are Σ^*-preserving, each $\pi_{MQ} \circ \sigma(W^{\bar{\nu}}_{\bar{M}})$ (resp. $\pi_{\bar{M}Q}(W^{\bar{\nu}}_{\bar{M}})$) is a generalized witness for $\pi_{MQ} \circ \sigma(\nu)$ (resp. $\pi_{\bar{M}Q}(\nu)$) w.r.t. $\pi_{MQ} \circ \sigma(p_{\bar{M}})$, $\bar{Q}$ (resp. $\pi_{\bar{M}Q}(p_{\bar{M}})$, Q) which yields the desired equality.

By the $(n+1)$-soundness of $\bar{M}$, every $x \in \bar{M}$ is of the form $\tilde{h}^{n+1}_{\bar{M}}(\xi, p_{\bar{M}})$ for some $\xi < \omega\varrho^{n+1}$; using the above equalities and the preservation properties of the maps in the above diagram we infer exactly as in the proof of Claim 3 of Theorem 5.2.1 that $\pi_{MQ} \circ \sigma = \pi_{\bar{M}Q}$. d) then follows immediately.

The first part of e) follows immediately from Lemma 5.1.7. The second part follows trivially if $\bar{\tau} = (\omega\varrho^{n+1})^{+\bar{M}} = \mathrm{ht}(\bar{M})$: In this case, $\bar{M}$ is not moved in the coiteration. Thus, $J^{\bar{E}}_{\bar{\tau}}$ is an initial segment of M (we set $\bar{E} = E^{\bar{M}}$ and $E = E^M$), so its height must be equal to $(\omega\varrho^{n+1})^{+M}$ since $\bar{M}$ and M contain the same subsets of $\omega\varrho^{n+1}$. Now suppose $\bar{\tau}$ is an element of $\bar{M}$. Then $\tau = (\omega\varrho^{n+1})^{+M} \in M$ and both $J^{\bar{E}}_{\bar{\tau}}, J^{E}_{\tau}$ are ZFC$^-$ models. We again apply the first part of e). Given $\xi < \bar{\tau}$, there is a bijection $f : \omega\varrho^{n+1} \to \bar{\tau}$ in $\bar{M}$. This bijection induces a binary relation R on $\omega\varrho^{n+1}$ so that f is an isomorphism between $\langle \omega\varrho^{n+1} \times \omega\varrho^{n+1}, R\rangle$ and $\langle \xi, \in\rangle$. We can thus encode ξ into a subset a of $\omega\varrho^{n+1}$ in the same manner we encoded the standard witness in the proof of the solidity theorem. Then $a \in \bar{M}$, so $a \in M$ and by acceptability, $a \in J^E_\tau$. It follows that $R \in J^E_\tau$ and we can collapse R inside J^E_τ; this transitive collapse is equal to ξ. Hence $\xi < \tau$. This proves that $\bar{\tau} \le \tau$; the converse follows by the symmetry of the argument. □(Lemma 5.3.4)

Remark. Clause c) holds even if $\alpha \neq \omega\varrho^{n+1}$, but only for mice for measures (not necessarily of order 0). We shall see later that this fails if we generalize the theory of mice for extenders. More precisely, the following argument will also go through for measures of higher order, but will fail in the case of extenders with more than one generator.

To prove c) for arbitrary α, we show that the M-side of the coiteration is above α. This is immediate if $\bar{\nu} = \mathrm{cr}(\sigma)$ is a limit cardinal in $\bar{M}$ (and therefore also in M). Thus, if the M-side is not above α, then it is easy to see (using the fact that both sides of the coiteration are simple and above κ) that $\mathcal{P}(\kappa) \cap (\bar{M}) = \mathcal{P}(\kappa) \cap M$, where κ is the critical point of the first measure used on the M-side. On the other hand, since $\sigma(\bar{\nu}) > \bar{\nu}$ and $\bar{M}$ agrees with M below $\bar{\nu} = \kappa^{+\bar{M}}$, M must contain more subsets of κ than $\bar{M}$. Contradiction.

Definition. Let M, N be mice. We say that N is a *derived iterate* of M iff there is a common simple iterate Q of M, N such that $\mathrm{rng}(\pi_{MQ}) \subset \mathrm{rng}(\pi_{NQ})$. The map $\sigma = \pi^{-1}_{NQ} \circ \pi_{MQ}$ is called the *derived iteration map*.

That the notion of derived iteration map is meaningful follows from the next lemma.

Lemma 5.3.5. *Let N be a derived iterate of M. The derived iteration map is uniquely determined, i.e. it does not depend on Q, π_{MQ} and π_{NQ}.*

Proof. Let Q, π_{MQ}, π_{NQ} witness that N is a derived iterate of M; set $\sigma = \pi^{-1}_{NQ} \circ \pi_{MQ}$. Coiterate M with N; let Q^* be the (not necessarily properly) shorter of the pair of mice at the end of the coiteration. We show:

$$M, N \text{ coiterate simply to } Q^*, \tag{5.3.1}$$

$$\operatorname{rng}(\pi_{MQ^*}) \subset \operatorname{rng}(\pi_{NQ^*}). \tag{5.3.2}$$

Setting $\sigma^* = \pi_{NQ^*}^{-1} \circ \pi_{MQ^*}$, we show that $\sigma = \sigma^*$. This will prove the lemma since σ^* depends merely on M and N.

To see (5.3.1), coiterate Q with Q^*. Let $\tilde{Q}$ be the shorter of the pair of mice at the end of this coiteration. Then

$$Q, Q^* \text{ coiterate simply to } \tilde{Q}. \tag{5.3.3}$$

The situation is illustrated by the following diagram:

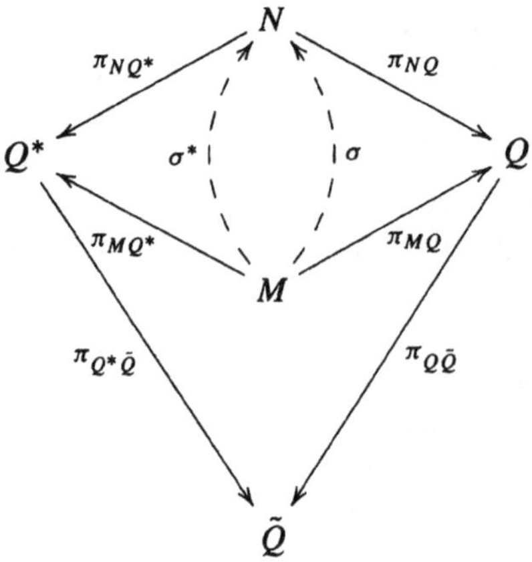

Suppose w.l.o.g. that the iteration from N to Q^* is non-simple. Then the iteration from M to Q^* must be simple. If the iteration from Q^* to $\tilde{Q}$ were non-simple, then $\tilde{Q}$ would be a non-simple iterate of N. On the other hand, the iteration $N \to Q \to \tilde{Q}$ would be simple, hence $\tilde{Q}$ would be both a simple and a non-simple iterate of N, which contradicts the Dodd–Jensen Lemma. It follows that the iteration from Q^* to $\tilde{Q}$ must be simple. Similarly, if the iteration from Q to $\tilde{Q}$ were non-simple, then $M \to Q^* \to \tilde{Q}$ would be a simple iteration from M to $\tilde{Q}$ and $M \to Q \to \tilde{Q}$ a non-simple one, which yields a contradiction again. Hence Q coiterates with Q^* simply to $\tilde{Q}$. To complete the proof of (5.3.1) we observe that $N \to Q^* \to \tilde{Q}$ is a non-simple iteration from N to $\tilde{Q}$ while $N \to Q \to \tilde{Q}$ is simple, which is again impossible, so Q^* must be a simple iterate of N after all. An argument of the same sort then yields (5.3.3).

By a straightforward application of the Dodd–Jensen Lemma we then conclude that all iteration maps in the above diagram commute. Hence

$$\operatorname{rng}(\pi_{Q^*\tilde{Q}} \circ \pi_{MQ^*}) = \operatorname{rng}(\pi_{Q\tilde{Q}} \circ \pi_{MQ}) \subset \operatorname{rng}(\pi_{Q\tilde{Q}} \circ \pi_{NQ}) = \operatorname{rng}(\pi_{Q^*\tilde{Q}} \circ \pi_{NQ^*})$$

(the equalities follow by the uniqueness of iteration maps), which immediately implies (5.3.2) and thus guarantees that σ^* is defined and that all maps in the above diagram commute. Using the latter we infer

$$\pi_{Q^*\tilde{Q}} \circ \pi_{NQ^*} \circ \sigma^* = \pi_{Q^*\tilde{Q}} \circ \pi_{MQ^*} = \pi_{Q\tilde{Q}} \circ \pi_{MQ} = \pi_{Q\tilde{Q}} \circ \pi_{NQ} \circ \sigma.$$

Then $\sigma = \sigma^*$ since $\pi_{Q^*\tilde{Q}} \circ \pi_{NQ^*} = \pi_{Q\tilde{Q}} \circ \pi_{NQ}$. □(Lemma 5.3.5)

Corollary 5.3.6. *Let M, N be mice. Then M, N have a common simple iterate iff they coiterate simply to a common mouse.* □(Corollary 5.3.6)

The following lemma shows that derived iteration maps are nothing more than iteration maps; this fact, as we shall see later, can only be generalized for mice for non-overlapping extenders, which corresponds to the level of at most one strong cardinal.

Lemma 5.3.7. *Let N be a derived iterate of M and $\sigma : M \to N$ the associated derived iteration map. Then N is a normal simple iterate of M and σ is the associated iteration map.*

Proof. By Corollary 5.3.6, M coiterates with N simply to a common mouse Q. Let π_{MQ}, π_{NQ} be the iteration maps, $\langle \nu_i;\ i < \theta \rangle$ the coiteration indices and $\langle \kappa_i;\ i < \theta \rangle$ the critical points of the coiteration. Set

$$\iota = \text{the least } i < \theta \text{ such that } E_{\nu_i}^{N_i} \neq \emptyset,$$

i.e. ι is the first stage of the coiteration at which N is moved. Then

$$\operatorname{rng}(\pi_{M_\iota Q}) \subset \operatorname{rng}(\pi_{NQ}).$$

To see this we recall that, by Lemma 4.2.4, every $x \in M_\iota$ is of the form

$$\pi_{MM_\iota}(f)(\alpha_1, \dots, \alpha_\ell),$$

where $f \in \Gamma(\kappa_0, M)$ and $\alpha_1, \dots, \alpha_\ell < \kappa_\iota$. Fixing n, p such that f is a good $\Sigma_1^{(n)}(M)$ function in p we obtain

$$\pi_{M_\iota Q}(x) = \pi_{MQ}(f)(\alpha_1, \dots, \alpha_\ell) = \pi_{NQ}(g)(\alpha_1, \dots, \alpha_\ell),$$

where $g \in \Gamma(\kappa_0, N)$ is $\Sigma_1^{(n)}(N)$ in $\pi_{NQ}^{-1} \circ \pi_{MQ}(p)$ by the same functionally absolute $\Sigma_1^{(n)}$-definition as f. These equalities follow from the fact that both the iteration from M_ι to Q as well as that from $N = N_\iota$ to Q is above κ_ι, which means that none of these iterations moves the ordinals $\alpha_1, \dots, \alpha_\ell$.

We next observe that ν_ι indexes an extender in M_ι, otherwise the above inclusion would yield

$$\kappa_\iota = \pi_{M_\iota Q}(\kappa_\iota) \in \operatorname{rng}(\pi_{NQ}),$$

which contradicts the fact that $\kappa_\iota = \operatorname{cr}(\pi_{NQ})$. Note also that $\mathcal{P}(\kappa_\iota) \cap M_\iota = \mathcal{P}(\kappa_\iota) \cap N$; denote this set by a. Hence both $E_{\nu_\iota}^{M_\iota}$ and $E_{\nu_\iota}^{N}$ are on a. Given $x \in a$, pick $\tilde{x} \in N$ so that $\pi_{M_\iota Q}(x) = \pi_{NQ}(\tilde{x})$. Then

$$x \in E_{\nu_\iota}^{M_\iota} \longleftrightarrow \kappa_\iota \in \pi_{M_\iota Q}(x) = \pi_{NQ}(\tilde{x}) \longleftrightarrow \kappa_\iota \in \pi_{NQ}(x) \longleftrightarrow x \in E_{\nu_\iota}^{N}.$$

The middle equivalence follows from the fact that $\tilde{x} \cap \kappa_\iota = x$. This computation shows that $E^{M_\iota}_{\nu_\iota} = E^N_{\nu_\iota}$, which is impossible, since ν_ι is a coiteration index.

It follows that the N-side of the coiteration is not moved. Consequently, $\pi_{NQ} = \mathrm{id}$, $N = Q$ and $\sigma = \pi_{NQ}$. □(Lemma 5.3.7)

Corollary 5.3.8. *Let M be a mouse and $\bar{M} = \mathrm{core}^n(M)$. Then M is a simple iterate of $\bar{M}$ above $\omega\varrho^n_{\bar{M}} = \omega\varrho^n_M$. In particular, every mouse is a simple iterate of its core above the ultimate projectum.*

Proof. This is an immediate consequence of Lemma 5.3.4 and Lemma 5.3.7 □(Corollary 5.3.8)

Corollary 5.3.9. *Let N be a simple iterate of M. Then N is a normal simple iterate of M.*

Proof. M coiterates with N simply to a common mouse, say Q. Let π be the iteration map from M to N and π_{MQ}, π_{NQ} the maps arising from the coiteration. By the Dodd–Jensen Lemma, $\pi_{MQ} = \pi_{NQ} \circ \pi$. It follows that π is a derived iteration map. By the proof of Lemma 5.3.7, the N-side of the coiteration is not moved, so $N = Q$ and $\pi = \pi_{MQ}$. □(Corollary 5.3.9)

5.4 The Canonical Ordering of Mice

The fact that one side of any mouse coiteration is simple leads to a natural ordering of mice, where the mouse which is truncated in the course of the coiteration is considered to be "larger". Intuitively, truncations correspond to "removing" a certain amount of information.

Definition. Let M, N be mice. Then

a) $M \sim^* N$ iff M, N have a common simple iterate;

b) $M <^* N$ iff there is a mouse which is a simple iterate of M and a non-simple iterate of N.

Lemma 5.4.1. *The relations $\sim^*$ and $<^*$ are transitive.*

Proof. We shall first focus on the transitivity of $\sim^*$. Suppose $M_1 \sim^* M_2$ and $M_2 \sim^* M_3$. Let Q_1 be a common simple iterate of M_1, M_2 and Q_2 be a common simple iterate of M_2, M_3. Coiterate Q_1 with Q_2; let Q be the shorter of the pair of mice at the end of the coiteration. The situation is depicted in the following diagram:

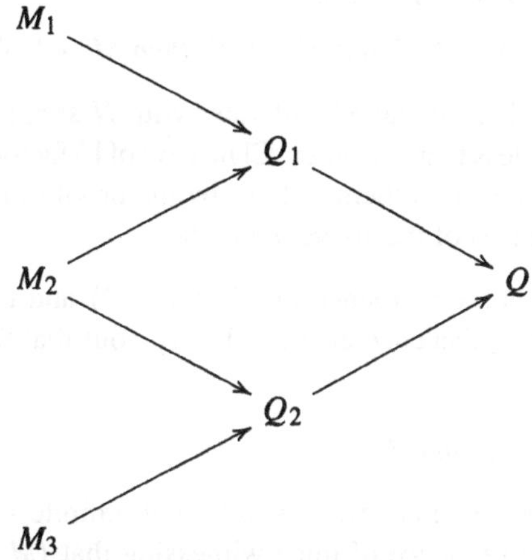

We observe that Q is a common simple iterate of both Q_1, Q_2. Suppose w.l.o.g. that the iteration from Q_1 to Q is non-simple. Then the iteration from Q_2 to Q must be simple. It follows that $M_2 \to Q_1 \to Q$ is a non-simple iteration from M_2 to Q and, at the same time, $M_2 \to Q_2 \to Q$ is a simple one, which contradicts the Dodd–Jensen Lemma. As an immediate consequence we have that Q is a common simple iterate of M_1, M_3, i.e. $M_1 \sim^* M_3$.

The transitivity of $<^*$ follows by a similar argument. Let $M_3 <^* M_2$ and $M_2 <^* M_1$; i.e. let the iteration from M_1 to Q_i $(i = 1, 2)$ be non-simple and that from M_j to Q_{j-1} $(j = 2, 3)$ be simple. This time we conclude that the iteration from Q_1 to Q is non-simple. □(Lemma 5.4.1)

Lemma 5.4.2. *The following are equivalent:*

a) $M <^* N$;

b) *M, N coiterate to a mouse Q such that the iteration from M to Q is simple and that from N to Q is non-simple. (Here we allow an extra truncation of the longer premouse at the end of the coiteration.)*

Proof. The proof proceeds as those of Lemma 5.3.7 and Corollary 5.3.8. □(Lemma 5.4.2)

Corollary 5.4.3.

a) *$\sim^*$ is an equivalence relation on mice;*

b) *$<^*$ is a strict linear quasi-ordering of mice relative to $\sim^*$, i.e.*

   i) $M \not<^* M$;

ii) $<^*$ *is transitive;*

iii) $M \not<^* N \;\&\; N \not<^* M \longrightarrow M \sim^* N$;

iv) *if* $M' \sim^* M$, $N' \sim^* N$ *and* $M <^* N$, *then* $M' <^* N'$.

Proof. To see iii) of b), observe that M coiterates with N simply to a common mouse, since none of them is $<^*$-less than the other. Clause iv) of b) factorizes to the inequalities $M' <^* N$ and $M <^* N'$; each of them follows by the proof of Lemma 5.4.1. The rest are immediate consequences of the above lemmata. □(Corollary 5.4.3)

In other words, $\sim^*$ is a congruence on $\langle \text{MICE}, <^* \rangle$ and the factorization yields a linear ordering on the equivalence classes. It turns out that this ordering is a well-ordering.

Lemma 5.4.4. $<^*$ *is well-founded.*

Proof. Suppose the contrary. Let $\langle M_i;\ i \in \omega \rangle$ be an infinite $<^*$-decreasing chain of mice and $\langle Q^1_i;\ i \in \omega \rangle$ a sequence of mice witnessing that $\langle M_i \rangle_i$ decreases, i.e. each Q^1_i is a non-simple iterate of M_i and a simple iterate of M_{i+1}. Given $j > 1$ and $i \in \omega$, define Q^j_i inductively to be the shorter of the pair of mice at the end of the coiteration of Q^{j-1}_i with Q^{j-1}_{i+1}. We obtain a system of mice and iteration maps as in the following diagram:

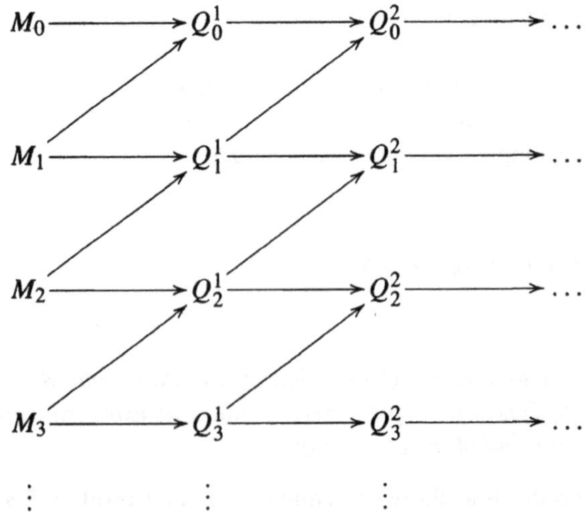

By the proof of clause b) in Lemma 5.4.1, the iteration from Q^j_i to Q^{j+1}_i is non-simple for every $j > 0$ and $i \in \omega$ (note that the upper left corner of the above diagram agrees with the diagram from the proof of that lemma). It follows that $M_0 \to Q^1_0 \to Q^2_0 \to \dots$ is a degenerate iteration of M_0. Contradiction. □(Lemma 5.4.4)

Corollary 5.4.5. $<^* / \sim^*$ *is a strict well-ordering of the class* MICE$/\sim^*$.

□(Corollary 5.4.5)

5.5 Condensation Lemma

Another consequence of solidity is the condensation lemma. Recall that in the case of L, the condensation lemma asserts that any Σ_1 substructure of J_α condenses to some J_β, that is, to another level of the hierarchy. In the more general context of mice, however, such substructure can condense to the core or a segment of an ultrapower of the original mouse.

Theorem 5.5.1. *Let $\bar{M}$ be a premouse, let M be a mouse and let*

$$\sigma : \bar{M} \underset{\Sigma_0^{(n)}}{\longrightarrow} M$$

be such that $\sigma \restriction \omega\varrho_{\bar{M}}^{n+1} = \mathrm{id}$. Then $\bar{M}$ is a mouse. Suppose moreover that $\bar{M}$ is sound above ν, where ν is the largest ordinal such that $\sigma \restriction \nu = \mathrm{id}$ (hence $\nu = \mathrm{cr}(\sigma)$ if $\nu \in \bar{M}$). Then one of the following holds:

a) *$\bar{M} = \mathrm{core}_\nu(M)$ and σ is the associated core map. Consequently, M is a simple iterate of $\bar{M}$.*

b) *$\bar{M}$ is a proper initial segment of M above ν.*

c) *$\bar{M}$ is a proper initial segment of $\mathrm{Ult}^*(M||\zeta, E_\nu^M)$ where ζ is maximal possible, i.e. maximal such that E_ν^M is a total measure in $M||\zeta$. If $\bar{M}$ is a proper segment of the above ultrapower, then $\bar{M}$ is in fact a proper segment of $\mathrm{Ult}(M||\nu, E_\nu^M)$.*

Proof. If $\nu = \mathrm{ht}(\bar{M})$, then either $\bar{M}$ does not have a top measure, in which case $\bar{M}$ is clearly an initial segment of M, or $\bar{M}$ does have a top measure, in which case this measure is, by Lemma 5.3.2, equal to E_ν^M and $\bar{M} = M||\nu$ as well. For the rest of the proof suppose that $\nu \in \bar{M}$, which means that $\nu = \mathrm{cr}(\sigma)$. That $\bar{M}$ is a mouse follows from Lemmata 5.3.2 and 5.1.6. Let Q be the shorter of the pair of premice at the end of the coiteration of M with $\bar{M}$. The above-mentioned lemmata also guarantee that the iteration from $\bar{M}$ to Q is simple and above ν.

Suppose also that the iteration from M to Q is simple. We show that in this case clause a) holds. First we observe that ν must be a limit cardinal in $\bar{M}$. Otherwise $\nu = \kappa^{+\bar{M}}$ for some κ, which would mean that $\sigma(\nu) = \kappa^{+M}$ and the iteration from M to Q, being beyond $\sigma(\nu)$ (this is the least ordinal $\geq \nu$ indexing a total measure in M), would be above κ. Then

$$\mathcal{P}(\kappa) \cap M = \mathcal{P}(\kappa) \cap Q = \mathcal{P}(\kappa) \cap \bar{M},$$

which would contradict the fact that M contains more subsets of κ than M (since $\sigma(\nu) > \nu$). Thus ν, being the critical point of σ and a limit cardinal of $\bar{M}$, must also be a limit cardinal in M; as an immediate consequence we have that the iteration from M to Q is above ν as well. Hence

- $\mathcal{P}(\nu) \cap M = \mathcal{P}(\nu) \cap Q = \mathcal{P}(\nu) \cap \bar{M}$;

- $\omega\varrho_M^{n+1} = \omega\varrho_Q^{n+1} = \omega\varrho_{\bar{M}}^{n+1} \stackrel{\text{def}}{=} \omega\varrho^{n+1}$.

The latter follows by the argument from the proof of Lemma 5.3.7. From this we infer that $\sigma''\omega\varrho_{\bar{M}}^n$ is cofinal in $\omega\varrho_M^n$. Here we assume w.l.o.g. that $\nu < \omega\varrho_{\bar{M}}^n$. For, if $\omega\varrho = \sup(\sigma''\omega\varrho_{\bar{M}}^n) < \omega\varrho_M^n$, then, fixing a $\Sigma_1(M^{n,p_M\restriction n})$ relation $\bar{B}(\eta, \xi)$ such that $b = \bar{B}(-, p_{\bar{M}}^n) \cap \omega\varrho^{n+1} \notin \bar{M}$, we would have that $b = B(-, \sigma(p_{\bar{M}}^n))$, where, setting $\tilde{M} = \langle |J_\varrho^E|, A^{n,\sigma(p_{\bar{M}}\restriction n)} \cap |J_\varrho^E| \rangle$ (here $E = E^M$), B is $\Sigma_1(\tilde{M})$ in $\sigma(p_M^n)$ by the same definition as $\bar{B}$. This follows from the fact that the map $\sigma : M^{n,p_M\restriction n} \to \tilde{M}$ is Σ_0-preserving and cofinal. Since $\tilde{M} \in M$, b would also be an element of M, which contradicts our earlier computations.

We thus conclude that σ is $\Sigma_0^{(n)}$-preserving and cofinal, and therefore $\Sigma_1^{(n)}$-preserving. Our aim is to prove that $\mathrm{rng}(\sigma) = \tilde{h}_M^{n+1}(\nu \cup \{p_M\})$. By the soundness of $\bar{M}$ above ν, every $x \in \bar{M}$ is of the form $\tilde{h}_{\bar{M}}^{n+1}(\eta, p_{\bar{M}})$ for some $\eta < \nu$, which together with the preservation properties of σ yields that $\mathrm{rng}(\sigma) = \tilde{h}_M^{n+1}(\nu \cup \{\sigma(p_{\bar{M}})\})$. It remains to verify the equality $\sigma(p_{\bar{M}} - \omega\varrho^{n+1}) = p_M - \omega\varrho^{n+1}$. That $\sigma(p_{\bar{M}} - \omega\varrho^{n+1}) \in P_M^{n+1}$ follows from the fact that both $\bar{M}$ and M contain the same subsets of $\omega\varrho^{n+1}$: given a $\Sigma_1^{(n)}(\bar{M})$ relation $\bar{A}(\eta, \xi)$ such that $a = \bar{A}(-, p_{\bar{M}} - \omega\varrho^{n+1}) \cap \omega\varrho^{n+1} \notin \bar{M}$ and a $\Sigma_1^{(n)}(M)$ relation A by the same definition, $A(-, \sigma(p_{\bar{M}} - \omega\varrho^{n+1})) \cap \omega\varrho^{n+1} = a \notin M$. To see that $\sigma(p_{\bar{M}} - \omega\varrho^{n+1}) = p_M - \omega\varrho^{n+1}$, we apply Corollary 1.12.4: $\bar{M}$, being iterable, is solid, hence $W_{\bar{M}}^\alpha \in \bar{M}$ for every $\alpha \in p_{\bar{M}}$. Since σ is $\Sigma_1^{(n)}$-preserving, $\sigma(W_{\bar{M}}^\alpha) \in M$ is a generalized witness for $\sigma(\alpha)$ w.r.t. $\sigma(p_{\bar{M}} - \omega\varrho^{n+1})$ for every $\alpha \in p_{\bar{M}} - \omega\varrho^{n+1}$. This completes the proof of clause a).

From now on suppose that the iteration from M to Q is non-simple. We first show that in this case $\bar{M}$ is not moved in the coiteration. Suppose the contrary. Then Q is not sound, since the iteration from $\bar{M}$ to Q is above the $(n+1)$-st projectum. It follows that the last mouse on the M side of the coiteration cannot be properly longer than Q, so there is no extra truncation at the end of the coiteration. This means that the last truncation point ι on the M-side is strictly less than the length of the coiteration and that the iteration from $M_\iota||\alpha_\iota$ to Q is simple. Exactly as above we show that $M_\iota||\alpha_\iota$, Q and $\bar{M}$ have the same $(n+1)$-st projectum. Note that $\pi_{\bar{M}Q}$, being an iteration map above ν, sends $p_{\bar{M}}$ to p_Q and is constant on ν. By the soundness of $\bar{M}$ above ν we infer exactly as in the previous paragraph that $\mathrm{rng}(\pi_{\bar{M}Q})$ is the $\Sigma_1^{(n)}$-hull of $\nu \cup \{p_Q\}$ over Q and that $\pi_{\bar{M}Q}$ is the uncollapsing map associated with this hull. If the iteration from $M_\iota||\alpha_\iota$ to Q is above ν then the same argument yields that $\mathrm{rng}(\pi_{M_\iota||\alpha_\iota,Q})$ is the same hull and that $\pi_{M_\iota||\alpha_\iota,Q}$ is the uncollapsing map. It follows that $M_\iota||\alpha_\iota = \bar{M}$ $(= \mathrm{core}_\nu(Q))$. For $\iota = 0$ this means that $\bar{M}$ is a proper initial segment of M. Thus, neither of $M, \bar{M}$ was moved in the coiteration, which contradicts our assumption. For $\iota > 0$ we have that $\nu_0 > \nu$ is a cardinal in M_ι and $\bar{M}$ is a proper initial segment of M_ι. Since the $(n+1)$-st projectum of $\bar{M}$ drops to ν, $\mathrm{ht}(\bar{M}) < \nu_0$ – otherwise $\tilde{h}_{\bar{M}}^{n+1} \in M_\iota$ would collapse ν_0 in M_ι. But M_ι agrees with M below ν_0, hence $\bar{M}$ is a proper initial segment of M and we have a contradiction again. Now consider the case where the iteration from $M_\iota||\alpha_\iota$ to

Q is not above ν. Then $\iota = 0$ and $\nu_0 = \nu$. The former is immediate; the latter follows from the fact that if $\nu_0 > \nu$, then ν would fail to be a cardinal in $M||\nu_0$ (because $\kappa_0 < \nu$ and there is no even local cardinal between κ_0 and ν_0) and therefore also in $\bar{M}$, since $\bar{M}$ agrees with M below ν_0. Set $\kappa = \kappa_0$. M_1, being the ultrapower of $M_0||\alpha_0$ by E_ν^M, is sound above $\kappa + 1$ since $M_1 = \tilde{h}_{M_1}^{n+1}(\kappa + 1 \cup \{p_{M_1}\})$; using this we infer as before that $\bar{M} = M_1$. By Lemma 5.3.2, $\bar{M}$ is not moved at the first stage in the coiteration in this case ($E^{\bar{M}} = \varnothing$, otherwise E_ν^M would not be the first measure used on the M-side of the coiteration), which means that the coiteration stops after the first step and the M-side is not moved at all, so we also have a contradiction to our assumption in this case.

At this point we know that $\bar{M} = Q$, i.e. $\bar{M}$ is a non-simple iterate of M. To finish the proof we mainly repeat the arguments from the previous paragraph. Let ι be the last truncation point in the iteration from M to $\bar{M}$. This time it might be the case that ι is equal to the length of the coiteration. If $\iota > 0$ then $\bar{M}$ is a proper initial segment of $M_{\iota+1}$ since $M_{\iota+1}$ is not sound above $\kappa_\iota > \nu$. For $\nu_0 > \nu$ we infer as above that $\bar{M}$ is a proper initial segment of $M||\nu_0$, so $\bar{M}$ is as in clause b). For $\nu_0 = \nu$ the same argument yields that $\bar{M}$ is a proper initial segment of M_1 (strictly speaking, if $\iota > 1$ then $\bar{M}$ will be a proper initial segment of $M_1||\nu_1$); that the first measure used in the iteration is E_ν^M follows exactly as in the previous paragraph. By the fact that $\omega\varrho_{\bar{M}}^{n+1} \leq \nu$, $\bar{M}$ is a proper initial segment of $M_1||\nu^{+M_1}$. Since $\mathrm{Ult}(J_\nu^E, E_\nu)$ is an initial segment of M_1 of length $\pi(\kappa)^{+M_1}$, where π is the associated ultrapower map, $\bar{M}$ must be a proper initial segment of that ultrapower; hence $\bar{M}$ is as in clause c) of the theorem. It remains to discuss the case where $\iota = 0$. Here we again use E_ν^M as the first measure in the iteration and conclude that either $\bar{M} = M_1$ or else $\bar{M}$ is a proper initial segment of M_1 and thus also of $\mathrm{Ult}(J_\nu^E, E_\nu)$; so M is again as in clause c). □(Theorem 5.5.1)

Remark. Clause c) of the condensation lemma cannot be omitted. We give two examples showing that $\bar{M}$ might happen to be a proper initial segment of $M_1 = \mathrm{Ult}^*(M||\alpha, E_\nu^M)$ as well as M_1 itself. We assume that there is a mouse N which contains a full measure $E_\tau^N \in N$ with critical point κ and that $\tau + 1 \in M$. We shall later see that such mice do exist if there are sufficiently large cardinals (one measurable, and even less, suffices).

Let $\bar{N}$ be the transitive collapse of $h_{N||(\tau+1)}(\kappa \cup \{\tau\})$ and $\vartheta : \bar{N} \to N||(\tau + 1)$ the uncollapsing map. Then $\bar{N}$ is an element of N and can be encoded in a subset of κ, so by acceptability, $\bar{N} \in N||\tau$. It follows that $\mathrm{ht}(\bar{N}) < \tau$; in particular, this holds for the σ-preimage ν of τ. The ordinal κ, being the cardinal predecessor of τ, is easily seen to be in the range of σ, so $\sigma(\kappa) = \kappa$. Hence $\nu = \mathrm{cr}(\sigma)$ (because there are no $\bar{N}$-cardinals between κ and ν). By Lemma 5.3.2, $E_\nu^{\bar{N}} = E_\tau^N \cap |J_\nu^E|$, where $E = E^N$.

The above construction of $\bar{N}$ also guarantees that $\omega\varrho_{\bar{N}}^1 \leq \kappa$. In fact, no projectum of $\bar{N}$ can be strictly smaller than κ since any bounded subset of κ definable over $\bar{N}$ is an element of $N||\tau$ and, since κ is a cardinal in $N||\tau$, of $N||\kappa = \bar{N}||\kappa$. The former itself implies that $\mathrm{Ult}^*(\bar{N}, E_\nu^{\bar{N}}) = \mathrm{Ult}(\bar{N}, E_\nu^{\bar{N}})$.

Let $Q = \mathrm{Ult}(N||(\tau + 1), E_\tau^N)$, $\bar{Q} = \mathrm{Ult}(\bar{N}, E_\nu^{\bar{N}})$ and let $\pi, \bar{\pi}$ be the associated ultrapower maps. Then Q agrees with N below $\tau = \kappa^{+Q}$ and $\bar{Q}$ agrees with $\bar{N}$ below

$\nu = \kappa^{+\bar{Q}}$. Define $\vartheta' : \bar{Q} \to Q$ in the usual way, i.e. $\vartheta'(\bar{\pi}(f)(\kappa)) = \pi \circ \vartheta(f)(\kappa)$ for $f : \kappa \to \bar{N}$, $f \in \bar{N}$. Standard arguments show that ϑ' is Σ_0-preserving, $\vartheta' \upharpoonright (\kappa+1) =$ id and $\vartheta(\nu) = \tau$. It follows that $\nu = \mathrm{cr}(\vartheta)$. Note also that $E_\nu^{\bar{Q}} = \varnothing$, while $E_\nu^{Q} = E_\nu^{N}$ is a partial measure.

We are now ready to give the first example. Pick $\zeta < \nu^{+\bar{Q}}$ such that ν is the largest cardinal in $\bar{Q}||\zeta$ (such a ζ can be obtained e.g. by a condensation argument inside $\bar{Q}||\bar{\pi}(\kappa)$). Let α be the collapsing level for ζ in $\bar{Q}$. Hence ζ is a cardinal in $\bar{Q}||\alpha$ but not in $\bar{Q}||(\alpha+1)$. Set $\bar{M} = \bar{Q}||\alpha$, $M = \vartheta'(\bar{M})$ and $\sigma = \vartheta' \upharpoonright \bar{M}$. Then $\omega\varrho^\omega_{\bar{M}} = \nu$: the inequality $\leq$ follows immediately from the definition of α, the reverse one from the fact that ν is a cardinal in $\bar{Q}$. The map $\sigma : \bar{M} \to M$ is fully elementary, i.e. much stronger than $\Sigma_0^{(n)}$-preserving, where n satisfies $\omega\varrho^{n+1}_{\bar{M}} \leq \nu < \omega\varrho^n_{\bar{M}}$. Setting β to be the collapsing level for ν in N, we show that $\bar{M}$ is a proper initial segment of $M_1 = \mathrm{Ult}^*(M||\beta, E^M_\nu)$ by excluding all other possibilities given by the condensation lemma. First of all, $\bar{M}$ cannot be the core of M since the ultimate projecta of the two mice differ: the projectum is ν in the case of $\bar{M}$ and τ in the case of M. To rule out clause b) note that $E^{\bar{M}}_\nu = \varnothing \neq E^N_\nu = E^M_\nu$. Hence c) holds. But $\bar{M} \neq M_1$ since the ultimate projectum of M_1 is κ.

Now we give the second example. Since the first projectum of $\bar{N}$ is κ, the same applies to $\omega\varrho^1_{\bar{Q}}$ (we have already mentioned above that $\bar{Q}$ is in fact a $*$-ultrapower of $\bar{N}$). Let $\bar{M} = \mathrm{core}_{\kappa+1}(\bar{Q})$ and $\bar{\sigma}$ be the core map. The critical point of $\bar{\sigma}$ must be greater than κ and since ν is the successor of κ both in $\bar{M}$ and $\bar{Q}$, $\mathrm{cr}(\bar{\sigma}) > \nu$. It follows that $E^{\bar{M}}_\nu = E^{\bar{Q}}_\nu \neq \varnothing$. Set $M = Q$ and $\sigma = \vartheta' \circ \bar{\sigma}$. Then $\sigma : \bar{M} \to M$ is Σ_0-preserving, $\omega\varrho^1_{\bar{M}} \leq \mathrm{cr}(\sigma) = \nu$ and $\bar{M}$ is sound above ν (in fact above $\kappa+1$). That $\bar{M}$ cannot be the core of M above ν follows e.g. from the fact that $\omega\varrho^\omega_{\bar{M}} = \kappa < \pi(\tau) \leq \omega\varrho^\omega_M$ (another reason is that $\bar{M} \in M$). That $\bar{M}$ is not an initial segment of M follows exactly as before: ν indexes a measure in $\bar{M}$, but not in M. Hence $\bar{M}$ is an initial segment of M_1. If $\bar{M}$ were a proper initial segment thereof, then, by the fact that $\omega\varrho^1_{\bar{M}} = \kappa$, M would contain a $\mathbf{\Sigma}_1(\bar{M})$-definable subset of κ which is not in $\bar{M}$. On the other hand, $\bar{M}$ and M contain the same subsets of κ since they agree below ν and ν is the successor of κ in both. Hence $\bar{M} = M_1$.

Lemma 5.5.2. *Let $\sigma : \bar{M} \to M$ be Σ_0-preserving, where $\bar{M}$, M are premice and M is 0-iterable. Suppose that $\mu = \mathrm{cr}(\sigma)$ is a limit cardinal in $\bar{M}$. Then $E^{\bar{M}} \upharpoonright \bar{\tau} = E^M \upharpoonright \bar{\tau}$, where $\bar{\tau} = \mu^{+\bar{M}}$.*

Proof. Given $\xi < \bar{\tau}$ such that μ is the largest cardinal in $\bar{M}||\xi$, let $\zeta = \zeta_\xi < \bar{\tau}$ be such that ξ is a cardinal in $\bar{M}||\zeta$, but not in $\bar{M}||(\zeta+1)$. We show that $\bar{M}||\zeta$ is an initial segment of M. Since ordinals of the form ζ_ξ are cofinal in $\bar{\tau}$, this will prove the lemma.

Any ξ as above is a cardinal successor of μ in $\bar{N} = \bar{M}||\zeta$ and is collapsed in $\bar{M}||(\zeta+1)$, so the latter structure contains a subset a of μ coding the ordinal ξ as in the proof of clause f) of Lemma 5.3.4. Such a set cannot be an element of $\bar{N}$: otherwise it would lie in $\bar{N}||\xi$ and, exactly as in that proof, we would be able to reconstruct ξ from a inside $\bar{N}||\xi$, which is impossible. Hence a, being an element of $\bar{M}||(\zeta+1) - \bar{M}||\zeta$, is definable and, by the soundness of $\bar{M}||\zeta$, Σ^*-definable over $\bar{M}||\zeta = \bar{N}$. It follows that $\omega\varrho^{\omega}_{\bar{N}} \leq \mu$.

We now apply the condensation lemma to the mouse $N = \sigma(\bar{N})$ (note that N is really fully iterable) and the map $\bar{\sigma} = \sigma \restriction \bar{N} : \bar{N} \to N$. That σ has all requisite properties follows from the previous paragraph and the fact that, being fully elementary, σ is certainly $\Sigma_0^{(m)}$-preserving for m such that $\omega\varrho^{m+1}_{\bar{N}} \leq \mu < \omega\varrho^{m}_{\bar{N}}$. We shall exclude clauses a) and c). The latter is immediate since c) would occur only if μ were an $\bar{N}$-successor. To see that a) fails, note that $\bar{N}$, being sound, is encodable into a subset b of μ, which is an element of $\bar{M}$. Using the preservation properties of σ together with the acceptability of M we infer that $b = \sigma(b) \cap \mu \in M||\tau$ and that $\bar{N}$ can be reconstructed inside $M||\tau$ from b. Hence $\bar{N} \in N$, since $M||\tau$ is an initial segment of N. This means that $\bar{N}$ cannot be the core of N. □(Lemma 5.5.2)

Corollary 5.5.3. *Let M be a mouse and $\bar{M} = \mathrm{core}^n(M)$. Then $E^{\bar{M}} \restriction \tau = E^M \restriction \tau$, where τ is the common successor of $\omega\varrho^n$.*

Proof. That $\bar{M}, M$ have the same n-th projectum and its successor was proved in Lemma 5.3.4. The conclusion of the corollary follows trivially if the critical point of the core map is greater than $\mu = \omega\varrho^n$ and from the previous lemma otherwise. In the latter case we show that μ is a limit cardinal in $\bar{M}$, which will guarantee that the assumptions of the previous lemma are satisfied.

If μ were a successor, say $\mu = \kappa^{+\bar{M}}$, then $\sigma(\mu) > \mu$ would be a successor of κ in M, so μ would fail to be a cardinal in M. But his contradicts the fact that μ is the m-th projectum. □(Corollary 5.5.3)

5.6 Upwards Extensions to Premice

Pseudoultrapowers, or upwards extensions of embeddings, constitute the heart of the proof of the covering lemma, but are also useful in other fine structural constructions. We introduced them in Chapter 3 in a very general context, but we are mainly concerned about upwards extensions to premice. Of course, our main concern is the well-foundedness and iterability of the target premice of such extensions. Unlike downwards extensions (which correspond to taking hulls), we cannot, in general, show that a pseudoultrapower of an iterable premouse is fully iterable, but we can show its sufficient iterability for further applications. Throughout this section we assume that we have a Σ_0 cofinal embedding

$$\sigma : J^E_\tau \to J^{\tilde{E}}_{\tilde{\tau}},$$

M is a premouse which end-extends J^E_τ and τ is regular in M. We shall occasionally use Q and $\tilde{Q}$ to denote J^E_τ and $J^{\tilde{E}}_{\tilde{\tau}}$, respectively.

Lemma 5.6.1. *Let $\tilde{\sigma} : M \to \tilde{M}$ be the k-extension of σ to M (so $\tilde{M}$ is transitive). Then $\tilde{M}$ is a premouse.*

Proof. By Lemma 3.6.3, the map $\tilde{\sigma}$ is Q-preserving. The conclusion then follows from the fact that being a premouse is a Q-statement. □(Lemma 5.6.1)

Lemma 5.6.2 (Interpolation Lemma). *Let M, M' be premice, $n = k(M)$ and let $\sigma : M \to M'$ be $\Sigma_0^{(n)}$-preserving. Set $\tilde{\tau} = \sup(\sigma''\tau)$. Then the canonical extension $\tilde{\sigma}$ of $\sigma \restriction Q$ exists and is $\Sigma_0^{(n)}$-preserving and cofinal. Hence $\tilde{M}$ is a premouse. Furthermore, there is a $\Sigma_0^{(n)}$-preserving map $\sigma' : \tilde{M} \to M'$ such that $\sigma' \restriction \tilde{\tau} = \mathrm{id}$ and $\sigma'(\tilde{\tau}) = \tau'$. It follows that if M' is iterable (above $\tilde{\tau}$), then so is $\tilde{M}$.*

Proof. The existence of $\tilde{M}, \tilde{\sigma}$ and σ' follows from the general version of the interpolation lemma (Lemma 3.6.10); that $\tilde{M}$ is a premouse follows from Lemma 5.6.1. By Lemma 3.6.7, $\omega\varrho_{\tilde{M}}^{n+1} \leq \tilde{\tau}$, so the conclusions on the iterability of $\tilde{M}$ are direct consequences of the condensation lemma (resp. its proof). □(Lemma 5.6.2)

Similarly, we can formulate a coarse version of the interpolation lemma for 0-iterable premice.

Lemma 5.6.3. *Let M, M' be premice and $\sigma : M \to M'$ be Σ_0-preserving. Set $\tilde{\tau} = \sup(\sigma''\tau)$. Then the 0-extension $\tilde{\sigma}$ of $\sigma \restriction Q$ exists and is Σ_0-preserving and cofinal. Hence $\tilde{M}$ is a premouse. Furthermore, there is a Σ_0-preserving map $\sigma' : \tilde{M} \to M'$ such that $\sigma' \restriction \tilde{\tau} = \mathrm{id}$ and $\sigma'(\tilde{\tau}) = \tau'$. It follows that if M' is 0-iterable, then so is $\tilde{M}$.*
□(Lemma 5.6.3)

We have seen that ω-complete embeddings do have upwards extensions. The possibility of mapping countable substructures of the pseudoultrapower $\tilde{M}$ back into the original structure M guarantees that $\tilde{M}$ inherits a certain amount of iterability from M.

Lemma 5.6.4. *Suppose σ is ω-complete.*

a) *If M is normally iterable then the pseudoultrapower $\tilde{M}$ is normally iterable above its $(n+1)$-st projectum, where $n = k(M)$.*

b) *If M is normally iterable above τ where τ is a limit (resp. above κ where $\tau = \kappa^{+M}$), then $\tilde{M}$ is normally iterable above $\tilde{\tau}$ (resp. above $\tilde{\kappa} = \tilde{\sigma}(\kappa)$).*

c) *If M is Σ_0-iterable, then so is the 0-pseudoultrapower $\tilde{M}$.*

Proof. We shall first focus on a). The canonical extension $\tilde{\sigma} : M \to \tilde{M}$ exists by Lemma 3.6.12 and is $\Sigma_0^{(n)}$-preserving and cofinal by Lemma 3.6.7. Suppose $\tilde{M}$ is not iterable above $\omega\varrho_{\tilde{M}}^{n+1}$ and pick an iteration $\mathfrak{I}$ witnessing this. Let θ be regular large enough that $\mathfrak{I} \in H_\theta$ and let $\bar{\sigma} : \bar{H} \to H_\theta$ be a countable fully elementary embedding, where $\bar{H}$ is transitive. Then $\bar{\mathfrak{I}} = \bar{\sigma}^{-1}(\mathfrak{I})$ witnesses that $\bar{M} = \bar{\sigma}^{-1}(\tilde{M})$ is not iterable above its $(n+1)$-st projectum. On the other hand, b) of Lemma 3.6.12 guarantees the existence of a $\Sigma_0^{(n)}$-preserving embedding $\pi : \bar{M} \to M$, so $\bar{M}$ must be normally iterable above $\omega\varrho_{\tilde{M}}^{n+1}$ after all, as follows from Lemma 4.3.7. Contradiction. This proves a). To see b), notice that $\omega\varrho_M^{n+1} \leq \tau$ and $\leq \kappa$ if $\tau = \kappa^{+M}$, and the corresponding inequalities hold in $\tilde{M}$. If $\bar{\sigma}(\bar{\tau}, \bar{\kappa}) = \tilde{\tau}, \tilde{\kappa}$ then $\pi(\bar{\tau}, \bar{\kappa}) = \tau, \kappa$ since $\bar{\sigma} = \tilde{\sigma} \circ \pi$, so the above argument goes through with "normally iterable above $\tilde{\tau}$, resp. $\tilde{\kappa}$" in place of "normally iterable above $\omega\varrho_{\tilde{M}}^{n+1}$". Clause c) follows by a similar, but much simpler argument.

□(Lemma 5.6.4)

Due to fine structure, the assumption on ω-completeness from the previous lemma can be considerably weakened.

Lemma 5.6.5. *Let* $\mathrm{cf}(\tau) > \omega$, $n = k(M)$, *let both* $M, \tilde{Q}$ *be iterable and let one of the following holds:*

- τ *is limit in* M *and* M *is sound above* τ*;*
- $\tau = \kappa^{+M}$ *and* M *is sound above* κ.

Let $\tilde{\sigma} : M \to \tilde{M}$ *be the canonical extension of* σ *guaranteed by Lemmata* 3.6.13 *and* 5.6.1. *Then* $\tilde{M}$ *is normally iterable above* $\tilde{\tau}$ *in the former case and above* $\tilde{\kappa} = \tilde{\sigma}(\kappa)$ *in the latter case.*

Proof. Let us focus on the case when τ is a cardinal successor of κ in M; the remaining case follows by the same argument. Suppose for a contradiction that $\tilde{M}$ is not normally iterable above $\tilde{\kappa}$. Using a combination of arguments from the proofs of Corollary 4.3.4 and Lemma 4.3.7 we construct a countable premouse $\bar{M}$ and an elementary embedding $\bar{\sigma} : \bar{M} \to \tilde{M}$ such that

- $\bar{M}$ is not normally iterable above $\bar{\kappa} = \bar{\sigma}^{-1}(\tilde{\kappa})$;
- $\omega\varrho_{\bar{M}}^{n+1} \leq \bar{\kappa} < \omega\varrho_{\bar{M}}^{n}$.

The latter is a consequence of the fact that the corresponding inequalities hold for $\tilde{M}$, as follows from Lemmata 3.6.7 and 3.6.9. On the other hand, we show that $\bar{M}$ is embeddable into a proper initial segment of $\tilde{Q}$ via a sufficiently preserving map guaranteeing such iterability, which will yield the desired contradiction.

Let $\langle x_i; i \in \omega\rangle$ be an enumeration of $\bar{M}$. To each i pick an $f_i \in \Gamma(\sigma, M)$ and an $\alpha_i < \tilde{\kappa}$ such that $\bar{\sigma}(x_i) = \tilde{\sigma}(f_i)(\alpha_i)$. Choose a regular θ large enough that $M, \langle f_i\rangle_i \in H_\theta$. Construct a countable elementary embedding $\bar{\sigma}^* : \bar{H} \to H_\theta$ with $M, \langle f_i\rangle_i \in \mathrm{rng}(\bar{\sigma})$ and $\bar{H}$ transitive. (As usual, this is done by taking appropriate hulls and collapsing.) Denote the preimages of these objects under $\bar{\sigma}^*$ by M^*, $\langle f_i^*\rangle_i$. Set

- $\sigma^* = \bar{\sigma}^* \restriction M^*$;
- $\kappa^* = (\sigma^*)^{-1}(\kappa)$, $\tau^* = (\sigma^*)^{-1}(\tau)$;
- $\tilde{\tau}^* = \sup[(\sigma^*)''\tau^*]$;

Since τ is not ω-cofinal, $\tilde{\tau}^* < \tau$. Apply the interpolation lemma (Lemma 3.6.10) to the triple M^*, M and σ^*. We obtain a premouse $\tilde{M}^*$ and a pair of Σ_0-preserving maps

$$\sigma_0^* : M^* \to \tilde{M}^*, \qquad \sigma_1^* : \tilde{M}^* \to M$$

such that $\sigma_1^* \circ \sigma_0^* = \sigma^*$, $\sigma_1^* \restriction \tilde{\tau}^* = \mathrm{id}$ and $\sigma_1^*(\tilde{\tau}^*) = \tau$.

By the elementarity of $\bar{\sigma}^*$ we have $n = k(M^*)$, so σ_0^* is $\Sigma_0^{(n)}$-preserving and cofinal. It follows from c) of the Interpolation lemma that σ_0^* is $\Sigma_0^{(n)}$-preserving. Since $\omega\varrho_M^{n+1} \leq \kappa$ and M is sound and solid (solidity follows from the iterability of M) above κ, the elementarity of $\bar{\sigma}^*$ yields that $\omega\varrho_{M^*}^{n+1} \leq \kappa^*$ and that M^* is sound and solid above κ. Furthermore, $\bar{\sigma}^*(p_{M^*}) = p_M$. This together with the preservation properties of σ_0^* guarantees that $\omega\varrho_{\tilde{M}^*}^{n+1} \leq \kappa$, $\sigma_0^*(p_{M^*}) - \kappa = p_{\tilde{M}^*} - \kappa$ and $\tilde{M}^*$ is sound and solid above κ, as follows from general preservation lemmata for pseudoultrapowers (Lemmata 3.6.7 and 3.6.9) (see also Lemma 1.12.5).

The triple $\tilde{M}^*$, M and σ_1^* has all requisite properties for the application of the condensation lemma, so one of the following holds:

i) $\tilde{M}^*$ is the core of M above $\tilde{\tau}^*$ and σ_1^* is the associated core map.

ii) $\tilde{M}^*$ is a proper initial segment of M.

iii) $\tilde{M}^*$ is an initial segment of $M' = \mathrm{Ult}^*(M||\zeta, E_{\tilde{\tau}^*}^M)$, where ζ is maximal such that $E_{\tilde{\tau}^*}^M$ is a total measure on $M||\zeta$.

We show that the second clause is the only possible. To exclude i), notice that σ_1^* is non-cofinal; otherwise it would be $\Sigma_1^{(n)}$-preserving and the fact that $\kappa \cup \{p_M\} \subset \mathrm{rng}(\sigma_1^*)$ together with the soundness of M above κ would imply that $\mathrm{rng}(\sigma_1^*) = M$ and $\sigma_1^* = \mathrm{id}$, which is false. So σ_1^* cannot be a core map, as core maps are cofinal. Now consider iii). Notice that $E_{\tilde{\tau}^*}^M$ is a partial measure and its critical point is κ. It follows that $\tilde{M}^*||\zeta$ projects to κ (i.e. its ultimate projectum is κ), so its $*$-ultrapower M' cannot be sound above κ. But $\tilde{M}^*$ *is* sound above κ, so the only way iii) can be satisfied is that $\tilde{M}^*$ is a proper initial segment of M'. Then $\tilde{\tau}^*$ fails to be a cardinal in M' since $\tilde{M}^*$ projects to κ. On the other hand, $\tilde{\tau}^* = \kappa^{+M'}$ since $\tilde{\tau}^*$ is a cardinal successor of κ in $M||\zeta$ (ζ was so chosen!) and $E_{\tilde{\tau}^*}^M$ is weakly amenable w.r.t. $M||\zeta$. Contradiction. It follows that ii) must apply, i.e. $\tilde{M}^*$ is a proper initial segment of M.

In fact, $\tilde{M}^*$ is a proper initial segment of $Q = J_\tau^{E^M}$ since $\tilde{\tau}^*$, the cardinal successor of κ in $\tilde{M}^*$, is strictly less than $\tau = \kappa^{+M}$. Set $N = \sigma(\tilde{M}^*)$. Then N is a proper initial segment of $\bar{Q}$, so N is iterable. The fine structural properties of $\tilde{M}^*$ are also preserved, so in particular $\omega\varrho_N^{n+1} \leq \bar{\kappa} < \omega\varrho_N^n$. We are finally in a position to define a $\Sigma_0^{(n)}$-preserving embedding $\sigma' : \bar{M} \to N$ sending $\bar{\kappa}$ to $\bar{\kappa}$. Since $\omega\varrho_{\bar{M}}^{n+1} \leq \bar{\kappa} < \omega\varrho_{\bar{M}}^n$, this will guarantee normal iterability of $\bar{M}$ above $\bar{\kappa}$ and thus yield the contradiction which we announced at the beginning of the proof. Return back to the sequence $\langle\bar{\sigma}(f_i)(\alpha_i);\ i \in \omega\rangle$ representing the elements of $\mathrm{rng}(\bar{\sigma})$ as well as to the $\bar{\sigma}^*$-preimages f_i^* of functions

f_i. Set $\tilde{f}_i^* = \sigma_0^*(f_i^*)$ for all i. The map σ' is defined by

$$\sigma'(x_i) = \sigma(\tilde{f}_i^*)(\alpha_i).$$

It remains to verify that σ' is correctly defined and sufficiently preserving. Given a $\Sigma_0^{(n)}$ formula $\varphi(v_1, \ldots, v_\ell)$, the statement

$$M \models \varphi(f_{i_1}(\xi_{i_1}), \ldots, f_{i_\ell}(\xi_{i_\ell}))$$

is $\Sigma_0^{(n)}$, so by the preservation properties of σ_1^* we have

$$u^{\varphi,M}_{f_{i_1},\ldots,f_{i_\ell}} = u^{\varphi,\tilde{M}^*}_{\tilde{f}^*_{i_1},\ldots,\tilde{f}^*_{i_\ell}} \tag{5.6.1}$$

(recall the definition of $u^{\varphi,M}_{f_1,\ldots,f_k}$ from Lemma 3.6.1). Then

$$\bar{M} \models \varphi(x_1, \ldots, x_\ell) \longleftrightarrow \tilde{M} \models \varphi(\tilde{\sigma}(f_{i_1})(\alpha_{i_1}), \ldots, \tilde{\sigma}(f_{i_\ell})(\alpha_{i_\ell}))$$

and the latter is, by Łos' Theorem for $\Sigma_0^{(n)}$ formulae, equivalent to

$$\prec\alpha_{i_1}, \ldots, \alpha_{i_\ell}\succ \in \sigma(u^{\varphi,M}_{f_{i_1},\ldots,f_{i_\ell}}).$$

By (5.6.1),

$$\sigma(u^{\varphi,M}_{f_{i_1},\ldots,f_{i_\ell}}) = \sigma(u^{\varphi,\tilde{M}^*}_{\tilde{f}^*_{i_1},\ldots,\tilde{f}^*_{i_\ell}}) = u^{\varphi,N}_{\sigma(\tilde{f}^*_{i_1}),\ldots,\sigma(\tilde{f}^*_{i_\ell})},$$

so the previous statement can be equivalently expressed as

$$N \models \varphi(\sigma(\tilde{f}^*_{i_1})(\alpha_{i_1}), \ldots, \sigma(\tilde{f}^*_{i_\ell})(\alpha_{i_\ell})).$$

This verifies the required preservation properties of σ' and thus completes the proof of the lemma. □(Lemma 5.6.5)

The proof of the previous lemma does not require the iterability of $\tilde{Q}$ itself, but merely the iterability of each of its proper initial segments. Thus, by Theorem 4.5.5, the iterability of all proper initial segments of $M||\tau$ already yields a sufficient amount of iterability for the above proof to go through. However, we still have to require a certain amount of iterability for M because of our application of the condensation lemma. One can prove an iterability criterion for pseudoultrapowers even without assuming soundness of M under the assumption that M is iterable above κ (the proof is based on a combination of the arguments from Lemma 5.6.5 and the next lemma), but such a criterion does not seem to be useful in the applications.

Finally we state an iterability criterion for 0-pseudoultrapowers. Of course, in this case we are interested in the 0-iterability. We stress that the full 0-iterability of M already guarantees the full 0-iterability of $\bar{M}$. This is due to the fact that 0-iterability can be ensured by a Σ_0-preserving embedding. (And the same applies to various weak versions of 0-iterability, as is easily seen.) In the previous lemma, where we dealt with the $*$-iterability, the situation is more delicate since the methods used in the proof yield a mere $\Sigma_0^{(n)}$-preservation degree for the embedding σ', which is too weak to guarantee more iterability for $\bar{M}$, even if M is fully $*$-iterable.

Lemma 5.6.6. *Let* $\mathrm{cf}(\tau) > \omega$, *let* Q *have a largest cardinal* κ *and let both* M, $\tilde{Q}$ *be* 0*-iterable. Let* $\tilde{\sigma} : M \to \tilde{M}$ *be the canonical* 0*-extension of* σ *guaranteed by Lemmata* 3.6.13 *and* 5.6.1. *Then* $\tilde{M}$ *is* 0*-iterable.*

Proof. We shall closely follow the proof of Lemma 5.6.5 and adopt the same notation. Similarly as before we construct a premouse $\bar{M}$, which this time fails to be 0-iterable, together with all other auxiliary structures and maps. We stress that the canonical extensions $\tilde{\sigma}$ and σ_0^* are in the present case 0-extensions, so σ_1^* is Σ_0-preserving, but not Σ_1-preserving. As before, $\tilde{\tau}^* < \tau$. Also, the functions f_i, f_i^* and $\tilde{f}_i^*$ are elements of the corresponding structures. Following the strategy of the above proof, we shall attempt to Σ_0-embed $\bar{M}$ into a 0-iterable premouse. Because of the absence of fine structure, the application of the condensation lemma has to be replaced by a coiteration argument.

Using the fact that $\sigma_1^* : \tilde{M}^* \to M$ is Σ_0-preserving, we infer:

- $\tilde{M}^*$ is 0-coiterable with M and the coiteration is above κ.
- Let $\tilde{M}_\vartheta^*$ and M_ϑ be the last premouse on the $\tilde{M}^*$-side and the M-side of the coiteration, respectively. Then the iteration from $\tilde{M}^*$ to $\tilde{M}_\vartheta^*$ is simple and $\tilde{M}_\vartheta^*$ is an initial segment of M_ϑ.

This follows from the proof of Lemma 5.1.6, where we apply the Dodd–Jensen Lemma for 0-iterations (Lemma 4.3.14) instead of Lemma 4.3.12. Notice that, since the iteration from $\tilde{M}^*$ to $\tilde{M}_\vartheta^*$ is above κ, the cardinal successor of κ remains unchanged in the course of the iteration, so $\tilde{\tau}^*$ is the cardinal successor of κ in $\tilde{M}_\vartheta^*$. Similarly, τ is the cardinal successor of κ in M_ϑ, provided that the M-side of the coiteration is beyond τ. We shall consider two cases.

Case 1. The first iteration index used on the M-side is at least τ or else M is not moved in the coiteration. Then $\tilde{M}_\vartheta^*$ must be a proper initial segment of M_ϑ since the two premice compute the cardinal successor of κ differently. As $\tilde{\tau}^*$ is not a cardinal in M_ϑ, $\tilde{M}_\vartheta^*$ must be a proper segment of $M_\vartheta||\tau = M||\tau$, and thus a proper segment of Q. Let $N = \sigma(\tilde{M}_\vartheta^*)$ and $\tilde{\pi}^* : \tilde{M}^* \to \tilde{M}_\vartheta^*$ be the iteration map. Define $\sigma' : \bar{M} \to N$ by

$$\sigma'(x_i) = \sigma \circ \tilde{\pi}^*(\tilde{f}_i^*)(\alpha_i).$$

Let us verify that σ' is Σ_0-preserving, which will complete the proof in this case. Given a Σ_0-formula $\varphi(v_1, \dots, v_\ell)$, we infer, as in the proof of Lemma 5.6.5, that

$$\bar{M} \models \varphi(x_1, \dots, x_\ell) \longleftrightarrow \prec\alpha_{i_1}, \dots, \alpha_{i_\ell}\succ \in \sigma(u^{\varphi,M}_{f_{i_1},\dots,f_{i_\ell}})$$

and

$$\sigma(u^{\varphi,M}_{f_{i_1},\dots,f_{i_\ell}}) = \sigma(u^{\varphi,\tilde{M}^*}_{\tilde{f}^*_{i_1},\dots,\tilde{f}^*_{i_\ell}}) = \sigma\left(\tilde{\pi}^*(u^{\varphi,\tilde{M}^*}_{\tilde{f}^*_{i_1},\dots,\tilde{f}^*_{i_\ell}}) \cap \kappa\right).$$

The second equality follows from the fact that $\tilde{\pi}^* \upharpoonright \kappa = \mathrm{id}$. Since each α_i is less than κ, the right side of the above equivalence can be rewritten as

$$\prec\alpha_{i_1}, \ldots, \alpha_{i_\ell}\succ \in \sigma \circ \tilde{\pi}^*(u^{\varphi, \tilde{M}^*}_{\tilde{f}^*_{i_1}, \ldots, \tilde{f}^*_{i_\ell}}),$$

so Łos' Theorem then yields

$$N \models \varphi\big(\sigma \circ \tilde{\pi}^*(\tilde{f}^*_{i_1})(\alpha_1), \ldots, \sigma \circ \tilde{\pi}^*(\tilde{f}^*_{i_\ell})(\alpha_\ell)\big).$$

Case 2. The first iteration index used on the M-side is strictly less than τ. It follows that a partial measure is used at the beginning of the iteration $\Im^M$ on the M-side. Since the application of such a measure involves a truncation of M at an ordinal strictly less than τ, this iteration can be identified with an iteration of Q. Let $\tilde{\Im}$ be the copy of $\Im^M$ via σ, let $\tilde{Q}_\vartheta$ be the last premouse of $\tilde{\Im}$ and let $\sigma_\vartheta : M_\theta \to \tilde{Q}_\vartheta$ be the copy map at the end of the copying process. The situation is depicted in the following diagram, where the dotted arrows indicate the iterations which might possibly be non-simple:

$$\begin{array}{ccccccc}
\tilde{M}^* & \longrightarrow & \tilde{M}^*_\vartheta & \hookrightarrow & M_\vartheta & \xrightarrow{\ \sigma_\vartheta\ } & \tilde{Q}_\vartheta \\
 & & & & \uparrow & & \uparrow \\
 & & & & \vdots & & \vdots \\
 & & & & Q & \xrightarrow{\ \sigma\ } & \tilde{Q}
\end{array}$$

Set

$$N = \begin{cases} \tilde{Q}_\vartheta & \text{if } \tilde{M}^*_\vartheta = M_\vartheta \\ \sigma_\vartheta(\tilde{M}^*_\vartheta) & \text{otherwise} \end{cases}$$

and define $\sigma' : \bar{M} \to N$ by

$$\sigma'(x_i) = \sigma_\vartheta \circ \tilde{\pi}^*(\tilde{f}^*_i)(\alpha_i),$$

where $\tilde{\pi}^* : \tilde{M}^* \to \tilde{M}^*_\vartheta$ is, as before, the iteration map. To verify that σ' is Σ_0-preserving, we can proceed as in Case 1, but we have to show that σ can be replaced by σ_ϑ. This is possible, as $\sigma \upharpoonright \tilde{\tau}^* = \sigma_\vartheta \upharpoonright \tilde{\tau}^*$, which guarantees that

$$\sigma\Big(\tilde{\pi}^*(u^{\varphi, \tilde{M}^*}_{\tilde{f}^*_{i_1}, \ldots, \tilde{f}^*_{i_\ell}}) \cap \kappa\Big) = \sigma_\vartheta\Big(\tilde{\pi}^*(u^{\varphi, \tilde{M}^*}_{\tilde{f}^*_{i_1}, \ldots, \tilde{f}^*_{i_\ell}}) \cap \kappa\Big).$$

The rest of the verification is then straightforward. □(Lemma 5.6.6)

This last criterion explicitly requires that Q has a largest cardinal κ. In the above proof, this is used for the conclusion that $\tilde{M}^*_\vartheta$ is an initial segment of an iterate of Q. However, if there are unboundedly many Q-cardinals in τ, it is in general not clear how this can be guaranteed. On the other hand, there are special situations, e.g. when $\tilde{\tau}^*$ is singular in Q or $\mathrm{cf}((\tilde{\tau}^*)^{+Q}) > \omega$ which can be handled by the above argument. Since these situations vary depending on the applications in which they arise, we refrain from stating a general lemma along these lines.

Notes

The solidity theorem, i.e. the fact that the solidity witnesses are elements of the structure, was discovered by Mitchell. Condensation lemma of the kind we considered in this book first appeared in [MiS94]. The canonical well-ordering of mice was introduced in the original work of Dodd and Jensen, but of course, the present version of this ordering formally differs from the original one.

Chapter 6
Extender Models

We are now ready to extend the theory of mice to proper class models, called *extender models* or *weasels*. This theory can be considered a generalization of the fine structure theory of L in the presence of measures. There is an obvious analogy between the hierarchies composing L and those composing extender models: while the J_α's are levels of L, mice are levels of $L[E]$. However, not every weasel plays the role analogous to that of L in the absence of $0^\#$. Some weasels can namely be "thinner" than others, i.e. they might omit some measure living in V. To demonstrate that the class of all weasels is rich, we shall introduce the *K^c-construction,* a method for constructing extender models which will provide us with weasels having certain "maximality" properties. These weasels reflect all relevant information about measures in V in the sense that they are *universal,* i.e. they absorb any coiterable premouse and, granted that the universe does not contain a mouse for measure of order 1 (this assumption substitutes the absence of $0^\#$), they compute successors of sufficiently large singular cardinals correctly.

6.1 Extender Models and Iterations

Definition. An *extender model,* or, equivalently, a *weasel,* is a model W of the form $L[E] = J^E_\infty$ such that $W||\alpha$ is a mouse for every $\alpha \in \mathbf{On}$.

The property of being a weasel is preserved by elementary maps; however, upward preservation already depends on a non-trivial result on iterability.

Lemma 6.1.1. *Let $\sigma : W \to W'$ be Σ_1-preserving, where W, W' are proper class J-models. If one of W, W' is a weasel, then so is the other one.*

Proof. The map σ is easily seen to be cofinal and therefore fully elementary, since both W, W' are ZFC models. That the property of being a weasel is preserved by σ backwards follows easily from Lemma 4.3.2. Now suppose that W is a weasel. Given an ordinal α, $W||\alpha$ is iterable in the sense of W. Hence $\sigma(W||\alpha)$ is iterable in the sense of W'. By Theorem 4.5.5, $\sigma(W||\alpha)$ is iterable in the sense of V. So W' is a weasel, as the mice of the form $\sigma(W||\alpha) = W'||\sigma(\alpha)$ are cofinal in the height of W'.
□(Lemma 6.1.1)

Weasel iterations and coiterations are natural extensions of the respective notions for premice. It is not hard to see that every weasel iteration can be approximated by a mouse iteration of some initial segment of the weasel in question. All relevant facts concerning coiterations generalize to weasels. However, a new phenomenon occurs: iterations might have length ∞ (that is, length **On**).

In the following, the pair $\langle W_i;\ i < \theta\rangle, \langle M_i;\ i < \theta\rangle$ denotes a coiteration of a weasel W with a premouse M with coiteration indices ν_i, critical points κ_i and iteration maps π_{ij}^W, π_{ij}^M.

Lemma 6.1.2. *Let W be a weasel and M a mouse coiterable with W. Then at least one side of the coiteration of W with M is simple.*

Proof. Suppose the contrary. Let i and j be the first truncation point on the W- and the M-side, respectively, and let $k = \max\{i, j\}$. The coiteration from k on is identical to the coiteration of *premice*, the W_{k+1} and M_{k+1}. This coiteration terminates, say at θ, and $W_\theta = M_\theta$ since neither of the latter two premice is sound. To get a contradiction we then argue exactly as in the proof of Lemma 5.3.1. □(Lemma 6.1.2)

Lemma 6.1.3. *Let W, M be as above and θ be the length of their coiteration.*

a) *If the W-side is non-simple, then θ is a successor ordinal, say $\delta + 1$, and M_δ is an initial segment of W_δ.*

b) *If the M-side is non-simple, then $\theta = \infty$.*

Proof. If the W-side is non-simple then, as in the proof of the previous lemma, we reduce the coiteration of W with M to a premouse coiteration; clause a) then follows immediately. If the M-side is non-simple then, by the previous lemma, the W-side must be simple. Hence, each W_i is a weasel. That $\theta < \infty$ would mean that the coiteration terminates; then θ is a successor, say $\delta + 1$, and M_δ, being a set size structure, is a proper initial segment of W_δ. But this is impossible, since M_β is not sound. □(Lemma 6.1.3)

The further results on weasel coiterations depend on the pressing down argument. As we also consider coiterations of length ∞, we shall need a version of Fodor's Theorem for proper classes. Recall that a class is any collection of sets which is boldface definable over V. A proper class S is stationary iff its intersection with any c.u.b. proper class is nonempty. Note that the statement "S is a stationary proper class" cannot be expressed by a single formula in the language of ZFC, but still can be expressed by an infinite scheme of formulae. If $\langle C_\alpha\,;\ \alpha < \infty\rangle$ is a sequence of c.u.b. proper classes which is itself a class (that is, the sequence as a whole is boldface definable over V), we can define the diagonal intersection $C = \Delta C_\alpha$ in the obvious way and prove that C is again a c.u.b. proper class.

Lemma 6.1.4 (Fodor's Theorem for classes). *Let S be a stationary class and let $f : S \to \mathbf{On}$ be a regressive function. Then there is an ordinal α such that $f(\xi) = \alpha$ for unboundedly many ξ.*

Proof. Suppose the contrary. Let $C_\alpha = [g(\alpha), \infty)$ where

$$g(\alpha) = \text{the l.u.b. of } (f^{-1})''\{\alpha\},$$

where the "l.u.b. of a set $A \subset \mathbf{On}$" is the least upper bound of A, i.e. the least ordinal which is *strictly* larger than all elements of A. Then each C_α is a c.u.b. proper class such that $f''C_\alpha$ does not contain α. Moreover, $\langle C_\alpha \rangle_\alpha$ is clearly a class. Let $C = \Delta C_\alpha$. Pick a $\xi \in S \cap C$. Then $f(\xi) \neq \alpha$ for every $\alpha < \xi$. But this is impossible, since f is regressive. □(Lemma 6.1.4)

Lemma 6.1.5. *Suppose the coiteration of a weasel W with a premouse M never terminates.*

a) *For every ordinal α there is an index i such that $\pi^W_{0,i}(\alpha) < \kappa_i$.*

b) *There is a c.u.b. class $C \subset \mathbf{On}$ such that $\pi^M_{ij}(\kappa_i) = \kappa_j$ for every $i < j$ from C.*

Proof. Suppose for a contradiction that $\kappa_i \leq \pi^W_{0,i}(\alpha)$ for some ordinal α and all i. Then $\nu_i = \kappa_i^{+W_i} < \pi^W_{0,i}(\alpha^{++W})$, so we can assume w.l.o.g. that α is regular in W and $\nu_i < \pi^W_{0,i}(\alpha) = \alpha_i$. We shall inductively verify that $W_i||\alpha_i$ agrees with the i-th iterate of $W||\alpha$ in the iteration using the same indices ν_i. For limit i, this follows immediately from the induction hypothesis and properties of direct limits; for successor i, we observe by the regularity of α_i in W_i that every $f : \kappa_i \to W_i$ with $\mathrm{rng}(f) \subset W_i||\alpha_i$ is in fact an element of $W_i||\alpha_i$, so that $W_{i+1}||\alpha_{i+1} = \mathrm{Ult}(W_i||\alpha_i, E^{W_i}_{\nu_i})$ and $\pi^W_{i,i+1} \restriction (W_i||\alpha_i)$ agrees with the associated ultrapower map. It follows that $\langle W_i||\alpha_i;\ i \in \mathbf{On}\rangle$, $\langle M_i;\ i \in \mathbf{On}\rangle$ is a coiteration of two *premice* $W||\alpha$ and M which does not terminate. Contradiction. This proves clause a).

Towards the proof of b), fix the last truncation point i_0 on the M-side. For every limit ordinal $i > i_0$ let $f(i)$ be the least iteration index $h \geq i_0$ such that $\kappa_i \in \mathrm{rng}(\pi^M_{hi})$. By Fodor's Theorem for proper classes, there is an $i^* \geq i_0$ such that $\kappa_i \in \mathrm{rng}(\pi^M_{i^*,i})$ for unboundedly many $i \in \mathbf{On}$. Since M_{i^*} is a set, there is a $\kappa^* \in M_{i^*}$ such that the indices i satisfying $\kappa_i = \pi^M_{i^*,i}(\kappa^*)$ still comprise a proper class, say C. Then clearly

$$\pi^M_{ij}(\kappa_i) = \kappa_j \text{ whenever } i < j \text{ and } i, j \in C.$$

We show that C is closed. Given a limit point j of C, note that $\tilde{\kappa} \overset{\text{def}}{=} \pi^M_{i^*,j}(\kappa^*) = \sup\{\kappa_i;\ i < j\}$: On the one hand, $\pi^M_{i^*,j}(\kappa^*) \geq \pi^M_{i^*,i}(\kappa^*) = \kappa_i$ for every $i \in C \cap j$, which yields the inequality $\geq$; on the other hand, each $\xi < \pi^M_{i^*,i}(\kappa^*)$ is of the form $\pi^M_{i,j}(\xi_i)$ for some $i \in C \cap j$ and $\xi_i < \kappa_i$, so that $\xi < \kappa_i$ since $\xi = \xi_i$. Now, by the normality of the iteration, $\kappa_j \geq \tilde{\kappa}$. But if $\kappa_j > \tilde{\kappa}$, then $\pi^M_{i^*,i}(\kappa^*) = \tilde{\kappa}$ for every $i \geq j$, so C would not be a proper class. Hence $\tilde{\kappa} = \kappa_j$. □(Lemma 6.1.5)

Lemma 6.1.6. *Let $\langle W_i\,;\ i < \infty\rangle$ be a simple normal iteration of a weasel W with indices ν_i, critical points κ_i and iteration maps π_{ij} such that*

$$(\forall\alpha)(\exists i)(\pi_{0,i}(\alpha) < \kappa_i). \tag{6.1.1}$$

Define W_∞ and $\pi_{i,\infty} : W_i \to W_\infty$ as follows:

$$W_\infty = \bigcup_{i<\infty} W_i||\kappa_i$$

$$\pi_{i,\infty}(x) = \pi_{ij}(x) \text{ where } j \text{ is large enough so that } \pi_{ij}(x) \in W_j||\kappa_j.$$

Then $\langle W_\infty, \langle \pi_{i,\infty} \rangle_i \rangle$ *is a direct limit of* $\langle W_i, \pi_{ij}\,;\, i \le j < \infty \rangle$.

Proof. The restrictions $\pi_{ij} \restriction (W_i||\kappa_i)$ of the iteration maps are identities, so each $W_i||\kappa_i$ is easily seen to be a transitive subset of W_∞. Since every fibre $\langle x_i;\ k \le i < \infty \rangle = \langle \pi_{ki}(x_k);\ k \le i < \infty \rangle$ is eventually constant with eventual value in some $W_j||\kappa_j$, the segments $W_j||\kappa_j$ cover all of W_∞. That the map defined in the statement of the lemma is the direct limit map, follows from the observation that we can replace each fibre by its eventual constant value. □(Lemma 6.1.6)

Lemma 6.1.7. *Let* $\langle W_i^h\,;\, i < \theta \rangle$, $h = 0, 1$ *be the coiteration of a pair of weasels* W^0, W^1 *with indices* ν_i, *critical points* κ_i *and iteration maps* π_{ij}^h.

a) *At least one side is simple.*

b) *If one side is non-simple, then* $\theta = \infty$.

c) *If* $\theta = \infty$, *then at least one side satisfies* (6.1.1).

d) *If* $\theta = \infty$ *and both sides satisfy* (6.1.1), *then* $W_\infty^0 = W_\infty^1$.

e) *If* $\theta = \infty$ *and the* W^h*-side does not satisfy* (6.1.1), *then there is a c.u.b. proper class* C *such that* $\pi^h(\kappa_i) = \kappa_j$ *whenever* $i, j \in C$ *and* $i < j$.

Proof. Clauses a), b) and d) follow easily from the previous lemmata. To see c), note that if (6.1.1) fails on both sides, then, as in the proof of Lemma 6.1.5(a), we could replace the coiteration of W^0 with W^1 by the coiteration of $W^0||\alpha_0$ with $W^1||\alpha_1$ of length ∞, where α^h is large enough that ν_i is strictly less than $\pi_{0,i}^h(\alpha^h)$ and regular in W^h $(h = 0, 1)$. This is, of course, impossible. Clause e) follows similarly: If (6.1.1) fails on the W^h-side, replace this side of the iteration by the iteration of $W^h||\alpha$ with the same iteration indices and suitably chosen α. This allows us to apply the proof of Lemma 6.1.5(b) to the mouse $W||\alpha$. □(Lemma 6.1.7)

6.2 The Canonical Ordering of Weasels

Given a coiteration of a weasel W with a premouse M that does not terminate, the direct limit W_∞ on the W-side is a weasel. We can also form the direct limit on the M-side, and obtain a well-founded structure M_∞ which cannot be transitivised, as it is not set-like, i.e. its ordinals are longer than **On**. However, the transitivized maximal set-like lower part of M_∞ is equal to the union $\bigcup M_i||\kappa_i$ and is easily seen to agree with W_∞, so W_∞ can be considered a proper initial segment of M_∞. For this reason, we consider W_∞ to be a simple iterate of W and a non-simple iterate of M.

Let W be a weasel or a premouse and let $\Im$ be a simple iteration of W of length ∞. Set $W_\infty = \bigcup W_i||\kappa_i$ and $\Im' = \Im^\frown\langle W_\infty \rangle$. We consider $\Im'$ to be an iteration of length $\infty + 1$, or, in other words, a *long* iteration of W. By Lemma 6.1.6, W_∞ is a direct limit along $\Im$ if (6.1.1) holds.

Definition. A long iteration $\mathfrak{I}$ of W is *simple* iff (6.1.1) holds, otherwise $\mathfrak{I}$ is *non-simple*.

Lemma 6.2.1. *Let $W = W^0$ be a weasel and W^j be a simple iterate of W^i with iteration maps π^{ij} where $i \leq j < \theta < \infty$. Then the direct limit W' of $\langle W^i, \pi^{ij};\ i \leq j < \theta \rangle$ is a simple iterate of W.*

The same conclusion holds for $\theta = \infty$ provided that the sequence $\langle \pi^{0,i}(\alpha);\ i < \infty \rangle$ is bounded for all ordinals α.

Proof. We give the proof for $\theta < \infty$; the case $\theta = \infty$ can be handled similarly and is left to the reader. Since the iterations we are dealing with are proper classes, the lemma should have been stated more carefully: We assume that we have a proper class $\mathfrak{I}$ representing a sequence of iterations $\langle \mathfrak{I}^k;\ k < \theta \rangle$ where each $\mathfrak{I}^k$ consists of weasels W^k_{ij} and maps π^k_{ij} for $i \leq j < \theta_k = \text{length}(\mathfrak{I}^k)$; the sequence $\mathfrak{I}$ is a concatenation of all $\mathfrak{I}^k$, i.e. $W = W^0_0$, $W^{k+1}_0 = W^k_{\theta_k}$ and W^k_0 is the direct limit of $\langle W^g_i, \pi^{gh}_{ij} \rangle$. In the above notation, W^k is equal to W^k_0 and $\pi^{k\ell}$ to the associated iteration map from W^k to W^ℓ.

Let α be regular in W and such that $\nu < \alpha$ whenever $\text{cr}(E^W_\nu) < \alpha$; there are certainly proper class many such ordinals α. For $k < \theta$ set $\alpha^k = \pi^{0,k}(\alpha)$. Since all $\mathfrak{I}^k$ are simple, we can assign an ordinal $\iota(k)$ to each k such that the iteration $\mathfrak{I}^k$ is above $\pi^k_{0,\iota(k)}(\alpha^k)$ from $\iota(k)$ on. Since α^k is regular in W^k, the iteration $\mathfrak{I}^k(\alpha^k)$ of $W^k||\alpha^k$, obtained by applying the first $\iota(k)$ indices of $\mathfrak{I}^k$, agrees with $\mathfrak{I}^k \upharpoonright \iota(k)$ in the sense that the mice and iteration maps of the former are initial segments of the corresponding objects of the latter. Let $\mathfrak{I}(\alpha)$ be the concatenation of all $\mathfrak{I}^k(\alpha^k)$ followed by a direct limit, if this concatenation is of limit length, and let $W(\alpha)$ be the last premouse in $\mathfrak{I}(\alpha)$. Then $W(\alpha)$ is a simple iterate of $W||\alpha$ and, by the proof of Lemma 5.3.7, $W(\alpha)$ is not moved in the coiteration against $W||\alpha$. Let $\mathfrak{I}^*_\alpha$ be the $W||\alpha$-side and $\vartheta(\alpha)$ its length. To each $\mathfrak{I}^*_\alpha$ we assign an iteration $\mathfrak{I}_\alpha$ of W which uses the same iteration indices as $\mathfrak{I}^*_\alpha$; informally, we simply "lengthen" the mice of $\mathfrak{I}^*_\alpha$ to the corresponding weasels. Again, $\mathfrak{I}_\alpha$ agrees with $\mathfrak{I}^*_\alpha$ in the above sense.

Now if $\beta > \alpha$ then $W(\beta)||\alpha = W(\alpha)$, so $\mathfrak{I}^*_\beta \upharpoonright \vartheta(\alpha)$ agrees with $\mathfrak{I}^*_\alpha$ in the sense of the previous paragraph. It follows that $\mathfrak{I}_\beta \upharpoonright \vartheta(\alpha) = \mathfrak{I}_\alpha$. Thus, the union $\mathfrak{I}'$ of all $\mathfrak{I}_\alpha$ is a simple normal iteration of W and if length$(\mathfrak{I}') = \infty$ then $\mathfrak{I}'$ satisfies (6.1.1) since $\mathfrak{I}'$ is above $\text{ht}(W(\alpha)) = \pi^{\mathfrak{I}'}_{0,\vartheta(\alpha)}(\alpha)$ from $\vartheta(\alpha)$ on. By Lemma 6.1.6, $W' = \bigcup W(\alpha)$ is the direct limit along $\mathfrak{I}'$. □(Lemma 6.2.1)

The boundedness requirement in the second part of the above lemma is essential. We can obtain a concatenation of ∞ many iterations of length 1 by iterating a total measure of a weasel W along the ordinals. By Dodd–Jensen Lemma for weasels (see below), such an iteration cannot be rearranged into an iteration satisfying (6.1.1).

Lemma 6.2.2. *Let W' be an iterate of a weasel or mouse W and W'' a non-simple iterate of W'. Then W'' is a non-simple iterate of W.*

Proof. Let $\mathfrak{I}$ be an iteration from W to W' and $\mathfrak{I}'$ a non-simple one from W' to W''. The fact that $\mathfrak{I}'$ is non-simple means that (6.1.1) fails; in other words, there is a regular

α in W' such that whenever $\Im' \restriction i$ is simple, $\nu_i < \pi^{\Im'}_{0,i}(\alpha)$. Thus, W'' is a non-simple iterate of the mouse $W||\alpha$ via the iteration $\Im''$ which uses the same iteration indices as $\Im'$. We next find a sufficiently large $\iota < \infty$ such that $\alpha < \kappa_\iota$ and $W_\iota||\kappa_\iota = W'||\kappa_\iota$. Then $\Im^* = (\Im \restriction \iota)^\frown \Im''$ constitutes a non-simple iteration of length $\infty + 1$ with the last model W''.

The existence of an ι as above in the non-trivial case where length$(\Im) = \infty + 1$ follows from the fact that for every α, $\kappa_i > \alpha$ from some i on. Otherwise we would obtain (using the pigeonhole principle) a proper class D of indices such that $\kappa_i = \kappa_j$ for all $i, j \in D$. Letting $\langle i_n;\ n \in \omega\rangle$ be the first ω elements of D and $i^* = \sup\{i_n\}_n$, $\langle \pi^{\Im}_{i_n,i^*}(\kappa_{i_n});\ i \in \omega\rangle$ would be an infinite decreasing sequence of ordinals in W_{i^*}.

□(Lemma 6.2.2)

In general, it is not clear that a long non-simple iteration followed by a simple iteration can be rearranged into a single non-simple iteration. Fortunately, we do not need this in order to develop the properties of the canonical ordering of weasels. We shall, however, need a more elaborate version of the Dodd–Jensen Lemma.

Lemma 6.2.3 (Dodd–Jensen Lemma for Weasels). *Let W' be an iterate of W, where W is a weasel or a mouse, and let $\sigma : W \to W'$ be Σ_1-preserving. Then any iteration from W to W' is simple and $\pi(\xi) \leq \sigma(\xi)$ for every ordinal ξ, where π is the associated iteration map. Consequently, W' cannot be both a simple and non-simple iterate of W and the iteration map does not depend on the actual iteration.*

More generally, let W_{i+1} be an iterate of W_i via an iteration $\Im_i$ for $i = 1, \ldots, n-1$. If there is a Σ_1-preserving map $\sigma : W_0 \to W_n$, then each $\Im_i$ is simple and, setting $\pi = \pi_{n-1} \circ \cdots \circ \pi_1 \circ \pi_0$ where π_i is the iteration map associated with $\Im_i$, $\sigma(\xi) \leq \pi(\xi)$ for every ordinal ξ.

Proof. We shall only prove the first part of the lemma and leave the slightly more elaborate proof of the second part to the reader (note, however, that the only thing to prove is that each $\Im_i$ is simple, as the minimality of π follows from the first part of the lemma). Furthermore, we shall only consider the case where W is a weasel and the iteration from W to W' is long, which is not automatically covered by the Dodd–Jensen Lemma for mice. Towards a contradiction, fix an iteration $\Im$ witnessing the failure of the lemma. If there is a truncation point $i < \infty$, pick an ordinal α regular in W such that $\Im \restriction i$ agrees with the iteration of $W||\alpha$ of length i with the same iteration indices. Fix $j > i$ such that $\sigma(\alpha) < \kappa_j$; then $W'||\sigma(\alpha)$ is a nonsimple iterate of $W||\alpha$ and at the same time $\sigma : W||\alpha \to W'||\sigma(\alpha)$ is fully elementary. This contradicts the Dodd–Jensen Lemma for mice. If the only truncation point in the iteration is ∞, then (6.1.1) fails, so we can fix an α regular in W and an ordinal j such that $\nu_i < \pi_{0,i}(\alpha)$ for all i and $\sigma(\alpha) < \kappa_j$, and obtain a contradiction as before. It follows that $\Im$ must be simple.

Now suppose $\sigma : W \to W'$ is a counterexample to the second part of the lemma, i.e. $\sigma(\xi) < \pi(\xi)$ for some ξ. We shall again reduce this to a contradiction to the Dodd–Jensen Lemma for mice. This time we fix j so that $\kappa_j > \pi_{0,j}(\xi)$ and set $\alpha = \xi^{++W}$. Then $\Im \restriction j$ can be replaced by the corresponding iteration of $W||\alpha$ and

$\sigma : W||\alpha \to W'||\sigma(\alpha)$ is fully elementary. But $\sigma(\alpha)$, being the double successor of $\sigma(\xi)$ in W', must, by our assumption, be less than or equal to $\pi(\xi)^{++W'} = \pi(\alpha)$. If it is strictly smaller, we obtain a contradiction as above. Otherwise $\sigma : W||\alpha \to W'||\pi(\alpha)$ and the contradiction follows from the minimality of the iteration map $\pi \restriction (W||\alpha) = \pi_{0,j} \restriction (W||\alpha)$. □(Lemma 6.2.3)

Lemma 6.2.4. *There is no countable sequence $\langle W_i;\ i \in \omega\rangle$ such that W_0 is a weasel or mouse and each W_{i+1} is a non-simple iterate of W_i.*

Proof. It suffices to focus on the case where each iteration $\Im_i$ from W_i to W_{i+1} is long. To each i we assign a $\vartheta(i)$ such that, setting $\kappa_i^* = \kappa_{\vartheta(i)}^{\Im_i}, \kappa_j^{\Im_i} = \pi_{ij}^{\Im_i}(\kappa_i^*)$ for unboundedly (and therefore c.u.b.-) many j. Note that the sequence $\langle\vartheta(i)\rangle_i$ can be correctly defined in ZFC. Let $\theta(i)$ be the least ordinal θ satisfying

$$\kappa_\theta^{\Im_i} = \pi_{\vartheta(i),\theta}^{\Im_i}(\kappa_i^*) > \kappa_{i+1}^*.$$

Then $\Im_0^* {}^\frown \Im_1^* {}^\frown \ldots {}^\frown \Im_i^* {}^\frown \ldots$, where $\Im_i^* = \Im_i \restriction (\theta(i) + 1)$, constitutes an iteration of W_0 which cannot be continued, i.e. the direct limit W^* along this iteration is not well-founded: Otherwise we could take W^* to be transitive; then $\pi_i^*(\kappa_i^*)$ would be an infinite decreasing chain of ordinals, where $\pi_i^* : W_i \to W^*$ are the direct limit maps. Given sufficiently large α regular in W, the same would apply to $W||\alpha$, contradicting the fact that $W||\alpha$ is a mouse. □(Lemma 6.2.4)

Using the above lemmata we can apply the proofs of Lemma 5.3.7 and Corollary 5.3.9 to weasels and obtain:

Lemma 6.2.5. *Let W be a weasel and $\sigma : W \to W'$ a derived iteration map, that is, there is a common simple iterate W^* of W, W' with iteration maps $\pi : W \to W^*$, $\pi' : W' \to W^*$ such that $\pi' \circ \sigma = \pi$. Then W' is a normal iterate of W and σ is the iteration map. In particular, if W' is a simple iterate of W, then W' is a simple normal iterate of W.* □(Lemma 6.2.5)

We define the canonical ordering of weasels in the obvious way; the above lemmata allow us to imitate the corresponding proofs for mice in the weasel context (Lemmata and Corollaries 5.4.1 – 5.4.4).

Definition. Let W, W' be weasels or mice.

a) $W \sim^* W'$ iff W, W' coiterate simply to a common weasel or mouse.

b) $W <^* W'$ iff there is a weasel or mouse which is a simple iterate of W and a non-simple iterate of W'

c) $W \leq^* W'$ iff $W <^* W'$ or $W \sim^* W'$.

Lemma 6.2.6.

a) *The relation* $<^*$ *is irreflexive, transitive, linear modulo* $\sim^*$, *i.e.*

$$(W \not<^* W' \;\&\; W' \not<^* W) \longrightarrow W \sim^* W'$$

and well-founded.

b) $\sim^*$ *is a congruence relation on* $\langle \mathcal{W}, <^* \rangle$, *where* $\mathcal{W}$ *constitutes the collection of all mice and weasels.*

□(Lemma 6.2.6)

Corollary 6.2.7. $<^*$ *is a well-ordering of mice and weasels modulo* $\sim^*$.

□(Corollary 6.2.7)

We stress that our statements about the "collection $\mathcal{W}$" are meant informally – as weasels are proper classes, no collection of weasels can be a class. Therefore, the formulations of the above lemmata and corollaries in the ZFC context (which are left to the reader) require some care. For instance, the last corollary can be understood as follows: Any class representing a collection of weasels is well-ordered by $<^*$ modulo $\sim^*$. This statement can obviously be expressed in ZFC.

Definition. A weasel W is *weakly universal* iff the coiteration of W with any mouse terminates.

It is not hard to see that any two weakly universal weasels coiterate simply to a common weasel; it suffices to observe that any non-simple weasel iteration can be substituted by an iteration of a sufficiently tall proper initial segments of the original weasels. Assuming that the W'-side of the coiteration of W against W' is non-simple, we can reduce the original coiteration to the coiteration of W with a mouse of the form $W'||\alpha$ which wouldn't terminate, and thus obtain a contradiction to weak universality. It follows that weakly universal weasels are precisely the maximal elements in the ordering $<^*$. The above argument in fact yields that, granting weak universality of W, there is no truncation on the W'-side no matter whether W' is weakly universal or not.

Lemma 6.2.8. *Let* $\sigma : W \to W'$ *be* Σ_0*-preserving, where* W, W' *are weasels and* W *is weakly universal. Then* W' *is also weakly universal. In particular, any simple iterate of a weakly universal weasel is again weakly universal.*

Proof. We first observe that, given any ordinal α, $W||\alpha \leq^* \sigma(W||\alpha)$ in the canonical ordering of mice. To see this note that, by the Dodd–Jensen Lemma, the last premouse $\tilde{W}$ on the $W||\alpha$-side of the coiteration of $W||\alpha$ with $\sigma(W||\alpha)$ must be a simple iterate of $W||\alpha$ as well as an iterate of $\sigma(W||\alpha)$. (See the proof of Lemma 5.1.7 for details).

Pick a mouse M. Let α be regular in W and large enough that $M \leq^* W||\alpha$. By the above observation, $M \leq^* W||\alpha \leq^* \sigma(W||\alpha)$. Hence the coiteration of $\sigma(W||\alpha)$ with M terminates. Since $\sigma(\alpha)$ is regular in W', the coiteration of W' with M agrees with that of $W'||\sigma(\alpha) = \sigma(W||\alpha)$ with M and therefore terminates. □(Lemma 6.2.8)

6.3 Universality

The notion of weak universality from the end of the previous section was a bit ad hoc; its only purpose was to help us understand the behaviour of the ordering $<^*$. We now introduce a strengthening of this notion in that we broaden the class of premice the weasels should absorb in coiterations to that of *all coiterable* premice. Considering this broader class of premice is somewhat more natural, since, when performing the coiteration of W with some premouse M, we do not really need to know that M is fully iterable, but merely that the coiteration can be carried out, which will be crucial in many arguments.

Definition. A weasel W is universal iff the coiteration of W with any coiterable premouse terminates, that is, its length is strictly less than ∞.

Obviously, every universal weasel is weakly universal. We shall later see that, under suitable smallness condition, the two notions agree. Given any premouse M coiterable with W, the M-side of the coiteration must be simple (this follows from the remarks above Lemma 6.2.8) and the last premouse on the M-side must be a proper initial segment of the last weasel or mouse on the W-side, which means that M iterates simply to a mouse. It follows that M itself is a mouse, i.e. fully iterable. Thus, proving the full iterability of a premouse merely amounts to proving its coiterability with a universal weasel.

Lemma 6.3.1. *Let W be a weasel that is not universal. There is a regular cardinal μ and a c.u.b. class of ordinals C such that for every $\beta \in C$,*

a) *β is inaccessible in W;*

b) $\mathrm{cf}(\beta^{+W}) = \mu$.

Proof. Let M be a premouse such that the coiteration of M with W does not terminate. By Lemma 6.1.5, the W-side satisfies (6.1.1). Let W' be the last weasel on the W-side, π_{ij} the iteration maps and $\pi = \pi_{0,\infty}$. For every β we have

$$\pi''\beta^{+W} \text{ is cofinal in } \pi(\beta^{+W}).$$

To see this, fix ι such that $\pi_{0,\iota}(\beta^{+W}) < \kappa_\iota$. Since $\pi_{\iota,\infty} \restriction (\pi_{0,\iota}(\beta^{+W}) + 1) = \mathrm{id}$, it suffices to prove the above statement with $\pi_{0,\iota}$ in place of π. We proceed by induction along the iteration from W to W_ι. For $i < \iota$ we have $\kappa_i < \pi_{0,i}(\beta^{+W})$, so the ultrapower map $\pi_{i,i+1}$ maps $\pi_{0,i}(\beta^{+W})$ cofinally to $\pi_{0,i+1}(\beta^{+W})$ since $\pi_{0,i}(\beta^{+W})$ is regular in W_i and, being a successor cardinal in W_i, differs from κ_i. This takes care of the successor step. For limit steps, note that direct limits automatically preserve cofinality of iteration maps.

On the M-side we "repeat the same critical point" on a c.u.b class, say C_0. In other words, $\pi^M_{ij}(\kappa_i) = \kappa_j$ for any $i < j$ from C_0. Let C_1 be the class of all $i \in C_0$ satisfying

$$\pi''_{h,i}\kappa_i \subset \kappa_i \text{ whenever } h < i.$$

Then C_1 is again a c.u.b. class. That C_1 is closed follows immediately. To see its unboundedness fix an ordinal α and for $n \in \omega$ pick $i(n) \in C_0$ such that $\kappa_{i(0)} > \alpha$ and $i(n+1)$ is the least $j \in C_0$ satisfying $\pi_{0,j}(\kappa_{i(n)}) < \kappa_j$. Then $i = \sup\{i(n)\,;\, n \in \omega\}$ is greater than α and $\pi''_{h,i}\kappa_i \subset \kappa_i$.

We next observe that $\mathcal{P}(\kappa_i) \cap W_i = \mathcal{P}(\kappa_i) \cap M_i$ for every $i \in C_1$. The inclusion $\supset$ follows from the fact that every subset of κ_i is a member of $M_i||\nu_i$ (since $E^{M_i}_{\nu_i}$ is a total measure in M_i, $\nu_i = \kappa_i^{+M_i}$) and W_i agrees with M_i below ν_i. To see the converse, pick an $a \in \mathcal{P}(\kappa_i) \cap W_i$ and an $\alpha > \kappa_i$ regular in W_i such that $a \in W_i||\alpha$. Let j be the least ordinal with the property $\pi_{0,j}(\alpha) < \kappa_j$. By the normality of the iteration from W_i to W_j we have $a = \pi_{ij}(a) \cap \kappa_i \in W_j$, hence $a \in W_j||\pi_{i,j}(\alpha)$ since $\pi_{i,j}(\alpha)$ is a cardinal in W_j. And since M_j agrees with W_j below κ_j, a is an element of M_j. Hence $a \in M_i$ by the normality of the iteration from M_i to M_j. (Note that this argument goes through for any $i \in C_0 - \min(C_0)$.)

By the above observations and the fact that κ_i is inaccessible in W_i, κ_i is inaccessible in both M_i and W_i and $\kappa_i^{+M_i} = \kappa_i^{+W_i}$ for $i \in C_1$; denote this common value by τ_i. From this we infer that each such κ_i must be regular in W_h for all $h < i$. Suppose this fails; then W_h contains an increasing cofinal function $f : \gamma \to \kappa_i$ for some $\gamma < \kappa_i$. Setting $\gamma' = \sup(\pi''_{h,i}\gamma)$, we obtain that $\gamma' \leq \pi_{0,i}(\gamma) < \kappa_i$ and $\pi_{h,i}(f)$ maps γ' cofinally into κ_i, which contradicts the regularity of κ_i in W_i.

Finally, we show that κ_i is a fixpoint of $\pi_{0,i}$. Let k be the least index such that $\kappa_i \in \operatorname{rng}(\pi_{ki})$. If $k > 0$, then $k = h+1$ for some h. Now π_{hk} is either the identity or the ultrapower map by the measure $E^{M_k}_{\nu_k}$, which is a member of $W_h||\kappa_i$ (the latter follows from the fact that κ_i is a cardinal in W_k). In either case, $\pi_{hk}(\kappa_i) = \kappa_i$ since κ_i is regular in W_k. This contradicts the definition of k, so $k = 0$, i.e. $\kappa_i \in \operatorname{rng}(\pi_{0,i})$. Since $i \in C_1$, this forces $\pi_{0,i}(\kappa_i) = \kappa_i$.

Let $C = \{\kappa_i\,;\, i \in C_1\}$. C is clearly a c.u.b. class and a) of the lemma holds for every $\beta \in C$. We now prove b). Pick $\beta \in C$, say $\beta = \kappa_i$ for an $i \in C_1$. By the argument from the beginning of this proof, $\pi_{0,i}$ maps κ_i^{+W} cofinally into $\pi_{0,i}(\kappa_i^{+W}) = \kappa_i^{+W_i} = \tau_i$. Setting i^* to be the least element of C_1, we have $\kappa^* = \kappa_{i^*}$ and $\tau^* = (\kappa^*)^{+M_{i^*}}$. The same argument as above yields that $\pi^M_{i^*,i}$ maps τ^* cofinally into $\kappa_i^{+M_i} = \tau_i$. Note that the above-mentioned argument is applicable to the iteration on the M-side, since if n satisfies $\omega\varrho^{n+1}_{M_{i^*}} \leq \kappa_{i^*} < \omega\varrho^n_{M_{i^*}}$, then $\omega\varrho^{n+1}_{M_i} \leq \kappa_i < \omega\varrho^n_{M_i}$ for all $i \in C_1$, so each κ_i is a member of $H^n_{M_i}$ and the $*$-ultrapower maps of the iteration restricted to the n-th reducts agree with the corresponding 0-ultrapower maps. Taken together, $\operatorname{cf}(\kappa_i^{+W}) = \operatorname{cf}(\tau_i) = \operatorname{cf}(\tau^*)$. Thus, $\mu = \operatorname{cf}(\tau^*)$ witnesses clause b). □(Lemma 6.3.1)

Corollary 6.3.2. *If $V = W$, where W is a weasel, then W is universal.*

□(Corollary 6.3.2)

The above results can be improved in the sense that we have a uniform upper bound for the cofinalities of *all* W-successors. This strengthened version is, however, only possible in the context of extender models for measures (or extenders with only very few generators) and fails in extender models with sufficiently strong extenders on the sequence. We shall return to this issue in the chapter on extender models below one strong cardinal.

Lemma 6.3.3. *Let M be an acceptable structure, let U be a weakly amenable measure on M and $M' = \mathrm{Ult}(M, U)$. Then for every τ' which is a successor cardinal in M' there is a successor cardinal τ in M that can be cofinally mapped into τ'.*

Proof. Let π be the ultrapower map, $\kappa = \mathrm{cr}(\pi)$ and let τ' be a successor cardinal in M'; say $\tau' = (\mu')^{+M'}$ for some M'-cardinal μ'. Pick $f \in M$, $f : \kappa \to \mathbf{On} \cap M$ such that $\mu' = \pi(f)(\kappa)$. Setting $f^+(\xi) = (f(\xi))^{+M}$ for all $\xi < \kappa$, we have $\tau' = \pi(f^+)(\kappa)$. In the following, we shall use the fact that f can be taken to be monotone. This follows from a Ramsey-like argument. Since U is κ-complete with respect to sequences from M, the standard proof of partition relations for measurable cardinals yields $\kappa \to (U)^2_2$ in the sense that the colourings are taken from M. Define a colouring e of $[\kappa]^2$ by

$$e(\{\xi_1, \xi_2\}) = \begin{cases} 0 & \text{if } \xi_1 < \xi_2 \text{ and } f(\xi_1) \le f(\xi_2); \\ 1 & \text{otherwise.} \end{cases}$$

Given a homogeneous set $x \in U$, e must be of colour 0 on x, which means that f is non-decreasing on x. Thus, the map $\tilde{f} : \kappa \to \kappa$ defined by $\xi \mapsto f(\min(x - \xi))$ is monotone and $\pi(\tilde{f})(\kappa) = \pi(f)(\kappa)$ since $\tilde{f}$ agrees with f on a set from U.

In the special case where f is eventually constant with value μ we have $\mu' = \pi(\mu)$, so it suffices to set $\tau = \mu^{+M}$ since π maps μ^{+M} cofinally to τ'. From now on assume that f is monotone, strictly increasing on a cofinal subset of κ, and $f(\xi) \ge \xi$ for every $\xi \in \kappa$. Recall that any α which lies strictly between μ' and τ' can be represented in the ultrapower as $\pi(f^*)(\kappa)$, where $f(\xi) < f^*(\xi) < f^+(\xi)$ for all $\xi \in \kappa$. Let X be the collection of all functions $f^* \in M$ satisfying the above inequality. Clearly, $X \in M$.

Let $\mu = \sup(\mathrm{rng}(f))$. Since we assume that μ' is the cardinal predecessor of τ' in M', there are unboundedly many $\xi < \kappa$ such that $f(\xi)$ is a cardinal in M (this follows by the Łos Theorem – the collection of all such ξ is a set from U). Hence μ is a limit cardinal in M. We show that $\tau = \mu^{+M}$ can be cofinally mapped into τ'. To every $\vartheta < \tau$ such that μ is the largest cardinal in J^A_ϑ (here $A = A^M$), pick the $<_M$-least enumeration $\langle f^\vartheta_i \,;\, i < \mu\rangle$ of $X \cap J^A_\vartheta$ and define $g^\vartheta : \kappa \to \mu$ by

$$g^\vartheta(\xi) = \sup\{f^\vartheta_i(\xi) + 1 \,;\, i < f(\xi)\}.$$

Then $g^\vartheta \in M$; moreover, $g^\vartheta \in X$ since each $f^+(\xi)$ is regular in M. From the monotonicity of f and the fact that its range is cofinal in μ we infer that g^ϑ eventually majorizes every function in $X \cap J^E_\vartheta$. Note also that if $\vartheta' < \vartheta$ are as above, then $g^{\vartheta'} \in J^A_\vartheta$ since the enumeration $\langle f^\vartheta_i \,;\, i < \kappa\rangle$ of $X \cap J^A_{\vartheta'}$ lies in J^A_ϑ. Hence, g^ϑ eventually majorizes $g^{\vartheta'}$.

Setting $\zeta_\vartheta = \pi(g^\vartheta)(\kappa)$, $\zeta_\vartheta > \pi(f^\vartheta_i)(\kappa)$ whenever $i < \mu$ because each f^ϑ_i is eventually dominated by g^ϑ. Since every $f \in X$ is of the form f^ϑ_i for some ϑ and i, the assignment $\vartheta \mapsto \zeta_\vartheta$ maps τ cofinally to τ'. In fact, this mapping is strictly increasing since each g^ϑ eventually majorizes all $g^{\vartheta'}$, where $\vartheta' < \vartheta$. □(Lemma 6.3.3)

Corollary 6.3.4. *Given a weasel W that is not universal, there is a cardinal μ such that* $\mathrm{cf}(\beta^{+W}) \leq \mu$ *for every* β.

Proof. Let M be a premouse whose coiteration with W does not terminate, W' be the last weasel on the W-side of the coiteration and π^M, π^W be the respective iteration maps. At the beginning of the proof of Lemma 6.3.1 we showed that π^W maps cardinal successors of W cofinally into their images. Therefore it suffices to show that $\mathrm{cf}(\beta^{+W'}) < \mu$ for every β. Let ι be greater than the last truncation point on the M-side (if there is any) and such that there is a fixed n satisfying $\omega\varrho^{n+1}_{M_i} \leq \kappa_i < \omega\varrho^n_{M_i}$ whenever $\iota \leq i$. Recall that the $*$-ultrapower maps from the iteration on the M-side restricted to the n-th reducts agree with the corresponding 0-ultrapower maps.

Fix β and pick the least iteration index $i \geq \iota$ such that $\beta < \kappa_i$; since κ_i is a limit cardinal in $W' = \bigcup M_j||\kappa_j$, $\beta^{+W'} = \beta^{+M_i} \subset H^n_{M_i}$. Granted that $i > \iota$, we find an $h < i$ and a successor cardinal δ in M_h, $\delta \subset H^n_{M_h}$, which can be cofinally mapped into β^{+M_i} via a strictly increasing map. If i is a limit, it suffices to set h to be the least $\bar{h} \geq \iota$ satisfying $\beta \in \mathrm{rng}(\pi^M_{\bar{h}i})$. Such an h is certainly strictly less than i; moreover, $\delta = (\pi^M_{h,i})^{-1}(\beta^{+M_i})$ is a cardinal successor in M_h, a subset of $H^n_{M_h}$ and $\pi^M_{hi} \restriction \delta$ is a strictly increasing mapping of δ cofinally into β^{+M_i}. If i is a successor ordinal, let $h = i - 1$. By Lemma 6.3.3, there is a successor cardinal δ in M_h and a cofinal strictly increasing mapping of δ into β^{+M_i}. A closer examination of the proof of that lemma yields that the cardinal predecessor of δ in M_h must be an element of the n-th reduct (since the same applies to β), hence $\delta \subset H^n_{M_h}$.

Set $i = i(0)$ and $\delta_0 = \beta^{+M_i}$. By iterating the algorithm from the previous paragraph we obtain a decreasing sequence of ordinals $i(n)$, cardinal successors δ_n in $M_{i(n)}$ and strictly increasing cofinal maps $g_n : \delta_n \to \delta_{n-1}$ for $n > 0$. After finitely many steps we reach ι, so say $i(m) = \iota$. Then $g_1 \circ \cdots \circ g_{i(n)}$ maps $\delta_{i(n)} \in M_\iota$ cofinally into β^{+M_i}. Hence, it suffices to set $\mu = \mathrm{card}(M)$. □(Corollary 6.3.4)

6.4 The Model K^c

The theory developed so far finally provides us with sufficient means for the construction of a canonical extender model. We shall introduce the ***K****ᶜ-construction,* which yields the so called *background certified core model* $\boldsymbol{K^c}$. In the course of the construction we inductively build approximations to the measure sequence of the desired model. The choice of the measures should guarantee both the iterability of the mouse at any stage of the construction as well as the universality of $\boldsymbol{K^c}$; the latter means that this model will satisfactorily reflect the information about mice which exist in $\boldsymbol{V}$. It turns out that ω-completeness can serve as a criterion on the choice of the next measure. The same criterion can be used to construct much larger models with overlapping extender sequences containing many strong cardinals. However, the construction of still larger models requires much more than mere ω-completeness, namely the existence of so called *background certificates*, which are – roughly speaking – sufficiently strong partial extenders in V; the extenders put on the sequence are then "fragments" of these certificates. We shall return to this more general version of the construction at the end of this book.

It follows that the outcome of the K^c-construction depends on the large cardinal properties of V. For instance, if $0^\#$ does not exist, then $K^c = L$. Otherwise K^c is distinct from L since, viewing $0^\#$ as the first nontrivial mouse (i.e. $0^\# \sim \langle J_\tau, U\rangle$, where U is the L-measure derived from an elementary embedding of L with the least possible critical point κ and $\tau = \kappa^{+L}$), $0^\#$ witnesses the non-universality of L.

Definition. The K^c-*construction* is a pair of sequences of mice

$$\langle N_\alpha; \alpha \in \mathbf{On}\rangle, \quad \langle M_\alpha; \alpha \in \mathbf{On}\rangle$$

such that $M_\alpha = \operatorname{core}(N_\alpha)$, inductively defined as follows.

$$N_0 = \langle J_0, \varnothing\rangle.$$

Setting $N_\alpha = \langle J_\zeta^{E^\alpha}, E^\alpha_{\omega\zeta}\rangle$ and $M_\alpha = \langle J_{\bar\zeta}^{\bar E^\alpha}, \bar E^\alpha_{\omega\bar\zeta}\rangle$,

$$N_{\alpha+1} = \begin{cases} \langle J_{\bar\zeta}^{\bar E^\alpha}, F\rangle & \text{if } \bar E^\alpha_{\omega\bar\zeta} = \varnothing \text{ and } F \text{ is the unique } \omega\text{-complete measure such that } \langle J_{\bar\zeta}^{\bar E^\alpha}, F\rangle \text{ is a premouse;} \\ \langle J_{\bar\zeta+1}^{\bar E^\alpha}, \varnothing\rangle & \text{otherwise.} \end{cases}$$

Let ϱ_ξ denote $\varrho^\omega_{N_\xi}$ ($= \varrho^\omega_{M_\xi}$ by Lemma 5.3.4) and, given α such that $\xi \le \alpha \le \infty$ (we exclude the case $\xi = \alpha = \infty$), set

$$\begin{aligned} \varrho_{\xi\alpha} &= \min\{\varrho_\eta;\ \xi \le \eta < \alpha\} \\ \tau_{\xi\alpha} &= (\omega\varrho_{\xi\alpha})^{+N_\xi} \\ \zeta &= \sup\{\tau_{\xi\alpha};\ \xi < \alpha\}. \end{aligned}$$

If $\omega\varrho_{\xi\alpha}$ is the largest cardinal in N_ξ or the height of N_ξ, then $\tau_{\xi\alpha}$ is understood to be $\operatorname{ht}(N_\xi)$. For limit α we set

$$N_\alpha = \langle J_\zeta^{E^\alpha}, \varnothing\rangle \quad \text{where} \quad E^\alpha = \bigcup_{\xi<\alpha} E^\xi \restriction \tau_{\xi\alpha}.$$

This defines a hierarchy of premice which, granting that the construction does not break down, stops at ∞. We set

$$K^c = N_\infty = \bigcup_{\xi<\infty} J^{E^\xi}_{\tau_{\xi\infty}}.$$

It is not immediately clear that the above described construction leads to an extender model. We have to prove, firstly, that the construction does not break down at any $\alpha < \infty$, and, secondly, that the height of K^c really is ∞. The potential difficulty at successor steps might be verifying the iterability of $N_{\alpha+1}$; this can be cleared up using our earlier iterability criterion. At limit stages, difficulties might occur due to the fact that the hierarchy $\langle N_\xi \rangle_\xi$ is not linear (that is, the measure sequences of the structures N_ξ might differ and $\mathrm{ht}(N_\xi)$ might drop for some large ξ) since we keep coring down to M_ξ's in the course of the construction; this coring down is important because we need to arrange that all proper initial segments of the final model are sound. The following lemma takes care of all of these questions.

Lemma 6.4.1. *The structure N_α belonging to the $\boldsymbol{K}^c$-construction is defined for every $\alpha < \infty$ and:*

a) *N_α is a premouse and each proper initial segment of N_α is iterable;*

b) *every total measure in N_α is ω-complete;*

c) $E^\xi \restriction \tau_{\xi\alpha} = E^\beta \restriction \tau_{\xi\alpha}$ *and* $\tau_{\xi\alpha} \leq (\omega\varrho_{\xi\alpha})^{+N_\beta}$ *whenever* $\xi \leq \beta \leq \alpha$.

Hence the $\boldsymbol{K}^c$-construction does not break down and each N_α is a mouse. Moreover, clause c) *also holds for $\alpha = \infty$, so $\boldsymbol{K}^c$ is well-defined.*

Proof. We proceed by induction on α. For $\alpha = 0$ the above conclusions follow trivially, so we shall focus on the induction step. The iterability of each N_α, which also guarantees the existence of M_α, is a consequence of a), b) and Theorem 4.5.1. Notice also that the latter part of c) is an immediate consequence of the former part.

Case 1. α is a successor ordinal, say $\alpha = \bar{\alpha} + 1$. By the induction hypothesis, $N_{\bar{\alpha}}$ is a mouse and $M_{\bar{\alpha}} = \mathrm{core}(N_{\bar{\alpha}})$. If the first possibility in the definition of N_α occurs, i.e. if the top predicate of $M_{\bar{\alpha}}$ is empty and there is a unique ω-complete measure such that adding this measure on top of $M_{\bar{\alpha}}$ results in a premouse, then $M_{\bar{\alpha}}$ is a ZFC^--model, so its ultimate projectum is $\mathrm{ht}(M_{\bar{\alpha}})$ and, consequently, $M_{\bar{\alpha}} = N_{\bar{\alpha}}$, as they coiterate simply above the ultimate projectum to a common mouse. Then a) and b) follow immediately from the definition of N_α and the induction hypothesis. To see c), pick any $\xi < \alpha$. Then $\varrho_{\xi\alpha} \leq \varrho_{\xi\bar{\alpha}}$, so $\tau_{\xi\alpha} \leq \tau_{\xi\bar{\alpha}}$ and, by the induction hypothesis, $N_{\bar{\alpha}}$ agrees with N_ξ below $\tau_{\xi\alpha}$. Since $N_{\bar{\alpha}}$ and N_α differ only by the top extender, $E^\alpha \restriction \tau_{\xi\alpha} = E^{\bar{\alpha}} \restriction \tau_{\xi\alpha} = E^\xi \restriction \tau_{\xi\alpha}$.

Now suppose the second possibility in the definition of N_α occurs. Then $M_{\bar{\alpha}} = \mathrm{core}(N_{\bar{\alpha}})$ is a sound mouse, so the same applies to every proper initial segment of N_α and a) follows immediately from the definition of premouse. In particular, the acceptability at the last level follows from the soundness of $M_{\bar{\alpha}}$. We now verify b). Recall that $E_\nu^{N_\alpha} = E_\nu^{M_{\bar{\alpha}}}$ for every $\nu < \mathrm{ht}(N_\alpha)$, so it suffices to establish the ω-completeness of the latter. Since $N_\alpha - M_{\bar{\alpha}}$ contains a subset of $\omega\varrho^\omega_{M_{\bar{\alpha}}} = \omega\varrho_{\bar{\alpha}}$, no $E_\nu^{M_{\bar{\alpha}}}$ with critical point greater or equal than $\omega\varrho_{\bar{\alpha}}$ is total in N_α. Thus, if $E_\nu^{N_{\bar{\alpha}}}$ is total, its critical point is strictly smaller than $\omega\varrho_{\bar{\alpha}}$, and $\nu \leq \omega\varrho_{\bar{\alpha}}$ because the latter is a cardinal in $M_{\bar{\alpha}}$. We observe that $E_\nu^{M_{\bar{\alpha}}} = E_\nu^{N_{\bar{\alpha}}}$, which will guarantee that $E_\nu^{N_{\bar{\alpha}}}$ is a total measure in $N_{\bar{\alpha}}$ and thus, by the induction hypothesis, its ω-completeness. The above equality follows from the fact that the core map $\sigma : M_{\bar{\alpha}} \to N_{\bar{\alpha}}$ is the identity on ν. This is trivial if $\nu < \omega\varrho_{\bar{\alpha}}$. Otherwise $\nu = \omega\varrho_{\bar{\alpha}}$ is a cardinal successor of some κ in $M_{\bar{\alpha}}$, so its image under σ is the cardinal successor of $\sigma(\kappa) = \kappa$ in $N_{\bar\alpha}$; then $\sigma(\nu) = \nu$ since $\omega\varrho_{\bar\alpha}$, being the ultimate projectum, is a cardinal in $N_{\bar\alpha}$. To see c), fix a $\xi < \alpha$. Notice that $\tau_{\xi\alpha} \le \tau_{\xi\bar\alpha}$, as $\omega\varrho_{\xi\alpha} \le \omega\varrho_{\xi\bar\alpha}$ and $\omega\varrho_{\xi\alpha}$ is the largest cardinal in $J^{E^\xi}_{\tau_{\xi\alpha}}$. This together with the induction hypothesis gives the agreement $E^\xi \restriction \tau_{\xi\alpha} = E^\beta \restriction \tau_{\xi\alpha}$ for all $\beta \le \bar\alpha$. Now as $\omega\varrho_{\xi\alpha} \le \omega\varrho_{\bar\alpha}$, we have $\tau_{\xi\alpha} \le \tau_{\bar\alpha} = (\omega\varrho_{\bar\alpha})^{+N_{\bar\alpha}}$. But $N_{\bar\alpha}$, $M_{\bar\alpha}$ and N_α agree below $\tau_{\bar\alpha}$ (the agreement between the former two follows from Corollary 5.5.3), which immediately yields the desired conclusion.

Case 2. α is a limit. We first verify that N_α is well-defined. For this it suffices to show that $E^\eta \restriction \tau_{\eta\alpha} = E^\xi \restriction \tau_{\eta\alpha}$ whenever $\eta \le \xi < \alpha$. By the induction hypothesis, $E^\eta \restriction \tau_{\eta\xi} = E^\xi \restriction \tau_{\eta\xi}$. Similarly as in Case 1 we observe that $\omega\varrho_{\eta\alpha} \le \omega\varrho_{\eta\xi}$ and $\tau_{\eta\alpha} \le \tau_{\eta\xi}$, which guarantees the required agreement between E^η and E^ξ. Notice that this argument goes through even if $\alpha = \infty$, so K^c is well-defined as well.

Clause a) follows immediately since any $J^{E^\xi}_{\tau_{\xi\alpha}}$ is an initial segment of N_ξ and therefore a mouse. To see b), pick ν such that E^α_ν is a total measure in N_α. Then $\nu < \tau_{\xi\alpha}$ for some ξ and by the agreement between N_ξ and N_α, $E^\alpha_\nu = E^\xi_\nu$ is a total measure in the sense of $N_\xi \| \tau_{\xi\alpha}$. But $\tau_{\xi\alpha}$ is a cardinal in N_ξ, therefore E^ξ_ν is a total measure in N_ξ. By the induction hypothesis, $E^\alpha_\nu = E^\xi_\nu$ is ω-complete. Clause c) follows as in the previous paragraph. □(Lemma 6.4.1)

Lemma 6.4.2. *K^c is an extender model, that is,* $\mathrm{ht}(K^c) = \infty$.

Proof. Recall that the sequence $\langle \varrho_{\alpha,\infty}\rangle_\alpha$ is non-decreasing. We show that this sequence can attain the same value only for *boundedly* many ordinals α; in other words, given any α, there is a $\beta > \alpha$ such that $\omega\varrho_{\beta,\infty} > \omega\varrho_{\alpha,\infty}$. This means that the values $\omega\varrho_{\alpha,\infty}$ converge to ∞. By c) of the previous lemma, each N_α agrees with K^c below $\omega\varrho_{\alpha,\infty}$, so the height of K^c is larger than any ordinal.

Fix α so that $\omega\varrho_\alpha = \omega\varrho_{\alpha,\infty}$. By c) of the previous lemma, $J^{E^\alpha}_{\tau_{\alpha,\infty}} = J^{E^\gamma}_{\tau_{\gamma,\infty}}$ for every $\gamma \ge \alpha$. In particular, $\mathcal{P}(\omega\varrho_{\alpha,\infty}) \cap N_\alpha \subset \mathcal{P}(\omega\varrho_{\alpha,\infty}) \cap N_\gamma$. If additionally $\omega\varrho_\gamma = \omega\varrho_{\alpha,\infty}$, then we can define a subset of $\omega\varrho_{\alpha,\infty}$ over N_γ which is distinct from all sets in N_γ and, by the above agreement between N_α and N_γ, also from all sets in N_α. This set is then an element of $N_{\gamma+1}$. Hence, whenever we reach a stage of the K^c-construction at which the ultimate projectum drops to $\omega\varrho_{\alpha,\infty}$, we add a new subset of $\omega\varrho_{\alpha,\infty}$ to K^c. Since there are at most $2^{|\omega\varrho_{\alpha,\infty}|}$ such subsets, there can only be boundedly (in **On**) many ordinals γ as above. Let β be the least upper bound for all such γ; then $\omega\varrho_{\beta'} > \omega\varrho_{\alpha,\infty}$ for all $\beta' \ge \beta$, that is, $\omega\varrho_{\beta,\infty} > \omega\varrho_{\alpha,\infty}$. □(Lemma 6.4.2)

The condition on putting a new measure on top of N_α in the K^c-construction might seem too restrictive at first glance. If, by chance, there were more candidates for adding to the measure sequence at stage α, we would not, according to the construction, add any of them. We might thus omit those measures which should definitely be included if we want to guarantee the universality of K^c. (In the extreme case, we might omit *all* measures, so that the outcome of the K^c-construction would just be L, a poor reward for the effort we put into developing this whole theory). Fortunately, this catastrophic scenario will not present itself since, by the bicephalus lemma, if there is a candidate for the top measure of N_α, then it is uniquely determined. The next lemma shows that the $\boldsymbol{K}^c$-construction builds the measure sequence very effectively.

Theorem 6.4.3. *Let $E = E^{\boldsymbol{K}^c}$.*

a) *If E_τ is a total measure in $\boldsymbol{K}^c$ then E_τ is ω-complete.*

b) *If F is an ω-complete total measure on $\boldsymbol{K}^c$ (we do* not *assume that $F \in \boldsymbol{K}^c$) and τ is such that $\langle J^E_\tau, F\rangle$ is a premouse, then $F = E_\tau$.*

Proof. To see a), suppose E_τ is a total measure in $\boldsymbol{K}^c$. By the proof of the previous lemma, there is an α such that $\omega\varrho_{\alpha,\infty} > \tau$. Then N_α agrees with $\boldsymbol{K}^c$ below $\tau_{\alpha,\infty} = (\omega\varrho_{\alpha,\infty})^{+N_\alpha}$, so τ is a cardinal in N_α. It follows that E_τ is a total measure in N_α. By b) of Lemma 6.4.1, E_τ is ω-complete.

For the proof of b) we will need the following fact.

Claim. Given a $\boldsymbol{K}^c$-cardinal τ, there is exactly one stage α of the $\boldsymbol{K}^c$-construction such that $N_\alpha = J^E_\tau$, namely

$$\alpha = \sup\{\gamma;\ \omega\varrho_\gamma < \tau\}.$$

Proof. To see this, we first observe that α is a limit ordinal. Indeed, if $\omega\varrho_\gamma < \tau$, then the same applies to $\omega\varrho_{\gamma+1}$ since $N_{\gamma+1}$ is obtained either by adding a top measure, so that the ultimate projectum cannot become larger, or else by passing on to the next level of constructibility. In the latter case, $N_{\gamma+1} = J^{E^\gamma}_{\zeta+1}$ where ζ is the height of M_γ, so M_γ is a proper initial segment of $N_{\gamma+1}$. By the soundness of M_γ, there is a definable function over M_γ, say f, mapping $\omega\varrho_\gamma$ onto M_γ. Such an f is an element of $N_{\gamma+1}$, so $\zeta = \mathbf{On} \cap \operatorname{rng}(f)$ is easily seen to be Σ_1-definable over $N_{\gamma+1}$ in the parameter f. Since the same also applies to every $\zeta + n$ where $n \in \omega$, every ordinal of $N_{\gamma+1}$ is Σ_1-definable over $N_{\gamma+1}$ from f and an ordinal less that $\omega\varrho_\gamma$. It follows that $N_{\gamma+1} \subset h_{N_{\gamma+1}}(\omega\varrho_\gamma \cup \{f\})$, so the first projectum of $N_{\gamma+1}$ is at most $\omega\varrho_\gamma$.

Notice that, by our definition of α, the height of $N_{\alpha'}$ for $\alpha' \geq \alpha$ is at least τ and $\omega\varrho_{\alpha,\infty} \geq \tau$, so that, by c) of Lemma 6.4.1, $J^{E^\alpha}_\tau = J^E_\tau$. Pick any $\gamma < \alpha$. Then $\omega\varrho_{\gamma\alpha} < \tau$ and, again by the definition of α, $\omega\varrho_{\gamma,\infty} = \omega\varrho_{\gamma\alpha}$. Hence $\tau_{\gamma\alpha} = \tau_{\gamma,\infty}$, so, again by c) of Lemma 6.4.1, $J^{E^\gamma}_{\tau_{\gamma\alpha}}$ is an initial segment of both N_α and $\boldsymbol{K}^c$. The fact that τ is a cardinal in $\boldsymbol{K}^c$ and $\omega\varrho_\gamma$ is the largest cardinal in $J^{E^\gamma}_{\tau_{\gamma\alpha}}$ then yields $\tau_{\gamma\alpha} < \tau$. Since the length of N_α is equal to the supremum of all ordinals $\tau_{\gamma\alpha}$, the height of N_α is at most τ, so $\operatorname{ht}(N_\alpha) = \tau$ and $N_\alpha = J^E_\tau$.

It remains to verify the uniqueness of α. Recall that $N_{\alpha+1}$ is either $J^E_{\tau+1}$ or $\langle J^E_\tau, F\rangle$ for some F. In the former case, $\omega\varrho_{\alpha+1,\infty} = \omega\varrho_{\alpha+1} = \tau$ and $\tau_{\alpha+1,\infty} = \tau + 1$. The agreement between the premice N_β given by c) of Lemma 6.4.1 guarantees that $\operatorname{ht}(N_\beta) \geq \tau + 1$ for all $\beta > \alpha$, which yields the desired uniqueness of α. In the latter case, this agreement applied to $N_{\alpha+1}$ instead of N_α yields that $\operatorname{ht}(N_\beta) > \tau$ for all $\beta > \alpha + 1$, and we have the uniqueness of α again. □(Claim)

Assume F is ω-complete, total on $\boldsymbol{K}^c$ and such that $\langle J^E_\tau, F\rangle$ is a premouse. Then $\tau = \kappa^{+\boldsymbol{K}^c}$ where κ is the critical point of E_τ. Let α be the level of the $\boldsymbol{K}^c$-construction such that $N_\alpha = J_\tau^E$. Then $N_{\alpha+1} = \langle J_\tau^E, F\rangle$ and, by the proof of the above claim, $N_{\alpha+1}$ is an initial segment of any N_δ for $\delta > \alpha$, and thus an initial segment of $\boldsymbol{K}^c$. This yields $F = E_\tau^{\alpha+1} = E_\tau$. □(Theorem 6.4.3)

The previous lemma sums up the crucial properties of $\boldsymbol{K}^c$ and in fact all information about $\boldsymbol{K}^c$ we shall need for the applications in the following. In particular, it enables us to avoid any further pain from examining of the $\boldsymbol{K}^c$-construction. At this point, we are ready to prove the universality of $\boldsymbol{K}^c$. We shall present a proof which shows that b) of Theorem 6.4.3 alone guarantees the universality. This fact motivates the following definition.

Definition. Let $\theta > \omega$ be a regular cardinal. A weasel W is *θ-full* iff given any ω-complete measure F, the following condition is satisfied: Setting $\kappa = \mathrm{cr}(F)$ and $\tau = \kappa^{+W}$, if

a) $\mathrm{cf}(\kappa) = \theta$ (in the sense of $\boldsymbol{V}$), and

b) $\langle J_\tau^{E^W}, F\rangle$ is a premouse,

then $F = E_\tau^W$.

Given a class A of regular cardinals, W is A-full iff W is θ-full for every $\theta \in A - \{\omega\}$.

$\boldsymbol{K}^c$ is obviously **On**-full. A straightforward generalization of the $\boldsymbol{K}^c$-construction yields a broad class of A-full weasels.

Definition. The A-full $\boldsymbol{K}^c$ over J_α^E is the model built by the modification of the $\boldsymbol{K}^c$-construction which starts from $N_0 = J_\alpha^E$ and only admits extenders whose critical points have cofinality in A. That is, having constructed $N_\alpha = \langle J_\zeta^{E^\alpha}, E_{\omega\zeta}^\alpha\rangle$, we set $N_{\alpha+1} = \langle J_\zeta^{E^\alpha}, F\rangle$ only if $E_{\omega\zeta}^\alpha = \varnothing$ and F is the unique ω-complete extender such that

- $\mathrm{cf}(\kappa) \in A$ where $\kappa = \mathrm{cr}(F)$;
- $\langle J_\zeta^{E^\alpha}, F\rangle$ is a premouse.

Otherwise we proceed as in the definition of $\boldsymbol{K}^c$.

One can also introduce a version of the model starting from a structure with a top predicate; but the above definition also covers this case (it suffices to pass on to a longer structure without a top predicate). The A-full $\boldsymbol{K}^c$-construction over $N_0 = \langle J_0, \varnothing\rangle$ always converges; however, this need not be the case for $N_0 = J_\alpha^E$ in general. Later we shall see that if J_α^E is an initial segment of a universal weasel, then the A-full $\boldsymbol{K}^c$ over J_α^E exists. This is as much as one can expect, since the weasel is designed to be universal, so there are no other conceivable candidates for N_0. A-full $\boldsymbol{K}^c$ models can be analyzed in the same way as $\boldsymbol{K}^c$; that they are really A-full in the sense of our above definition follows from the proof of Theorem 6.4.3; in fact, the following statements completely characterize A-full $\boldsymbol{K}^c$ over J_α^E:

- $J_\alpha^{E^W} = J_\alpha^E$.

- If $\nu \geq \alpha$, E_ν^W is a total measure in W with critical point κ, then $\mathrm{cf}(\kappa) \in A$ and E_ν^W is ω-complete.

- If $\nu \geq \alpha$, F is a total measure on W, $\langle J_\nu^{E^W}, F\rangle$ is a premouse and $\mathrm{cf}(\kappa) \in A$ where $\kappa = \mathrm{cr}(F)$, then $F = E_\nu^W$.

This follows from the universality (see the next theorem). If W_1, W_2 satisfy the above clauses, they are universal, so they coiterate simply beyond α to a common weasel, say Q. This means that if ν is the first point of disagreement between W_1, W_2 and $E_\nu^{W_i} \neq \varnothing$, then $E_\nu^{W_i}$ is a total measure in W_i, so by the second clause, $E_\nu^{W_i}$ is ω-complete, $\mathrm{cf}(\kappa) \in A$ where $\kappa = \mathrm{cr}(E_\nu^{W_i})$ and $\langle J_\nu^{E'}, F\rangle$ is a premouse, where $E^{W_1} \restriction \nu = E' = E^{W_2} \restriction \nu$. As the above coiteration is simple on both sides and above κ (the latter can be easily seen), the weasels W_1, Q, W_2 contain the same subsets of κ. The last clause above then implies that $E_\nu^{W_i} = E_\nu^{W_{3-i}}$; so there cannot be any disagreement between W_1 and W_2 after all.

There are other variations on the $\boldsymbol{K}^c$ construction, for instance one can attempt to build a version of the A-full $\boldsymbol{K}^c$-construction over J_α^E by admitting only extenders with critical points greater than α. Models arising from such constructions are also useful in applications.

Theorem 6.4.4. *Any θ-full weasel is universal. In particular, $\boldsymbol{K}^c$ and the A-full $\boldsymbol{K}^c$ over any J_α^E (if it exists) is universal.*

Proof. Suppose for a contradiction that W is not universal and that M is a premouse whose coiteration with W does not terminate. Let ν_i be the coiteration indices, κ_i the critical points and $\tau_i = \kappa_i^{+W}$. Recall that $\nu_i = \kappa_i^{+M_i} = \kappa_i^{+W_i}$. By Lemma 6.3.1, there is a c.u.b. class C of ordinals such that given any $i < j$ from C,

a) $\kappa_i = \mathrm{cr}(\pi_{ij}^M)$ and $\pi_{ij}^M(\kappa_i) = \kappa_j$;

b) $(\pi_{ij}^M)''\nu_i$ is cofinal in ν_j;

c) $\pi_{0,i}^W(\kappa_i) = \kappa_i$;

d) $\kappa_i^{+W_i} = \nu_i$ and $(\pi_{0,i}^W)''\tau_i$ is cofinal in $\nu_i = \pi_{0,i}^W(\tau_i)$;

e) $\pi_{ij}^W \restriction \kappa_i = \mathrm{id}$, so that $\pi_{0,j}^W \restriction \kappa_i = \pi_{0,i}^W \restriction \kappa_i$.

Let $E = E^W$; we first show the existence of a c.u.b. class C^* such that each $i \in C^*$ satisfies

$$E_{\tau_i} = \varnothing \tag{6.4.1}$$

Setting $\tilde{W}_i = J_{\tau_i}^E$, we shall construct a c.u.b. class $D \subset C^*$ and a commutative system of cofinal Σ_0-preserving maps

$$\sigma_{ij} : \tilde{W}_i \to \tilde{W}_j \quad \text{where} \quad i, j \in D \;\&\; i < j$$

such that

$$\operatorname{cr}(\sigma_{ij}) = \kappa_i \text{ and } \sigma_{ij}(\kappa_i) = \kappa_j \text{ for all } i, j.$$

We note that the first step, i.e. proving that $E_{\tau_i} = \emptyset$ for c.u.b.-many i, is, surprisingly, superfluous if we generalize the whole proof for larger mice, e.g. mice for non-overlapping extenders. The whole problem is connected with the indexing of measures, resp. extenders. (One could, of course do an alternative argument instead of proving $E_{\tau_i} = \emptyset$ for c.u.b.-many i, but some additional argument would seem to be necessary.) An alternative which would enable us to skip over the above-mentioned additional argument would be a restrictive, so called *smallness* assumption on V; in our case such an assumption would rule out the existence of a measure of order one. In terms of mice we say that the mouse called $0^\ddagger$ (see the next section) does not exist. Since all applications of our theory (including the construction of the core model K) apply only to the region below $0^\ddagger$, the corresponding weaker result on the universality is sufficiently general. (So an inpatient reader can skip over the first step of this proof).

With the above system of maps we shall complete the proof of the theorem as follows. To each $i \in D$ we assign the measure F_i on $\mathcal{P}(\kappa_i) \cap W$ defined by

$$x \in F_i \longleftrightarrow \kappa_i \in \sigma_{ij}(x)$$

(since the system σ_{ij} commutes, this definition does not depend on the choice of $j > i$). F_i is clearly a measure on $\tilde{W}_i$ and therefore also on W. We obtain a contradiction by showing that at least one F_i must be on the W-sequence, i.e. $F_i = E_{\tau_i}$, which is impossible by (6.4.1). In order to do so we apply b) of Theorem 6.4.3.

To see that $\langle J^E_{\tau_i}, F_i\rangle$ is a premouse, we shall verify the weak amenability and the coherency condition for the top measure; all other requirements follow immediately. The ultrapower $\operatorname{Ult}(\tilde{W}_i, F_i)$ exists since we can Σ_0-embed the corresponding term model $\mathbb{D}(\tilde{W}_i, F_i)$ into $\tilde{W}_j$ by $[f] \mapsto \sigma_{ij}(f)(\kappa_i)$. Denote this ultrapower by W'_i and the associated ultrapower map by π'_i. Let $\sigma'_{ij} : W'_i \to \tilde{W}_j$ be the embedding defined by $\sigma'_{ij}(\pi'_i(f)(\kappa_i)) = \sigma_{ij}(f)(\kappa_i)$. Then $\sigma'_{ij}(\kappa_i) = \sigma'_{ij}(\pi'_i(\mathrm{id})(\kappa_i)) = \sigma_{ij}(\mathrm{id})(\kappa_i) = \mathrm{id}(\kappa_i) = \kappa_i$. Similarly, σ'_{ij} does not move any ordinal less than κ_i, so $\sigma'_{ij} \restriction (\kappa_i + 1) = \mathrm{id}$. Given any $x \subset \kappa_i$ from W'_i we have $\sigma'_{ij}(x) = x$, so x is an element of W and therefore of $\tilde{W}_i$. This proves that the ultrapower by F_i does not increase the power set of κ_i and thus the weak amenability of F_i. Note also that we have in fact obtained the equality $\mathcal{P}(\kappa_i) \cap W'_i = \mathcal{P}(\kappa_i) \cap W$. In particular, this means that $\kappa^{+W'_i} = \kappa_i^{+\tilde{W}_i} = \tau_i$. By the above computation, $\operatorname{cr}(\sigma'_{ij}) > \kappa_i$. It follows that $\operatorname{cr}(\sigma'_{ij}) \geq \tau_i$, as the critical point must be regular in W'_i. But

$$\sigma'_{ij}(\tau_i) = \sigma'_{ij}(\kappa^{+W'_i}) = (\sigma'_{ij}(\kappa_i))^{+\tilde{W}_j} = \kappa_i^{+\tilde{W}_j} = \tau_j,$$

so $E^{W'_i}_{\tau_i} = \emptyset$ since the same applies to $E_{\tau_i} = \sigma'_{ij}(E^{W'_i}_{\tau_i})$. This proves the coherency condition for F_i.

In the following we show that at least one (in fact, stationarily many) F_i is ω-complete; this is sufficient for the application of b) of Theorem 6.4.3 and thus for the above-mentioned contradiction. Suppose the contrary; say $S \subset D$ is a stationary class of limit points of D with uncountable cofinality such that for no $i \in S$ is F_i ω-complete. To each such i pick a witness to this, i.e. a sequence $\langle x_n^i;\ n \in \omega\rangle$ of sets from F_i with empty intersection. Define a function $f : S \to \mathbf{On}$ by setting $f(i)$ to be the least ι such that $\{x_n^i;\ n \in \omega\} \subset \mathrm{rng}(\sigma_{\iota,i})$. For every n there is an $\iota(n) < i$ such that $x_n^i \in \mathrm{rng}(\sigma_{\iota(n),i})$; the supremum of all $\iota(n)$ is then strictly less that i since $\mathrm{cf}(i) > \omega$. It follows that f is regressive. By Fodor's Theorem for classes, there is an unbounded class $S^* \subset S$ such that $f \restriction S^*$ is constant with value, say, i^*. This means that each sequence $\langle x_n^i\rangle_n$ has a (pointwise!) preimage in $\tilde{W}_{i^*}$. Since there are at most $|\tilde{W}_{i^*}|^\omega$ such possible preimages, there is a fixed sequence $\langle y_n;\ n \in \omega\rangle$ and a proper class $S' \subset S^*$ such that for every $n \in \omega$ and $i \in S'$ we have $\sigma_{i^*,i}(y_n) = x_n^i$. Pick $i < j$ from S'; then $\sigma_{ij}(x_n^i) = x_n^j$ for every n. The fact that each $x_n^i \in F_i$ yields $\kappa_i \in \sigma_{ij}(x_n^i) = x_n^j$, so the intersection of all $\{y_n^i;\ n \in \omega\}$ is nonempty after all (it contains κ_i). Contradiction.

This completes the proof of the theorem modulo the verification of (6.4.1) on a c.u.b. class and the construction of the maps σ_{ij}. We shall first focus on the former. Suppose for a contradiction that $E_{\tau_i} \neq \varnothing$ on a stationary class $S \subset C$ of limit points of C. Then $E_{\nu_i}^{W_i} = \pi_{0,i}^W(E_{\tau_i}) \neq \varnothing$. Since $E_{\nu_i}^{W_i}$, $E_{\nu_i}^{M_i}$ is a pair of distinct measures (because they are used in the coiteration), we can pick $y \subset \kappa_i$ which is an element of exactly one of them; for such y we have $\pi_{ij}^W(y) \cap \nu_i \neq \pi_{ij}^M(y) \cap \nu_i$ (here $j > i$ can be arbitrary). By d), there is a $\bar{\zeta} < \tau_i$ such that, setting $\zeta = \pi_{0,i}^W(\bar{\zeta})$, we have $y \in W_i||\zeta$. Pick an $x \subset \kappa_i$, $x \in W$ coding a κ_i-enumeration of $\mathcal{P}(\kappa_i) \cap W||\bar{\zeta}$. Such an x exists since there is an enumeration g of $W||\bar{\zeta}$ with $\mathrm{dom}(g) = \kappa_i$ in W and we can think of x as the isomorphic image of the graph of g under the Gödel pairing function. Then $\pi_{0,i}^W(x)$ encodes a κ_i-enumeration of $\mathcal{P}(\kappa_i) \cap W_i||\zeta$, so there is an $\eta < \kappa_i$ such that y is the η-th element of $W_i||\zeta$ in the coding by $\pi_{0,i}^W(x)$, briefly $y = [\pi_{0,i}^W(x)]_\eta$. We claim that

$$\pi_{ij}^W \circ \pi_{0,i}^W(x) \cap \nu_i \neq \pi_{ij}^M \circ \pi_{0,i}^W(x) \cap \nu_i. \tag{6.4.2}$$

Otherwise

$$\begin{aligned} \pi_{ij}^W(y) \cap \nu_i &= \pi_{ij}^W([\pi_{0,i}^W(x)]_\eta) \cap \nu_i \\ &= [\pi_{ij}^W \circ \pi_{0,i}^W(x)]_\eta \cap \nu_i = [\pi_{ij}^M \circ \pi_{0,i}^W(x)]_\eta \cap \nu_i \\ &= \pi_{ij}^M([\pi_{0,i}^W(x)]_\eta) \cap \nu_i \\ &= \pi_{ij}^M(y) \cap \nu_i, \end{aligned}$$

which would contradict our choice of y. Thus, to each $i \in S$ we can pick an $x_i \subset \kappa_i$ such that (6.4.2) holds with x_i in place of x. Since $\pi_{0,i}^W(x_i) \in M_i$, we can define a function $f : S \to \mathbf{On}$ by setting $f(i)$ to be the least ι such that $\pi_{0,i}^W(x_i) \in \mathrm{rng}(\pi_{\iota,i}^M)$. f is clearly regressive, so there is a proper class $S^* \subset S$ such that $f \restriction S^*$ is constant with some value i^*. This means that each $\pi_{0,i}^W(x_i)$ has a $\pi_{i^*,i}^M$-preimage in M_{i^*}. But M_{i^*} is a set, so we can find a $z^* \in M_{i^*}$ and a proper class $S' \subset S^*$ such that $\pi^W_{0,i}(x_i) = \pi^M_{i^*,i}(z^*)$ for all $i \in S'$. Pick $i < j$ from S'; then

$$\pi^W_{0,j}(x_j) = \pi^M_{ij} \circ \pi^W_{0,i}(x_i).$$

In particular, $\pi^W_{0,i}(x_i) = \pi^W_{0,j}(x_j) \cap \kappa_i = \pi^W_{0,i}(x_j) \cap \kappa_i$; the former follows from the above equality and the fact that $\mathrm{cr}(\pi^M_{ij}) = \kappa_i$; the latter then follows from the former since $\mathrm{cr}(\pi^W_{ij}) = \kappa_i$. Using this we infer

$$\pi^W_{ij} \circ \pi^W_{0,i}(x_i) \cap \nu_i = \pi^W_{ij}(\pi^W_{0,i}(x_j) \cap \kappa_i) \cap \nu_i = \pi^W_{0,j}(x_j) \cap \nu_i = \pi^M_{ij} \circ \pi^W_{0,i}(x_i) \cap \nu_i,$$

which contradicts our choice of x_i.

It remains to construct the system of maps σ_{ij}. First we shall find a proper class of ordinals D^* such that, given any $i < j$ from D^*, we have

$$(\pi^M_{ij} \circ \pi^W_{0,i})''\tilde{W}_i \subset (\pi^W_{0,j})''\tilde{W}_j, \tag{6.4.3}$$

which will enable us to define σ_{ij} by

$$\sigma_{ij} = (\pi^W_{0,j})^{-1} \circ \pi^M_{ij} \circ \pi^W_{0,i}.$$

This again involves a pressing down argument. By b) and d), all ordinals ν_i and τ_i have the same cofinality, say γ. Let $D' \subset C^*$ be a stationary class of limit points of C^* of cofinality greater than γ. To each $i \in D'$ pick an increasing sequence $\langle \xi^i_\eta;\ \eta < \gamma \rangle$ cofinal in τ_i. We define a function $f : D' \to \mathbf{On}$ by setting $f(i)$ to be the least ι such that $(\pi^W_{0,i})''\{\xi^i_\eta;\ \eta < \gamma\} \subset \mathrm{rng}(\pi^M_{\iota,i})$. Then f is regressive since $\gamma < \mathrm{cf}(i)$; this follows exactly as in the pressing down argument from the above proof of the ω-completeness of F_i. Proceeding as in that argument, we arrive at a proper class $D^* \subset D'$, an ordinal i^* and a sequence $\langle \zeta^i_\eta;\ \eta < \gamma \rangle$ such that for $i \in D^*$ and $\eta < \gamma$ we have $\pi^W_{0,i}(\xi^i_\eta) = \pi^M_{i^*,i}(\zeta^i_\eta)$. Thus, given $i < j$ from D^*,

$$\pi^M_{ij} \circ \pi^W_{0,i}(\xi^i_\eta) = \pi^W_{0,j}(\xi^j_\eta). \tag{6.4.4}$$

Let f^i_η be the $<_W$-least surjection of κ_i onto $W||\xi^i_\eta$ in W. Given $x \in \tilde{W}_i$, there is an $\eta < \gamma$ such that $x \in W||\xi^i_\eta$, so $x = f^i_\eta(\delta)$ for some $\delta < \kappa_i$. Then

$$\begin{aligned}\pi^M_{ij} \circ \pi^W_{0,i}(x) &= \pi^M_{ij} \circ \pi^W_{0,i}(f^i_\eta(\delta)) = \pi^M_{ij} \circ \pi^W_{0,i}(f^i_\eta)(\pi^M_{ij} \circ \pi^W_{0,i}(\delta)) \\ &= \pi^W_{0,j}(f^j_\eta)(\pi^W_{0,i}(\delta)) = \pi^W_{0,j}(f^j_\eta)(\pi^W_{0,j}(\delta)) = \pi^W_{0,j}(f^j_\eta(\delta)) = \pi^W_{0,j}(x),\end{aligned}$$

which proves (6.4.3) on D^*. The third equality follows from (6.4.4) and the fact that the f^i_η have a uniform definition in W from ξ^i_η and that $\mathrm{cr}(\pi^M_{ij}) = \kappa_i$. To see the fourth equality, notice that $\pi^W_{0,j} \restriction \kappa_i = \pi^W_{0,i} \restriction \kappa_i$ since $\pi^W_{ij} \restriction \kappa_i = \mathrm{id}$.

Finally we show that this diagram can be continuously extended to a diagram indexed by $D = D^* \cup \lim(D^*)$, i.e. that for any limit point ι of D^*, the direct limit of $\langle \tilde{W}_i, \sigma_{ij};\ i \leq$

$j < \iota\rangle$ is equal to $\tilde{W}_\iota$. For the moment, denote this direct limit by W_ι^* and the direct limit maps by $\sigma_{i,\iota}$. Pick an $h \in D^*$ which is strictly greater than ι (such an h exists since D^* is a proper class). Then W_ι^* can be Σ_0-embedded into $\tilde{W}_h$ by $\sigma_{i,\iota}(x) \mapsto \sigma_{ih}(x)$, so W_ι^* is well-founded. Since W_ι^* is also extensional, we can take it to be transitive; let $\sigma_{\iota,h} : W_\iota^* \to \tilde{W}_h$ be the above defined embedding. A straightforward computation gives us that $\mathrm{cr}(\sigma_{i,\iota}) = \kappa_i$ for every i. Let $\kappa_\iota^* = \sigma_{i,\iota}(\kappa_i)$; this value is clearly independent of the choice of $i \in D^* \cap \iota$; moreover, κ_ι is easily seen to be the largest cardinal in W_ι^*. Since all maps σ_{ij} for $i, j < \iota$ are cofinal, so is each $\sigma_{i,\iota}$: Given $\zeta \in W_\iota^*$, pick $j \in D^* \cap \iota$ such that $\zeta = \sigma_{j,\iota}(\zeta')$ for some $\zeta' \in \tilde{W}_j$; by choosing $\bar{\zeta} \in \tilde{W}_i$ satisfying $\zeta' < \sigma_{ij}(\bar{\zeta})$ we obtain $\zeta < \sigma_{j,\iota} \circ \sigma_{ij}(\bar{\zeta}) = \sigma_{i,\iota}(\bar{\zeta})$. Recall that $D^* \subset C^*$, i.e. κ_ι, being a limit point of C^*, is a critical point of the M-side of the coiteration of W against M. We show that

$$\kappa_\iota^* = \sup\{\kappa_i;\ i \in D^* \cap \iota\} = \kappa_\iota = \mathrm{cr}(\sigma_{\iota,h}) \quad \text{and} \quad \sigma_{\iota,h}(\kappa_\iota) = \kappa_h.$$

That κ_ι is at least the above supremum, is an immediate consequence of the fact that κ_i is the critical point of $\sigma_{i,\iota}$. To see the reverse inequality, pick a $\zeta < \kappa_\iota$. For some $j \in D^* \cap \iota$ we have $\zeta = \sigma_{j,\iota}(\zeta_j)$ and, by the elementarity of $\sigma_{j,\iota}$, $\zeta_j < \kappa_j$. But κ_j is the critical point of $\sigma_{j,\iota}$, so $\zeta = \zeta_j < \kappa_j$. As for the critical point of $\sigma_{\iota,h}$, note that $\sigma_{\iota,h}(\kappa_\iota) = \sigma_{\iota,h}(\sigma_{i,\iota}(\kappa_i)) = \kappa_h$, so it suffices to show that $\sigma_{\iota,h} \restriction \kappa_\iota = \mathrm{id}$. But for any $\zeta < \kappa_\iota$ there is an i such that $\zeta < \kappa_i$; so $\sigma_{\iota,h}(\zeta) = \sigma_{ij}(\zeta) = \zeta$.

Applying Lemma 5.5.2 to W_ι^*, $\tilde{W}_h$ and $\sigma_{\iota,h}$ we infer that W_ι^* is an initial segment of $\tilde{W}_h$, so W_ι^* is an initial segment of $\tilde{W}_\iota$ since κ_ι is the largest cardinal in W_ι^*. Our aim is to show that, in fact, $W_\iota^* = \tilde{W}_\iota$, so it suffices to rule out the possibility that W_ι^* is a proper initial segment of $\tilde{W}_\iota$. Let $\bar{W}_i = J_{\nu_i}^{E^i}$ for $i \in C$; then $\pi_{0,i}^W$ embeds $\tilde{W}_i$ cofinally into $\bar{W}_i$. By the definition of σ_{ij} we have $\pi_{0,j}^W \circ \sigma_{ij} = \pi_{ij}^M \circ \pi_{0,i}^W$ whenever $i < j$ are from $D^* \cap \iota$, so we can define an embedding $\pi^* : W_\iota^* \to \bar{W}_\iota$ by

$$\pi^*(\sigma_{i,\iota}(x)) = \pi_{i,\iota}^M \circ \pi_{0,i}^W(x).$$

Since W_ι^* is the direct limit of $\langle \tilde{W}_i, \sigma_{ij}\rangle_{i,j}$, the values $\sigma_{i,\iota}(x)$ cover all of W_ι^*, so π^* is total. That π^* is a well-defined Σ_0-preserving map follows by a standard computation: Given a Σ_0-formula $\varphi(v_1, \dots, v_\ell)$ and an ℓ-tuple $x_1, \dots, x_\ell \in W_\iota^*$, fix $i \in D^* \cap \iota$ such that $x_k = \sigma_{i,\iota}(x_k^i)$ for some $x_k^i \in \tilde{W}_i$, where $i = 1, \dots, \ell$. Then

$$\begin{aligned}
\bar{W}_\iota \models \varphi(\pi^*(x_1), \dots, \pi^*(x_\ell)) &\longleftrightarrow \bar{W}_\iota \models \varphi(\pi_{i,\iota}^M \circ \pi_{0,i}^W(x_1^i), \dots, \pi_{i,\iota}^M \circ \pi_{0,i}^W(x_\ell^i)) \\
&\longleftrightarrow \tilde{W}_i \models \varphi(x_1^i, \dots, x_\ell^i) \\
&\longleftrightarrow W_\iota^* \models \varphi(x_1, \dots, (x_\ell)).
\end{aligned}$$

A straightforward computation yields $\pi^*(\kappa_\iota) = \kappa_\iota$. Notice also that $\pi^* \restriction J_{\kappa_\iota}^E = \pi_{0,\iota}^W \restriction J_{\kappa_\iota}^E$: Given any set $a \in J_{\kappa_\iota}^E$, fix an $i \in D^* \cap \iota$ such that $a \in J_{\kappa_i}^E$; then

$$\pi^*(a) = \pi^*(\sigma_{i,\iota}(a)) = \pi_{i,\iota}^M \circ \pi_{0,i}^W(a) = \pi_{0,i}^W(a) = \pi_{0,\iota}^W(a).$$

Furthermore, π^* is cofinal since the same applies to all maps involved in the definition of π^* (the verification of this fact is easy and in the same spirit as the above verification of the cofinality of $\sigma_{i,\iota}$ above). It follows that every $x \in \tilde{W}_\iota$ is of the form $\pi^*(f)(\delta)$ for some $f \in W_\iota^*$ with $\mathrm{dom}(f) \in \kappa_\iota$ and a $\delta \in \pi^*(\mathrm{dom}(f))$. Indeed, by the cofinality of π^*, we can find a $\zeta \in W_\iota^*$ such that $x \in \pi^*(J_\zeta^E)$; if $g \in W_\iota^*$ is a surjection of κ_ι onto J_ζ^E, then $x = \pi^*(g)(\delta)$ for some $\delta < \kappa_\iota$, so certainly $x = \pi^*(g \restriction (\delta+1))(\delta)$. Thus, it suffices to set $f = g \restriction \delta$.

Assuming that W_ι^* is a *proper* initial segment of $\tilde{W}_\iota$, say $W_\iota^* = J_\zeta^E$ for some $\zeta < \vartheta_\iota$, we shall derive a contradiction by showing that $\tilde{W}_\iota$ can be embedded into its proper initial segment. Recall that $\pi_{0,\iota}^W$ embeds $\tilde{W}_\iota$ cofinally into $\tilde{W}_\iota$. Define $\sigma : \tilde{W} \to \pi_{0,\iota}^W(W_\iota^*)$ by

$$\sigma(\pi^*(f)(\delta)) = \pi_{0,\iota}^W(f)(\delta)$$

whenever f, δ are as above. It suffices to verify that σ is a well-defined Σ_0-preserving map. Given a Σ_0-formula $\varphi(v_1, \dots, v_\ell)$ and $x_i = \pi^*(f_i)(\delta_i) \in \tilde{W}_i$ for $i = 1, \dots, \ell$,

$$\begin{aligned}\tilde{W}_\iota \models \varphi(x_1, \dots, x_\ell) &\longleftrightarrow \tilde{W}_\iota \models \varphi(\pi^*(f_1)(\delta_1), \dots, \pi^*(f_\ell)(\delta_\ell))\\ &\longleftrightarrow \prec\delta_1, \dots, \delta_\ell\succ \in u^{\tilde{W}_\iota,\varphi}_{\pi^*(f_1),\dots,\pi^*(f_\ell)}\end{aligned}$$

We recall that

$$u^{W_\iota^*,\varphi}_{f_1,\dots,f_\ell} = \{\prec\eta_1, \dots, \eta_\ell\succ \in \kappa_\iota;\ W_\iota^* \models \varphi(f_1(\eta_1), \dots, f_\ell(\eta_\ell))\}.$$

Since

$$u^{\tilde{W}_\iota,\varphi}_{\pi^*(f_1),\dots\pi^*(f_\ell)} = \pi^*(u^{W_\iota^*,\varphi}_{f_1,\dots,f_\ell}) = \pi_{0,\iota}^W(u^{W_\iota^*,\varphi}_{f_1,\dots,f_\ell}) = u^{\pi_{0,\iota}^W(W_\iota^*),\varphi}_{\pi_{0,\iota}^W(f_1),\dots,\pi_{0,\iota}^W(f_\ell)},$$

the above statement can be equivalently expressed as

$$\pi_{0,\iota}^W(W_\iota^*) \models \varphi(\pi_{0,\iota}^W(f_1)(\delta_1), \dots, \pi_{0,\iota}^W(f_\ell)(\delta_\ell)),$$

that is,

$$\pi_{0,\iota}^W(W_\iota^*) \models \varphi(\sigma(x_1), \dots, \sigma(x_\ell)).$$

The above computations follow easily using the identities obtained in the previous paragraph. We have arrived at a contradiction, so $W_\iota^* = \tilde{W}_\iota$ after all and we have the desired commutative system of cofinal embeddings $\sigma_{ij} : \tilde{W}_i \to \tilde{W}_j$ for $i < j$ from D. This completes the proof of the universality of W. □(Theorem 6.4.4)

Lemma 6.4.5. *Let W be an extender model. Then $(K^c)^W = W$, i.e. W believes that it is K^c.*

Proof. Work in W. Then W is universal by Corollary 6.3.2 and $\bar{K} = (K^c)^W$ by the previous theorem. Hence, the coiteration of W with $\bar{K}$ is simple on both sides. Let ν be the least index of disagreement between W and $\bar{K}$, i.e. $\nu = \nu_0$. It follows that ν can index only total measures on both the W- and $\bar{K}$-sequences, so if E_ν^W is a measure, then it is ω-complete (in fact, κ-complete, where $\kappa = \mathrm{cr}(E_\kappa^W)$, since we work in W). Now $\langle J_\nu^{E^{\bar{K}}}, E_\nu^W\rangle = W||\nu$ is a premouse, so $E_\nu^{\bar{K}} = E_\nu^W$ by Theorem 6.4.3. This contradicts the choice of ν, so $E_\nu^W = \varnothing$ after all. Consequently, $E_\nu^{\bar{K}}$ must be a measure.

The iteration from W_1, resp. $\bar{K}_1$, on are both above κ_1, so $W = W_1$ and $\bar{K}_1$ contain the same subsets of κ_1. This means that κ_1 is inaccessible in both W and $\bar{K}_1$. By acceptability, $E_\nu^{\bar{K}} \in W||\kappa_1 = \bar{K}_1||\kappa_1$, so we have a contradiction to the fact that $E_\nu^{\bar{K}} \notin \bar{K}_1$ (recall that, after forming an ultrapower, the applied measure vanishes).

□(Lemma 6.4.5)

Lemma 6.4.6. *Let κ be measurable. Then κ is measurable in $\boldsymbol{K}^c$.*

Proof. Let U be a measure on κ in $\boldsymbol{V}$, let $M = \mathrm{Ult}(\boldsymbol{V}, U)$ and let π be the associated ultrapower map. Then $K' = \bigcup\{\pi(\boldsymbol{K}^c||\zeta);\ \zeta \in \mathbf{On}\}$ is the model $\boldsymbol{K}^c$ constructed inside M and $\pi \restriction \boldsymbol{K}^c : \boldsymbol{K}^c \to K'$ is Σ_1-preserving. By Lemma 6.2.8, K' is weakly universal, so $\boldsymbol{K}^c$ coiterates with K' simply to a common weasel. Since both weasels agree up to κ, their coiteration is above κ. It follows that they contain the same subsets of κ, hence the cardinal successors of κ in $\boldsymbol{K}^c$ and K' agree. Let τ be this common successor. By applying Lemma 5.5.2 to (the appropriate initial segments of) $\boldsymbol{K}^c$, K' and π we infer that $E^{\boldsymbol{K}^c} \restriction \tau = E^{K'} \restriction \tau$; denote this common part of the two measure sequences by E.

Suppose first that $E_\tau^{K'} \neq \varnothing$. Then $E_\tau^{K'}$ is ω-complete. To see this notice that, by the fact that K' is $\boldsymbol{K}^c$ in the sense of M, $E_\tau^{K'}$ is ω-complete in M, as follows from a) of Theorem 6.4.3. But M, being an ultrapower by a total measure on $\boldsymbol{V}$, is closed under ω-sequences (in fact, under κ-sequences), so any witness to the failure of the ω-completeness of $E_\tau^{K'}$ in $\boldsymbol{V}$ would be in M. Since $\langle J_\tau^E, E_\tau^{K'}\rangle$ is a premouse, $E_\tau^{\boldsymbol{K}^c} = E_\tau^{K'}$ by b) of Theorem 6.4.3.

Now suppose that $E_\tau^{K'} = \varnothing$. Let $F = U \cap J_\tau^E$. We show that F is weakly amenable w.r.t. J_τ^E. Given any $f \in J_\tau^E$ such that $f : \kappa \to \mathcal{P}(\kappa) \cap \boldsymbol{K}^c$ we have

$$f(\xi) \in F \longleftrightarrow \kappa \in \pi(f)(\xi),$$

so $\{\xi < \kappa;\ f(\xi) \in F\} = \{\xi < \kappa;\ \kappa \in \pi(f)(\xi)\} \in K'$. By acceptability, this set is an element of J_τ^E. We would like to show that $\langle J_\tau^E, F\rangle$ is a premouse; the only requirement on F we have to verify is the coherency condition. But this follows by the argument from the proof of Theorem 6.4.4, which we used to verify the premousehood of $\langle J_{\tau_i}^E, F_i\rangle$. Finally, F is ω-complete since it is a subset of a full measure in $\boldsymbol{V}$, so $E_\tau^{\boldsymbol{K}^c} = F$ again.

□(Lemma 6.4.6)

6.5 $0^{\ddagger}$

The class of mice we have defined contains only measures of order 0. Indeed, the coherency condition guarantees that, after forming an ultrapower by E_ν, the ordinal ν cannot index a measure; by the Łos Theorem this means that the set of all

$\xi < \kappa = \mathrm{cr}(E_\nu)$ which are critical points of some E_ζ has E_ν-measure 0. We now introduce the object $0^{\ddagger}$, called 0-*sword,* which plays the role of the "first mouse" containing a measure of order 1. Its relationship to extender models is analogous to that of $0^{\#}$ to L.

Definition. An *s-premouse* is a structure $M = \langle J_\nu^E, E_\nu, E_{\nu+1}\rangle$ such that $E_\nu \neq \varnothing \neq E_{\nu+1}$ and

a) $\langle J_\nu^E, E_\nu\rangle$ is a premouse;

b) $M' = \langle J_\nu^E, E_{\nu+1}\rangle$ satisfies all requirements on a premouse except that the coherency condition is modified as follows: if $N = \mathrm{Ult}(M', E_{\nu+1})$ then $\nu + 1 \subset \mathrm{wfc}(N)$, $E^N \upharpoonright \nu = E \upharpoonright \nu$ and $E_\nu^N = E_\nu$.

An *s-mouse* is a 0-iterable *s*-premouse.

Notice that $\mathrm{cr}(E_{\nu+1}) = \mathrm{cr}(E_\nu)$. By the proof of Lemma 4.1.2, being an *s*-premouse is a Q-condition. The measure $E_{\nu+1}$ has order 1 since, by the Łos Theorem, there are E_ν-measure 1 many $\xi < \kappa = \mathrm{cr}(E_{\nu+1})$ that are critical points of some E_ζ. We could weaken the requirement $E_\nu^N = E_\nu$ to $E_\nu \neq \varnothing$, but the difference is only cosmetic: If M satisfies this weaker coherency condition and $N = \mathrm{Ult}(M, E_{\nu+1})$, then $M^* = \langle J_\nu^E, E_\nu^N, E_{\nu+1}\rangle$ is an *s*-premouse. Moreover, if M is 0-iterable, there is a 0-iterable *s*-premouse: iterate the measure $E_{\nu+1}$ ω_1-times to obtain M_1 with an ω-complete $E_{\nu_1+1}^{M_1}$. Then perform the iteration from the proof of Theorem 4.5.6; by the proof of this theorem, all measures of the resulting structure M_2, except possibly $E_{\nu_2}^{M_2}$, will be ω-complete. Let

$$N = \mathrm{Ult}(M_2, E_{\nu_2+1}^{M_2}) \quad \text{and} \quad P = \langle J_{\nu_2}^{E^N}, E_{\nu_2}^N, E_{\nu_2+1}^{M_2}\rangle;$$

then P is an *s*-premouse all of whose measures are ω-complete. The argument from the proof of Lemma 4.2.2 then shows that P is 0-iterable. One can further weaken the coherency condition to "$\mathrm{cr}(E_{\nu+1})$ is measurable in N". This follows from the results of the previous section and the fact that $J_\lambda^{E^N}$ believes it is K^c, where λ is the largest cardinal in N (we leave the verification of this to the reader).

In the following we develop the fine structure theory for *s*-mice. It is clear that any *s*-premouse which is Σ_0-embeddable into an *s*-mouse is itself an *s*-mouse. Notice also that any nonsimple iterate of an *s*-mouse is a mouse. Furthermore, we can attempt to coiterate an *s*-premouse against a premouse; in this case we perform 0-iterations on the *s*-mouse side and $*$-iterations on the mouse side. The arguments from the proofs of lemmata concerning the comparison process (in particular the Comparison Lemma and Lemma 5.3.1) go through in the context of *s*-mice, so we have:

Lemma 6.5.1. *Let M, resp. N, be either an s-mouse or a mouse. Then their coiteration terminates below* $(\max\{\mathrm{card}(M), \mathrm{card}(N)\})^+$, *one side of the coiteration is simple and the last structure on the simple side is an initial segment of the last structure on the side which involves a truncation.* □(Lemma 6.5.1)

It follows that in any coiteration of a mouse against an s-mouse, the last model on the mouse side is a proper initial segment of the last model on the s-mouse side. This is clear if the s-mouse side involves a truncation. In the remaining case notice that the mouse side cannot involve a truncation since, by the previous lemma, the s-mouse side would be simple and its last structure would be an initial segment of the last model on the mouse side. However, this is impossible, since the former is an s-mouse (by the fact that the corresponding iteration is simple), while the latter a mouse. A similar argument goes through if both sides of the coiteration are simple. Moreover, it generalizes in a straightforward way to s-mice, so we have:

Lemma 6.5.2. *Any two s-mice coiterate simply to a common s-mouse.* □(Lemma 6.5.2)

Pick an s-mouse N. Let $X = h_M(\varnothing)$, let M be the transitive collapse of $M|X$ and σ the uncollapsing map. Then $\omega\varrho^1_M = \omega$, $p_M = \varnothing$ and M is sound. Furthermore, σ is Σ_1-preserving, so M is an s-mouse. By the previous lemma, M, N coiterate simply to a common mouse. It follows that $\mathcal{P}(\omega) \cap M = \mathcal{P}(\omega) \cap N$. If A_M is $\Sigma_1(M)$ such that $a = A_M \cap \omega \notin M$ and A_N is $\Sigma_1(N)$ by the same definition, then the preservation properties of σ guarantee that $A_N \cap \omega = a$, so $a \notin N$. Hence $\omega\varrho^1_N = \omega$ and $p_N = \varnothing$. As an immediate consequence we have that 0-iterations of s-mice are in fact $*$-iterations, since the first projectum drops below the critical point of any measure on the sequence.

Pick s-mice M, N such that $\omega\varrho^1_M = \omega = \omega\varrho^1_N$, $p_M = \varnothing = p_N$ and both M, N are sound. Since M, N have a common simple iterate, we can employ the argument from the proof of Lemma 5.1.4 and infer that $\pi_M = \pi_N$, where π_M and π_N is the iteration map on the M- and the N-side, respectively. Hence $M =$ the transitive collapse of $\operatorname{rng}(\pi_M) =$ the transitive collapse of $\operatorname{rng}(\pi_N) = N$. We have thus established:

Lemma 6.5.3. *There is a unique sound s-mouse M such that $\omega\varrho^1_M = \omega$ and $p_M = \varnothing$.* □(Lemma 6.5.3)

Definition. The unique s-mouse from the previous lemma is called 0-*sword* and is denoted by $0^\ddagger$.

Hence, $0^\ddagger$ exists iff there is an s-mouse iff there is a measure of order 1. In the following we shall develop some of its basic properties.

Lemma 6.5.4. *Suppose that $0^\ddagger$ exists. Given any weasel W, the coiteration of W with $0^\ddagger$ is simple and has length ∞, so the last weasel on the W-side is a non-simple iterate of $0^\ddagger$. Furthermore, $\operatorname{cf}(\kappa^{+W}) = \omega$ for all κ.*

Proof. The first part of the lemma follows by an analysis similar to that which yields the proof of Lemma 6.5.2 (see the paragraph preceding the lemma). Towards the proof of the second part, let M_∞ be the weasel resulting from the coiteration. By the proof of Corollary 6.3.4, to each cardinal successor ϑ in M_∞ there is a cardinal successor in $0^\ddagger$ which can be cofinally mapped into ϑ, so $\operatorname{cf}(\vartheta) = \omega$, as $0^\ddagger$ is countable. The rest follows from the fact that the iteration map on the W-side maps successors cofinally into their images. □(Lemma 6.5.4)

Thus, in the presence of $0^\ddagger$ no weasel computes cardinal successors correctly: No cardinal successor of W is a V-cardinal. W appears to be "thin", that is, it contains considerably less sets than V. This is caused by the fact that any weasel omits a substantial fact about V, namely the existence of a measure of higher order. As one might anticipate from the remarks preceding Lemma 6.3.3, the above result cannot be directly generalized for weasels with extenders; there we can conclude merely that the cofinalities of W-successors are ω-cofinal on a c.u.b. class and bounded otherwise.

It is also worthwhile to note that, given an s-mouse M, the weasel M^* obtained by iterating the measure of order 1 (that is, the measure $E^M_{\tau+1}$, where $\tau = \mathrm{ht}(M)$) is universal. To see this, notice that if N is a premouse coiterable with M^*, then N is coiterable with M since the M-side of the coiteration can be viewed as an iteration of M^* using the measures $E^{M_i}_{\nu_i}$. Indeed, the height τ_i of M_i is a successor cardinal in M^*, $M^*_i||\tau_i = M_i||\tau_i$ for all i, and applying the measure of order one at the i-th step corresponds to "unfolding" a longer initial segment of M^*_i – the N-side of the coiteration is not moved at this step. But both M, N are set-size structures, so the argument of the comparison lemma guarantees that their coiteration terminates, say in θ steps. Because of the close relationship between M and M^*, it is easy to see that the same would apply to the coiteration of M^* with N.

The above features of $0^\ddagger$ resemble the corresponding features of $0^\#$: if we view the latter as the first mouse with a measure, that is, the unique sound mouse of the form $Q = \langle J_\tau, U\rangle$ such that $\omega\varrho^1_Q = \omega$ and $p_Q = \varnothing$, then iterating the measure U out of the universe (i.e. ∞ many times) yields L, and cardinal successors of L are ω-cofinal. It is also easy to see that the critical points of this iteration constitute a c.u.b. class of indiscernibles for L; the proof easily generalizes for K and $0^\ddagger$ in place of L and $0^\#$, where K is obtained by iterating the measure of order one out of the universe. The truth of K is then encoded in $0^\ddagger$ in the same way the truth of L is encoded in $0^\#$.

We shall establish two more properties of $0^\ddagger$ concerning its complexity, which are obvious reformulations of the corresponding properties of $0^\#$. Let h be the Σ_1-Skolem function of $0^\ddagger$; recall that h is uniformly Σ_1 over $0^\ddagger$. Since the range of h is the whole $0^\ddagger$, we can encode $0^\ddagger$ into a countable structure $\langle D, E, I\rangle$ via h in the same way we encoded premice in the discussion preceding Theorem 4.3.6. Using the recursive isomorphism r from that construction we encode $0^\ddagger$ into a real. Note that if M is a ZF^- model containing $0^\ddagger$, then $0^\ddagger$ is automatically countable in M, and, as Lemma 4.3.5 – Theorem 4.3.6 easily generalize to s-mice, we have:

Lemma 6.5.5. *$0^\ddagger$ is Π_1-definable over H_{ω_1}. Thus, $0^\ddagger$ is a Π^1_2-singleton and the relation $n \in 0^\ddagger$ is Δ^1_3.*

Proof. Let r be the recursive isomorphism fixed in the discussion preceding Theorem 4.3.6. Then $r \in H_{\omega_1}$ and the same applies to $0^\ddagger$, if it exists. We shall verify that

$$y = 0^\ddagger \longleftrightarrow H_{\omega_1} \models \varphi(y),$$

where $\varphi(y)$ is the formula $(\forall M)\varphi_0(M, y)$ for $\varphi_0(M, y)$ of the form

$$(M \text{ a transitive } \mathrm{ZF}^-\text{-model } \& \ y \in M) \longrightarrow (\exists N \in M)(M \models \psi(y, M, N))$$

and where $\psi(y, M, N)$ is the conjunction of the formulae

- N is a sound s-mouse such that $\omega\varrho_N^1 = \omega$ and $p_N = \varnothing$;
- $(\exists D, E, I)(\exists\sigma)$
  - $\langle D, E, I\rangle = r''y$ &
  - $\sigma : \langle D, E\rangle/I \to \langle N, \in\rangle$ is a structure isomorphism.

This will prove the lemma since $\varphi(v)$ is obviously a Π_1-formula. ("$M \models \mathrm{ZF}^-$" can be expressed as a Σ_0 statement with the parameter M and the same applies to "$M \models \psi(y, M, N)$".)

Suppose $0^\ddagger$ exists and y is the real identified with $0^\ddagger$. Given any transitive ZFC$^-$ model, r is certainly an element of M, so if $y \in M$ then we can decode the structure $\langle D, E, I\rangle$ inside M and transitivize it there; we thus obtain the s-mouse representation of $0^\ddagger$. Since $0^\ddagger$ is iterable in V, it is iterable in M by the absoluteness of well-foundedness. Hence $H_{\omega_1} \models \varphi(y)$.

On the other hand, if $y \in M$ and $M \models \psi(y, M, N)$ for some N, then N is the s-premouse encoded in y and is iterable in M. Notice that N does not depend on the choice of M, since decoding N from y is a "local" procedure. Since M can be arbitrary, the proof of Lemma 4.3.8 verifies that N is a countably iterable s-premouse having all the requisite fine structural properties. It follows that N is fully iterable, i.e. $N = 0^\ddagger$. □(Lemma 6.5.5)

Notice that, if $M \subset V$ is a ZF$^-$ model containing ω_1 and $M \models \varphi(y)$ for some y, then M thinks it contains $0^\ddagger$, i.e. an s-premouse Q with the first projectum equal to 1 and $p_Q = \varnothing$ such that Q is iterable in the sense of M. By the remarks preceding Theorem 4.3.6, M is iterable in any ZF$^-$ model $M' \subset V$ containing M, so Q is equal to $0^\ddagger$ in the sense of any such M'. In particular, Q is the genuine $0^\ddagger$ (i.e. in the sense of V). It follows that any M' as above contains $0^\ddagger$.

Lemma 6.5.6. *$0^\ddagger$ cannot be generically added by a set-size forcing. In other words, if $0^\ddagger$ does not exist, $\mathbb{P} \in V$ is a forcing notion and G a $\mathbb{P}$-generic filter over V, then $0^\ddagger$ does not exist in $V[G]$ either.*

Proof. Suppose $0^\ddagger \in V[G]$. There is a $p \in \mathbb{P}$ such that $p \Vdash_{\mathbb{P}} (\exists x)\varphi(x)$, where $\varphi(v)$ is the formula defining $0^\ddagger$ from the previous lemma. Let H be $\mathrm{Col}(\omega, \mathcal{P}(\mathbb{P}))$-generic over V. Since there are only countably many dense subsets of $\mathbb{P}$ in $V[H]$, there is a $\mathbb{P}$-generic filter $\bar{G}$ in $V[H]$ such that $p \in \bar{G}$. Then $V[\bar{G}] \models (\exists x)\varphi(x)$, that is, $V[\bar{G}]$ believes it contains $0^\ddagger$. Pick y such that $V[\bar{G}] \models \varphi(y)$; by the remarks preceding this lemma, y is the unique real such that $V[H] \models \varphi(y)$, so for every $n \in \omega$,

$$n \in y \longleftrightarrow V[H] \models (\exists x)(\varphi(x)\ \&\ n \in x) \longleftrightarrow\ \Vdash (\exists x)(\varphi(x)\ \&\ \check{n} \in x);$$

here we mean the forcing relation associated with $\mathrm{Col}(\omega, \mathcal{P}(\mathbb{P}))$ over V. The latter equivalence is a consequence of the homogeneity of the collapse. It follows that $y \in V$. So $V \models \varphi(y)$, as $\varphi(y)$ holds in $V[H]$ and in fact $V \models \varphi(y)$, where $V = H^V_{\omega_1}$, since $y \in V$. The above argument goes through for any $\mathbb{P}$-generic H over V and y is obviously independent of the choice of H, so we have $\Vdash (\check{V} \models \varphi(\check{y}))$, which proves that $\varphi(y)$ holds in V. This means that $0^\ddagger$ exists in V. □(Lemma 6.5.6)

In the absence of $0^\ddagger$, we obtain the following strengthening of Lemma 6.4.6: If κ is measurable in V and U is a normal measure on κ, then $E^{K^c}_\tau = U \cap K^c$ where τ is the cardinal successor of κ in K^c. Going back to the proof of Lemma 6.4.6, under $\neg 0^\ddagger$ we can show that $E^{K'}_\tau = \varnothing$, so $\langle J^E_\tau, U \cap J^E_\tau \rangle$ is a premouse. To see this, notice that otherwise $\langle J^E_\tau, E^{K'}_\tau, U \cap J^E_\tau \rangle$ would be an s-mouse, as all its total measures are ω-complete.

6.6 Weak Covering

Jensen's Covering Lemma for L asserts that, in the absence of $0^\#$, every uncountable subset of ordinals can be covered by a constructible set of the same size. Our aim is to find a generalization of this theorem for weasels. We stress that there is no hope of having the full result for weasels containing total measures since such weasels have generic Prikry sequences of small size which are cofinal in the associated measurables. Measures of higher order give rise to Prikry "systems" of higher complexity and there is a rich theory with a variety of covering results for the true core model K arising from the analysis of such systems. This theory was developed for applications in cardinal arithmetic and infinitary combinatorics and lead to the complete solution of one of the most prominent questions in the area, the Singular Cardinal Hypothesis.

The current section is devoted to the proof of the so-called *Weak Covering Theorem* for K^c, which asserts that, in the absence of $0^\ddagger$, K^c computes the successors of sufficiently large singular cardinals correctly. Notice that in the case of L and $0^\#$, this is an easy consequence of Jensen's Covering Lemma. The purpose of this theorem is mainly technical: It will be needed for the construction of the true core model K. Later, after we have built K, we shall prove the Weak Covering Theorem for K, which will be of much more interest. For these reasons, we shall focus mainly on the following version of the Weak Covering Theorem, whose proof is technically less demanding than that of the full version and which still suffices for the construction of K.

Definition. A cardinal κ is *ω-closed* iff $\lambda^\omega < \kappa$ for all $\lambda < \kappa$.

Theorem 6.6.1. *Assume that $0^\ddagger$ does not exist; put briefly $\neg 0^\ddagger$. Then for every ω-closed cardinal κ,*

$$\mathrm{cf}(\kappa^{+K^c}) \geq \kappa.$$

In particular, if κ is singular in V, we have

$$\kappa^{+K^c} = \kappa^+.$$

It is worthwhile to point out a sort of dichotomy, which was already obvious when dealing with $0^\#$ and L. In the present more general context of weasels, Lemma 6.5.4

asserts that the presence of $0^\ddagger$ causes that cardinal successors are computed by extender models "as incorrectly as possible", that is, their cofinality is countable (or, as we shall see later for non-overlapping extenders, at least bounded). On the other hand, the absence of $0^\ddagger$ guarantees that the cardinal successors of $\boldsymbol{K}^c$ are computed correctly. Notice finally that it is essential that weak covering is possible only in the case of "rich" weasels; weasels which omit too many measures need not be universal and, by the proof of Lemma 6.3.1, any premouse iterating past such weasels would force low cofinality for cardinal successors in these models.

For the sake of completeness we also state the full version of the Weak Covering Theorem. Its proof follows the same strategy, but builds on a technical means called the *Frequent Extensions of Embeddings Lemma.* We shall introduce the Frequent Extensions of Embeddings Lemma in the next chapter where we will need it for the proof of the Weak Covering Theorem for $\boldsymbol{K}$. There we will also give hints as to which amendments should be made to the argument in order to obtain a proof of the full version of the Weak Covering Theorem for $\boldsymbol{K}^c$; the details will nevertheless be left to the reader.

Theorem 6.6.2. *Assume $\neg 0^\ddagger$. Let $\alpha \geq \omega_2$ be an ω-closed cardinal. Then for every $\boldsymbol{K}^c$-cardinal $\beta > \alpha$,*

$$\mathrm{cf}(\beta^{+\boldsymbol{K}^c}) \geq \mathrm{card}(\beta).$$

In particular, $\kappa^{+\boldsymbol{K}^c} = \kappa^+$ for every singular cardinal (in the sense of V) $\kappa > \alpha$.

The rest of this section, apart form the remarks at the very end, is devoted to the proof of Theorem 6.6.1. Pick a cardinal κ with all requisite properties and suppose that $\gamma = \mathrm{cf}(\tau) < \kappa$, where

$$\tau = \kappa^{+\boldsymbol{K}^c}.$$

From this we derive a contradiction by showing that τ must be collapsed inside $\boldsymbol{K}^c$. Pick a sequence of ordinals b of order type γ cofinal in τ and a regular θ larger than κ^+. Construct an elementary substructure X of H_θ satisfying:

- $b \subset X$ and $J_\tau^E \in X$ where $E = E^{\boldsymbol{K}^c}$;
- ${}^\omega X \subset X$;
- $\mathrm{card}(X) < \kappa$.

This can be done by an easy induction, setting

$$\begin{aligned} X_0 &= b \cup \{J_\tau^E\} \\ X_{\xi+1} &= \text{the hull of } {}^\omega X_\xi \text{ in } H_\theta \\ X_\lambda &= \bigcup_{\xi<\lambda} X_\xi \quad \text{for limit } \lambda \\ X &= X_{\omega_1}. \end{aligned}$$

The set X is easily seen to be closed under countable sequences and, since $\mathrm{card}(X_\xi) = \gamma^\omega$ for all countable $\xi > 0$, we have $\mathrm{card}(X) = \omega_1 \cdot \gamma^\omega < \kappa$ by our assumptions on κ.

Let

- $\bar{H}$ be the transitive collapse of X and $\sigma : \bar{H} \to H_\theta$ the uncollapsing map;
- $\sigma(\bar{\tau}, \bar{\kappa}) = \tau, \kappa$;
- $\sigma(J^{\bar{E}}_{\bar{\tau}}) = J^E_\tau$; we shall often write $\bar{\boldsymbol{K}}$ instead of $J^{\bar{E}}_{\bar{\tau}}$;
- $\alpha = \operatorname{cr}(\sigma)$.

Notice that, by construction, $\bar{H}$ is closed under ω-sequences, i.e. any countable sequence $\langle x_i;\ i \in \omega\rangle$ satisfying $x_i \in \bar{H}$ for all $i \in \omega$ is itself an element of $\bar{H}$. Furthermore,

$$\sigma \restriction J^{\bar{E}}_{\bar{\tau}} : J^{\bar{E}}_{\bar{\tau}} \to J^E_\tau \quad \text{is } \Sigma_0\text{-preserving and cofinal.} \tag{6.6.1}$$

Lemma 6.6.3. $\mathcal{P}(\alpha) \cap \bar{\boldsymbol{K}} \subsetneqq \mathcal{P}(\alpha) \cap \boldsymbol{K}^c$.

Proof. This is the only place in the proof of weak covering where we use the assumption $\neg 0^\ddagger$. The inclusion $\subset$ in the statement of the lemma follows trivially, so our task is to show that the two sets are different. Suppose for a contradiction that they are equal. Then the cardinal successors of α in $J^{\bar{E}}_{\bar{\tau}}$ and J^E_τ agree; denote this common value by ϑ. This follows from the standard coding argument: any ordinal less than the cardinal successor of α in one of the structures can be encoded into a subset of α and decoded in the other structure by collapsing; the collapsing map then constitutes a bijection between the ordinal in question and α. By Lemma 5.5.2 we have $\bar{E} \restriction \vartheta = E \restriction \vartheta$, so the structures $J^{\bar{E}}_{\bar{\tau}}$ and J^E_τ agree up to ϑ.

Let U be a measure on J^E_ϑ derived from σ, i.e.

$$U = \{x \in J^E_\vartheta;\ \alpha \in \sigma(x)\}.$$

U is easily seen to be an α-complete ultrafilter over J^E_τ; we show that U is normal. Pick a regressive function $f : a \to \alpha$ from J^E_ϑ, where $a \in U$. Then $\alpha \in \sigma(a)$; set $\delta = \sigma(f)(\alpha)$. This value is strictly less than α, as $\sigma(f)$ is regressive. For $\bar{a} = \{\xi \in a;\ f(\xi) = \delta\}$ we have $\alpha \in \sigma(\bar{a})$, so $\bar{a}$ is a set from U on which f is constant.

The measure U is furthermore weakly amenable w.r.t. J^E_ϑ and ω-complete. To see the former, pick any $f : \alpha \to \mathcal{P}(\alpha)$ from J^E_ϑ. We show that the set a of all $\xi < \alpha$ satisfying $f(\xi) \in U$ is an element of J^E_ϑ. But for $\xi < \alpha$ we have

$$f(\xi) \in U \longleftrightarrow \alpha \in \sigma(f(\xi)) = \sigma(f)(\xi),$$

so $a \in J^E_\tau$. Hence $a \in J^E_\vartheta$ by acceptability. To see the latter, pick (in V !) a countable sequence $\langle x_i;\ i \in \omega\rangle$ of sets from U. We intend to show that these sets have nonempty intersection. Each $\sigma(x_i)$ is an element of X and since X is closed under ω-sequences, the whole sequence $\langle \sigma(x_i);\ i \in \omega\rangle$ is in X. It follows that this sequence is in the range of σ and in fact, $\sigma(\langle x_i;\ i \in \omega\rangle) = \langle \sigma(x_i);\ i \in \omega\rangle$. Using this we conclude that the intersection of all $\sigma(x_i)$ must be nonempty, as

$$\alpha \in \bigcap_{i \in \omega} \sigma(x_i) = \sigma\Big(\bigcap_{i \in \omega} x_i\Big).$$

By the elementarity of σ, the intersection of all sets x_i must be nonempty as well.

Let $J^{E'}_{\vartheta'} = \mathrm{Ult}(J^E_\vartheta, U)$ and let $\bar{\sigma}$ be the associated ultrapower map. Since the restriction $\sigma \restriction J^E_\vartheta : J^E_\vartheta \to J^E_{\sigma(\vartheta)}$ is elementary, this ultrapower exists and can be embedded into $J^E_{\sigma(\vartheta)}$ via the Σ_0-preserving map $\sigma' : \bar{\sigma}(f)(\alpha) \mapsto \sigma(f)(\alpha)$ (this can be verified by a standard argument like that in the proof of Claim 3 of the proof of Theorem 5.2.1). Then $\sigma'(\alpha) = \alpha$ and the same applies to every $\xi < \alpha$, so the critical point of σ' must be greater than α. By the weak amenability of U, the structures J^E_ϑ and $J^{E'}_{\vartheta'}$ contain the same subsets of α, so the same argument as above yields that ϑ is the cardinal successor of α in $J^{E'}_{\vartheta'}$ and that the two structures agree up to ϑ. Since ϑ is also the cardinal successor of α in $J^E_{\sigma(\vartheta)}$, we conclude that

$$\mathrm{Ult}(J^E_\vartheta, U)||\vartheta = \langle J^E_\vartheta, E_\vartheta\rangle. \tag{6.6.2}$$

We are finally ready to derive the desired contradiction. Suppose first that $E_\vartheta = \varnothing$. Then $\langle J^E_\vartheta, U\rangle$ is a premouse: weak amenability of the top measure (and thus amenability of the structure) and the coherency condition are guaranteed by the previous two paragraphs; the remaining clauses of the definition of premouse follow immediately. Since E is the measure sequence of $\boldsymbol{K}^c$ and U is ω-complete, $U = E_\vartheta$ by Theorem 6.4.3, so E_ϑ cannot be empty after all. But then $M = \langle J^E_\vartheta, E_\vartheta, U\rangle$ is an s-premouse, as follows from (6.6.2). Moreover, all total measures in M are ω-complete, so M is an s-mouse by the proof of Lemma 4.5.2. This means that $0^\ddagger$ exists after all. □(Lemma 6.6.3)

Let $\bar{\vartheta} = \alpha^{+\bar{K}}$. By the previous lemma, $\bar{\vartheta}$ is strictly less than ϑ, the cardinal successor of α in the sense of $\boldsymbol{K}^c$. Hence there is a $\zeta < \vartheta$ such that $\bar{\vartheta}$ is a cardinal in $\boldsymbol{K}^c||\zeta$, but not in $\boldsymbol{K}^c||(\zeta + 1)$. Let

$$M_0 = \boldsymbol{K}^c||\zeta.$$

Then

- $\bar{\vartheta} = \alpha^{+M_0}$ and $\bar{E} \restriction \bar{\vartheta} = E^{M_0} \restriction \bar{\vartheta}$;
- $\omega\varrho^\omega_{M_0} \leq \alpha$;
- M_0 is sound.

The coiteration of $\bar{\boldsymbol{K}}$ with $\boldsymbol{K}^c$ is above α on the $\bar{\boldsymbol{K}}$-side since the two structures agree up to $\bar{\vartheta}$ and the $\bar{\boldsymbol{K}}$-side is simple by the universality of $\boldsymbol{K}^c$. It follows that any structure on the $\bar{\boldsymbol{K}}$-side of the coiteration contains the same subsets of α as $\bar{\boldsymbol{K}}$. As M_0 projects to α, the first iteration index used on the $\boldsymbol{K}^c$-side (if there is any) must be at most ζ, so this coiteration can be identified with the coiteration of $\bar{\boldsymbol{K}}$ with M_0. The following lemma is rather surprising.

Lemma 6.6.4. *$\bar{\boldsymbol{K}}$ is not moved in the coiteration with $\boldsymbol{K}^c$, that is, no measure is applied on the $\bar{\boldsymbol{K}}$-side of the coiteration.*

Proof. Let ν_i be the iteration indices, κ_i the associated critical points, and let M_i and $\bar{K}_i$ be the structures on the M_0-side and the $\bar{K}$-side of the coiteration, respectively. Suppose for a contradiction that $\bar{K}$ is moved and let ι be the first index such that $E_{\nu_\iota}^{\bar{K}_\iota} \neq \varnothing$. Then $\bar{K}_i = \bar{K}$ and $E_{\nu_\iota}^{\bar{K}_\iota} = \bar{E}_{\nu_\iota}$. Let ζ_ι be the maximal ζ' such that ν_i is the cardinal successor of κ_i in $M_\iota||\zeta'$, or, in other words, maximal such that $\bar{E}_{\nu_\iota}$ is a total measure over the corresponding truncate. Then

$$M_\iota^* = M_\iota||\zeta_\iota \text{ projects to } \kappa_\iota \text{ and is sound above } \kappa_\iota.$$

This is obvious if $\zeta_\iota < \mathrm{ht}(M_\iota)$. Otherwise the above statement follows from:

$$\text{Each mouse } M_i \text{ projects to } \kappa_i \text{ and is sound above } \kappa_i. \tag{6.6.3}$$

We shall prove this by induction on the length of the iteration. For $i = 0$ there is nothing to prove. To handle successor steps, suppose M_i satisfies (6.6.3). Then $M_{i+1} = \mathrm{Ult}^*(M_i^*, E_{\nu_i}^{M_i})$ where $M_i^* = M_i||\zeta_i$ and ζ_i is defined in the same way as ζ_ι above. Clearly, M_i^* satisfies (6.6.3) as well. Let n be such that $\omega\varrho_{M_i^*}^{n+1} \leq \kappa_i < \omega\varrho_{M_i^*}^{n}$. Each element x of the ultrapower M_{i+1} is of the form $\pi_{i,i+1}^M(f)(\kappa_i)$, where $f \in \Gamma(\kappa_i, M_i^*)$ and $\pi_{i,i+1}^M$ is the associated ultrapower map. The function f can be expressed as $f(\xi) \simeq g(\xi, p)$ for some good $\Sigma_1^{(n-1)}(M_i^*)$ function g and a parameter $p \in M_i^*$. So

$$x = g'(\kappa_i, \pi_{i,i+1}^M(p)) = g'(\kappa_i, \tilde{h}_{M_{i+1}}^{n+1}(\xi, \pi_{i,i+1}^M(p_{M_i})))$$

where g' is the good $\Sigma_1^{(n-1)}(M_{i+1})$ function with the same functionally absolute definition as g and $p = \tilde{h}_{M_i}^{n+1}(\xi, p_{M_i})$ for some $\xi < \kappa_i$. That p can be written in this form follows from the soundness of M_i^* above κ_i. As in the proof of Lemma 3.2.4 we infer that x is $\Sigma_1^{(n)}(M)$-definable from $\pi_{i,i+1}^M(p_{M_i^*}) = p_{M_{i+1}}$ (recall that iteration maps preserve standard parameters) and parameters less than $\kappa_i + 1$, so

$$M_{i+1} = \tilde{h}_{M_{i+1}}^{n+1}((\kappa_i + 1) \cup \{p_{M_{i+1}}\}). \tag{6.6.4}$$

That the above computation goes through follows from the fact that $\pi_{i,i+1}^M$ is $\Sigma_1^{(n)}$-preserving (it is in fact Σ^*-preserving).

To handle a limit case λ, we follow the same strategy; the main point is to obtain the analogue to (6.6.4) with λ in place of $i + 1$. This is easy, as each $x \in M_\lambda$ is of the form $\pi_{i,\lambda}^M(x_i)$ for a sufficiently large i and some $x_i \in M_i$. By the induction hypothesis, x_i can be expressed as $\tilde{h}_{M_i}^{n+1}(\xi, p_{M_i})$, where $\xi < \kappa_i$. Let us make precise that i is chosen so that there are no truncations between i and λ and that $\omega\varrho_{M_j^*}^{n+1} \leq \kappa_j < \omega\varrho_{M_j^*}^{n}$ whenever $i \leq j \leq \lambda$. Similarly as above, the preservation properties of the iteration maps enable us to infer that $M_\lambda = \tilde{h}_{M_\lambda}^{n+1}(\kappa_\lambda \cup \{p_{M_\lambda}\})$, which completes the proof of (6.6.3). □(6.6.3)

Since the $\bar{K}$-side of the coiteration is simple, the measure $\bar{E}_{\nu_\iota}$ is a total measure in $\bar{K}$. Its image $E_{\sigma(\nu_\iota)}$ under σ is therefore a total measure in K^c and thus, by Theorem 6.4.3,

ω-complete. Then $\bar{E}_{\nu_\iota}$ is ω-complete in the sense of $\bar{H}$ by the elementarity of σ. Since $\bar{H}$ is closed under ω-sequences, $\bar{E}_{\nu_\iota}$ is ω-complete in the sense of V. Let

$$\pi : M_\iota^* \xrightarrow[\bar{E}_{\nu_\iota}]{*} \tilde{M}.$$

By the ω-completeness of $\bar{E}_{\nu_\iota}$, this ultrapower exists and $\tilde{M}$ is a premouse normally iterable above κ_ι. The latter follows from the ω-completeness of $\bar{E}_{\nu_\iota}$ by the argument from the proof clause b) of Lemma 5.6.4. (In other words, the above ultrapower can be understood as the canonical extension of the ω-complete embedding $\pi \restriction J_{\nu_\iota}^{E_\iota}$, where $E_\iota = E^{M_\iota^*}$.) Since M_ι^* agrees with $\tilde{M}$ above κ_ι and κ_ι is a cardinal in both structures, M_ι^* coiterates with $\tilde{M}$ above κ_ι, so the two premice are coiterable. Furthermore, π is $\Sigma_0^{(m)}$-preserving and cofinal for m satisfying $\omega\varrho_{M_\iota^*}^{m+1} \le \kappa_\iota < \omega\varrho_{M_\iota^*}^{n}$; the cofinality follows from the weak amenability of $\bar{E}_{\nu_\iota}$ w.r.t. M_ι^*. We can now repeat the argument from the proof of Claim 3 of the proof of Theorem 5.2.1 and infer that

- M_ι^* coiterates with $\tilde{M}$ simply above κ_ι to a common mouse Q;
- $\tilde{\pi} \circ \pi = \pi^*$, where $\tilde{\pi}$ and π^* is the iteration map on the $\tilde{M}$-side and the M_ι^*-side, respectively;
- $E_{\nu_\iota}^{M_\iota^*} = \bar{E}_{\nu_\iota}$.

But the measures $E_{\nu_\iota}^{M_\iota^*}$, $\bar{E}_{\nu_\iota}$ come from the coiteration of M_ι^* with $\bar{K}$, so they must disagree. Contradiction. □(Lemma 6.6.4)

The last model on the M_0-side of the coiteration of M_0 with $\bar{K} = J_{\bar{\tau}}^{\bar{E}}$ is thus a lengthening of $J_{\bar{\tau}}^{\bar{E}}$. All iteration indices and critical points of the coiteration are either below $\bar{\kappa}$, or else there is a last ultrapower

$$\pi_{\lambda,\lambda+1}^{M} : M_\lambda^* \xrightarrow[\bar{E}_{\bar{\tau}}]{*} M_{\lambda+1},$$

that is, $\nu_\lambda = \bar{\tau}$. In the former case, let M_λ be the last mouse on the M_0-side of the coiteration. In both cases, $E^{M_\lambda} \restriction \bar{\tau} = \bar{E} \restriction \bar{\tau}$ (recall: M_λ^* is either M_λ or the result of the last truncation). Set

$$M = M_\lambda || \zeta \text{ where } \zeta \text{ is the maximal } \zeta' \text{ such that } \bar{\tau} = \bar{\kappa}^{+M_\lambda||\zeta'}.$$

Then

- M projects to $\bar{\kappa}$; let $n \in \omega$ be such that $\omega\varrho_M^{n+1} \le \bar{\kappa} < \omega\varrho_M^{n}$;
- M is sound above $\bar{\kappa}$;
- $\bar{\tau} = \bar{\kappa}^{+M}$;

- $J_{\bar\tau}^{E^M} = J_{\bar\tau}^{\bar E}$.

The former two clauses are immediate if $\zeta < \mathrm{ht}(M_\lambda)$; otherwise they follow from (6.6.3). Let

$$\tilde\sigma : M \to N$$

be the canonical extension of $\sigma \restriction J_{\bar\tau}^{\bar E} : J_{\bar\tau}^{\bar E} \to J_\tau^E$ to M. This canonical extension exists by (6.6.1) and the fact that σ is ω-complete. Using the above fine structural properties of M together with Lemmata 3.6.7 and 3.6.9 we infer:

- $\omega\varrho_N^{n+1} \leq \kappa < \omega\varrho_N^n$;
- N is sound above κ;
- $\tilde\sigma(\bar\kappa) = \kappa, \tilde\sigma(\bar\tau) = \tau$ and $\tau = \kappa^{+N}$.

Furthermore, the ω-completeness of σ guarantees that N is normally iterable above κ, as follows from clause b) in Lemma 5.6.4. Since N agrees with $\boldsymbol{K}^c$ above κ and κ is a cardinal both in N and $\boldsymbol{K}^c$, the premouse N coiterates with $\boldsymbol{K}^c$ above κ. As $\boldsymbol{K}^c$ is universal, this coiteration terminates, say after θ^* many steps, the iteration on the N-side is simple and the last premouse N_{θ^*} on the N-side is an initial segment of the last premouse W_{θ^*} on the $\boldsymbol{K}^c$-side.

The above fine structural properties of N enable us to encode N into a subset b of κ which is $\boldsymbol{\Sigma}_1^{(n)}(N)$ in the same way we encoded the witness W in the proof of Theorem 5.2.1. Since the iteration on the N-side is simple and above κ, we have $b \in \boldsymbol{\Sigma}_1^{(n)}(N_{\theta^*})$. Let k be the last truncation point on the $\boldsymbol{K}^c$-side if $W_{\theta^*} = N_{\theta^*}$ and $k = \theta^*$ otherwise. In the former case, using the fact that the iteration from W_k^* to W_{θ^*} is simple and above κ (recall that W_k^* is the result of the last truncation), we infer that $b \in \boldsymbol{\Sigma}_1^{(n)}(W_k^*)$, so $b \in W_k$. In the latter case, the conclusion $b \in W_k$ follows immediately. Finally, the fact that the iteration from $\boldsymbol{K}^c$ to W_k is above κ yields $b \in \boldsymbol{K}^c$. Now, exactly as in the proof of Theorem 5.2.1, we can reconstruct N from b by forming a transitive collapse inside $\boldsymbol{K}^c$, so $N \in \boldsymbol{K}^c$. The above derived fine structural properties of N guarantee that $N = \tilde h_N^{n+1}(\kappa \cup \{p_N\})$. As $\tilde h_N^{n+1}$ is definable over N, we have $\tilde h_N^{n+1} \in \boldsymbol{K}^c$. Thus, in $\boldsymbol{K}^c$ there is a partial function $\xi \mapsto \tilde h_N^{n+1}(\xi, p_N)$ which maps κ onto $\tau = \kappa^{+\boldsymbol{K}^c}$. Contradiction. □(Theorem 6.6.1)

We have similar covering results for A-full $\boldsymbol{K}^c$ over J_β^E. In fact, one can show that Theorem 6.6.2 holds for such models and sufficiently large $\alpha > \beta$. However, as we have already mentioned, there is little use for such a strong result in applications, so we shall rather state the simpler analogue of Theorem 6.6.1.

Theorem 6.6.5. *Assume* $\neg 0^\ddagger$. *Let* W *be the* A*-full* $\boldsymbol{K}^c$ *over* J_α^E. *Then for every* ω*-closed cardinal* $\kappa > \alpha \cdot \min(A)$,

$$\mathrm{cf}(\kappa^{+W}) \geq \kappa.$$

In particular, if κ *is singular in* V, *then*

$$\kappa^{+W} = \kappa^+.$$

Proof. The proof of Theorem 6.6.1 goes through with a simple amendment to that of Lemma 6.6.3. There we used the **On**-fullness of $\boldsymbol{K}^c$ to argue that the measure U actually agrees with E_ϑ. In the present more general context we have mere A-fullness at our disposal, so we need to arrange for $\mathrm{cf}(\mathrm{cr}(U)) \in A$. This can be easily done in that we modify the construction of X from the beginning of the proof of Theorem 6.6.1. This time we construct X in ε many steps, where ε is the least uncountable element of A. We start from X_0 such that $\varepsilon \cup \alpha \subset X_0$. Fix a regular cardinal $\delta \leq \kappa$ strictly greater than $\mathrm{card}(X_0)^\omega$ and for every $\xi < \varepsilon$ set $\alpha_\xi = \sup(\delta \cap X_\xi)$. Clearly $\alpha_\xi < \delta$ and $\alpha_\xi \notin X_\xi$. At successor steps we let

$$X_{\xi+1} = \text{ the hull of } {}^\omega X_\xi \cup \{\alpha_\xi\} \text{ in } H_\theta;$$

it follows that α_ξ is a strictly increasing sequence of length ε and its supremum α^* is less than δ. As $X \cap \delta = \alpha^*$, the critical point of the associated uncollapsing map $\sigma : \bar{H} \to H_\theta$ agrees with α^*, so its cofinality is ε. □(Theorem 6.6.5)

Using the Weak Covering Theorem, we can show that every weakly universal weasel is in fact universal. Here, of course we need a smallness condition on V, in our case we use $\neg 0^\ddagger$. This result generalizes to weasels with strong cardinals. As we shall see later, in the presence of strong cardinals, the notions of universality and weak universality require a reformulation.

Lemma 6.6.6. *Assume $\neg 0^\ddagger$. Then every weakly universal weasel computes the successors of V-cardinals correctly on a stationary class. Consequently, every weakly universal weasel is universal.*

Proof. Let $\bar{W}$ be the θ-full $\boldsymbol{K}^c$ for a fixed regular uncountable cardinal θ (in the sense of V). Given a weakly universal weasel W, the two weasels coiterate simply to a common weasel, say Q. Denote the weasels on the $\boldsymbol{K}^c$-side and the W-side by $\bar{W}_i$ and W_i, respectively, and the associated iteration maps by $\bar{\pi}_{ij}$ and π_{ij}. By (6.1.1), the class

$$C = \{\alpha \in \mathbf{On};\ \bar{\pi}''_{0,\infty}\alpha \subset \alpha \ \&\ \pi''_{0,\infty}\alpha \subset \alpha\}$$

is closed and proper, as can be proved by the standard closure argument (see e.g. the proof of Lemma 6.3.1), and the class

$$S = \{\alpha \in C;\ \alpha \text{ is } \omega\text{-closed and } \mathrm{cf}(\alpha) = \omega\}$$

is stationary. We show that $\alpha^{+W} = \alpha^+$ for all $\alpha \in S$.

The crucial point is the observation that $\bar{\pi}_{0,i}(\alpha) = \alpha$ for all $\alpha \in S$ and $i \in \mathbf{On}$. This follows by induction on i: As $\bar{W}_j \models \mathrm{cf}(\pi_{0,j}(\alpha)) \neq \kappa_j$ for all $j < i$, the image $\bar{\pi}''_{0,i}\alpha \subset \alpha$ is cofinal in $\bar{\pi}_{0,i}(\alpha)$, which yields the above conclusion. Hence $\bar{\pi}(\alpha) = \alpha$ and, applying the weak covering to $\bar{W}$, $\alpha^{+Q} = \bar{\pi}(\alpha^{+\boldsymbol{K}^c}) = \bar{\pi}(\alpha^+) \geq \alpha^+$. This guarantees that Q computes the cardinal successors correctly for all $\alpha \in S$.

Using the iteration map π we now pull the situation down to W. Let ϑ be the least index such that the iteration on the W-side is above α from ϑ on. Then obviously

$\alpha^{+W_\vartheta} = \alpha^{+Q} = \alpha^+$. If there is no $\iota < \vartheta$ satisfying $\mathrm{cf}^{W_\iota}(\alpha) = \kappa_\iota$, then the argument from the previous paragraph yields that $\pi_{0,\vartheta}(\alpha) = \alpha$ and α^{+W} is mapped by $\pi_{0,\vartheta}$ cofinally into $\pi_{0,\vartheta}(\alpha^{+W}) = \alpha^{+W_\vartheta} = \alpha^+$, so $\alpha^{+W} = \alpha^+$. Now suppose there is an ι as above. As the iteration is normal, there is exactly one such ι. Regarding the iteration from W to W_ι and that from $W_{\iota+1}$ to W_ϑ, we can apply the above reasoning and infer that α^{+W} is mapped cofinally into α^{+W_ι} and $\alpha^{+W_{\iota+1}}$ cofinally into $\alpha^{+W_\vartheta} = \alpha^+$. It follows that $\alpha^{+W_{\iota+1}} = \alpha^+$ and, to establish that W computes the cardinal successor of α correctly, it only remains to prove that $\alpha^{+W_\iota} = \alpha^+$. Suppose this fails. Setting $\delta = \alpha^{+W_\iota}$, we have $\delta < \alpha^+$. As $W_{\iota+1} = \mathrm{Ult}(W_\iota, E_{\nu_\iota}^{W_\iota})$ and $\pi_{\iota,\iota+1}$ is the associated ultrapower map,

$$\pi_{\iota,\iota+1}(\alpha) \leq \mathrm{card}(\mathrm{Ult}(W_\iota||\delta, E_{\nu_\iota}^{W_\iota})) \leq \mathrm{card}((^{\kappa_\iota}(W_\iota||\delta))^{W_\iota}) = \mathrm{card}(W_\iota||\delta) = \alpha.$$

So $\alpha < \pi_{\iota,\iota+1}(\alpha) < \alpha^+ = \alpha^{+W_{\iota+1}}$, which is impossible, as $\pi_{\iota,\iota+1}(\alpha)$ is a cardinal in $W_{\iota+1}$. This completes the proof. □(Lemma 6.6.6)

Notes

The notion of universality was first considered by Mitchell [Mi1]. The $\boldsymbol{K}^c$-construction, as presented in this book, is just Steel's $\boldsymbol{K}^c$-construction adapted to ω-complete measures. Jensen [Je88] used a slightly different construction, which yields essentially the same model once we are below $0^\P$. The proof of the universality of $\boldsymbol{K}^c$ presented here is due to Jensen [Je98], where he uses an additional assumption that **On** is inaccessible. The ZFC version of the proof is due to Schindler and the author. Steel [St96] does not prove the universality of $\boldsymbol{K}^c$ directly, but uses a so called "cheap covering lemma" instead. $0^\ddagger$ was introduced in [Je88] as the first mouse containing a measure of order 1; $0^\P$ was already considered by Dodd [D82]. The proof of the weak covering lemma is a generalization of that of Jensen's covering lemma for $\boldsymbol{L}$. Two important aspects of the proof that we have presented, namely the fact that $\bar{K}$ is not moved in the coiteration with $\boldsymbol{K}^c$ as well as the iterability of the liftup, are due to Mitchell and all proofs of weak covering for higher models follow the same scenario. In those models, however, that there are many other new aspects, that have to be considered.(see e.g. [MSS94] and [Sch1]).

Chapter 7

The Core Model

At last we have developed means sufficient to give a definition of the core model $\boldsymbol{K}$, prove its existence and discuss its basic properties. The idea behind the core model is to provide an extender model with a "maximal" measure sequence (or, more generally, an extender sequence) in the sense that ordinals indexing measures in such a sequence are distributed as densely as possible. There are various ways of stating this rigorously. One way is to observe that if we form a Skolem hull of a sufficiently large class inside a weasel and collapse it, the outcome is a weasel with a "denser" measure sequence. Notice the parallel between considering such collapsed hulls of weasels and the cores of mice. An alternative way of determining maximal measure sequences is to observe that no sequence associated with an iterate of some extender model can be maximal, since the corresponding iteration indices cannot index measures in the iterate, but *do* index measures in the weasel at the beginning of the iteration. In this case, "maximal" would mean that sequence whose associated weasel is not an iterate of any other weasel, or more generally, the one which does not omit any measure belonging to some universal weasel. We shall see that under a certain smallness condition the two approaches are equivalent. (Intuitively, this looks plausible since, after forming an ultrapower of any model W, the collapsed range of the associated ultrapower map is precisely W.) The former one leads to Steel's definition of $\boldsymbol{K}$, which appears to demand the use second-order set theory at first glance, but can be reformulated in ZFC. This approach turns out to be more convenient for proving the existence and uniqueness of $\boldsymbol{K}$ as well as developing its basic properties. The latter approach leads to the inductive definition of $\boldsymbol{K}$, which has a "local" nature and which can be stated without developing any theory in advance. After having established the existence, uniqueness and universality of the core model, we shall prove its fundamental properties: the rigidity lemma, generic absoluteness and weak covering.

7.1 Inductive Definition of $\boldsymbol{K}$

The "maximality" of the measure sequence E of the core model $\boldsymbol{K}$ should guarantee, as has already been indicated in the above introduction, that any measure belonging to the sequence of an arbitrary universal weasel is absorbed by E. A natural attempt at such a sequence would be to proceed inductively along the ordinals and at each step add a measure belonging to a universal weasel cohering with the sequence constructed so far, if such a measure exists.

Definition. A mouse $M = \langle J_\alpha^E, E_{\omega\alpha}\rangle$ is *strong* iff there is a universal weasel W such that $M = W||\alpha$. If $E_{\omega\alpha} = \varnothing$, we say briefly that the structure $J_\alpha^{E^M}$ is strong.

Notice that if $E_{\omega\alpha} \neq \varnothing$ and $M = \langle J_\alpha^E, E_{\omega\alpha}\rangle$ is strong, then so is J_α^E, since the latter is equal to $\mathrm{Ult}(W, E_{\omega\alpha})||\alpha$, being thus an initial segment of a universal weasel. Trivially, any initial segment of a strong mouse is again strong.

We are now able to give a rigorous definition of $\boldsymbol{K}$. This definition appears to make use of second-order set theory, as the definition of strong mouse does. In the rest of this section we shall show that strongness can be equivalently described as a ZFC condition and that it in fact has a "local" character.

Definition. The measure sequence $E = E^{\boldsymbol{K}}$ is defined inductively as follows:

$$E_{\omega\nu} \simeq \begin{cases} \langle J_\nu^E, F\rangle & \text{if } F \text{ is the unique measure such that } \langle J_\nu^E, F\rangle \text{ is strong;} \\ \varnothing & \text{if } J_\nu^E \text{ is strong and for no measure } F \text{ is } \langle J_\nu^E, F\rangle \text{ strong.} \end{cases}$$

If neither of the above conditions holds for ν, the object $E_{\omega\nu}$ is undefined and the construction of E meets a premature end. If E_α is defined for all ordinals α (of course, $E_\alpha = \varnothing$ whenever α is a successor ordinal) we set

$$\boldsymbol{K} = \boldsymbol{L}[E] = J_\infty^E = \bigcup_{\alpha \in \mathbf{On}} J_\alpha^E.$$

The extender model $\boldsymbol{K}$ is the *core model.*

Lemma 7.1.1. *A mouse* $M = \langle J_\alpha^E, E_{\omega\alpha}\rangle$ *is strong iff for any premouse* N *satisfying* $M = N||\alpha$, *if* N *is normally iterable beyond* $\alpha + 1$ *(i.e. with indices strictly greater than* α*), then:*

a) N *is iterable;*

b) $M = \mathrm{core}(N)||\alpha$.

Proof. We first show that if M is strong, then M satisfies a) and b). Let W be a universal weasel witnessing the strongness of M. Pick any premouse N which is a lengthening of M. Since the measure sequences of N and W agree below $\alpha + 1$, the coiteration of N with W is beyond $\alpha + 1$, that is, N is coiterable with W. Since W is universal, this coiteration terminates and is simple on the N-side. The last premouse N' on the N-side is an initial segment of the last structure W' on the W-side, which is either a weasel or a mouse, so N' is iterable. As the iteration map $\pi : N \to N'$ on the N-side is total and Σ^*-preserving, N is iterable. This proves a).

To see b), suppose first that the iteration from N to N' is not above $\omega\varrho_N^\omega$. Since we work with measures of order zero, and the iteration only uses total measures, this is possible only if $\omega\varrho_N^\omega \geq \alpha$. But then b) follows from the fact that $\mathrm{core}(N)$ agrees with N below the successor of the ultimate projectum (Corollary 5.5.3). Now suppose the iteration from N to N' is above the ultimate projectum. Then N' is not sound, so there must be a truncation on the W-side of the coiteration, W' is a mouse and $W' = N'$. Let i be the last truncation point on the W-side and W_i^* the result of this last truncation; then

$$W_i^* = \mathrm{core}(W') = \mathrm{core}(N') = \mathrm{core}(N).$$

But since the coiteration is beyond $\alpha+1$, we have $W_i^*||\alpha = W||\alpha = M$. This completes the proof of b).

We shall now turn to the proof of the converse, i.e. assuming that the conclusion of the lemma holds for any premouse N as above, we construct a universal weasel which is a lengthening of M. This weasel will be the $\boldsymbol{K}^c$ over M (see the definition of A-full $\boldsymbol{K}^c$ over J_α^E) in the previous chapter. That the construction will yield the desired lengthening of M will follow from a) and b).

Define hierarchies N_β, M_β exactly as in the $\boldsymbol{K}^c$-construction with only one difference – we set

$$N_0 = M.$$

We then follow the inductive argument of Lemma 6.4.1 and show:

- All total measures in N_β with indices strictly greater than α are ω-complete.

The point here is that if $N_{\beta+1} = \langle J_{\zeta+1}^{\bar{E}^\beta}, \varnothing\rangle$ (recall $\bar{E}^\beta = E^{M_\beta}$ where $M_\beta = \text{core}(N_\beta)$), then all total measures of $N_{\beta+1}$ have critical points *below* the ultimate projectum of N_β, so the measure sequences E^β, $\bar{E}^\beta$ agree up to this projectum. Repeating the proof of Lemma 4.5.2, we obtain that each N_β is normally iterable beyond $\alpha+1$. Condition a) then guarantees that N_β is in fact fully iterable, so it does make sense to form cores and the construction does not break down. Finally, M_β agrees with M up to α due to condition b), so M remains an initial segment of both N_β and M_β for all ordinals β and thus also of the resulting model $\boldsymbol{K}^c[M]$. That $\boldsymbol{K}^c[M]$ exists (i.e. that the construction converges) and is universal follows by the same arguments as in the case of $\boldsymbol{K}^c$. □(Lemma 7.1.1)

The same proof yields a somewhat weaker condition of strongness for structures without top predicates: J_α^E is strong iff a) and b) from the statement of the previous lemma follow for any lengthening N of J_α^E which is normally iterable beyond α.

7.2 Steel's Definition of $\boldsymbol{K}$

The heart of Steel's definition of $\boldsymbol{K}$ is the notion of an *incompressible structure*. Such a structure, roughly speaking, constitutes an initial segment of some universal weasel which is contained in the Skolem hull of any reasonably rich class of ordinals, and thus cannot be "compressed" by further collapsing. It turns out that any two incompressible structures are compatible in the sense that one is an initial segment of the other, so we can form the union of all of them; this union will be a candidate for $\boldsymbol{K}$. However, we have to show that there are incompressible structures of arbitrary length. The key point here is the construction of a decreasing sequence of so-called *thick* classes of ordinals; formalizing this construction in ZFC makes the issues somewhat technical.

Definition. Let W be a weasel and X a class of ordinals. We say that X is *thick* in W iff X is unbounded in **On** and for all sufficiently large regular ε (that is, there is a ε_0 such that for all $\varepsilon \geq \varepsilon_0$) we have:

a) X is closed under singular limits of ε-sequences;

b) $\beta^{+W} = \beta^+$ and $X \cap \beta^+$ is unbounded in β^+ for all large enough $\beta \in X$ which are cardinals in V;

c) $\mathrm{cf}^W(\beta) \neq \mathrm{cr}(E_\nu^W)$ whenever $\beta \in X$ and E_ν^W is total in W.

We shall see below that if $0^\ddagger$ does not exist, many universal weasels have thick classes, so the above definition is not vacuous.

Definition. The structure J_α^E is *incompressible* iff there is a universal weasel W and a thick class X in W such that

a) $J_\alpha^E = J_\alpha^{E^W}$;

b) $\alpha \subset h_W(Y)$ for any thick class Y in W, where $h_W(Y)$ is the Σ_1-Skolem hull of Y in W. (Thus, h_W is the Σ_1-Skolem function for W.)

We say that the pair W, X *witness* the incompressibility of J_α^E.

Clearly, if J_α^E is incompressible and $\beta < \alpha$ then J_β^E is incompressible as well. The above definition again makes use of second-order set theory. It can be formulated in Gödel-Bernays set theory, which is a conservative extension of ZF. We shall, however, work in ZFC, where classes can be identified with pairs consisting of formulae and parameters. We shall use the above definition only in ways which can be handled directly in ZFC. That is, to prove the incompressibility of a given structure we shall *define* a witnessing pair W, X; condition b) can then be expressed by a scheme of formulae and verifying this condition amounts to giving a *uniform* proof which can be applied to all formulae of the scheme.

Theorem 7.2.1. *Given a pair of incompressible structures, one of them is an initial segment of the other.*

Theorem 7.2.2 ($\neg 0^\ddagger$). *For any ordinal α there is an incompressible structure of height α.*

Definition. $K^* = \bigcup\{J_\alpha^E;\ \alpha \in \mathbf{On} \text{ and } J_\alpha^E \text{ is incompressible}\}$

Our strategy is, assuming $\neg 0^\ddagger$, to prove the existence of K^* and then verify that K^* actually agrees with K. We shall develop basic properties of thick classes and prove Theorem 7.2.1 below; the proof of Theorem 7.2.2 is more demanding and we devote the whole next section to this task.

Lemma 7.2.3 ($\neg 0^\ddagger$). *Let A be a set of regular cardinals. The class*

$$X = \{\beta \in \mathbf{On};\ \mathrm{cf}(\beta) > \mathrm{l.u.b.}(A)\ \&\ \kappa \leq \beta < \kappa^+ \text{ for some strong limit cardinal } \kappa\}$$

is thick in the A-full K^c.

Proof. We verify the clauses in the definition of a thick class. Both a) and the second part of b) follow by the standard closure argument, here it suffices to set $\varepsilon_0 = \text{l.u.b.}(A)$. The first part of b) follows from the weak covering theorem. Finally, c) follows from the conctruction of the A-full $\boldsymbol{K}^c$. □(Lemma 7.2.3)

More generally, the class X from the previous lemma is thick in $\boldsymbol{K}^c$ over arbitrary J^E_α.

Lemma 7.2.4. *Let W be a weasel and X a class of ordinals.*

a) *If X is thick in W, then W is universal.*

b) *If $\langle X_\xi;\ \xi < \theta\rangle$ is a sequence of thick classes in W and $\theta < \infty$, then $\bigcap_\xi X_\xi$ is thick in W.*

c) *If F is a measure (or, more generally, an extender) on W and $W' = \text{Ult}(W, F)$ with the ultrapower map π, then*

$$X' = \{\beta \in X;\ \pi(\beta) = \beta\}$$

is thick in both W and W'.

d) *If X is thick in W and W' is a simple iterate of W with the iteration map π, then*

$$X' = \{\beta \in X;\ \pi(\beta) = \beta\}$$

is thick in both W and W'. (The iteration from W to W' can be long.)

e) *If $\sigma : \bar{W} \to W$ is Σ_1-preserving and $X \subset \text{rng}(\sigma)$ is thick in W, then*

$$\bar{X} = \{\beta \in X;\ \sigma(\beta) = \beta\}$$

is thick in both $\bar{W}$, W.

f) *($\neg 0^\ddagger$) If W is (weakly) universal, then W has a simple iterate W' such that there is a class Y thick in W'.*

Proof. Clause a) follows from Lemma 6.3.1. If W were not universal, then $\beta^{+W} < \beta^+$ for a c.u.b. class C of ordinals. On the other hand, if we pick an increasing ε-sequence of ordinals alternating between X and C for sufficiently large regular ε, the associated supremum κ is a limit point of both X and C, and therefore a member of $C \cap X$. As X is thick, W must compute the cardinal successor of κ correctly, a contradiction.

To see b) it suffices to prove the unboundedness of the intersection in **On** and in β^+ for $\beta > \theta$ (both follow by the same standard closure argument); the remaining requirements on thickness follow immediately. Let us briefly describe the argument. Given an ordinal β_0, let ζ^ξ_0 be the least member of X_ξ greater than β_0 and let β_1 be some upper bound for all ζ^ξ_0. Inductively construct ζ^ξ_η and $\beta_{\eta+1}$ for all $\eta < \varepsilon$ in the same manner. Then $\beta = \sup\{\beta_\eta;\ \eta \in \varepsilon\} = \sup\{\zeta^\xi_\eta;\ \eta < \varepsilon\}$ whenever $\xi < \theta$, so $\beta \in \bigcap X_\xi$.

For c), we stress that F need not be an element of W. We first verify a) from the definition of thickness. Setting $\kappa = \mathrm{cr}(F)$, the ultrapower embedding π maps any β of cofinality larger than κ cofinally into its image, i.e. $\pi''\beta$ is cofinal in $\pi(\beta)$. If furthermore $\pi''\beta \subset \beta$, we then have $\pi(\beta) = \beta$. But the collection of such $\beta \in X$ is unbounded and closed under limits of sufficiently large cofinalities, as follows by the standard closure argument: Given β_0, we pick an increasing sequence β_ξ such that $\beta_\xi > \pi(\beta_{\bar{\xi}})$ whenever $\bar{\xi} < \xi$; then $\beta = \sup \beta_\xi \in X$ by the thickness of X and $\pi''\beta \subset \beta$. The same argument also shows the unboundedness of $X' \cap \beta^+$ in β^+; notice that $\beta^{+W'} = \pi(\beta^{+W}) = \beta^+$ whenever $\beta \in X'$ is a cardinal in V. Clause c) in the definition of thickness follows by the elementarity of π: If $\mathrm{cf}(\mathrm{cr}(E_{\nu'}^{W'})) = \beta$ for some $\beta \in X'$ and ν', then $\mathrm{cf}(\mathrm{cr}(E_\nu^W)) = \beta$ for some ν, contradicting the thickness of X in W.

We now verify d). If the iteration is of length less than ∞, then, by the proof of Lemma 6.3.1, any β which is regular in W and larger than the supremum of the critical points of the iteration is mapped by π cofinally into $\pi(\beta)$, so we can proceed as in c). If the iteration is of length ∞, we use the fact that the iteration is simple, that is, (6.1.1). Let W_i be the weasels of the iteration and π_{ij} the iteration maps. We amend the definition of β_ξ from above as follows: for each ξ pick i_ξ so that $\pi_{0,i_\xi}(\beta_\xi) < \kappa_{i_\xi}$ and set $\beta_{\xi+1}$ to be the least element of X strictly greater than $\pi_{0,i_\xi}(\beta_\xi)$; take limits at limit stages. Setting $\beta = \sup \beta_\xi$ and $i = \sup i_\xi$, we have $\pi_{0,i}''\beta \subset \beta$. Since $\kappa_i > \pi_{0,i}(\beta)$, it suffices to show that $\pi_{0,i}(\beta) = \beta$ to see that $\pi(\beta) = \beta$. Suppose this fails. Let j be the least counterexample, i.e. the least ordinal with $\pi_{0,j}(\beta) > \beta$. Then j must be a successor, say $j = h + 1$, as otherwise $\beta = \pi_{kj}(\bar{\beta})$ for some $\bar{\beta} < \beta$ (since $\pi_{kj}(\beta) > \beta$); but then for any β' which lies between $\bar{\beta}$ and β we would have $\pi_{0,j}(\beta') \geq \pi_{kj}(\beta') > \pi_{kj}(\bar{\beta}) = \beta$, contradicting the fact that $\pi_{0,i}''\beta \subset \beta$. So $W_j = \mathrm{Ult}(W_h, E_{\nu_h}^{W_h})$ and $\pi_{hj}(\beta) > \beta$. It follows that $\mathrm{cf}^{W_h}(\beta) = \kappa_h = \mathrm{cr}(E_{\nu_h}^{W_h})$; otherwise $\pi_{hj}''\beta \subset \beta$ would be cofinal in $\pi_{hj}(\beta)$ and, consequently, $\pi_{hj}(\beta) = \beta$. By the elementarity of $\pi_{0,h}$ and the minimality of j, we then have $\mathrm{cf}^W(\beta) = \mathrm{cr}(E_\nu^W)$ for some ν, contradicting the thickness of X in W. This verifies unboundedness of X'; its closedness under limits of sufficiently large cofinalities follows by a similar argument. The clauses b) and c) of the definition of thickness follow exactly as in the proof of c) above.

To see e), we prove the unboundedness of $\bar{X}$; the rest follows by similar arguments or arguments from the proof of c). Given an ordinal β_0 construct β_ξ and β in the same manner as in the proof of c). Then $\beta \in X$ by the thickness of X. In particular, $\beta \in \mathrm{rng}(\sigma)$. Thus there is a $\bar{\beta}$ such that $\sigma(\bar{\beta}) = \beta$, and $\bar{\beta} \geq \beta$ as $\sigma''\beta \subset \beta$. It follows that $\bar{\beta} = \beta$.

Finally we prove f). We know that $\boldsymbol{K}^c$ has a thick class X. Coiterate $\boldsymbol{K}^c$ with W. By the universality of W, both sides of the coiteration are simple and there is a common simple iterate W' of $\boldsymbol{K}^c$, W (as follows from Lemma 6.1.6). Let $\pi : \boldsymbol{K}^c \to W'$ be the iteration map on the $\boldsymbol{K}^c$-side. By c), the class $Y = \{\beta \in X;\ \pi(\beta) = \beta\}$ is thick in W'.

□(Lemma 7.2.4)

Proof of Theorem 7.2.1. By the remarks after the definition of incompressibility, it clearly suffices to show that if $J_\alpha^{E_1}$, $J_\alpha^{E_2}$ are incompressible, then $E_1 = E_2$. Let W_i, X_i

witness the incompressibility of $J_\alpha^{E_i}$ $(i = 1, 2)$. By universality, W_1 coiterates with W_2 simply to a common weasel W; let π_1, π_2 be the associated iteration maps and

$$X = \{\zeta \in X_1;\ \pi_1(\zeta) = \zeta\} \cap \{\zeta \in X_2;\ \pi_2(\zeta) = \zeta\}.$$

By Lemma 7.2.4, X is thick in W_1, W_2 and W. Let $Y = h_W(X)$ and $Y_i = h_{W_i}(X)$ for $i = 1, 2$. Notice that Y, Y_1, Y_2 are mutually isomorphic, as witnessed by the assignments

$$h_W(j, \prec\zeta_1, \dots, \zeta_\ell\succ) \mapsto h_{W_i}(j, \prec\zeta_1, \dots, \zeta_\ell\succ) \qquad (i = 1, 2).$$

If $\bar{W}$ is the transitive hull of X and $\sigma : \bar{W} \to W$ the associated uncollapsing map, these assignments allow us to define Σ_1-preserving maps $\sigma_i : \bar{W} \to W_i$ in the natural way, so that $\pi_i \circ \sigma_i = \sigma$ and $\mathrm{rng}(\sigma_i) = Y_i$. As X is thick in W_i, we have $\alpha \subset Y_i$, so $\sigma_i \restriction \alpha = \mathrm{id}$. It follows that

$$E_1 = E^{W_1} \restriction \alpha = E^{\bar{W}} \restriction \alpha = E^{W_2} \restriction \alpha = E_2. \qquad \square\text{(Lemma 7.2.1)}$$

7.3 The Existence of *K*

Giving an incompressible structure J_α^E amounts to producing a pair W, X witnessing its incompressibility. We shall present an inductive definition of a "sequence" of thick classes in $\boldsymbol{K}^c$, each of them witnessing incompressibility of a sufficiently tall structure (in some weasel embeddable into $\boldsymbol{K}^c$). The analysis of thick classes involves collapsing arguments, and measures generated by the associated uncollapsing embeddings also turn out to be useful when dealing with general properties of $\boldsymbol{K}$, so we shall list the basic fact about such measures at the very beginning.

Definition. Let W be a universal weasel. A measure U on W with critical point κ is *W-correct* iff

- setting $\nu = \kappa^{+W}$, either $\langle J_\nu^{E^W}, U\rangle$ is a premouse or $\langle J_\nu^{E^W}, E_\nu^W, U\rangle$ is an s-premouse;
- $\mathrm{Ult}(W, U)$ exists.

We stress that we *do not* assume that $U \in W$. The structure $\langle J_\nu^{E^W}, U\rangle$ (resp. $\langle J_\nu^{E^W}, E_\nu^W, U\rangle$) is frequently called a *beaver* (resp. *s-beaver*) for W.

Lemma 7.3.1. *Let W be a universal weasel and let U be a W-correct measure with critical point κ. Set $\nu = \kappa^{+W}$. Suppose one of the following holds:*

a) $W' = \mathrm{Ult}(W, E_\alpha^W)$ *for some* $\alpha \geq \nu$;

b) *If W' is a simple iterate of W beyond ν;*

c) *W' is a universal weasel such that $E^{W'} \restriction \nu = E^W \restriction \nu$.*

Then $\mathrm{Ult}(W', U)$ *exists. Consequently, if* $M = \langle J_\nu^{E^W}, U\rangle$ *is a premouse, then* U *is* W'*-correct. This also holds if* M *is an s-premouse and* $E^W, E^{W'}$ *agree up to* $\nu + 1$.

Proof. Let $\tilde{W} = \mathrm{Ult}(W, U)$ and $\sigma : W \to \tilde{W}$ be the associated ultrapower map; furthermore, let $\mu = \mathrm{cr}(E_\alpha^W)$ and $\sigma(\alpha, \mu) = (\alpha', \mu')$.

For a), let $\pi : W \to W'$ be the associated ultrapower map by E_α^W. Form $\tilde{W}' = \mathrm{Ult}(\tilde{W}, E_{\alpha'}^{W'})$ with the ultrapower map $\tilde{\pi}$ and define $\sigma' : W' \to \tilde{W}'$ to complete the diagram, i.e.

$$\sigma'(\pi(f)(\mu)) = \tilde{\pi} \circ \sigma(f)(\mu').$$

By coherency, the weasels W, W' agree up to α. We show that the same applies to the maps σ, σ', so that $\sigma' \restriction (W'||\alpha) = \sigma \restriction (W||\alpha)$. Clearly it suffices to show that the two maps agree on α. From the proof of Lemma 4.1.2 we know that any $\xi < \alpha$ in W' can be expressed in the form $\pi(f^*)(\mu)$ where $f : \mu \to \xi$ is an arbitrary surjection and $f^*(\eta) = \mathrm{otp}(f''\eta)$. Fixing such a surjection we have

$$\sigma'(\xi) = \sigma'(\pi(f^*)(\mu)) = \tilde{\pi}(\sigma(f)^*)(\mu') = \sigma(\xi),$$

where the last equality follows from the fact that $\sigma(f)$ is a surjection of μ' onto $\sigma(\xi)$. As $\alpha \geq \nu$, the maps σ, σ' agree on $W||\nu = W'||\nu$, so in particular $\sigma'(x) = \sigma(x)$ for every $x \in \mathcal{P}(\kappa) \cap W'$. Thus,

$$x \in U \longleftrightarrow \kappa \in \sigma'(x)$$

for every such x. This already guarantees that $\mathrm{Ult}(W', U)$ exists, since the standard computation shows that the term model $\mathbb{D}(W', U)$ can be embedded into $\tilde{W}'$ via $[f] \mapsto \sigma'(f)(\kappa)$.

To see b), let $\mathfrak{I}$ be the iteration from W to W' and $\tilde{\mathfrak{I}}$ the copy of $\mathfrak{I}$ via σ with the last weasel $\tilde{W}'$. Then the last copy map $\sigma' : W' \to \tilde{W}'$ agrees with σ up to ν. This can be shown by induction along the length of $\mathfrak{I}$ using a). We first observe that the agreement applies to the copy map $\sigma_1 : W_1 \to \tilde{W}_1$ after forming the initial ultrapower; however, the rest of $\mathfrak{I}$ is above ν, so the induction trivializes. Having the desired agreement, we can proceed as in the proof of a) and embed $\mathbb{D}(W', U)$ into $\tilde{W}'$.

For c), notice that the weasels coiterate simply beyond ν to a common weasel Q. By b), U is Q-correct; let $\sigma_Q : Q \to Q' = \mathrm{Ult}(Q, U)$ be the associated ultrapower embedding. Denote the iteration map from W' to Q by π'. As the iteration is beyond ν, and therefore above κ, the critical point of π' is at least κ. Then $\mathbb{D}(W', U)$ can be embedded into Q' by $[f] \mapsto \sigma_Q \circ \pi'(f)(\kappa)$, as can be verified following the standard computation: Given a formula $\varphi(v_1, \ldots, v_\ell)$ and functions $f_1, \ldots, f_\ell$ from ${}^\kappa W' \cap W'$,

$$\mathbb{D}(W', U) \models \varphi([f_1], \ldots, [f_\ell]) \longleftrightarrow u_{f_1,\ldots,f_\ell}^{\varphi,W'} \in U \longleftrightarrow \kappa \in \sigma_Q(u_{f_1,\ldots,f_\ell}^{\varphi,W'}).$$

But

$$u_{f_1,\ldots,f_\ell}^{\varphi,W'} = \{\eta < \kappa;\ Q \models \varphi(\pi'(f_1)(\eta), \ldots, \pi'(f_\ell)(\eta))\}$$

by the elementarity of π' and the fact that $\mathrm{cr}(\pi') \geq \kappa$, so the rightmost expression can be equivalently rewritten as

$$Q' \models \varphi(\sigma_Q \circ \pi'(f_1)(\kappa), \ldots, \sigma_Q \circ \pi'(f_\ell)(\kappa)) \qquad \square(\text{Lemma 7.3.1})$$

The next two lemmata show a close relationship between correct measures and thick classes. The first one establishes the uniqueness of correct measures, the second one guarantees that any thick class gives rise to a correct measure and vice versa, which enables us to replace quantifying over thick classes by quantifying over set-size objects, namely correct measures.

Lemma 7.3.2. *Let U, U' be normal measures with a common critical point κ. Suppose U is W-correct and U' is W'-correct, where W, W' are universal weasels which agree up to κ. Set $M = \langle J_\nu^E, U\rangle$ and $M' = \langle J_{\nu'}^{E'}, U'\rangle$, where E, E' are measure sequences of W, W' and ν, ν' are cardinal successors of κ in W, W'. Then $\nu = \nu'$ and, if both M, M' are either premice or s-premice, $U = U'$.*

Proof. As W, W' agree up to κ and are universal, they coiterate simply above κ to a common weasel Q. (Here we use the observation that κ is an inaccessible cardinal both in W and W', so no measure with index larger than κ can have critical point below κ. It follows that W, Q and W' contain the same subsets of κ, which implies $\nu = \nu'$. By the previous lemma, $\mathrm{Ult}(W, U')$ exists. Let W'' be $\boldsymbol{K}^c$ over J_ν^E. Then once again, W coiterates with W'' simply above κ to a common weasel Q^*, which has a thick class by Lemma 7.2.3 and Lemma 7.2.4, and which is extendible by both U, U'. So w.l.o.g. we can suppose that $Q^* = W$ and W has a thick class, say X.

Let $W_1 = \mathrm{Ult}(W, U)$, $W_2 = \mathrm{Ult}(W, U')$ and π_1, π_2 be the associated ultrapower maps. Then W_1, W_2 are universal weasels which agree up to κ (in fact, up to ν), so they coiterate simply above κ to a common weasel W^*; denote the associated iteration maps by π_1^*, π_2^*. Set

$$\bar{X} = \text{the class of all } \xi \in X \text{ satisfying } \pi_1^* \circ \pi_1(\xi) = \xi = \pi_2^* \circ \pi_2(\xi).$$

$\bar{X}$ is obviously thick in W, W_1, W_2 and W^*. Let $Y = h_W(\kappa \cup \bar{X})$, let $\bar{W}$ be the transitive collapse of Y and $\sigma : \bar{W} \to W$ the uncollapsing map. Then $\bar{W}$ is universal, as $\{\xi \in \bar{X};\ \sigma(\xi) = \xi\}$ is thick in $\bar{W}$. By construction, $\bar{W}$ agrees with W up to κ, so the argument from the previous paragraph yields $\mathcal{P}(\kappa) \cap W \subset \bar{W}$. Hence, any $x \subset \kappa$ from W is in $\mathrm{dom}(\sigma)$ and $\sigma(x) = h_W(\xi, \xi_1, \ldots, \xi_\ell)$ for some $\xi < \kappa$ and $\xi_1, \ldots, \xi_\ell \in \bar{X}$. It follows that

$$\pi_1^* \circ \pi_1(\sigma(x)) = h_{W^*}(\xi, \xi_1, \ldots, \xi_\ell) = \pi_2^* \circ \pi_2(\sigma(x)).$$

We claim that W_1, W_2 coiterate above ν. This is immediate if both M, M' are premice, as $E_\nu^{W_1} = \varnothing = E_\nu^{W_2}$ in this case. Now suppose M, M' are s-premice. If the coiteration were not above ν then $E_\nu^{W_1} \neq E_\nu^{W_2}$, where both $E_\nu^{W_1}, E_\nu^{W_2}$ are measures with critical point κ and $E_\nu^{W_i}$ is the first measure used on the i-th side. Note that W_1, W_2 and W contain the same subsets of κ. So given any such set x we have

$$\begin{aligned} x \in E_\nu^{W_1} \longleftrightarrow \kappa \in \pi_1^*(x) &\longleftrightarrow \kappa \in \pi_1^* \circ \pi_1(\sigma(x)) = \pi_2^* \circ \pi_2(\sigma(x)) \\ &\longleftrightarrow \kappa \in \pi_2^*(x) \longleftrightarrow x \in E_\nu^{W_2}. \end{aligned}$$

The second and third equivalence follows from the fact that $\pi_i(\sigma(x)) \cap \kappa = x$ for $i = 1, 2$ and from the equality we established in the previous paragraph. Consequently, $E_\nu^{W_1} = E_\nu^{W_2}$, a contradiction.

Now pick any $x \in \mathcal{P}(\kappa) \cap W$. Arguing as above, we obtain:

$$\begin{aligned} x \in U \longleftrightarrow \kappa \in \pi_1(x) &\longleftrightarrow \kappa \in \pi_1(\sigma(x)) \\ &\longleftrightarrow \kappa = \pi_1^*(\kappa) \in \pi_1^* \circ \pi_1(\sigma(x)) = \pi_2^* \circ \pi_2(\sigma(x)) \\ &\longleftrightarrow \kappa \in \pi_2(\sigma(x)) \longleftrightarrow \kappa \in \pi_2(x) \longleftrightarrow x \in U'. \end{aligned}$$

Thus, $U = U'$, which proves the lemma. □(Lemma 7.3.2)

Lemma 7.3.3. *Let W be a weasel and let X be a thick class in W. The following are equivalent for every ordinal α:*

a) *There is a thick class $X' \subset X$ such that $\alpha \notin h_W(\alpha \cup X')$.*

b) *There is a W-correct measure U with critical point α such that $U \neq E_\nu$, where $E = E^W$ and $\nu = \alpha^{+W}$. Moreover, $\langle J_\nu^E, U\rangle$ is a premouse if $E_\nu = \varnothing$ and $\langle J_\nu^E, E_\nu, U\rangle$ is an s-premouse if $E_\nu \neq \varnothing$.*

This lemma should be understood as follows: Given a definition of a thick class X, we produce the corresponding measure U and vice versa: Given a measure U as in b), we produce a definition of the thick class X'. Moreover, the whole procedure is uniform.

Proof. Suppose X' is as in a). Let $\bar{W}$ be the transitive collapse of $h_W(\alpha \cup X')$ and $\sigma : \bar{W} \to W$ the associated uncollapsing map. Then α is the least ordinal not in the range of σ, so $\alpha = \text{cr}(\sigma)$.

We show that $\mathcal{P}(\alpha) \cap \bar{W} = \mathcal{P}(\alpha) \cap W$, which will guarantee that the measure derived from σ is a measure on W. As rng(σ) contains the thick class X', the weasel $\bar{W}$ is universal. It follows that $\bar{W}$ coiterates with W simply to a common weasel, say Q. It suffices to show that they coiterate above α. By the fact that $\bar{W}$ agrees with W below α, this will follow immediately once we have shown that α is a limit cardinal both in W and $\bar{W}$. But if the latter were not the case, then the above coiteration would still be above the cardinal predecessor κ of α in W, since κ is a cardinal in both weasels by the elementarity of σ. Consequently, W, Q and $\bar{W}$ contain the same subsets of κ. In particular, W contains some $b \subset \kappa$ coding α, as $\alpha < \sigma(\alpha) = \kappa^{+W}$, and this b must also be an element of $\bar{W}$. However, this is impossible as α, being the critical point of σ, is regular in $\bar{W}$.

Now let $U \subset \mathcal{P}(\alpha)$ be the measure derived from σ, that is,

$$x \in U \longleftrightarrow \alpha \in \sigma(x).$$

By the previous paragraph, U is a measure on W. Also, $\alpha^{+\bar{W}} = \nu$, where $\nu = \alpha^{+W}$. The measure U is weakly amenable w.r.t. J_ν^E (here $E = E^W$), as given any $e \in J_\nu^E$ such that $e : \alpha \to J_\nu^E$, we have $\{\xi < \alpha;\ e(\xi) \in U\} = \{\xi < \alpha;\ \alpha \in \sigma(e)(\xi)\} \in J_\nu^E$. Lemma 5.5.2 guarantees that $\bar{W}$ agrees with W below ν. It is easy to verify that $\langle J_\nu^E, U\rangle$ satisfies all requirements on a premouse except possibly the coherency condition, which we verify now. As U is derived from σ, the ultrapower $\bar{W}' = \mathrm{Ult}(\bar{W}, U)$ exists and can be embedded into W via the standard map $k : \pi(f)(\kappa) \mapsto \sigma(f)(\kappa)$, where π is the associated ultrapower embedding. Now $k(\kappa) = \kappa$, so $\mathrm{cr}(k)$ is at least the cardinal successor $\bar{\nu}'$ of κ in $\bar{W}'$. Furthermore, $k(\bar{\nu}') = \nu$ and $\bar{\nu}' = \nu$; the inequality $\geq$ follows from the fact that $\mathcal{P}(\alpha) \cap W \subset \mathcal{P}(\alpha) \cap \bar{W} \subset \bar{W}'$. Thus, $\mathrm{cr}(k) > \nu$ and $\bar{W}$, $\bar{W}'$ and W agree below ν. It follows immediately that $\langle J_\nu^E, U\rangle$ is a premouse if $E_\nu^W = \varnothing$ and that $\langle J_\nu^E, E_\nu, U\rangle$ is an s-premouse otherwise. Consequently, $U \neq E_\nu^W$.

We now prove the converse, i.e. given a measure U as in b), we produce the definition of some thick $X' \subset X$. Let $W' = \mathrm{Ult}(W, U)$ and $\pi_U : W \to W'$ be the associated ultrapower map. Then W' is universal, so W coiterates with W simply to a common weasel, say Q. Denote the iteration maps from W and W' to Q by π and π', respectively. Setting

$$X' = \text{the class of all } \xi \in X \text{ satisfying } \pi' \circ \pi_U(\xi) = \xi = \pi(\xi),$$

X' is thick in W, W' and Q by Lemma 7.2.4. We show that $\alpha \notin h_W(\alpha \cup X')$, thus completing the proof.

Suppose for a contradiction that $\alpha \in h_W(\alpha \cup X')$. Let $\bar{W}$ be the transitive collapse of $h_W(\alpha \cup X')$ and $\sigma : \bar{W} \to W$ the associated uncollapsing map. Then $\alpha + 1 \subset \mathrm{rng}(\sigma)$, so $\mathrm{cr}(\sigma) > \alpha$. As in the proof of a) $\to$ b) we show that $\bar{W}$ and W contain the same subsets of α. These two facts imply that $\sigma(x) = x$ for every $x \in \mathcal{P}(\alpha) \cap W$, which means that $\mathcal{P}(\alpha) \cap W \subset h_W(\alpha \cup X')$. Given an $x \in \mathcal{P}(\alpha) \cap W$, pick ordinals $\eta_1, \dots, \eta_k < \alpha$ and $\xi_1, \dots, \xi_\ell \in X'$ such that $x = h_W(\eta_1, \dots, \eta_k, \xi_1, \dots, \xi_\ell)$; then

$$\pi' \circ \pi_U(x) = h_Q(\eta_1, \dots, \eta_k, \xi_1, \dots, \xi_\ell) = \pi(x).$$

That $\pi' \circ \pi_U(\eta) = \eta = \pi(\eta)$ for $\eta < \alpha$ is a consequence of the fact that the coiteration of W with W' is above α, which follows by the argument from the proof of the first part of the lemma. Setting $x = \alpha$, we see that $\pi(\alpha) > \alpha$, so that $\alpha = \mathrm{cr}(\sigma)$ and $E_\nu^W \neq \varnothing$ is the first measure used on the W-side of the coiteration. As a consequence we have that $E_\nu^{W'} = \varnothing$, as otherwise $E_\nu^{W'} = E_\nu^W$. It follows that the W'-side of the coiteration is above ν, which yields $\pi'(\alpha) = \alpha$. Now given any $x \in \mathcal{P}(\alpha) \cap W$ we have

$$x \in E_\nu^W \longleftrightarrow \alpha \in \pi(x) \longleftrightarrow \pi'(\alpha) \in \pi' \circ \pi_U(x) \longleftrightarrow \alpha \in \pi_U(x) \longleftrightarrow x \in U,$$

so $U = E_\nu^W$. Contradiction. □(Lemma 7.3.3)

We are now ready to begin with the construction of $\boldsymbol{K}$. Let $W = \boldsymbol{L}[E]$ be the ω_1-full $\boldsymbol{K}^c$. In the following we would like to define a class $X^* \subset \mathbf{On} \times \mathbf{On}$ coding a sequence of witnesses for incompressible structures. More precisely, to any $\bar{\alpha}$ we want to have an α such that $X_\alpha = \{\xi;\ \langle\alpha, \xi\rangle \in X^*\}$ is thick in W and $J^{\bar{E}}_{\bar{\alpha}}$ is incompressible, where $\bar{E}$ is the measure sequence of the transitive collapse of $h_W(X_\alpha)$. Intuitively, we shall define X_α by recursion as follows. At limit stages we form the intersection of all thick classes constructed so far. At successor stages $\alpha + 1$, where $\alpha \in h_W(X_\alpha)$, we check whether $\alpha \in h_W(Z)$ for all thick $Z \subset X_\alpha$. If this is *not* the case we let $X_{\alpha+1}$ be the class X' constructed as in the previous lemma. In all remaining cases we set $X_{\alpha+1} = X_\alpha$. Notice that this construction, provided that it does not break down, yields exactly what we need, as each $\alpha \cap h_W(X_\alpha)$ is an "uncollapsed" copy of an incompressible structure. There are, however, two issues to be cleared up. Firstly, we have to ensure that the above procedure really generates witnesses for arbitrarily long incompressible structures, in other words, that $\alpha \in X_{\alpha+1}$ for arbitrarily large α. Secondly, each X_α is a proper class, so the above recursion requires certain care if we want to carry it out in ZFC. A natural strategy would be to define a sequence representing all proper initial segments of X_α's, which will enable us to work with set-size objects, and use *this* sequence for the definition of X^* and $\boldsymbol{K}^*$. However, in order to determine even a proper initial segment of X_α, say $X_\alpha \cap \delta$, we have to refer to the respective initial segments $h_W(X_\beta) \cap J^E_\delta$ of the hulls of the *whole* X_β's for $\beta < \alpha$, as follows from the argument of the previous lemma. For this reason, it is convenient to generate auxiliary initial segments of the hulls (or sets containing enough information about such segments) in the course of the recursion together with the initial segments of X_α's.

Recall that W is the ω_1-full $\boldsymbol{K}^c$ and $E = E^W$. Let X be a thick class in W, say the one given by Lemma 7.2.3. We define matrices of sets Y^δ_α, X^δ_α for $\delta, \alpha \in \mathbf{On}$ as follows. For δ which is not a limit cardinal in the sense of V we leave Y^δ_α and X^δ_α undefined. Given a limit cardinal δ, we set

$$Y^\delta_0 = J^E_\delta, \qquad X^\delta_0 = X \cap \delta.$$

Having defined Y^δ_β, X^δ_β for $\beta \leq \alpha$, we consider two cases. If $\alpha \notin Y^\delta_\alpha$, we set

$$Y^\delta_{\alpha+1} = Y^\delta_\alpha, \qquad X^\delta_{\alpha+1} = X^\delta_\alpha.$$

If $\alpha \in Y^\delta_\alpha$, let $\bar{W}^\delta_\alpha$ be the transitive collapse of Y^δ_α and let $\sigma^\delta_\alpha : \bar{W}^\delta_\alpha \to J^E_\delta$ be the associated uncollapsing map. We define $X^\delta_{\alpha+1}$ and $Y^\delta_{\alpha+1}$ only in the case where $\bar{X}^\delta_\alpha = \{\xi \in X^\delta_\alpha;\ \sigma^\delta_\alpha(\xi) = \xi\}$ is of size δ. We let $\bar{\alpha}$ be the preimage of α under σ^δ_α and $\bar{\nu}$ be the cardinal successor of $\bar{\alpha}$ in $\bar{W}^\delta_\alpha$, and ask whether there is a unique measure U on $\bar{W}^\delta_\alpha$ with critical point $\bar{\alpha}$ such that setting $\bar{E}^{\alpha,\delta}$ to be the measure sequence of $\bar{W}^\delta_\alpha$,

- $U \neq E^{\delta,\alpha}_{\bar{\nu}}$;

- $\langle J^{E^{\delta,\alpha}}_{\bar{\nu}}, U\rangle$ is a premouse provided that $E^{\delta,\alpha}_{\bar{\nu}} = \varnothing$ and $\langle J^{E^{\delta,\alpha}}_{\bar{\nu}}, E^{\delta,\alpha}_{\bar{\nu}}, U\rangle$ is an s-premouse otherwise;

- $\tilde{W}^{\delta}_{\alpha} = \mathrm{Ult}(\bar{W}^{\delta}_{\alpha}, U)$ exists and $\bar{W}^{\delta}_{\alpha}$ coiterates with $\tilde{W}^{\delta}_{\alpha}$ simply to a common mouse.

Granted that U as above exists, let $\pi^{\delta}_{\alpha} : \bar{W}^{\delta}_{\alpha} \to \tilde{W}^{\delta}_{\alpha}$ be the associated ultrapower map by U, and let $\bar{\pi}^{\delta}_{\alpha}$ and $\tilde{\pi}^{\delta}_{\alpha}$ be the iteration map on the $\bar{W}^{\delta}_{\alpha}$-side and the $\tilde{W}^{\delta}_{\alpha}$-side of the coiteration, respectively. We set

$$X^{\delta}_{\alpha+1} = \{\xi \in \bar{X}^{\delta}_{\alpha};\ \tilde{\pi}^{\delta}_{\alpha} \circ \pi^{\delta}_{\alpha}(\xi) = \xi = \bar{\pi}^{\delta}_{\alpha}\},$$

provided that the set on the right side is of size δ, and define

$$\begin{aligned} \bar{Y}^{\delta}_{\alpha+1} &= \{x \in \bar{W}^{\delta}_{\alpha};\ \bar{\pi}^{\delta}_{\alpha}(x) \in \mathrm{rng}(\tilde{\pi}^{\delta}_{\alpha} \circ \pi^{\delta}_{\alpha})\} \\ Y^{\delta}_{\alpha+1} &= (\sigma^{\delta}_{\alpha})''\bar{Y}^{\delta}_{\alpha}. \end{aligned}$$

Otherwise we leave $X^{\delta}_{\alpha+1}$ and $Y^{\delta}_{\alpha+1}$ undefined. This completes the construction of $X^{\delta}_{\alpha+1}$ and $Y^{\delta}_{\alpha+1}$.

For limit α we set

$$\begin{aligned} X^{\delta}_{\alpha} &= \bigcap_{\beta<\alpha} X^{\delta}_{\beta} \\ Y^{\delta}_{\alpha} &= \bigcap_{\beta<\alpha} Y^{\delta}_{\alpha} \end{aligned}$$

provided that these intersections are of size δ and leave X^{δ}_{α} and Y^{δ}_{α} undefined otherwise.

Given an ordinal α, set

$$D_{\alpha} = \text{ the class of all } \delta \text{ such that } X^{\delta}_{\alpha} \text{ is defined.}$$

We will show below that for a fixed α, the sequences Y^{δ}_{α}, X^{δ}_{α} stabilize in the sense that Y^{γ}_{α}, X^{γ}_{α} are proper segments of Y^{δ}_{α}, X^{δ}_{α} for sufficiently large $\gamma < \delta$, say $\gamma, \delta > \vartheta(\alpha)$. This fact enables us to define X^*, Y^* and the associated projections along the α-th coordinate X_{α}, Y_{α} in the obvious way:

$$\begin{aligned} X_{\alpha} &= \bigcup\{X^{\delta}_{\alpha};\ \delta \in D_{\alpha} - \vartheta(\alpha)\} \\ Y_{\alpha} &= \bigcup\{Y^{\delta}_{\alpha};\ \delta \in D_{\alpha} - \vartheta(\alpha)\} \\ X^* &= \text{ the class of all } \langle \alpha, \xi \rangle \text{ such that } \xi \in X^{\delta}_{\alpha} \text{ for some } \delta > \vartheta(\alpha) \\ Y^* &= \text{ the class of all } \langle \alpha, x \rangle \text{ such that } x \in Y^{\delta}_{\alpha} \text{ for some } \delta > \vartheta(\alpha). \end{aligned}$$

Given a fixed $\delta \in D_{\alpha} - \vartheta(\alpha)$, the sets Y^{δ}_{α} and X^{δ}_{α} constitute a decreasing sequence with respect to the inclusion, so the same applies to Y_{α} and X_{α}.

The following three lemmata guarantee that the above recursion yields objects with the desired properties. The first of them is the means which keeps the recursion going. The second lemma makes it clear that the classes X_{α} and Y_{α} collapse to a witness for the incompressible structure of height $\mathrm{otp}(\alpha \cap Y_{\alpha})$, and the third lemma ensures that the whole construction does not trivialize, in other words, that it produces witnesses for incompressible structures of arbitrary height.

Lemma 7.3.4. *Given W and X as above, the following holds for the objects constructed in the above-described recursion.*

a) *D_α is a c.u.b. class on a tail-end for every ordinal α.*

b) *$Y_\alpha^\delta \prec_{\Sigma_1} J_\delta^E$ and $X_\alpha^\delta \subset Y_\alpha^\delta$ whenever $\delta \in D_\alpha$.*

c) *There is a $\vartheta = \vartheta(\alpha)$ such that $Y_\alpha^\gamma = J_\gamma^E \cap Y_\alpha^\delta$ and $X_\alpha^\gamma = X_\alpha^\delta \cap \gamma$ whenever $\gamma, \delta \in D_\alpha$ and $\vartheta < \gamma < \delta$.*

d) *$Y_\alpha \prec_{\Sigma_1} W$ and X_α is thick in W.*

e) *If $\alpha \in Y_\alpha$ and there is a thick class Z in W such that $\alpha \notin h_W(Z)$, then, letting*

- *$\bar{W}_\alpha$ be the transitive collapse of Y_α and $\sigma_\alpha : \bar{W}_\alpha \to W$ be the corresponding uncollapsing map;*
- *$\bar{\alpha} = (\sigma_\alpha)^{-1}(\alpha)$, $\bar{\nu} = \bar{\alpha}^{+\bar{W}_\alpha}$, $\bar{X}_\alpha = \{\xi \in X_\alpha;\ \xi = \sigma_\alpha(\xi)\}$ and $\bar{E} = E^{\bar{W}_\alpha}$;*
- *$\tilde{W}_\alpha = \mathrm{Ult}(\bar{W}_\alpha, U)$ and π_α be the associated ultrapower map for the unique $\bar{W}_\alpha$-correct measure $U \neq E_{\bar{\nu}}^{\bar{W}_\alpha}$ with critical point $\bar{\alpha}$ such that $\langle J_{\bar{\nu}}^{\bar{E}}, U\rangle$ is a premouse if $\bar{E}_{\bar{\nu}} = \varnothing$ and $\langle J_{\bar{\nu}}^{\bar{E}}, \bar{E}_\nu, U\rangle$ is an s-premouse otherwise;*
- *$\bar{\pi}_\alpha$ and $\tilde{\pi}_\alpha$ be the iteration maps of the coiteration of $\bar{W}_\alpha$ with $\tilde{W}_\alpha$ (they coiterate simply to a common weasel),*

we have

$$X_{\alpha+1} = \{\xi \in \bar{X}_\alpha;\ \tilde{\pi}_\alpha \circ \pi_\alpha(\xi) = \xi = \bar{\pi}_\alpha(\xi)\}$$
$$Y_{\alpha+1} = \sigma_\alpha'' \bar{Y}_\alpha,$$

where

$$\bar{Y}_\alpha = \{x \in \bar{W}_\alpha;\ \bar{\pi}_\alpha(x) \in \mathrm{rng}(\tilde{\pi}_\alpha \circ \pi_\alpha)\}.$$

In all other cases,

$$X_{\alpha+1} = X_\alpha \quad \text{and} \quad Y_{\alpha+1} = Y_\alpha.$$

f) *For limit α we have*

$$X_\alpha = \bigcap_{\beta<\alpha} X_\beta \quad \text{and} \quad Y_\alpha = \bigcap_{\beta<\alpha} Y_\beta.$$

Lemma 7.3.5. *For any ordinal α and arbitrary thick class $Z \subset X_\alpha$ we have $Y_\alpha \cap \alpha = h_W(Z) \cap \alpha$. Consequently, $\bar{W}_\alpha$, $\bar{X}_\alpha$ from* e) *and* f) *of Lemma* 7.3.4 *witness the incompressibility of $J_{\bar{\alpha}}^{\bar{E}}$.*

It follows from the above construction that the classes Y_α and X_α constitute decreasing sequences. By Lemma 7.3.5, the initial segments $Y_\beta \cap \alpha$ remain the same for every $\beta \geq \alpha$.

Lemma 7.3.6. *There are unboundedly many ordinals α such that $\alpha \in Y_{\alpha+1}$.*

Deferring the proofs of these lemmata, we shall focus on the definition of K^*. By Lemma 7.3.5 and the fact that Y_α^δ is a proper segment of Y_α for $\alpha > \vartheta(\alpha)$, being an incompressible structure is a property expressible in the language of ZFC:

$$J_{\bar{\alpha}}^{\bar{E}} \text{ is incompressible} \longleftrightarrow (\exists \alpha, \delta)(\bullet\ Y_\alpha^\delta \text{ is defined and } \delta > \vartheta(\alpha),\ \bullet\ \mathrm{otp}(Y_\alpha^\delta \cap \alpha) = \bar{\alpha} \text{ and }\ \bullet\ J_{\bar{\alpha}}^{\bar{E}} \text{ is a proper segment of the transitive collapse of } Y_\alpha^\delta)$$

We stress that the above formula contains no parameters, as the recursive definition of X_α^δ and Y_α^δ refers only to the definitions W and X, which are *lightface* definable in V, and as $\vartheta(\alpha)$ is definable from the sequences Y_α^δ and X_α^δ. It follows immediately that K^* is *lightface* definable in V.

The next lemmata sum up some basic properties of K^*. In particular, we obtain that $K^* = K$, i.e. the inductive definition yields the same model as Steel's definition and that the measure sequence of K is *maximal* in the sense that it absorbs all measures which are possible candidates for elements of the E^K-sequence.

Lemma 7.3.7. *K^* is θ-full for any regular uncountable θ. Consequently, K^* is universal.*

Proof. The latter follows from the former by Theorem 6.4.4. Suppose U is an ω-complete measure on K^* with critical point κ of cofinality θ and $\langle J_\nu^{E^*}, U\rangle$ is a premouse, where E^* is the measure sequence of K^* and ν is the cardinal successor of κ in K^*. Pick a pair W, X witnessing the incompressibility of $J_{\nu+1}^{E^*}$. Then U is easily seen to be W-correct. If $U \neq E_\nu^W$, there would be a thick class $Z \subset X$ such that $\kappa \notin h_W(Z)$, contradicting the incompressibility of $J_{\nu+1}^{E^*}$. Thus, $U = E_\nu^W = E_\nu^*$. □(Lemma 7.3.7)

Having seen that K^* is universal, an elaboration on the above argument shows that K^* absorbs *all* correct measures and not only those which are ω-complete.

Lemma 7.3.8. *Let U be K^*-correct. Then $U = E_\nu^*$ where ν, κ and E^* are as in the previous lemma.*

Proof. Let W, X witness the incompressibility of $J_{\nu+1}^{E^*}$. Then W agrees with K^* below $\nu+1$, so U is W-correct by Lemma 7.3.1. As in the proof of the previous lemma we then infer that $U = E_\nu^W = E_\nu^*$. □(Lemma 7.3.8)

Lemma 7.3.9. $K^* = K$.

Proof. We show that the measure sequence E^* of K^* satisfies the conditions from the inductive definition of E^K. This will prove the lemma, as the above conditions determine E^K uniquely.

Fix an ordinal ν. If $\langle J_\nu^{E^*}, U\rangle$ is a strong mouse for some measure U, then this is witnessed by some universal weasel W which agrees with $\boldsymbol{K}^*$ below $\nu = \omega\nu$ and such that $U = E_\nu^W$. Then U is $\boldsymbol{K}^*$-correct by Lemma 7.3.1, so $U = E_\nu^*$.

If $E_{\omega\nu}^* = \varnothing$, the above observation implies that $\langle J_\nu^{E^*}, U\rangle$ is not strong for any U. If $E_{\omega\nu}^* \neq \varnothing$, it implies that $E_{\omega\nu}^*$ is the unique measure U such that $\langle J_\nu^{E^*}, U\rangle$ is strong.

□(Lemma 7.3.9)

Finally we give the proofs of Lemmata 7.3.4 – 7.3.6 and thus complete the proof of the existence of $\boldsymbol{K}^*$.

Proof of Lemma 7.3.4. We proceed by induction on α. The case $\alpha = 0$ is trivial except perhaps for the first part of d). Notice that it suffices to verify that J_τ^E is a Σ_1 substructure of W whenever τ is a cardinal in W. This follows by an easy condensation argument. Given a Σ_1-formula $\varphi(v) \longleftrightarrow (\exists z)\psi(z, v)$ where ψ is Σ_0 and an $a \in J_\tau^E$, pick x witnessing that $\varphi(a)$ holds in W, i.e. $W \models \psi(x, a)$. Let $\xi < \tau$ and ζ be such that $a \in J_\xi^E$ and $x \in J_\zeta^E$. If M is the transitive collapse of the Σ_1-hull of $\xi \cup \{x\}$ in J_ζ^E, then $M \models \psi(\bar{x}, a)$ where $\bar{x}$ is the image of x under the collapsing map. Moreover, $M = h_M(\xi \cup \{\bar{x}\})$, which means that M can be encoded in a subset of ξ. Applying acceptability, this subset is an element of J_τ^E and can be decoded there, so $M \in J_\tau^E$, which implies $\bar{x} \in J_\tau^E$ and $J_\tau^E \models \psi(\bar{x}, a)$.

We next verify the induction step at successor ordinals. Granting that the lemma holds for α, we verify a) – f) for $\alpha + 1$. This is trivial if $\alpha \notin Y_\alpha^\delta$ for some (all) $\delta > \vartheta(\alpha)$ and, in fact, $D_{\alpha+1}$ agrees with D_α above $\vartheta(\alpha)$. Now suppose that $\alpha \in Y_\alpha$. For $\delta \in D_\alpha$, $\delta > \vartheta(\alpha)$, we have that $Y_\alpha^\delta, X_\alpha^\delta$ are proper segments of Y_α, X_α, so $\bar{W}_\alpha^\delta, \bar{X}_\alpha^\delta$ are proper segments of $\bar{W}_\alpha, \bar{X}_\alpha$ and $\sigma_\alpha^\delta = \sigma_\alpha \restriction \bar{W}_\alpha^\delta$. If Z is a thick class in W satisfying $\alpha \notin h_W(Z)$ then the same applies to $Z \cap \bar{X}_\alpha$, so we can assume w.l.o.g. that $Z \subset \bar{X}_\alpha$. Then Z is thick both in W and W_α and $\bar{\alpha} \notin h_{W_\alpha}(Z)$, so the existence of such a Z is equivalent to the existence of a $\bar{W}_\alpha$-correct measure with critical point $\bar{\alpha}$ by Lemma 7.3.3.

Let us make an observation which will be useful in what follows. If $R = \boldsymbol{L}[F]$ is a universal weasel and δ a cardinal in R, then for any measure U on R, the ultrapower $\tilde{R}^\delta = \mathrm{Ult}(J_\delta^F, U)$ is an initial segment of $\tilde{R} = \mathrm{Ult}(R, U)$ of height $\pi''\delta$ (provided that these ultrapowers exist) and $\pi^\delta = \pi \restriction J_\delta^F$, where π and π^δ are the associated ultrapower maps. This follows immediately from the fact that δ is a cardinal in R, so that any function from R bounded in δ, which is used in the ultrapower construction is actually in J_δ^F. Similarly, if $\tilde{R}$ is a normal iterate of R with iteration indices ν_i and $\tilde{\pi}$ is the associated iteration map, then the iterate of J_δ^F using the indices ν_i is the initial segment of $\tilde{R}$ of height $\tilde{\pi}''\delta$. In particular, if δ is a limit cardinal in R and the i-th critical point κ_i is less than $\tilde{\pi}_{0,i}''\delta$, then $\nu_i < \tilde{\pi}_{0,i}''\delta$. Consequently, if δ_h is a limit cardinal in $R_h = \boldsymbol{L}[F_h]$ for $h = 1, 2$ then the coiteration of $J_{\delta_1}^{F_1}$ with $J_{\delta_2}^{F_2}$ is an initial segment of the coiteration of R_1 with R_2.

Now suppose that there is a $\bar{W}_\alpha$-correct measure U with critical point $\bar{\alpha}$. Such a measure is uniquely determined in the sense of Lemma 7.3.2, $\langle J_{\bar{\nu}}^{\bar{E}}, U\rangle$ is a premouse if

$\bar{E}_{\bar{\nu}} = \varnothing$ and $\langle J_{\bar{\nu}}^{\bar{E}}, \bar{E}_{\bar{\nu}}, U\rangle$ is an s-premouse otherwise. The class

$$\begin{aligned} D'_\alpha = {} & \text{the class of all } \delta \in D_\alpha - \vartheta(\alpha) \text{ satisfying} \\ & \bullet\ \operatorname{card}(\{\bar{X}_\alpha \cap \delta);\ \tilde{\pi}_\alpha \circ \pi_\alpha(\xi) = \xi = \bar{\pi}_\alpha(\xi)\}) = \delta, \\ & \bullet\ \sup[(\tilde{\pi}_\alpha \circ \pi_\alpha)''\delta] = \sup(\bar{\pi}''\delta) \end{aligned}$$

is closed and unbounded by the standard closure argument and, as δ is a limit cardinal both in $\bar{W}_\alpha$ and $\tilde{W}_\alpha = \operatorname{Ult}(\bar{W}_\alpha, U)$, the premice $\bar{W}_\alpha^\delta = J_\delta^{\bar{E}}$ and $\tilde{W}_\alpha^\delta = J_\delta^{\tilde{E}}$ coiterate simply to $J_{\delta'}^{E'}$ where $Q_\alpha = L[E']$ is the common coiterate of $\bar{W}_\alpha$ and $\tilde{W}_\alpha$ and

$$\sup[(\tilde{\pi}_\alpha \circ \pi_\alpha)''\delta] = \delta' = \sup(\bar{\pi}''\delta).$$

Furthermore, $\pi_\alpha^\delta = \pi_\alpha \restriction \bar{W}_\alpha$, $\tilde{\pi}_\alpha^\delta = \tilde{\pi}_\alpha \restriction \tilde{W}_\alpha$ and $\bar{\pi}_\alpha^\delta = \bar{\pi}_\alpha \restriction \bar{W}_\alpha$. Applying the above observation, D'_α is the class of all δ's from $D_\alpha - \vartheta(\alpha)$ such that U has all those properties with respect to Y_α^δ and X_α^δ which are required in the definition of $Y_{\alpha+1}^\delta$ and $X_{\alpha+1}^\delta$.

We prove that, on a tail-end of ordinals, U is the unique such measure for $\delta \in D'_\alpha$ and no such measure exists for the rest of δ's. For, if $U_\delta \neq U$ were another such measure, then $U_\delta = U' \neq U$ for a proper class of $\delta \in D_\alpha - \vartheta(\alpha)$, where U' is fixed. As each $\bar{W}_\alpha^\delta$ is an initial segment of $\bar{W}_\alpha$, this would mean that $U' \neq U$ is $\bar{W}_\alpha$-correct, hence $U' = U$ by Lemma 7.3.2, a contradiction. For $\delta \in D'_\alpha$, the common fixpoints of σ_α^δ, $\tilde{\pi}_\alpha^\delta \circ \pi_\alpha^\delta$ and $\bar{\pi}_\alpha^\delta$ constitute a set of size δ, so $Y_{\alpha+1}^\delta$ and $X_{\alpha+1}^\delta$ are defined. Hence D'_α is a tail-end of $D_{\alpha+1}$, which proves a) of the lemma. Furthermore, $X_{\alpha+1}^\delta = X_{\alpha+1} \cap \delta$ and $Y_{\alpha+1}^\delta = Y_{\alpha+1} \cap J_\delta^E$, so that c) – e) of the lemma hold. To see b), notice that $Y_{\alpha+1}^\delta$ is the intersection of the ranges of $\tilde{\pi}_\alpha^\delta \circ \pi_\alpha^\delta$ and $\bar{\pi}_\alpha^\delta$, which are both Σ_1-substructures of J_δ^E.

In the case where there is no $\bar{W}_\alpha$-correct measure U with critical point $\bar{\alpha}$, the above argument shows that, on a tail-end of ordinals δ, no U_δ satisfying the requirements from the recursion step can exist. It follows that $D_{\alpha+1}$ agrees with D_α on that tail-end, $X_{\alpha+1}^\delta = X_\alpha^\delta$ and $Y_{\alpha+1}^\delta = Y_\alpha^\delta$. This completes the proof at successor steps.

Finally, if α is a limit and the lemma holds for all $\beta < \alpha$, let ϑ be a common upper bound for all $\vartheta(\beta)$. Given $\gamma < \delta$ from the intersection $\bar{D}_\alpha$ of all classes $D_\beta - \vartheta$, each X_β^γ, Y_β^γ is an initial segment of X_β^δ, Y_β^δ, so the same also applies to the corresponding intersections X_α^δ, Y_α^δ and X_α^δ, Y_α^δ. Then $\bar{D}_\alpha$ is a tail-end of D_α and the verification of a) – d) and f) is a routine. □(Lemma 7.3.4)

Proof of Lemma 7.3.5. As Y_α is a Σ_1-substructure of W, the hull $h_W(Z)$ is contained in Y_α whenever $Z \subset X_\alpha$. Thus, it suffices to show that $Y_\alpha \cap \alpha \subset h_W(Z)$ for any thick $Z \subset X_\alpha$. Suppose this fails and α is the least with this property. Then α is a successor ordinal, say $\alpha = \beta+1$, as for limit α we have the inclusions $Y_\alpha \cap \bar{\alpha} \subset Y_{\bar{\alpha}} \cap \bar{\alpha} \subset h_W(Z)$ for $\bar{\alpha} < \alpha$ by the minimality of α and monotonicity of the Y_α-sequence. So $Y_\beta \cap \beta \subset h_W(Z)$ and $\beta \in Y_{\beta+1} - h_W(Z)$.

Following the notation from Lemma 7.3.4, the class $\bar{Z} = \{\xi \in Z;\ \sigma_\beta(\xi) = \xi\}$ is thick both in W and W_β and $\sigma_\beta^{-1}(\beta) = \bar{\beta} \notin h_{\bar{W}_\beta}[(\sigma_\beta^{-1})''Z] \supset h_{\bar{W}_\beta}(\beta \cup \bar{Z})$; the inclusion here follows again by the minimality of α. By Lemma 7.3.2, there is a $\bar{W}_\beta$-correct measure $U \neq \bar{E}_{\bar{\nu}}$ (here $\bar{E}$ is the measure sequence of $\bar{W}_\beta$ and $\bar{\nu}$ the cardinal successor of $\bar{\beta}$ in $\bar{W}_\beta$) such that $\bar{E}_{\bar{\nu}} = \tilde{E}_{\bar{\nu}}$, where $\tilde{W}_\beta = \mathrm{Ult}(\bar{W}_\beta, U)$. So $\bar{W}_\beta$ coiterates with $\tilde{W}_\beta$ simply beyond $\bar{\nu}$, i.e. above $\bar{\beta}$, to a common weasel. It follows that $\bar{\pi}_\beta(\bar{\beta}) = \bar{\beta}$ and $\mathrm{cr}(\tilde{\pi}_\beta \circ \pi_\beta) = \bar{\beta}$, so that $\bar{\pi}_\beta(\bar{\beta}) \notin \mathrm{rng}(\tilde{\pi}_\beta \circ \pi_\beta)$. Consequently, $\beta \notin Y_{\beta+1}$, a contradiction. □(Lemma 7.3.5)

Proof of Lemma 7.3.6. Consider the class

$$D = \text{the class of all } \omega\text{-closed cardinals } \alpha \text{ satisfying } \mathrm{cf}(\alpha) = \omega_1.$$

D is a proper class closed under limits of ω_1-sequences.

Claim. Given an $\alpha \in D$ and a thick class Z in W, we have $\alpha \in h_W(\alpha \cup Z)$.

Proof of Claim. Suppose the contrary; let α and Z witness this. By Lemma 7.3.3, there is a W-correct measure $U \neq E_\nu$ with critical point α, where $\nu = \alpha^{+W}$. As α is ω-closed, the weak covering lemma yields $\alpha^{+W} = \alpha^+$, so $\mathrm{cf}(\alpha^{+W}) > \omega$. This guarantees that U is ω-complete: Given a sequence $\langle x_i;\ i \in \omega\rangle$ of sets from U, this sequence is bounded in α^{+W}, so pick a $\nu^* < \nu$ such that $x_i \in J^E_{\nu^*}$ for all i. Then $U \cap J^E_{\nu^*} \in W$, as $\langle J^E_\nu, U\rangle$ is amenable. Let $f : \alpha \to U \cap J^E_{\nu^*}$ enumerate $U \cap J^E_{\nu^*}$ in W. The diagonal intersection x of f is in U and is almost contained in any x_i, i.e. there is a $\beta_i < \alpha$ such that $x - \beta_i \subset x_i$. As α is not ω-cofinal, $\beta = \sup \beta_i < \alpha$, so $\varnothing \neq x - \beta \subset x_i$ for every $i \in \omega$.

Now $E_\nu = \varnothing$, otherwise $M = \langle J^E_\nu, E_\nu, U\rangle$ would be an s-premouse by Lemma 7.3.3, and, as all total measures in M are ω-complete, M would be iterable, thus witnessing the existence of $0^\ddagger$. Consequently, $\langle J^E_\nu, U\rangle$ is a premouse. As U is ω-complete, $\mathrm{cf}(\alpha) = \omega_1$ and W is ω_1-full, we have $U = E_\nu$, a contradiction. □(Claim)

Suppose for a contradiction that there is only a set of ordinals α satisfying $\alpha \in Y_{\alpha+1}$; let ϑ be the least upper bound for such α's (in the sense that $\vartheta \notin Y_{\vartheta+1}$). Using the remark after Lemma 7.3.5, we conclude that $Y_{\beta+1} \cap (\beta+1) = Y_{\beta+1} \cap \vartheta = Y_{\vartheta+1} \cap \vartheta$ for every $\beta \geq \vartheta$, as otherwise there would be a $\beta' \geq \vartheta$ such that $\beta' \in Y_{\beta+1} \cap (\beta+1)$, hence $\beta' \in Y_{\beta'+1}$ by the monotonicity of the sequence $\langle Y_\alpha\rangle_\alpha$. For $\beta \geq \vartheta$ let

$$\theta_\beta = \min(Y_{\beta+1} - \beta) = \min(Y_{\beta+1} - \vartheta).$$

Clearly, $\theta_\beta > \beta$. Pick any $\gamma \in D, \gamma > \vartheta$ such that $\theta_\beta < \gamma$ whenever $\beta < \gamma$. By the usual closure argument, there is a proper class of such γ's. Applying the above claim, fix $\xi_1, \ldots, \xi_k < \gamma$ and $\zeta_1, \ldots, \zeta_\ell \in X_{\gamma+1}$ such that $\gamma = h_W(\xi_1, \ldots, \xi_k, \zeta_1, \ldots, \zeta_\ell)$. Fix a $\beta < \gamma$ strictly larger than ϑ and all of $\xi_1, \ldots, \xi_k$. Then

$$(\exists \xi_1, \ldots, \xi_k < \theta_\beta)(\theta_\beta < h_W(\xi_1, \ldots, \xi_k, \zeta_1, \ldots, \zeta_\ell) < \theta_\gamma).$$

As $Y_{\gamma+1} \subset Y_{\beta+1}$ and $Y_{\beta+1}$ is a Σ_1-substructure of W, there are $\xi_1, \ldots, \xi_k \in Y_{\beta+1} \cap \theta_\beta$ witnessing the above statement. Such ξ_i are in $Y_{\beta+1} \cap \vartheta = Y_{\vartheta+1} \cap \vartheta = Y_{\gamma+1} \cap \vartheta$ by the discussion at the beginning of this paragraph. But then we also have

$$h_W(\xi_1, \ldots, \xi_k, \zeta_1, \ldots, \zeta_\ell) \in Y_{\gamma+1},$$

so $Y_{\gamma+1} \cap (\theta_\beta, \theta_\gamma) \neq \varnothing$ after all. Contradiction. □(Lemma 7.3.6)

7.4 Embeddings of K and Generic Absoluteness

The main result of this section is the *Rigidity Lemma* asserting that there are no nontrivial embeddings of $\boldsymbol{K}$ into itself. This is a generalization of the well-known fact that, under $\neg 0^{\#}$, there are no nontrivial embeddings of $\boldsymbol{L}$, as $\boldsymbol{K} = \boldsymbol{L}$ in this case. If $0^{\#}$ exists, $\boldsymbol{L}$ can embed, of course, only to $\boldsymbol{L}$. The same is, however, not literally true for $\boldsymbol{K}$ below $0^{\ddagger}$. Nevertheless, we can show that embeddings of $\boldsymbol{K}$ are nothing more than iteration maps. Throughout this section, E stands for the measure sequence of $\boldsymbol{K}$.

Lemma 7.4.1. *Let $\sigma : \boldsymbol{K} \to W$ be a nontrivial embedding with critical point κ. Then W is universal and, setting $\nu = \kappa^{+\boldsymbol{K}} = \kappa^{+W}$, we have*

$$E_\nu = \{x \in \mathcal{P}(\kappa) \cap \boldsymbol{K};\ \kappa \in \sigma(x)\}.$$

This means that, letting $\boldsymbol{K}' = \mathrm{Ult}(\boldsymbol{K}, E_\nu)$ and π be the associated ultrapower map, we can define an embedding $\sigma' : \boldsymbol{K}' \to W$ by $\sigma'(\pi(f)(\kappa)) = \sigma(f)(\kappa)$ for $f \in {}^{\kappa}\boldsymbol{K}$. It follows that $\kappa' = \mathrm{cr}(\sigma') > \nu = \kappa^{+\boldsymbol{K}'}$.

Proof. W is weakly universal by Lemma 6.2.8 and therefore universal, as follows from Lemma 6.6.6. It follows that $\boldsymbol{K}$ and W coiterate simply above κ to a common weasel (for this argument see e.g. the proof of Lemma 7.3.3), so $\kappa^{+\boldsymbol{K}} = \kappa^{+W}$. Denote this value by ν. Let

$$U = \{x \in \mathcal{P}(\kappa) \cap \boldsymbol{K};\ \kappa \in \sigma(x)\},$$

$\boldsymbol{K}' = \mathrm{Ult}(\boldsymbol{K}, U)$ and let $\sigma' : \boldsymbol{K}' \to W$ be defined as above. From the existence of the ultrapower embedding $\pi_U : \boldsymbol{K} \to \boldsymbol{K}'$ we infer exactly as above that $\nu = \kappa^{+\boldsymbol{K}'}$. A standard computation yields that $\sigma' \restriction (\kappa + 1) = \mathrm{id}$, so $\mathrm{cr}(\sigma) \geq \nu$ and in fact the strict inequality holds, since ν is the common successor of κ in $\boldsymbol{K}'$ and W. The argument from the proof of Lemma 7.3.3 shows that the models $\boldsymbol{K}$, $\boldsymbol{K}'$ and W agree below ν and that either $\langle J_\nu^E, U\rangle$ is a premouse or $\langle J_\nu^E, E_\nu^{\boldsymbol{K}'}, U\rangle$ is an s-premouse. If the former holds, then U is $\boldsymbol{K}$-correct, so $U = E_\nu$ by Lemma 7.3.8. Now suppose the latter holds. As $\boldsymbol{K}$ and $\boldsymbol{K}'$ coiterate simply above κ to a common weasel, $E^{\boldsymbol{K}'}$ is $\boldsymbol{K}$-correct by Lemma 7.3.1, so $E_\nu^{\boldsymbol{K}'} = E_\nu$ and, consequently, U is $\boldsymbol{K}$-correct. As before it follows that $U = E_\nu$. (Notice that this shows that the latter case in fact cannot appear.) □(Lemma 7.4.1)

Theorem 7.4.2 (Rigidity Lemma). *There are no nontrivial elementary embeddings*

$$\sigma : \boldsymbol{K} \to \boldsymbol{K}.$$

Proof. Suppose the contrary; let $\sigma : \boldsymbol{K} \to \boldsymbol{K}$ be an elementary embedding with critical point κ. By the previous lemma, there is an embedding $\sigma' : \boldsymbol{K}' \to \boldsymbol{K}$ with critical point strictly larger than $\nu = \kappa^{+\boldsymbol{K}} = \kappa^{\boldsymbol{K}'}$. As $E_\nu^{\boldsymbol{K}'} = \varnothing$, this would also mean that $E_\nu = \varnothing$, a contradiction. (Notice that this shows that the latter case in fact cannot occur □(Theorem 7.4.2)

In the following, we strengthen the above results. This involves, first of all, generalizing the notion of incompressibility.

Definition. Given ordinals $\alpha \geq \gamma$, we say that $J_\alpha^{\bar{E}}$ is *incompressible over* γ iff there is a weasel W and a class X thick in W satisfying:

a) $\bar{E} \upharpoonright \alpha = E^W \upharpoonright \alpha$;

b) if Y is a thick class in W, then $\delta \subset h_W(\gamma \cup Y)$.

Exactly as in the case of incompressibility we can show that if $J_{\alpha_1}^{E_1}$ and $J_{\alpha_2}^{E_2}$ are incompressible over γ and such that $E_1 \upharpoonright \gamma = E_2 \upharpoonright \gamma$, then one of these structures is an initial segment of the other. The proof of Lemma 7.3.8 adapted to the context of γ-incompressibility, yields:

Lemma 7.4.3. *Suppose $\tilde{K}$ is a weasel with measure sequence F such that each J_α^F is incompressible over γ, where γ is fixed. If U is $\tilde{K}$-correct with critical point κ and $\nu \geq \gamma$, where ν is the cardinal successor of κ in $\tilde{K}$, then $U = F_\nu$.* □(Lemma 7.4.3)

The model $\tilde{K}$ from the previous lemma is the *core model over* J_γ^F; we denote it by $K[J_\gamma^F]$. An obvious generalization of the proof of Lemma 7.3.9 yields the inductive definition of $K[J_\gamma^F]$:

Lemma 7.4.4. *Suppose $\alpha \geq \gamma$ and $\langle J_\alpha^F, U\rangle$ is strong, where $U \neq \varnothing$. Then $U = F_\alpha$.*
□(Lemma 7.4.4)

The next lemma tells us that iterates of K are core models over appropriate structures. It generalizes in a straightforward way to iterates of $K[J_\gamma^F]$.

Lemma 7.4.5. *Let W be a simple iterate of K and $\gamma = \text{l.u.b.}\{\nu_i;\ i \leq \theta\}$, where ν_i are the iteration indices and $\theta + 1$ is the length of the iteration. Then $W = K[J_\gamma^{E^W}]$.*

Proof. We show that each $J_\delta^{E^W}$ is incompressible over γ for $\delta \geq \gamma$. Notice that it suffices to prove this for *unboundedly* many δ's; so we can assume w.l.o.g. that $\delta = \pi^W(\bar{\delta})$, where $\bar{\delta}$ is regular (in V) and π^W is the iteration map from K to W. Fix Q, X witnessing the incompressibility of $J_{\bar{\delta}}^E$ over γ. Let $W_i = L[E^i]$ be the i-th model in the iteration from K to W, $Q_i = L[F^i]$ the i-th model in the iteration of Q using the indices ν_i, and $\pi_i^W : K \to W_i$ and $\pi_i^Q : Q \to Q_i$ the corresponding iteration maps. As $\bar{\delta}$ is regular both in K and Q, the i-th iterate $J_{\delta_i}^{E^i}$ of $J_{\bar{\delta}}^E$ by the iteration indices ν_i is an initial segment of both W_i and Q_i and

$$\pi_i^W(\bar{\delta}) = \sup((\pi_i^W)''\bar{\delta}) = \delta_i = \sup((\pi_i^Q)''\bar{\delta}) = \pi_i^Q(\bar{\delta}).$$

This follows from the fact that those elements of the i-th iterate, which are in the initial segment determined by $\pi_i^W(\bar{\delta})$, are of the form $\pi_i^W(f)(\xi)$ for some $f \in J_{\bar{\delta}}^E$ and $\xi < \text{l.u.b.}\{\nu_j;\ j < i\}$. See the proof of Lemma 6.1.5 and Lemmata 4.2.3 and 4.2.4.

As a consequence of the above discussion we have that $J_\delta^{E^W}$ is and initial segment of Q_θ and $\pi_\theta^Q \upharpoonright (\bar{\delta} + 1) = \pi_\theta^W \upharpoonright (\bar{\delta} + 1)$, so that $\pi_\theta^Q(\bar{\delta}) = \delta$. Set

$$X' = \{\xi;\ \xi \in X \ \&\ \pi_\theta^Q(\xi) = \xi\};$$

the class X' is thick both in Q and Q_θ. We show that Q_θ, X' witness the incompressibility of $J_\delta^{E^W}$ over γ. Let Z be a class thick in Q_θ; suppose w.l.o.g. that $Z \subset X'$. Then Z is thick in Q. We want to prove that $\delta \subset h_{Q_\theta}(\gamma \cup Z)$, so fix any $\zeta < \delta$. As we pointed out in the previous paragraph, there is an $f \in J_{\bar\delta}^E$ and a $\xi < \gamma$ such that $\zeta = \pi_\theta^Q(f)(\xi)$. As Q, X witness the incompressibility of $J_{\bar\delta}^E$, there are $\xi_1, \ldots, \xi_\ell \in Z$ such that $f = h_Q(\xi_1, \ldots, \xi_\ell)$. Then $\zeta = \pi_\theta^Q(f)(\xi) = h_{Q_\theta}(\xi_1, \ldots, \xi_\ell)(\xi)$, so $\zeta \in h_{Q_\theta}(\gamma \cup Z)$, as desired. □(Lemma 7.4.5)

Lemma 7.4.1 and Theorem 7.4.2 can be the reformulated for $\boldsymbol{K}[J_\gamma^F]$ as follows:

Lemma 7.4.6. *Let $\sigma : \boldsymbol{K}[J_\gamma^F] \to W$ be an embedding with critical point κ and $\gamma \leq \nu$, where ν is the cardinal successor of κ in $\boldsymbol{K}[J_\gamma^F]$. Then W is universal, $\nu = \kappa^{+W}$ and*

$$E_\nu^F = \{x \in \mathcal{P}(\kappa) \cap \boldsymbol{K}[J_\gamma^F];\ \kappa \in \sigma(x)\},$$

where F is the measure sequence of $\boldsymbol{K}[J_\gamma^F]$. Let $\boldsymbol{K}' = \mathrm{Ult}(\boldsymbol{K}[J_\gamma^F], E_\nu^F)$ and let π be the associated ultrapower map. Then $\boldsymbol{K}' = \boldsymbol{K}[J_{\nu+1}^{F'}]$ where $\boldsymbol{K}' = \boldsymbol{L}[F']$ and the map $\sigma' : \pi(f)(\kappa) \mapsto \sigma(f)(\kappa)$ embeds $\boldsymbol{K}'$ into W and its critical point is strictly greater than ν. □(Lemma 7.4.6)

Theorem 7.4.7 (Rigidity Lemma for $\boldsymbol{K}[J_\gamma^F]$). *Suppose $\sigma : \boldsymbol{K}[J_\gamma^F] \to \boldsymbol{K}[J_\gamma^F]$ is such that $\sigma \restriction \gamma^- = \mathrm{id}$, where γ^- is the cardinal predecessor of γ in $\boldsymbol{K}[J_\gamma^F]$. Then $\sigma = \mathrm{id}$.* □(Theorem 7.4.7)

Using the above results we can now prove the following strengthenings of the Rigidity Lemma. Analogous strengthenings also hold for the core model up to one strong cardinal, as we shall see later, but require a careful reformulation in the case of larger models.

Theorem 7.4.8. *Let $\sigma : \boldsymbol{K} \to W$. Then W is a simple normal iterate of $\boldsymbol{K}$ and σ is the iteration map.*

Proof. The models $\boldsymbol{K}$ and W coiterate simply to a common weasel, as W is universal. Moreover, W is not moved in the coiteration, as otherwise Lemma 7.3.1 would guarantee that $\langle J_{\nu_i}^{E^i}, E_{\nu_i}^{W_i}\rangle$ is $\boldsymbol{K}_i$-correct, where i is the first step of the iteration such that $E_{\nu_i}^{W_i} \neq \emptyset$ and E^i the measure sequence of $\boldsymbol{K}_i$. By Lemma 7.4.5, $\boldsymbol{K}_i = \boldsymbol{K}[J_{\tilde\nu_i}^{E^i}]$ where $\tilde\nu_i = \mathrm{l.u.b.}\{\nu_h;\ h < i\} \leq \nu_i$, so by applying Lemma 7.4.4 we could infer that $E_{\nu_i}^{W_i} = E_{\nu_i}^i$, a contradiction. Hence W agrees with the last model on the $\boldsymbol{K}$ side of the coiteration; therefore it is a simple iterate of $\boldsymbol{K}$.

To see that σ is the iteration map from $\boldsymbol{K}$ to W we inductively define embeddings $\sigma_i : \boldsymbol{K}_i \to W$ such that

$$\sigma_i \restriction \tilde\nu_i = \mathrm{id} \tag{7.4.1}$$

and $\sigma_j \circ \pi_{ij} = \sigma_i$ whenever $i \leq j$, where π_{ij} is the iteration map from $\boldsymbol{K}_i$ to $\boldsymbol{K}_j$. We set $\sigma_0 = \sigma$. At successor steps $i+1$ we use the induction hypothesis $\sigma_i \restriction \tilde\nu_i = \mathrm{id}$ and the fact that $K_i = K[J^{E^i}_{\tilde{\nu}_i}]$ to infer, using Lemma 7.4.6, that the measure derived from σ_i is equal to $E^i_{\nu_i}$, the map $\sigma_{i+1} : \pi_{i,i+1}(f)(\kappa_i) \mapsto \sigma_i(f)(\kappa_i)$ embeds K_{i+1} into W and its critical point is strictly larger than $\nu_i = \kappa_i^{+W}$, so that (7.4.1) holds with $i+1$ in place of i, and $\sigma_{i+1} \circ \pi_{i,i+1} = \sigma_i$. At limit steps i we define σ_i so that the whole diagram consisting of σ_h and π_{hj} commutes; i.e. $\sigma_i(\pi_{hi}(x)) = \sigma_h(x)$. Since the iteration of K we are dealing with is normal, each $\xi < \tilde{\nu}_i$ is less than κ_h for some $h < i$, and $\pi_{hi}(\xi) = \xi$ and $\sigma_i(\xi) = \sigma_h(\xi) = \xi$ by the induction hypothesis, so (7.4.1) holds at limit steps as well.

If the length of the iteration from K to W is $\theta < \infty$, then $W = K[J^{E^\theta}_{\tilde{\nu}_\theta}]$, $\sigma_\theta : W \to W$ and $\sigma_\theta \restriction \tilde{\nu}_\theta = \mathrm{id}$. Theorem 7.4.7 guarantees that $\sigma_\theta = \mathrm{id}$, so $\sigma = \sigma_\theta \circ \pi_{0,\theta} = \pi_{0,\theta}$. In the case where the length of the iteration is ∞ define $\sigma_\infty : W = K_\infty \to W$ by $\sigma_\infty : \pi_{i,\infty}(x) \mapsto \sigma_i(x)$. Then $\sigma_\infty \restriction \kappa_i = \sigma_i \restriction \kappa_i = \mathrm{id}$, for all i, so $\sigma = \mathrm{id}$, as the κ_i's are unbounded in **On**. □(Theorem 7.4.8)

Theorem 7.4.9. *Any universal weasel is a simple normal iterate of K.*

Proof. See the first part of the proof of the previous theorem. □(Theorem 7.4.9)

Corollary 7.4.10. *Let W be universal and $\sigma : W \to K$. Then $W = K$ and $\sigma = \mathrm{id}$.*

Proof. By the previous theorem, W is a simple normal iterate of K. Let $\pi : K \to W$ be the iteration map. Then $\sigma \circ \pi : K \to K$, so the Rigidity Lemma implies that $\sigma \circ \pi = \mathrm{id}$ and the conclusion follows immediately. □(Corollary 7.4.10)

The last property of K we describe in this section is *absoluteness,* i.e. its independence of the universe in which K is constructed. This need not be true for an arbitrary universe, but it *is* true for a universe which, roughly speaking, contains mice of the same strength as V. Let us first focus on *generic absoluteness* of K, that is, the fact that K constructed in set generic extensions of V agrees with the K constructed in V.

Theorem 7.4.11. *Let $\mathbb{P}$ be a set-size forcing notion and let G be a $\mathbb{P}$-generic filter over V. Then $K^{V[G]} = K$.*

Proof. The forcing $\mathbb{P}$ is of set-size, so it does not add $0^\ddagger$. Hence $K^{V[G]}$ exists and each of its proper initial segments is incompressible in $V[G]$. Setting $E = E^K$, we show that each J^E_α is incompressible in the sense of $V[G]$, which yields the desired equality.

Let W, X witness the incompressibility of J^E_α in V. Theorem 4.5.5 guarantees that W is a weasel in the sense of $V[G]$, as it suffices to verify the iterability of proper initial segments of W. Furthermore, since $\mathbb{P}$ is a small forcing, X is thick in W in the sense of $V[G]$. To see this, it is enough to verify a) and b) below for $V[G]$ in place of V. Fix a regular ε_0 larger than the saturation of $\mathbb{P}$ and such that the following hold in V:

a) X is closed under singular limits of length at least ε_0;

b) $\beta^+ = \beta^{+W}$ and $X \cap \beta^+$ is unbounded for any V-cardinal $\beta \in X$ that is larger than ε_0.

Suppose also w.l.o.g. that $X \cap \varepsilon_0 = \varnothing$. $\mathbb{P}$ preserves cardinals and cofinalities larger than ε_0, so given any β as in b), we have $\beta^{+W} = \beta^+ = \beta^{+V[G]}$; the rest of b) follows trivially. For a) we use the fact that if $\langle \gamma_\xi;\ \xi < \varepsilon \rangle \in V[G]$ is an increasing ε-sequence in X of regular length $\varepsilon \geq \varepsilon_0$ and γ is its supremum, then there exists an increasing sequence $\langle \gamma'_\xi;\ \xi < \varepsilon \rangle$ in V with supremum γ. Such a sequence can be constructed using the standard boundedness argument. (Fix $\langle a_\xi;\ \xi < \varepsilon \rangle \in V$ such that $\gamma_\xi \in a_\xi \subset X \cap \gamma$ and $\operatorname{card}^V(a_\xi) < \varepsilon_0$ for all $\xi < \varepsilon$; then define γ'_ξ recursively to be the least $\gamma' \in X \cap \gamma$ satisfying $\gamma' > \sup(a_\xi \cup \{\gamma'_{\bar{\xi}};\ \bar{\xi} < \xi\})$.) It follows that $\gamma \in X$.

Our aim is to show that W, X witness the incompressibility of J^E_α in the sense of $V[G]$. For this it suffices to prove that any class $Z \subset X$ thick in W in the sense of $V[G]$ contains a proper class $\bar{Z}$ in the sense of V satisfying a) and b); such class will automatically be thick in W in the sense of V, so $\alpha \subset h_W(\bar{Z}) \subset h_W(Z)$, as desired. The class $\bar{Z}$ is defined in the obvious way using the above mentioned boundedness argument. Assume Z is the class of all ζ satisfying $V[G] \models \varphi(\zeta, a)$ for a parameter $a = \tau_G \in V[G]$ and that $\psi(\zeta, \eta, a)$ determines the enumeration of Z by ordinals in $V[G]$. Setting

$$a_\eta = \{\zeta;\ (\exists p \in \mathbb{P});\ p \Vdash \psi(\check{\zeta}, \check{\eta}, \tau)\},$$

$\langle a_\eta;\ \eta \in \mathbf{On} \rangle$ is a class in the sense of V and is proper, as the η-th element of Z is in a_η. Define ζ_η for $\eta \in \mathbf{On}$ by recursion as follows

$$\begin{aligned} \zeta_0 &= \text{l.u.b.}(a_0); \\ \zeta_{\eta+1} &= \text{l.u.b.}(a_{\zeta_\eta}); \\ \zeta_\eta &= \sup_{\bar{\eta}<\eta} \zeta_{\bar{\eta}}, \end{aligned}$$

and define a V-class $\bar{Z}'$ by

$$\bar{Z}' = \{\zeta_\eta;\ \varepsilon^Z \leq \operatorname{cf}(\eta) < \zeta_\eta \ \&\ \zeta_\eta \text{ a cardinal in } V\},$$

where $\varepsilon^Z \geq \varepsilon_0$ is fixed so that Z is closed under singular limits of cofinality at least ε^Z. Then $\bar{Z}'$ is a class in the sense of V, as $\langle \zeta_\eta \rangle_\eta$ is, and it follows easily that $\bar{Z}'$ is closed under singular ε-limits whenever $\varepsilon^Z \leq \varepsilon$. By construction, the ζ_η-th element of Z lies between ζ_η and $\zeta_{\eta+1}$, so if η is a limit of cofinality ε, then there is an increasing sequence of elements of Z with supremum ζ_η. From this we immediately obtain that $\bar{Z}' \subset Z$. Consequently, $\beta^{+W} = \beta^+$ for every $\beta \in \bar{Z}'$. We would like to arrange b) for $\bar{Z}$; as it is not clear that the above construction will guarantee the second part of b), we simply add enough ordinals between β and β^+ for any $\beta \in \bar{Z}'$. To each such β define ζ^β_η for $\eta < \beta^+$ as before with the exception that this time we set $\zeta^\beta_0 = \beta$ and $\zeta^\beta_{\eta+1} = \text{l.u.b.}(a_{\zeta_\eta} \cap \beta^+)$; it is straightforward to verify by induction that

$$\bar{Z}^\beta = \{\zeta^\beta_\eta;\ \varepsilon^Z \leq \operatorname{cf}(\eta) < \zeta^\beta_\eta\}$$

is a subset of Z cofinal in β^+ and closed under singular limits of cofinality at least ε^Z. Then

$$\bar{Z} = \bar{Z}' \cup \bigcup_{\beta \in Z'} Z^\beta$$

satisfies a) and b) and is clearly a subclass of Z, which completes the proof.

□(Theorem 7.4.11)

One can also have absoluteness of $\boldsymbol{K}$ constructed in a sufficiently large inner model.

Theorem 7.4.12. *Let M be an inner model such that $\boldsymbol{K} \subset M$. Then $\boldsymbol{K}^M = \boldsymbol{K}$.*

Proof. Suppose for a contradiction that $\boldsymbol{K}^M \neq \boldsymbol{K}$. Notice that we do *not* assume that $\boldsymbol{K}$ is a class in the sense of M. Let $\bar{E}$ be the measure sequence of $\boldsymbol{K}^M$. Clearly, each level of $\boldsymbol{K}$ is iterable in M and conversely, Theorem 4.5.5 guarantees that each initial segment of $\boldsymbol{K}^M$ is iterable in the sense of V. Also, $\boldsymbol{K}^M$ is a class in V as M is, so $\boldsymbol{K}^M$ is a weasel in V. The two weasels are coiterable in V and they coiterate simply to a common weasel. To see the latter notice that in the case of a truncation on the $\boldsymbol{K}$-side we could replace the above coiteration by a coiteration of $\boldsymbol{K}^M$ with a sufficiently long proper initial segment of $\boldsymbol{K}$, which is in M by our assumption. This would contradict the universality of $\boldsymbol{K}^M$ in M. Similarly, the existence of a truncation on the $\boldsymbol{K}^M$-side would yield a contradiction to the universality of $\boldsymbol{K}$ in V. Now let ν be the least index of disagreement between $\boldsymbol{K}^M$ and $\boldsymbol{K}$. By the standard argument (see e.g. the proof of Lemma 7.4.1), there is a κ which is a cardinal in both $\boldsymbol{K}$ and $\boldsymbol{K}^M$ such that ν is the cardinal successor of κ in both weasels; moreover, $\boldsymbol{K}$ agrees with $\boldsymbol{K}^M$ below ν. Suppose $\bar{E}_\nu \neq \varnothing$; then $\bar{E}_\nu$ is $\boldsymbol{K}$-correct by Lemma 7.3.1, so $\bar{E}_\nu = E_\nu$ by Lemma 7.3.8, a contradiction. It follows that $\bar{E}_\nu = \varnothing$ and $E_\nu \neq \varnothing$. As $E_\nu \in M$, we conclude as above that E_ν is $\boldsymbol{K}^M$-correct in M, so Lemma 7.3.8 applied inside M yields $\bar{E}_\nu = E_\nu$ again. Contradiction. □(Theorem 7.4.12)

7.5 Weak Covering for $\boldsymbol{K}$

The weak covering theorem for an extender model W asserts – roughly speaking – that W correctly computes successors of certain singular cardinals. In the chapter on extender models we proved a version of the weak covering theorem for $\boldsymbol{K}^c$ and successors of ω-closed singular cardinals and stated a stronger version which guarantees that the successors of *all sufficiently large* singular cardinals are computed correctly. However, the lower bound determining the exact meaning of "sufficiently large" here depends on the cardinal arithmetic in V; furthermore, $\boldsymbol{K}^c$ is not absolute, so this version of the weak covering theorem does not prove very useful for applications. In this section we prove the strongest possible weak covering theorem for $\boldsymbol{K}$ under the assumption $\neg 0^\ddagger$. As an immediate consequence we obtain that $\boldsymbol{K}$ correctly computes successors of *all* singular cardinals.

Theorem 7.5.1 (Weak Covering Theorem for $\boldsymbol{K}$). *Let $\kappa \geq \omega_2^V$ be a cardinal in $\boldsymbol{K}$. Then*

$$\mathrm{cf}(\kappa^{+\boldsymbol{K}}) \geq \mathrm{card}(\kappa).$$

The above conclusion is false for $\kappa < \omega_2$ even in the simplest case where $0^{\#}$ does not exist; a canonical counterexample is given by Bukovský-Namba forcing. The proof of the above theorem closely follows that of the corresponding Theorem 6.6.1 for $\boldsymbol{K}^c$, so we shall focus mainly on the new issues occurring in the present context. There are two main points in the whole argument which require a new approach. Firstly, we have to apply different criteria for absorbing measures. In the case of $\boldsymbol{K}^c$ we used the ω-completeness; here we will use the maximality of the $E^{\boldsymbol{K}}$-sequence. Secondly, we need to refine the upwards extension techniques in order to make the arguments from the earlier proof applicable for all κ's. For this purpose we formulate and prove the *Frequent Extensions of Embeddings Lemma*. As in the case of single extensions, there are more variants of the Frequent Extensions of Embeddings Lemma, depending on the inner structure of the models which constitute the domains of the original embeddings and on the fact whether we are interested in coarse or fine structural extensions. Here we shall merely prove the versions of the lemma needed for the proof of Theorem 7.5.1, which correspond to the case where the domains of the original embeddings have a largest cardinal. The remaining case is left to the reader as a demanding exercise.

Theorem 7.5.2 (Frequent Extensions of Embeddings Lemma, fine structural version). *Consider a linear system of structures and maps with the following properties:*

a) *$\alpha > \omega_1$ is a regular cardinal;*

b) *$Q = J_\tau^E$ is a mouse with the largest cardinal $\kappa \geq \alpha$ (here E need not be the measure sequence of $\boldsymbol{K}$, contravening our earlier convention);*

c) *$\langle \alpha_\xi;\ \xi \leq \alpha\rangle$ is a normal sequence satisfying $\alpha_\alpha = \alpha$;*

d) *$Q_\xi = J_{\tau_\xi}^{E^\xi}$ for $\xi \leq \alpha$ is a system of premice such that $Q_\alpha = Q$ and $\mathrm{card}(Q_\xi) < \alpha$ for $\xi < \alpha$; moreover, κ_ξ is the largest cardinal in Q_ξ;*

e) *$\sigma_{\xi\xi'} : Q_\xi \to Q_{\xi'}$ is a commutative system of Σ_0-preserving maps such that $\sigma_{\xi\xi'}(\kappa_\xi) = \kappa_{\xi'}$ and, for limit $\lambda \leq \alpha$, $\langle Q_\lambda, \sigma_{\xi\lambda}\rangle_\xi$ is the direct limit of the diagram $\langle Q_\xi, \sigma_{\xi\xi'}\rangle_{\xi\leq\xi'<\lambda}$, that is, $Q_\lambda = \bigcup\{\mathrm{rng}(\sigma_{\xi\lambda});\ \xi < \lambda\}$.*

Setting $\tilde{\tau}_\xi = \sup(\sigma''_{\xi\alpha}\tau_\xi)$ and $\tilde{Q}_\xi = J_{\tilde{\tau}_\xi}^E$, we have

$$\sigma_{\xi\alpha} : Q_\xi \underset{\Sigma_0}{\longrightarrow} \tilde{Q}_\xi \quad \textit{cofinally}.$$

Fix a stationary set $S \subset \alpha$ of ordinals of uncountable cofinality and to each $\xi \in S$ assign a mouse M_ξ such that

A) *M_ξ is a lengthening of Q_ξ, i.e. $E^\xi = E^{M_\xi} \restriction \tau_\xi$;*

B) τ_ξ *is the cardinal successor of* κ_ξ *in* M_ξ*;*

C) *One of the following holds:*

1) $\omega\varrho^{\omega}_{M_\xi} > \kappa_\xi$*;*

2) $\omega\varrho^{\omega}_{M_\xi} \leq \kappa_\xi$ *and* M_ξ *is sound above* κ_ξ*.*

Under the above assumptions, for all but nonstationarily many ξ *we have:*

i) *The canonical extension* $\tilde{\sigma}_\xi : M_\xi \to \tilde{M}_\xi$ *exists;*

ii) $\tilde{M}$ *is normally iterable above* κ_ξ *if* C2) *holds and fully iterable otherwise.*

The coarse version of the Frequent Extensions of Embeddings Lemma is also worth stating.

Theorem 7.5.3 (Frequent Extensions of Embeddings Lemma, coarse version). *Consider* J^E_τ, α, κ *and the system of structures* Q_ξ *and maps* $\sigma_{\xi\xi'}$ *satisfying* a) – e) *from the previous theorem. Define* $\tilde{Q}_\xi$ *as before, fix a stationary set* $S \subset \alpha$ *of ordinals of uncountable cofinality and to each* $\xi \in S$ *assign a* Σ_0*-mouse* M_ξ *satisfying* A) *and* B) *of the previous theorem.*

Then for all but nonstationarily many ξ *we have:*

i) *The* 0*-extension* $\tilde{\sigma}_\xi : M_\xi \to \tilde{M}_\xi$ *exists;*

ii) $\tilde{M}_\xi$ *is* Σ_0*-iterable.*

Proof of Theorem 7.5.2. We shall first prove i) of the theorem. Suppose for a contradiction that the conclusion fails, i.e. that we have a stationary $S_1 \subset S$ such that for $\xi \in S_1$, the canonical extension of $\sigma_{\xi\alpha}$ does not exist. We can w.l.o.g. assume that $S_1 = S$. Given $\xi \in S$, we can further suppose that

- τ_ξ is minimal in the sense that if $\tau'_\xi < \tau_\xi$ and M'_ξ is a mouse satisfying A) – C) with τ'_ξ, $Q'_\xi = J^{E_\xi}_{\tau'_\xi}$ and M'_ξ in place of τ_ξ, Q_ξ and M_ξ, then the canonical extension of $\sigma_{\xi\alpha} \restriction Q'_\xi$ to M'_ξ exists.

This normalizing condition is not necessary for the whole argument, but it yields a certain simplification. To see that it can be adopted, fix $\tau^1_\xi \leq \tau_\xi$ minimal in the above sense together with a mouse M^1_ξ witnessing that the canonical extension of $\sigma_{\xi\alpha} \restriction Q^1_\xi$ to M^1_ξ does not exist, where $Q^1_\xi = J^{E_\xi}_{\tau^1_\xi}$, and such that A) – C) hold. If $\tau''_\xi = \tau_\xi$ on some stationary set $S' \subset S$, we just replace S by S'. Otherwise, using the pressing down argument (each τ^1_ξ is in the range of $\sigma_{g(\xi),\xi}$ for some $g(\xi) < \xi$) and the fact that each Q_ξ is of size strictly less than α, we find a $\bar{\xi} < \alpha$, a stationary set $S' \subset S$ such that $\min(S') > \bar{\xi}$ and a $\bar{\tau} \in Q_{\bar{\xi}}$ such that $\tau^1_\xi = \sigma_{\bar{\xi}\xi}(\bar{\tau})$ whenever $\xi \in S'$. Setting

$\tau_\xi^2 = \sigma_{\bar\xi\xi}(\bar\tau)$, $Q_\xi^2 = J_{\tau_\xi^2}^{E^\xi}$ and $\sigma_{\bar\xi\xi}^2 = \sigma_{\bar\xi\xi} \restriction J_{\bar\tau}^{E^{\bar\xi}}$ we obtain a system satisfying a) – e). If we replace S by S', the mice M_ξ^1 witness the minimality of $\tau_\xi^2 = \tau_\xi^1$ for all $\xi \in S'$.

We can further restrict the set S so that if 2) of C) holds, then there is a fixed $n \in \omega$ such that

- $\omega\varrho_{M_\xi}^{n+1} \leq \kappa_\xi < \omega\varrho_{M_\xi}^n$ whenever $\xi \in S$.

To each $\xi \in S$ fix ω-sequences $f_i^\xi \in \Gamma(\sigma_{\xi\alpha}) = \Gamma(\kappa_\xi, M_\xi)$ and $\beta_i^\xi < \kappa$ witnessing that $\mathbb{D}(M_\xi, \sigma_{\xi\alpha})$ is ill-founded, that is,

$$\prec\beta_{i+1}^\xi, \beta_i^\xi\succ \in \sigma_{\xi\alpha}(u_{i+1,i}^\xi),$$

where $u_{i+1,i}^\xi = \{\prec\eta_1, \eta_0\succ;\ f_{i+1}^\xi(\eta_1) \in f_i^\xi(\eta_0)\}$. Also fix parameters $q_i^\xi \in M_\xi$ such that each f_i^ξ is a good $\Sigma^*(M_\xi)$-function in q_i^ξ; in the case of C2), f_i^ξ is a good $\Sigma_1^{(n-1)}(M_\xi)$-function in q_i^ξ. Given $\xi \in S$, let

- X_ξ be the smallest elementary substructure of M_ξ containing q_i^ξ for all $i \in \omega$,
- M_ξ^* be the transitive collapse of $M_\xi | X_\xi$ and
- $\sigma_\xi^* : M_\xi^* \to M_\xi$ be the uncollapsing map.

Then X_ξ and M_ξ^* are countable and $p_{M_\xi} \in X_\xi$. As σ_ξ^* is fully elementary, M_ξ^* is a premouse, $\sigma_\xi^*(p_{M_\xi^*}) = p_{M_\xi}$ and each initial segment of M_ξ^* is solid. Also, granting that C2) holds, $\omega\varrho_{M_\xi^*}^{n+1} \leq \kappa_\xi^* < \omega\varrho_{M_\xi^*}^n$ and M_ξ^* is sound above κ_ξ^*, where $\sigma_\xi^*(\kappa_\xi^*) = \kappa_\xi$. Let $f_i^{*\xi}$ denote that good $\Sigma^*(M_\xi^*)$ function in $q_i^{*\xi} = (\sigma_\xi^*)^{-1}(q_i^\xi)$ which has the same functionally absolute definition as f_i^ξ; then $\sigma_\xi^*(f_i^{*\xi}) = f_i^\xi$.

Claim 1. σ_ξ^* maps $\tau_\xi^* = (\kappa_\xi^*)^{+M_\xi^*}$ cofinally into τ_ξ. Thus, τ_ξ is ω-cofinal.

Proof of Claim 1. This is a consequence of the above minimality assumption on τ_ξ. Suppose $\tilde\tau_\xi^* = \sup((\sigma_\xi^*)''\tau_\xi^*) < \tau_\xi$. Using the Interpolation Lemma (Lemma 5.6.2), we infer that the canonical extension $\tilde\sigma_\xi^* : M_\xi^* \to \tilde M_\xi^*$ of

$$\sigma_\xi^* \restriction J_{\tau_\xi^*}^{E^{*\xi}} : J_{\tau_\xi^*}^{E^{*\xi}} \to J_{\tilde\tau_\xi^*}^{E^\xi}$$

exists (here $E^{*\xi}$ is the measure sequence of M_ξ^*) and that there is a Σ_0-preserving map $\sigma_\xi^{*\prime} : \tilde M_\xi^* \to M_\xi$ such that $\sigma_\xi^{*\prime} \circ \tilde\sigma_\xi^* = \sigma_\xi^*$. It follows that $\sigma_\xi^{*\prime} \restriction \tilde\tau_\xi^* = \mathrm{id}$ and $\sigma_\xi^{*\prime}(\tilde\tau_\xi^*) = \tau_\xi$ if $\tilde\tau_\xi^* \in \tilde M_\xi^*$. Our aim is to show that $\tilde M_\xi^*$ is a counterexample to the minimality of τ_ξ. Using general facts about pseudoultrapowers together with the soundness and solidity of M_ξ^* above κ_ξ^*, we can argue as in the proof of Lemma 5.6.5 to infer:

- C1) implies that $\omega\varrho^{\omega}_{\tilde{M}^*_\xi} > \kappa_\xi$ and $\sigma^{*\prime}_\xi$ is Σ^*-preserving;
- C2) implies that $\omega\varrho^{n+1}_{\tilde{M}^*_\xi} \leq \kappa_\xi < \omega\varrho^{n}_{\tilde{M}^*_\xi}$, $\tilde{M}^*_\xi$ is sound above κ_ξ and that $\sigma^{*\prime}_\xi$ is $\Sigma^{(n)}_0$-preserving.

The preservation properties of $\sigma^{*\prime}_\xi$ guarantee the full iterability of $\tilde{M}^*_\xi$: In the former case the conclusion is immediate; in the latter case we apply the Condensation Lemma.

It remains to prove that the canonical extension of $\sigma_{\xi\alpha} \restriction J^{E^\xi}_{\tilde{\tau}^*_\xi}$ to $\tilde{M}^*_\xi$ does not exist. Let $\tilde{f}^{*\xi}_i = \tilde{\sigma}^*_\xi(f^{*\xi}_i)$. Then $\sigma^{*\prime}_\xi(\tilde{f}^{*\xi}_i) = f^\xi_i$ and

$$\{\prec\eta_1, \eta_0\succ;\ \tilde{f}^{*\xi}_{i+1}(\eta_1) \in \tilde{f}^{*\xi}_i(\eta_0)\} = u^\xi_{i+1,i},$$

as can be easily verified using the preservation properties of $\sigma^{*\prime}_\xi$. This guarantees that $[\beta^\xi_i, \tilde{f}^{*\xi}_i]$ represent an infinite decreasing sequence with respect to the membership relation in $\mathbb{D}(\tilde{M}^*_\xi, \sigma_{\xi\alpha})$. □(Claim 1)

By the countability of $X_\xi = \mathrm{rng}(\sigma^*_\xi)$ and the fact that $\mathrm{cf}(\xi) > \omega$, there is a $g(\xi) < \xi$ such that $X_\xi \cap Q_\xi \subset \mathrm{rng}(\sigma_{g(\xi),\xi})$. The pressing down argument yields a stationary $S' \subset S$ and an ordinal $\bar{\xi}$ such that $g \restriction S'$ is constant with the value $\bar{\xi}$. Again, we can w.l.o.g. suppose that $S' = S$ and $\bar{\xi} < \min(S)$. Setting $Q^*_\xi = J^{E^{*\xi}}_{\tau^*_\xi}$ for $\xi \in S$, we define a Σ_0-preserving map $\sigma_\xi : Q^*_\xi \to Q_\xi$ by

- $\sigma_\xi = \sigma^{-1}_{\bar{\xi}\xi} \circ \sigma^*_\xi$.

Claim 1 guarantees that both σ_ξ and $\sigma_{\bar{\xi}\xi}$ are cofinal. Using the arguments from the proof of the Interpolation Lemma and Lemma 5.6.5 we show:

- The canonical extension $\bar{\sigma}_\xi : M^*_\xi \to \bar{M}_\xi$ of σ_ξ exists and there is a Σ_0-preserving map $\sigma'_\xi : \bar{M}_\xi \to M_\xi$ extending $\sigma_{\bar{\xi}\xi}$, defined by

$$\sigma'_\xi(\bar{\sigma}_\xi(f)(\zeta)) = \sigma^*_\xi(f)(\sigma_{\bar{\xi}\xi}(\zeta))$$

for $f \in \Gamma(\kappa^*_\xi, M^*_\xi)$ and $\zeta < \kappa_{\bar{\xi}}$. Moreover, $\bar{\sigma}_\xi$ is $\Sigma^{(k)}_1$-preserving whenever $\omega\varrho^{k+1}_{M^*_\xi} > \kappa^*_\xi$.

- C1) implies that $\omega\varrho^{\omega}_{\bar{M}_\xi} > \kappa_{\bar{\xi}}$ and σ'_ξ is Σ^*-preserving.
- C2) implies that $\omega\varrho^{n+1}_{\bar{M}_\xi} \leq \kappa_{\bar{\xi}} < \omega\varrho^{n}_{\bar{M}_{\bar{\xi}}}$ and $\bar{M}_{\bar{\xi}}$ is both sound and solid above $\kappa_{\bar{\xi}}$, so $\bar{\sigma}_\xi$ is $\Sigma^{(n)}_0$-preserving and cofinal. The cofinality of $\bar{\sigma}_\xi$ guarantees that σ'_ξ is $\Sigma^{(n)}_0$-preserving. (In fact, σ'_ξ is also cofinal, as it maps $\tau_{\bar{\xi}}$ cofinally into τ_ξ.)

The definition of σ'_ξ is correct, as $\Gamma(\sigma_\xi, M^*_\xi) = \Gamma(\kappa^*_\xi, M^*_\xi) = \Gamma(\sigma^*_\xi, M^*_\xi)$. The verification of the preservation properties of σ'_ξ is standard: Given a Σ_0-formula $\varphi(v_1, \ldots, v_\ell)$ and $[\beta_j, f_j]$ representing objects in $\mathbb{D}^*(M^*_\xi, \sigma_\xi)$ for $j = 1, \ldots, \ell$,

$$\mathbb{D}^*(M^*_\xi, \sigma_\xi) \models \varphi([\beta_1, f_1], \ldots, [\beta_\ell, f_\ell]) \longleftrightarrow \prec\beta_1, \ldots, \beta_\ell\succ \in \sigma_\xi(u^{M^*_\xi,\varphi}_{f_1,\ldots,f_\ell})$$

and the latter can be equivalently rewritten as

$$\prec\sigma_{\bar{\xi}\xi}(\beta_1), \ldots, \sigma_{\bar{\xi}\xi}(\beta_\ell)\succ \in \sigma_{\bar{\xi}\xi} \circ \sigma_\xi(u^{M^*_\xi,\varphi}_{f_1,\ldots,f_\ell}) = \sigma^*_\xi(u^{M^*_\xi,\varphi}_{f_1,\ldots,f_\ell}) = u^{M_\xi,\varphi}_{\sigma^*_\xi(f_1),\ldots,\sigma^*_\xi(f_\ell)},$$

so the former holds iff

$$M_\xi \models \varphi[\sigma^*_\xi(f_1)(\sigma_{\bar{\xi}\xi}(\beta_1)), \ldots, \sigma^*_\xi(f_\ell)(\sigma_{\bar{\xi}\xi}(\beta_\ell))].$$

This computation also goes through for $\Sigma^{(k)}_0$-formulae whenever $\omega\varrho^k_{M^*_\xi} > \kappa^*_\xi$. If C2) holds, the cofinality of $\bar{\sigma}_\xi$ yields that σ'_ξ maps $H^n_{\bar{M}_\xi}$ *into* $H^n_{M_\xi}$, which is needed for the proof that σ'_ξ is $\Sigma^{(n)}_0$-preserving. As an immediate consequence of the preservation properties of σ'_ξ we obtain:

- C1) implies that $\bar{M}_\xi$ is fully iterable;
- C2) implies that $\bar{M}_\xi$ is normally iterable above $\kappa_{\bar{\xi}}$.

We are now ready to derive a contradiction. We begin with the case C2), which permits simpler argumentation. Given $\xi, \xi' \in S$, the premice $\bar{M}_\xi$ and $\bar{M}_{\xi'}$ coiterate *above* $\kappa_{\bar{\xi}}$, so the above iterability result guarantees that the two mice are coiterable. As $\bar{M}_\xi$ and $\bar{M}_{\xi'}$ agree below $\tau_{\bar{\xi}}$ and $\tau_{\bar{\xi}}$ is the cardinal successor of $\kappa_{\bar{\xi}}$ in both, $\mathcal{P}(\kappa_{\bar{\xi}}) \cap \bar{M}_\xi = \mathcal{P}(\kappa_{\bar{\xi}}) \cap \bar{M}_{\xi'}$. It follows that the two premice coiterate simply to a common premouse, say M', since both project to $\kappa_{\bar{\xi}}$. Using the properties of pseudoultrapowers we have seen that both $\bar{M}_\xi$ and $\bar{M}_{\xi'}$ are sound and solid above $\kappa_{\bar{\xi}}$, so using the standard argument we infer that

- $\bar{M}_\xi =$ the transitive collapse of $\tilde{h}^{n+1}_{M'}(\kappa_{\bar{\xi}} \cup \{p_{M'}\}) = \bar{M}_{\xi'}$,

as the iteration maps send $p_{\bar{M}_\xi} - \kappa_{\bar{\xi}}$ (resp. $p_{\bar{M}_{\xi'}} - \kappa_{\bar{\xi}}$) to $p_{M'} - \kappa_{\bar{\xi}}$. Denote this common assignment by $\bar{M}$.

Pick $\xi < \xi'$ in S such that $\beta^\xi_i \in \mathrm{rng}(\sigma_{\xi'\alpha})$ for all $i \in \omega$. Set $\bar{f}^\xi_i = \bar{\sigma}_\xi(f^{*\xi}_i)$. The map $\bar{\sigma}_\xi$ is sufficiently preserving so that this definition does make sense and each $\bar{f}^\xi_i$ is an element of $\Gamma(\kappa_{\bar{\xi}}, \bar{M})$. Recalling that $\prec\beta^\xi_{i+1}, \beta^\xi_i\succ \in \sigma_{\xi\alpha}(u^\xi_{i+1,i})$, we obtain

$$\prec\sigma^{-1}_{\xi'\alpha}(\beta^\xi_{i+1}), \sigma^{-1}_{\xi'\alpha}(\beta^\xi_i)\succ \in \sigma^{-1}_{\xi'\alpha} \circ \sigma_{\xi\alpha}(u^\xi_{i+1,i}) = \sigma_{\xi\xi'}(u^\xi_{i+1,i}).$$

The preservation properties of $\sigma'_\xi : \bar{M} \to M_\xi$ guarantee that $\sigma'_\xi(\bar{f}^\xi_i) = f^\xi_i$, so $\sigma'_{\bar{\xi}\xi}(\bar{u}^\xi_{i+1,i}) = u^\xi_{i,i+1}$, where

$$\bar{u}^\xi_{i+1,i} = \{\prec\eta_1, \eta_0\succ \in \kappa_{\bar{\xi}};\ \bar{f}^\xi_{i+1}(\eta_1) \in \bar{f}^\xi_i(\eta_0)\}.$$

As each $\sigma'_\zeta : \bar{M} \to M_\zeta$ extends $\sigma_{\bar{\xi}\zeta}$ and is sufficiently preserving, we have

$$\sigma_{\xi\xi'}(u^\xi_{i+1,i}) = \sigma_{\xi\xi'} \circ \sigma'_{\bar{\xi}\xi}(\bar{u}^\xi_{i+1,i}) = \sigma_{\xi\xi'} \circ \sigma_{\bar{\xi}\xi}(\bar{u}^\xi_{i+1,i}) = \sigma_{\bar{\xi}\xi'}(\bar{u}^\xi_{i+1,i}) = \sigma'_{\bar{\xi}\xi'}(\bar{u}^\xi_{i+1,i}),$$

and the set on the rightmost side is equal to

$$\{\prec\eta_1, \eta_0\succ \in \kappa_{\xi'};\ \sigma'_{\xi'}(\bar{f}^\xi_{i+1})(\eta_1) \in \sigma'_{\xi'}(\bar{f}^\xi_i)(\eta_0)\}.$$

It follows that

$$\sigma'_{\xi'}(\bar{f}^\xi_0)(\sigma^{-1}_{\xi'\alpha}(\beta^\xi_0)) \ni \sigma'_{\xi'}(\bar{f}^\xi_1)(\sigma^{-1}_{\xi'\alpha}(\beta^\xi_1)) \ni \cdots \ni \sigma'_{\xi'}(\bar{f}^\xi_i)(\sigma^{-1}_{\xi'\alpha}(\beta^\xi_i)) \ni \ldots$$

is an infinite decreasing $\in$-sequence. Contradiction.

In the case of C1), the proof is similar in spirit, but a bit more complicated, as this time we cannot prove that all $\bar{M}_\xi$ are equal. Instead, we shall employ the idea of the proof of Lemma 5.6.6. We know that all M_ξ are fully iterable, so we can pick a $\xi \in S$ for which M_ξ is minimal in the mouse order. Then, as before, we choose $\xi' > \xi$ in S such that $\beta^\xi_i \in \operatorname{rng}(\sigma_{\xi'\alpha})$ for all $i \in \omega$. The minimality of $\bar{M}_\xi$ guarantees that $\bar{M}_\xi$ and $\bar{M}_{\xi'}$ coiterate to a common mouse, say $\bar{M}'$, so that the $\bar{M}_\xi$-side is simple. As before, this coiteration is above $\kappa_{\bar{\xi}}$. Since $\sigma'_{\xi'}$ is Σ^*-preserving, we can copy the iteration on the $\bar{M}_{\xi'}$-side onto an iteration of $M_{\xi'}$ via $\sigma'_{\xi'}$. Let $\bar{\pi}$ be the iteration map from $\bar{M}_\xi$ to $\bar{M}'$, let M' be the last model in the iteration of $M_{\xi'}$ and let $\sigma' : \bar{M}' \to M'$ be the last copy map. The situation is depicted in the following diagram:

$$\begin{array}{ccccc} \bar{M}_\xi & \xrightarrow{\bar{\pi}} & \bar{M}' & \xrightarrow{\sigma'} & M' \\ & & \uparrow & & \uparrow \\ & & \bar{M}_{\xi'} & \xrightarrow{\sigma'_{\xi'}} & M_{\xi'} \end{array}$$

The dotted arrows indicate the iterations which might be nonsimple. As the iteration from $\bar{M}_{\xi'}$ to $\bar{M}'$ is above $\kappa_{\bar{\xi}}$, the iteration from $M_{\xi'}$ to M' is above $\kappa_{\xi'}$, so all premice of the former iteration agree below $\tau_{\bar{\xi}}$ and those of the latter iteration below $\tau_{\xi'}$. It follows that $Q_{\bar{\xi}}$ is an initial segment of every premouse of the former iteration and $\tau_{\bar{\xi}}$ is the cardinal successor of $\kappa_{\bar{\xi}}$ in each of these premice. Also, the same holds of the latter iteration with $Q_{\xi'}$, $\tau_{\xi'}$ and $\kappa_{\xi'}$ in place of $Q_{\bar{\xi}}$, $\tau_{\bar{\xi}}$ and $\kappa_{\bar{\xi}}$. Furthermore, all copy maps are Σ^*-preserving and agree below $\tau_{\bar{\xi}}$, as follows from the copying lemma (Lemma 4.3.1). In particular,

- $\tau_{\bar{\xi}}$ is the cardinal successor of $\kappa_{\bar{\xi}}$ in $\bar{M}'$;
- $\sigma' \upharpoonright Q_{\bar{\xi}} = \sigma_{\bar{\xi}\xi'}$ and σ' is Σ^*-preserving;
- $\bar{\pi} \upharpoonright \kappa_{\bar{\xi}} = \mathrm{id}$.

Arguing as in the previous case,

$$\sigma_{\xi\xi'}(u^{\xi}_{i+1,i}) = \sigma_{\bar{\xi}\xi'}(\bar{u}^{\bar{\xi}}_{i+1,i}) = \sigma'(\bar{u}^{\bar{\xi}}_{i+1,i}) = \sigma'(\bar{\pi}(\bar{u}^{\bar{\xi}}_{i+1,i}) \cap \kappa_{\bar{\xi}}) = \sigma' \circ \bar{\pi}(\bar{u}^{\bar{\xi}}_{i+1,i}) \cap \kappa_{\xi'},$$

where the set on the rightmost side is an initial segment of

$$\{\prec\eta_1, \eta_0\succ \in \kappa_{\xi'};\ \sigma' \circ \bar{\pi}(\bar{f}^{\bar{\xi}}_{i+1})(\eta_1) \in \sigma' \circ \bar{\pi}(\bar{f}^{\bar{\xi}}_{i})(\eta_0)\}.$$

This yields an infinite decreasing sequence of the form

$$\sigma' \circ \bar{\pi}(\bar{f}^{\bar{\xi}}_0)(\sigma^{-1}_{\bar{\xi}'\alpha}(\beta^{\xi}_0)) \ni \sigma' \circ \bar{\pi}(\bar{f}^{\bar{\xi}}_1)(\sigma^{-1}_{\bar{\xi}'\alpha}(\beta^{\xi}_1)) \ni \cdots \ni \sigma' \circ \bar{\pi}(\bar{f}^{\bar{\xi}}_i)(\sigma^{-1}_{\bar{\xi}'\alpha}(\beta^{\xi}_i)) \ni \cdots$$

and completes the proof of i) of the theorem.

Remark. In the case C2) one could argue roughly as in C1), thus avoiding the use of the fact that all $\bar{M}_{\xi}$ are equal. When listing the properties of σ'_{ξ}, we mentioned that σ'_{ξ} is in fact $\Sigma^{(n)}_0$-preserving and cofinal. This means that σ'_{ξ} is strongly preserving above $\kappa_{\bar{\xi}}$ in the sense of Lemma 3.4.8, where "above $\kappa_{\bar{\xi}}$ has the obvious meaning. As a consequence we have that *any* (i.e. not only a normal one) iteration of $\bar{M}_{\xi}$ above $\kappa_{\bar{\xi}}$ can be copied onto an iteration of M_{ξ}, which establishes the full iterability of $\bar{M}_{\xi}$ above $\kappa_{\bar{\xi}}$. It follows that the mouse ordering is well-founded on the collection of all M_{ξ}, so that we can choose a ξ such that M_{ξ} is minimal. This suffices to guarantee that the argument of C1) goes through.

The proof of ii) closely follows the above argument. Suppose for a contradiction that the conclusion fails on a stationary set; as before we can w.l.o.g. identify this set with S. Furthermore, we can take τ_{ξ} to be minimal in the sense that if $\tau'_{\xi} < \tau_{\xi}$ and M'_{ξ} extends Q'_{ξ}, then the canonical extension of $\sigma_{\xi\alpha} \upharpoonright Q'_{\xi}$ to M'_{ξ} exists and the associated target structure of this extension is normally iterable above κ if C2) holds and fully iterable otherwise. This can be shown by an argument similar that in the proof of i). (This time we set τ^1_{ξ} to be minimal such that either the corresponding extension of embedding does not exist or the target structure is not normally iterable above κ, resp. not fully iterable, and use i) of the theorem to exclude the former.) Finally, suppose that $n_{\xi} = n$ for a fixed n if C2) holds. To each $\xi \in S$ pick a countable premouse $M^{\star}_{\xi}$ together with a Σ^*-preserving map $\sigma^{\star}_{\xi} : M^{\star}_{\xi} \to \tilde{M}_{\xi}$ witnessing the failure of iterability for $\tilde{M}_{\xi}$, that is:

- $M^{\star}_{\xi}$ is not normally iterable above $\kappa^{\star}_{\xi} = (\sigma^{\star}_{\xi})^{-1}(\kappa)$ if C2) holds and not fully iterable otherwise.

Such $M_\xi^\star$ and $\sigma_\xi^\star$ can be constructed by a condensation argument as in the proof of Lemma 4.3.3. To each ξ, let $\langle x_i^\xi;\ i \in \omega\rangle$ be an enumeration of $M_\xi^\star$ and $f_i^\xi \in \Gamma(\sigma_{\xi\alpha}, M_\xi)$ and $\beta_i^\xi < \kappa$ be such that

$$\sigma_\xi^\star(x_i^\xi) = \tilde{\sigma}_{\xi\alpha}(f_i^\xi)(\beta_i^\xi)$$

for all $\xi \in S$ and $i \in \omega$. Let $M_\xi^*, \sigma^*, \kappa_\xi^*$ and τ_ξ^* be as before, but this time constructed using the new functions f_i^ξ. Also, let $f_i^{*\xi}$ be the σ_ξ^*-preimage of f_ξ^*, i.e. the function from $\Gamma(\kappa_\xi^*, M_\xi^*)$ satisfying $\sigma_\xi^*(f_i^{*\xi}) = f_i^\xi$. As before we obtain:

Claim 1′. σ_ξ^* maps τ_ξ^* cofinally to τ_ξ.

Proof of Claim 1′. We construct $\tilde{M}_\xi^*, \tilde{\sigma}_\xi^*$ and $\sigma^{*\prime}_\xi$ as in the proof of Claim 1. Assuming that $\tilde{\tau}_\xi^* = \sup((\sigma_\xi^*)''\tau_\xi) < \tau_\xi$, we infer that $\tilde{M}_\xi$ is a lengthening of $\tilde{Q}_\xi^* = J_{\tilde{\tau}_\xi^*}^{E^\xi}$ having all requisite properties and, by the minimality of τ_ξ, the canonical extension

$$\tilde{\sigma}_{\xi\alpha}^* : \tilde{M}_\xi^* \to \tilde{M}_\xi^\star$$

of $\sigma_{\xi\alpha} \restriction \tilde{Q}_\xi^*$ exists. We show that $\tilde{M}_\xi^\star$ fails to satisfy the required iterability condition, which contradicts the minimality of τ_ξ.

To prove the failure of the above specified amount of iterability for $\tilde{M}_\xi^\star$, it suffices to embed $M_\xi^\star$ into $\tilde{M}_\xi^\star$ via a sufficiently preserving map. Given $i_1, \ldots, i_\ell \in \omega$, an integer $k \in \omega$ such that $\omega\varrho_{M_\xi}^k > \kappa$ and a $\Sigma_0^{(k)}$-formula $\varphi(v_1, \ldots, v_\ell)$, we have

$$\tilde{M}_\xi^\star \models \varphi(\tilde{\sigma}_{\xi\alpha}^*(\tilde{f}_{i_1}^{*\xi})(\beta_{i_1}^\xi), \ldots, \tilde{\sigma}_{\xi\alpha}^*(\tilde{f}_{i_\ell}^{*\xi})(\beta_{i_\ell}^\xi)) \longleftrightarrow \prec\beta_{i_1}^\xi, \ldots, \beta_{i_\ell}^\xi\succ \in \sigma_{\xi\alpha}(u),$$

where $\tilde{f}_i^{*\xi}$ is as in the proof of Claim 1 and

$$u = u_{\tilde{f}_{i_1}^{*\xi}, \ldots, \tilde{f}_{i_\ell}^{*\xi}}^{\tilde{M}_\xi^*, \varphi} = u_{f_{i_1}^\xi, \ldots, f_{i_\ell}^\xi}^{M_\xi, \varphi},$$

as $\sigma_\xi^{*\prime}$ is sufficiently preserving, maps $\tilde{f}_i^{*\xi}$ to f_i^ξ and $\sigma_\xi^{*\prime} \restriction \kappa_\xi = \mathrm{id}$. Then the right side of the above equivalence can be rewritten as

$$\tilde{M}_\xi \models \varphi(\tilde{\sigma}_{\xi\alpha}(f_1^\xi)(\beta_1^\xi), \ldots, \tilde{\sigma}_{\xi\alpha}(f_\ell^\xi)(\beta_\ell^\xi))$$

and, recalling that $\tilde{\sigma}_{\xi\alpha}(f_i^\xi)(\beta_i^\xi) = \sigma_\xi^\star(x_i^\xi)$, this is equivalent to

$$M_\xi^\star \models \varphi(x_1^\xi, \ldots, x_\ell^\xi).$$

It follows that the map

$$\tilde{\sigma}_\xi^\star : x_i^\xi \mapsto \tilde{\sigma}_{\xi\alpha}^*(\tilde{f}_i^{*\xi})(\beta_i^\xi)$$

is a $\Sigma_0^{(k)}$-preserving embedding of $M_\xi^\star$ into $\tilde{M}_\xi^\star$ sending $\kappa_\xi^\star$ to κ. If C1) holds, $\tilde{\sigma}_\xi^\star$ is easily seen to be Σ^*-preserving, so $\tilde{M}_\xi^\star$ cannot be fully iterable since $M_\xi^\star$ fails to be. If C2) holds, then both $\kappa_\xi^\star$ and κ are in the interval $[\omega\varrho^{n+1}, \omega\varrho^n)$ of the respective premice and $\tilde{\sigma}_\xi^\star$ is $\Sigma_0^{(n)}$-preserving, so $\tilde{M}_\xi^\star$ cannot be normally iterable above κ, as $M_\xi^\star$ fails to be normally iterable above $\kappa_\xi^\star$. □(Claim 1′)

The above claim enables us to find a $\bar{\xi} < \alpha$ and construct premice $\bar{M}_\xi$ extending $Q_{\bar{\xi}}$, and embeddings $\bar{\sigma}_\xi : M_\xi^* \to \bar{M}_\xi$ and $\sigma_\xi' : \bar{M}_\xi \to M_\xi$ as in the proof of i). We then derive a contradiction by embedding $M_\xi^\star$ into some $M_{\xi'}$ via a sufficiently preserving embedding. It will be clear that the construction of such an embedding is a mere refinement of the method from the proof of i). Again, we begin with the case C2).

If C2) holds, then, as before, $\bar{M}_\xi = \bar{M}$ for all ξ. We fix $\xi, \xi' \in S$ such that $\xi < \xi'$ and all β_i^ξ are in the range of $\sigma_{\xi'\alpha}$. Then the assignment

$$x_i \mapsto \sigma_{\xi'}'(\bar{f}_i^\xi)(\sigma_{\xi'\alpha}^{-1}(\beta_i^\xi))$$

where $\bar{f}_i^\xi = \bar{\sigma}_\xi(f_i^{*\xi})$, is a $\Sigma_0^{(n)}$-preserving embedding of $M_\xi^\star$ into $M_{\xi'}$ sending $\kappa_\xi^\star$ to $\kappa_{\xi'}$. This yields the desired contradiction, as $M_{\xi'}$ is fully iterable, but $M_\xi^\star$ fails to be normally iterable above $\kappa_\xi^\star$. To verify the preservation properties of the above embedding, notice that for all $i_1, \ldots, i_\ell$ and any $\Sigma_0^{(n)}$-formula $\varphi(v_1, \ldots, v_\ell)$ we have

$$M_{\xi'} \models \varphi(\sigma_{\xi'}'(\bar{f}_1^\xi)(\bar{\beta}_1^\xi), \ldots, \sigma_{\xi'}'(\bar{f}_\ell^\xi)(\bar{\beta}_\ell^\xi)) \longleftrightarrow \prec\bar{\beta}_1^\xi, \ldots, \bar{\beta}_\ell^\xi\succ \in u^{M_{\xi'},\varphi}_{\sigma_{\xi'}'(\bar{f}_1^\xi),\ldots,\sigma_{\xi'}'(\bar{f}_\ell^\xi)},$$

where $\bar{\beta}_i^\xi = \sigma_{\xi'\alpha}^{-1}(\beta_i^\xi)$. Using the fact that the maps σ_ζ' extend the original maps $\sigma_{\bar{\xi}\zeta}$ together with the preservation properties of $\sigma_{\xi'}'$ we infer:

$$u^{M_{\xi'},\varphi}_{\sigma_{\xi'}'(\bar{f}_1^\xi),\ldots,\sigma_{\xi'}'(\bar{f}_\ell^\xi)} = \sigma_{\xi'}'(u^{\bar{M}_\xi,\varphi}_{\bar{f}_1^\xi,\ldots,\bar{f}_\ell^\xi}) = \sigma_{\xi\xi'} \circ \sigma_\xi'(u^{\bar{M}_\xi,\varphi}_{\bar{f}_1^\xi,\ldots,\bar{f}_\ell^\xi}) = \sigma_{\xi\xi'}(u^{M_\xi,\varphi}_{f_1^\xi,\ldots,f_\ell^\xi}).$$

The second equality follows exactly as in the proof of i). By applying $\sigma_{\xi'\alpha}$ to these equalities, the above proves to be equivalent to

$$\prec\beta_1^\xi, \ldots, \beta_\ell^\xi\succ \in \sigma_{\xi\alpha}(u^{M_{\xi_i},\varphi}_{f_1^\xi,\ldots,f_\ell^\xi}),$$

and this, in turn, is equivalent to $\bar{M}_\xi \models \varphi(\bar{\sigma}_{\xi\alpha}(f_1^\xi)(\beta_1^\xi), \ldots, \bar{\sigma}_{\xi\alpha}(f_\ell^\xi)(\beta_\ell^\xi))$ and thus to $M_\xi^\star \models \varphi(x_1, \ldots, x_\ell)$.

If C1) holds, then each $\bar{M}_\xi$ is fully iterable, as $\bar{\sigma}_\xi$ and σ_ξ' are Σ^*-preserving. As before we fix $\xi < \xi'$ such that $\bar{M}_\xi$ is minimal among all $\bar{M}_\zeta$ in the mouse ordering and all β_i^ξ are in the range of $\sigma_{\xi'\alpha}$. Finally, define $\bar{M}'$, M', $\bar{\pi}$ and σ' exactly as before. We show that

$$x_i \mapsto \sigma' \circ \bar{\pi}(\bar{f}_i^\xi)(\sigma_{\xi'\alpha}^{-1}(\beta_i^\xi))$$

is a Σ^*-preserving embedding of M_ξ^* into M'. Indeed, for any $i_1, \ldots, i_\ell \in \omega$ and a Σ^*-formula $\varphi(v_1, \ldots, v_\ell)$ we have:

$$M' \models \varphi(\sigma' \circ \bar{\pi}(f_1^\xi)(\bar{\beta}_1^\xi), \ldots, \sigma' \circ \bar{\pi}(f_\ell^\xi)(\bar{\beta}_\ell^\xi))$$

holds if and only if

$$\prec \bar{\beta}_1^\xi, \ldots, \bar{\beta}_\ell^\xi \succ \in u^{M',\varphi}_{\sigma'\circ\bar{\pi}(\bar{f}_1^\xi),\ldots,\sigma'\circ\bar{\pi}(\bar{f}_\ell^\xi)} \cap \kappa_{\xi'} = \sigma_{\bar{\xi}\xi'}(u^{\bar{M}_\xi,\varphi}_{\bar{f}_1^\xi,\ldots,\bar{f}_\ell^\xi}) = \sigma_{\xi\xi'}(u^{M_\xi,\varphi}_{f_1^\xi,\ldots,f_\ell^\xi}),$$

where $\bar{\beta}_i^\xi$ and $\bar{f}_i^\xi$ are as in the case C2) above. The first equality holds because $\sigma' \circ \bar{\pi}$ is Σ^*-preserving and agrees with $\sigma_{\bar{\xi}\xi'}$ on $\kappa_{\bar{\xi}}$. So equivalently,

$$\prec \beta_1^\xi, \ldots, \beta_\ell^\xi \succ \in \sigma_{\xi\alpha}(u^{M_\xi,\varphi}_{f_1^\xi,\ldots,f_\ell^\xi}),$$

and as above we see that this is equivalent to $M_\xi^* \models \varphi(x_1^\xi, \ldots, x_\ell^\xi)$, which completes the proof in the case C1).

Finally, we note that the remark after the proof of i) also applies to ii), i.e. using the idea mentioned there we could apply the argument from the case C1) to the case C2) as well, and thus unify the proof of the theorem. □(Theorem 7.5.2)

The proof of Theorem 7.5.3 is left to the reader, as it virtually agrees with the proof of the case C1) of Theorem 7.5.2. The only difference is that this time all extensions are coarse, so we have to work with 0-iterability instead of $*$-iterability and use lemmata on coarse extensions, which might slightly differ from the fine structural ones. In fact, the special case of Theorem 7.5.2, where all M_ξ are ZFC$^-$-models, agrees with the coarse version of the Frequent Extensions of Embeddings Lemma, as there is no difference between fine structural and coarse extensions there.

Proof of Theorem 7.5.1. Suppose the theorem fails and κ is the minimal counterexample, i.e. the least K-cardinal larger than or equal to ω_2 whose cardinal successor in K has cofinality strictly smaller than the size of κ. Set

$$\tau = \kappa^{+K}, \quad \gamma = \mathrm{cf}(\tau).$$

We derive a contradiction as in the proof of the weak covering theorem for K^c (Theorem 6.6.1), i.e. by showing that τ is actually collapsed inside K. We shall closely follow the proof of Theorem 6.6.1. As κ need not even be a cardinal in V, we cannot work with ω-closed structures anymore; instead, we apply the Frequent Extensions of Embeddings Lemma. Fix a regular α satisfying $\max(\gamma, \omega_1) < \alpha \leq \kappa$. We shall work in $V[G]$, where G is $\mathrm{Col}(\alpha, \tau)$-generic over V. Here $\mathrm{Col}(\alpha, \tau)$ is the usual poset for collapsing τ to α consisting of partial functions of size less than α, so $V[G]$ does not contain any new maps from $\bar{\alpha}$ to V for $\bar{\alpha} < \alpha$. Also, $0^\ddagger$ does not exist in $V[G]$ and K is the core model in the sense of $V[G]$ by Lemma 6.5.6 and Theorem 7.4.11.

In $V[G]$, fix a surjective mapping $f : \alpha \xrightarrow{\text{onto}} J_\tau^E$ where, as usually, E is the measure sequence of $\boldsymbol{K}$. As $\mathrm{cf}(\tau) < \alpha$ and α remains regular in $V[G]$, $f''\alpha^*$ is cofinal in τ for some $\alpha^* < \alpha$. Consider the set $C \subset \alpha$ of all $\bar{\alpha} > \alpha^*$ satisfying

$$(f''\bar{\alpha}) \cap \alpha = \bar{\alpha} \text{ and } J_\tau^E|(f''\bar{\alpha}) \text{ is an elementary substructure of } J_\tau^E.$$

Then C is obviously a c.u.b. subset of α; let

- $\langle \alpha_\xi;\ \xi < \alpha \rangle$ be the normal enumeration of C;
- $Q_\xi = J_{\tau_\xi}^{E^\xi}$ be the transitive collapse of $J_\tau^E|(f''\bar{\alpha})$ and let κ_ξ be the largest cardinal in Q_ξ;
- $\sigma_{\xi\alpha} : Q_\xi \to J_\tau^E$ be the uncollapsing map and $\sigma_{\xi\xi'} = \sigma_{\xi'\alpha}^{-1} \circ \sigma_{\xi\alpha}$ for $\xi' > \xi$;
- $\alpha_\xi = \mathrm{cr}(\sigma_{\xi\alpha}) = \mathrm{cr}(\sigma_{\xi\xi'})$ and $\vartheta_\xi = \kappa_\xi^{+Q_\xi}$.

We thus obtain a commutative diagram satisfying a) – e) of the Frequent Extensions of Embeddings Lemma. Finally, let S be the set of all ordinals less than α of uncountable cofinality. The following is an analogue to Lemma 6.6.3.

Claim 1. For all but nonstationarily many $\xi \in S$,

$$\mathcal{P}(\alpha_\xi) \cap Q_\xi \subsetneqq \mathcal{P}(\alpha_\xi) \cap \boldsymbol{K}. \tag{7.5.1}$$

Proof. Suppose this fails. Then α_ξ is an inaccessible cardinal both in Q_ξ and $\boldsymbol{K}$ and $\vartheta_\xi = \alpha_\xi^{+\boldsymbol{K}}$ on some stationary set $S' \subset S$ of ξ's. Also, $\boldsymbol{K}$ agrees with Q_ξ below ϑ_ξ by the condensation lemma. As $\sigma_{\xi\alpha}(\alpha_\xi) = \alpha$, the system of structures and maps $\langle \bar{Q}_\xi, \sigma_{\xi\xi'} \restriction \bar{Q}_\xi;\ \xi < \xi' \leq \alpha \rangle$, where $\bar{Q}_\xi = J_{\vartheta_\xi}^{E^\xi}$, satisfies a) – e) of the Frequent Extensions of Embeddings Lemma. Furthermore, $M_\xi = \boldsymbol{K}$ is a proper lengthening of $\bar{Q}_\xi$ for $\xi \in S'$ and the assumptions of the coarse version of the Frequent Extensions of Embeddings Lemma are satisfied, so for all but nonstationarily many $\xi \in S'$, the canonical 0-extension

$$\tilde{\sigma}_{\xi\alpha}^* : \boldsymbol{K} \to \tilde{\boldsymbol{K}}_\xi$$

of $\sigma_{\xi\alpha} \restriction \bar{Q}_\xi$ exists and $\tilde{\boldsymbol{K}}_\xi$ is a weasel. By Theorem 7.4.8, $\tilde{\boldsymbol{K}}_\xi$ is a simple iterate of $\boldsymbol{K}$ and $\tilde{\sigma}_{\xi\alpha}^*$ is the associated iteration map. As $\alpha = \mathrm{cr}(\tilde{\sigma}_{\xi\alpha}^*)$, the first iteration index used in this iteration must be ϑ_ξ. It follows that $E_{\vartheta_\xi} \neq \varnothing$ and $E_{\vartheta_\xi}^{\tilde{\boldsymbol{K}}_\xi} = \varnothing$. But $\alpha_\xi < \alpha$, so $\vartheta_\xi \leq \alpha$ and $E_{\vartheta_\xi}^{\tilde{\boldsymbol{K}}_\xi} = E_{\vartheta_\xi}$, as $\tilde{\boldsymbol{K}}_\xi$ is a lengthening of $J_{\tilde{\vartheta}_\xi}^E$, where $\tilde{\vartheta}_\xi = \sup(\sigma_{\xi\alpha}''\vartheta_\xi)$. Contradiction. □(Claim 1)

In the following, we can w.l.o.g. assume that (7.5.1) holds for all $\xi \in S$. Exactly as in the proof for $\boldsymbol{K}^c$ we observe that, in the coiteration of Q_ξ with $\boldsymbol{K}$, we truncate on the $\boldsymbol{K}$-side at the very first step. Letting $M_{\xi,0}$ be the result of this truncation, the coiteration of Q_ξ with $\boldsymbol{K}$ essentially agrees with the coiteration of Q_ξ with $M_{\xi,0}$. Furthermore,

- $\vartheta_\xi = \alpha_\xi^{+M_{\xi,0}}$ and $E^\xi \restriction \vartheta_\xi = E^{M_{\xi,0}} \restriction \vartheta_\xi$;
- $\omega\varrho^\omega_{M_{\xi,0}} \le \kappa_{\xi,0}$, where $\kappa_{\xi,0}$ is the first critical point in the coiteration;
- $M_{\xi,0}$ is sound.

Denote the iteration indices of the above coiteration by $\nu_{\xi,i}$, critical points by $\kappa_{\xi,i}$ and mice on the $M_{\xi,0}$-side by $M_{\xi,i}$.

Claim 2. Q_ξ is not moved in the coiteration for all but nonstationarily many $\xi \in S$.

Proof. This corresponds to Lemma 6.6.4 and the proof is essentially the same, so we shall discuss only the new issues which arise here. Suppose for a contradiction that Q_ξ is moved for stationarily many ξ and $\iota(\xi)$ is the first index i such that $E^\xi_{\nu_{\xi,i}} \neq \varnothing$. Set $\nu(\xi) = \nu_{\xi,\iota(\xi)}$. As before we truncate $M_{\xi,\iota(\xi)}$ to some M^*_ξ, so $\nu(\xi)$ is the cardinal successor of $\kappa(\xi) = \kappa_{\xi,\iota(\xi)}$ in M^*_ξ. It follows that M^*_ξ projects to $\kappa(\xi)$, is sound above $\kappa(\xi)$, and $E^\xi_{\nu(\xi)}$ is a total measure over M^*_ξ. In order to make sure that the argument from the proof of Lemma 6.6.4 goes through, we have to verify that $\tilde{M}_\xi = \mathrm{Ult}(M^*_\xi, E^\xi_{\nu(\xi)})$ exists and is normally iterable above $\kappa(\xi)$.

Notice that $\nu(\xi) \in Q_\xi$, as Q_ξ does not have a top measure. Using a pressing down argument we find a $\bar{\xi} < \alpha$ and a stationary set $S' \subset S$ such that $\nu(\xi) \in \mathrm{rng}(\sigma_{\bar{\xi}\xi})$. By applying the fact that the size of $Q_{\bar{\xi}}$ is strictly less than α, there is a $\bar{\nu} \in Q_{\bar{\xi}}$ satisfying $\nu(\xi) = \sigma_{\bar{\xi}\xi}(\bar{\nu})$ for stationarily many $\xi \in S'$. We can again w.l.o.g. assume that this holds for all such ξ. Now $\bar{\nu}$ is the cardinal successor of some $\bar{\kappa}$ in $Q_{\bar{\xi}}$, so $\kappa(\xi) = \sigma_{\bar{\xi}\xi}(\bar{\kappa})$. Setting $(\kappa', \nu') = \sigma_{\bar{\xi}\alpha}(\bar{\kappa}, \bar{\nu})$ and $Q^*_\xi = J^{E^\xi}_{\nu(\xi)}$, each $\sigma^*_{\xi\alpha} = \sigma_{\xi\alpha} \restriction Q^*_\xi$ maps Q^*_ξ elementarily into $Q^* = J^E_{\nu'}$. The system $\langle Q^*_\xi, \sigma^*_{\xi\xi'};\ \xi < \xi' \le \alpha\rangle$ then satisfies a) – e) of the Frequent Extensions of Embeddings Lemma and each M^*_ξ is a lengthening of Q^*_ξ satisfying A) – C), where C2) is the case. Also, w.l.o.g. $n_\xi = n$ for some fixed n and all ξ. Hence, for all but nonstationarily many $\xi \in S'$:

- The canonical extension $\tilde{\sigma}^*_{\xi\alpha} : M^*_\xi \to M'_\xi$ of $\sigma^*_{\xi\alpha}$ exists, and is $\Sigma^{(n)}_0$-preserving and cofinal;
- $\omega\varrho^{n+1}_{M'_\xi} \le \kappa' < \omega\varrho^n_{M'_\xi}$ and M'_ξ is sound above κ';
- M'_ξ is normally iterable above κ'.

As κ' is a cardinal in K, the coiteration of M'_ξ with K is above κ', so the two structures are coiterable. A standard argument yields:

- M'_ξ is an initial segment of K.

Otherwise one of the structures M'_ξ, K is moved in the coiteration. If M'_ξ is not moved, then M'_ξ is a proper initial segment of the last model K' on the K-side. This is clear in the case where there is no truncation on the K-side; in the remaining case it suffices to observe that $\boldsymbol{K}'$ is not sound above κ'. Notice that the first coiteration index ν_0' is a cardinal in all structures on the $\boldsymbol{K}$-side with the possible exception of $\boldsymbol{K}$, and is at least as large as $\tilde{\nu}(\xi) = \sup((\sigma^*_{\xi\alpha})''\nu(\xi))$, as M_ξ' agrees with $\boldsymbol{K}$ below $\tilde{\nu}(\xi)$. Then M_ξ' is a proper initial segment of $\boldsymbol{K}'||\nu_0'$ since ν_0' is a cardinal in $\boldsymbol{K}'$ and M_ξ' projects to $\kappa' < \nu_0'$. But $\boldsymbol{K}'$ agrees with $\boldsymbol{K}$ below ν_0', so in fact, M_ξ' is an initial segment of $\boldsymbol{K}$. Now suppose that M_ξ' is moved in the coiteration. Then $\boldsymbol{K}$ must also be moved and there must be a truncation on the $\boldsymbol{K}$-side, as the last model on the M_ξ'-side is not sound, but is an initial segment of $\boldsymbol{K}'$. The argument from the proof of Lemma 5.1.4 yields that M_ξ' agrees with the result of the last truncation on the $\boldsymbol{K}$-side and it follows from the above discussion that M_ξ' must actually be an initial segment of $\boldsymbol{K}$. (Instead of ν_0' we now consider the index ν_ℓ' of the first extender used on the $\boldsymbol{K}$-side.)

$E^\xi_{\nu(\xi)}$ is a total measure on Q_ξ, as the Q_ξ-side of the coiteration of Q_ξ with $\boldsymbol{K}$ is simple, so $E_{\nu'}$ is a total measure in $\boldsymbol{K}$ with critical point κ'. Let

$$\pi' : \boldsymbol{K} \underset{E_{\nu'}}{\longrightarrow} \tilde{\boldsymbol{K}}.$$

Then $\tilde{M}_\xi' = \pi'(M_\xi')$, being an initial segment of $\tilde{\boldsymbol{K}}$, is a mouse. As $\tilde{\sigma}^*_{\xi\alpha}$ is an extension of $\sigma^*_{\xi\alpha}$, for all $x \subset \kappa(\xi)$ we have

$$x \in E^\xi_{\nu(\xi)} \quad\longleftrightarrow\quad \tilde{\sigma}^*_{\xi\alpha}(x) \in E_{\nu'}.$$

A standard computation then yields that $[f] \mapsto \pi' \circ \tilde{\sigma}^*_{\xi\alpha}(f)(\kappa')$ is a structure preserving embedding of $\mathbb{D}^*(M^*_\xi, E^\xi_{\nu(\xi)})$ into $\tilde{M}_\xi'$, so $\tilde{M}_\xi$ exists. Moreover, as

$$\omega\varrho^{n+1}_{\tilde{M}_\xi'} \leq \pi'(\kappa') < \omega\varrho^n_{\tilde{M}_\xi'},$$

the same computation shows that the above assignment is in fact a $\Sigma_0^{(n)}$-preserving embedding of $\tilde{M}_\xi$ into $\tilde{M}_\xi'$ sending $\kappa(\xi)$ to κ', as is depicted below:

$$\begin{array}{ccc}
M_\xi' & \xrightarrow{\ \pi'\ } & \tilde{M}_\xi' \\
\Big\uparrow{\scriptstyle \tilde{\sigma}^*_{\xi\alpha}} & & \Big\uparrow \\
M^*_\xi & \xrightarrow[E^\xi_{\nu(\xi)}]{\ *\ } & \tilde{M}_\xi
\end{array}$$

This guarantees the normal iterability of $\tilde{M}_\xi$ above $\kappa(\xi)$. The rest of the proof is just a literal repetition of the argument from the proof of Lemma 6.6.4 □(Claim 2)

In the rest of the proof of Theorem 7.5.1 we can w.l.o.g. suppose that Q_ξ is not moved in the coiteration with $\boldsymbol{K}$ for all $\xi \in S$. Exactly as in the proof of Theorem 6.6.1 we find an iterate $M_{\xi,\theta(\xi)}$ of $M_{\xi,0}$ which lengthens Q_ξ and an initial segment M_ξ of

$M_{\xi,\theta(\xi)}$ projecting to κ_ξ and such that τ_ξ is the cardinal successor of κ_ξ in M_ξ. Such an M_ξ clearly satisfies A) – C) of the Frequent Extensions of Embeddings Lemma, so for all but nonstationarily many $\xi \in S$:

- the canonical extension $\tilde{\sigma}_{\xi\alpha} : M_\xi \to N_\xi$ of $\sigma_{\xi\alpha} : Q_\xi \to J_\tau^E$ exists and is $\Sigma_0^{(n)}$-preserving and cofinal;
- $\omega\varrho_{N_\xi}^{n+1} \leq \kappa < \omega\varrho_{N_\xi}^n$ and N_ξ is sound above κ;
- N_ξ is normally iterable above κ;

As in the proof of Claim 2 we observe that N_ξ is coiterable with $\boldsymbol{K}$, as their coiteration is above κ. We then follow the end of the proof of Theorem 6.6.1 and show that $N_\xi \in \boldsymbol{K}$. (Also notice that the argument from the proof of Claim 2 in fact shows that N_ξ is an initial segment of $\boldsymbol{K}$.) Since N_ξ projects to κ and lengthens J_τ^E, $\boldsymbol{K}$ must contain a subset of κ which lies outside J_τ^E, so τ must be collapsed to κ in $\boldsymbol{K}$. Contradiction. □(Theorem 7.5.1)

In the rest of this section we give a hint on how to prove Theorem 6.6.2. Work again in $\boldsymbol{V}[G]$, where G is $\mathrm{Col}(\alpha, J_\tau^E)$-generic over $\boldsymbol{V}$. Build the system $\langle Q_\xi, \sigma_{\xi\xi'}\rangle_{\xi,\xi'}$ as before. To prove Claim 1, define an ultrafilter U_ξ on α_ξ by

$$x \in U_\xi \longleftrightarrow \alpha_\xi \in \sigma_{\xi\alpha}(x)$$

and show that U_ξ is ω-complete for all but nonstationarily many $\xi \in S$. Assume the contrary and for stationarily many $\xi \in S$ pick a sequence $\langle x_i^\xi;\ i \in \omega\rangle$ of elements of U_ξ with empty intersection. Using a pressing down argument, find a $\bar{\xi} < \alpha$ satisfying $x_i^\xi \in \mathrm{rng}(\sigma_{\bar{\xi}\xi})$ for all i and stationarily many ξ and then, using the fact, that α is ω-closed, a sequence $\langle \bar{x}_i;\ i \in \omega\rangle$ of elements of $Q_{\bar{\xi}}$ and a stationary set S' of ξ's such that $x_i^\xi = \sigma_{\bar{\xi}\xi}(\bar{x}_i)$ for all i. Given any pair $\xi < \xi'$ from S' and $i \in \omega$, we have $\sigma_{\xi\xi'}(x_i^\xi) = x_i^{\xi'}$. Use this to show that the intersection of all sets $x_i^{\xi'}$ contains α_ξ, contradicting our assumption.

Having proved the ω-completeness of U_ξ, notice that U_ξ, being a subset of Q_ξ, is an element of $\boldsymbol{V}$, which permits us to apply the argument from the proof of Lemma 6.6.3 to complete the proof of Claim 1. The rest of the proof of the theorem is a literal repetition of the proof of Theorem 7.5.1.

Notes

The notion of thick class is due to Mitchell. The inductive definition of $\boldsymbol{K}$ is originally due to Jensen; the definition of $\boldsymbol{K}$ based on the hull property is due to Steel [St96] and was abstracted from the construction of Mitchell [Mi1]. Steel [St96] also has an inductive definition of $\boldsymbol{K}$ below a Woodin cardinal. The construction of $\boldsymbol{K}$ we presented is basically due to Steel, however, his original construction seems to require second-order set theory (e.g. Morse–Kelley). The adaptation of the construction to the ZFC-context is due to Neeman. The frequent extension lemma and its generalizations to the core model context are due to Jensen. Jensen also proved the Σ^1_3-correctness of $\boldsymbol{K}$ below one measurable cardinal. Steel proved the same result for the core model up to a Woodin cardinal using an additional assumption on measurable cardinals in $\boldsymbol{V}$. In fact, he proved a much stronger result than a mere Σ^1_3-correctness. A proof of Σ^1_3-correctness for $\boldsymbol{K}$ below one strong cardinal under a weaker hypothesis is due to and Steel and Welch [SW98].

Chapter 8

One Strong Cardinal

The aim of the present chapter is to generalize the core model theory developed so far to a theory for models constructed relative to coherent extender sequences which can contain one strong cardinal. This theory can be considered a cosmetic improvement of the general theory under the assumption "no sharp for an inner model with a strong cardinal", so the reader interested in the general theory of models for overlapping extenders might skip over the present chapter. However, one strong cardinal constitutes a natural dividing line between "small" and "large" core models in that below this line, the comparison process can be carried out using *linear* iterations, which considerably simplifies both the theory and applications. For this reason, it is convenient to use the core model theory based on linear iterations whenever the situation permits. As an example we mention the singular cardinal hypothesis, whose failure has the consistency strength well below one strong cardinal, but there are many other examples related to changing cofinality of cardinals or infinitary combinatorics. Based on linear iterations, the core model theory below one strong cardinal looks very much like that for measures of order 0 and there are only very few places which require qualitatively new ideas. For this reason, we give only a sketch of the theory, stating merely those theorems, which require reformulations and leaving all standard proofs to the reader.

There is one important point in which our core model theory below a strong cardinal differ from that for measures of order 0, namely that the premice are no longer amenable structures. For this reason, whenever we want to form ultrapowers or do fine structural arguments, we replace them by amenable structures coding the same amount of information. This coding and decoding process constitutes an inconvenience which still pays, as the price for giving up the linear comparison process would be much higher.

8.1 Premice

As we have already mentioned in the introduction to this chapter, we shall introduce two kinds of premice. Firstly, the so called *bookkeeping* premice, which will determine the extender sequences and which need not be necessarily amenable. Secondly, the so called *expansions* of the former, which will be amenable, although larger than is necessary for coding the same amount of information. As most of the arguments will be done using the expansions, we shall call them *premice*. However, it will be convenient to be somewhat sloppy with the terminology and use the term "premouse" for both bookkeeping premice and expansions, as in most cases, the meaning will be clear from the context.

Definition. A structure M is a *bookkeeping premouse* iff M is of the form $\langle J_\alpha^E, F\rangle$ and satisfies the clauses a) – f) below.

a) J_α^E is acceptable.

b) $E \subset \omega\alpha \times M \times \omega\alpha$ and $F \subset M \times \omega\alpha$.

Set

$$E_\nu = \begin{cases} \{\langle x, \xi\rangle;\ \langle \nu, x, \xi\rangle \in E\} & \text{if } \nu < \omega\alpha \text{ and} \\ F & \text{if } \nu = \omega\alpha, \end{cases}$$

and

$$M||\nu = \langle J_\nu^E, E_{\omega\nu}\rangle \text{ for } \nu \leq \omega\alpha.$$

c) $E_\nu = \varnothing$ whenever ν is *not* primitive recursively closed and, if $E_\nu \neq \varnothing$, then $E_\nu \subset J_\nu^E$ (notice that $\omega\nu = \nu$ in this case) and $\langle E_{\nu,\xi};\ \xi < \nu\rangle$ is a weakly amenable hypermeasure on J_ν^E of length ν with critical point $\kappa < \nu$, where

$$E_{\nu,\xi} = \{x \subset \kappa;\ \langle x, \xi\rangle \in E_\nu\}.$$

Informally, we shall call E_ν a hypermeasure or an extender.

d) Let $E_\nu \neq \varnothing$ and $\pi : J_\nu^E \underset{E_\nu}{\longrightarrow} M'$ weakly. Then

i) $E^{M'} \restriction \nu = E \restriction \nu$;

ii) $\nu \in \text{wfcore}(M')$;

iii) $E_\nu^{M'} = \varnothing$.

For $M = \langle J_\alpha^E, F\rangle$ satisfying a) – d) we define the so called *expansion* $\widehat{M}$ of M by

$$\widehat{M} \simeq \begin{cases} M & \text{if } F = \varnothing \text{ and} \\ \langle J_{\hat{\alpha}}^{\widehat{E}}, \widehat{F}\rangle & \text{if } F \neq \varnothing, \end{cases}$$

where $J_{\hat{\alpha}}^{\widehat{E}} = \text{Ult}(J_\tau^E, F)$, π is the associated ultrapower map and $\widehat{F} = \pi \restriction \mathcal{P}(\kappa) \cap M$ for $\tau = \kappa^{+M}$ and $\kappa = \text{cr}(F)$. The original structure M is then called the *cutback* of $\widehat{M}$ and can be reconstructed from $\widehat{M}$ in the obvious way, as $E = \widehat{E} \restriction \alpha$ and F is the hypermeasure coding the extender $\widehat{F} \restriction \alpha$.

e) If $\nu < \alpha$ and $\widehat{M||\nu}$ exists then it is sound.

f) *(Initial segment condition)* Let $E_\nu \neq \varnothing$ and $\kappa = \text{cr}(E_\nu)$. If $\kappa^{+M||\nu} \leq \bar{\nu} < \nu$ and $\bar{\nu}$ is primitive recursively closed, then $E_{\bar{\nu}} \neq \varnothing$.

Although E represents a sequence hypermeasures, we shall be somewhat sloppy with the notation and call the objects E_ν extenders. In the same spirit, if F is the top extender of $\widehat{M}$, then we consider $F|\alpha$ to be the top hypermeasure (or, in our more liberal terminology, the top extender) of M. Notice also that $\widehat{E} \restriction \alpha = E$, as follows from d).

Definition. M is a *premouse* iff $M = \widehat{M}$ for some bookkeeping premouse M.

Definition. Given a bookkeeping premouse M and an ordinal $\kappa \in M$, we define:

$$\begin{aligned} \mathrm{O}^M(\kappa) &= \{\nu \geq \kappa^{+M};\ E^M_\nu \neq \varnothing \ \&\ \mathrm{cr}(E^M_\nu) = \kappa\} \\ \mathrm{o}^M(\kappa) &= \text{the least primitive recursively closed } \beta \geq \text{l.u.b.}(\mathrm{O}^M(\kappa)) \\ \mathrm{o}^M_*(\kappa) &= \mathrm{otp}(\mathrm{O}^M(\kappa)) \end{aligned}$$

The ordinal $\mathrm{o}^M(\kappa)$ is called the *order* of κ in M and $\mathrm{o}^M_*(\kappa)$ *Mitchell order* of κ in M.

Our definition of Mitchell order is not fully consistent with the standard definition as not all extenders in M which are below $E_\beta \in M$ in the Mitchell order must be on the sequence E. On the other hand, if $\mathrm{o}^M_*(\kappa) \geq \lambda$ then the Mitchell order of κ in the standard sense is at least λ, and this is what is needed in the applications.

There are other points where we shall be sloppy with the terminology. This will prove useful, as the notation and continuous switching between bookkeeping premice and the corresponding expansion might obscure the exposition.

- We shall omit the "hat" operator whenever there is no danger of confusion. This in particular applies to various fine structural parameters. We shall e.g. write $\omega\varrho^n_M$, p_M, $h^{n,p\restriction n}_M$, ... etc. instead of $\omega\varrho^n_{\widehat{M}}$, $p_{\widehat{M}}$, $h^{n,p\restriction n}_{\widehat{M}}$, ... etc.
- On the other hand, we shall use "hat" in certain situations which allow to drop it; usually to stress that we work with the expansion. Another typical case when the "hat" will not be omitted is talking about maps between structures. Thus, $\sigma : M \to N$ will always mean a map between the bookkeeping premice and *not* between the corresponding expansions.
- We shall avoid using the term "bookkeeping premouse" unless there is a danger of confusion. So we shall talk about "premice" as it will be clear from the context which structures we refer to.

As the ordinals indexing extenders are primitive recursively closed, the following fact will be useful for the computations below

Fact. The statement "ξ is primitive recursively closed" is uniformly Δ_1 over any J^A_α.

We now prove that the set of indices of E^M extenders with critical point κ is essentially an interval, and that the extenders of E do not overlap, that is, the ordinals indexing extenders with a critical point μ cannot be cofinal in any measurable cardinal (in M) larger than μ.

Lemma 8.1.1. *Let M be a bookkeeping premouse and $E = E^M$.*

a) *If $\kappa^+ \leq \gamma < \mathrm{o}^M(\kappa)$ and γ is primitive recursively closed then* $\mathrm{cr}(E_\gamma) = \kappa$.

b) *If $\mu < \kappa$ and there is a total extender E_β in M with critical point κ then* $\mathrm{o}^M(\mu) < \kappa$.

Proof. To see a), let E_β be an extender with critical point κ and $\tau \leq \gamma < \beta$, where $\tau = \kappa^{+M}$. We show that $\mathrm{cr}(E_\gamma) = \kappa$. Suppose for contradiction that $\mathrm{cr}(E_\gamma) = \mu < \kappa$. Notice that κ is inaccessible in M as E_β is weakly amenable w.r.t. M. So $\vartheta = \mu^{+M} < \kappa$. Now if

$$\pi : J^E_\beta \underset{E_\beta}{\longrightarrow} M' \quad \text{weakly}$$

then $\pi(\kappa) \geq \beta$ in M', so by d) in the definition of premouse,

$$M' \models (\exists\xi)(\vartheta \leq \xi \leq \pi(\kappa)\ \&\ \xi \text{ is primitive recursively closed }\ \&\ E_\xi = \varnothing),$$

as β witnesses the existence of such a ξ. Using the elementarity of π we infer that there is a primitive recursively closed ordinal $\bar{\xi}$ such that $\vartheta \leq \bar{\xi} \leq \kappa$ such that $E_{\bar{\xi}} = \varnothing$. As $\kappa < \gamma$, this contradicts the initial segment condition for γ.

If $\mu > \kappa$, the proof is similar; the only difference is that we form an ultrapower of M by E_γ. Also, it is easy to see that the above proof yields also b). □(Lemma 8.1.1)

Next we need to know that premousehood can be internally characterized by a simple sentence. Unlike the measure of order 0 case, no Q-statement will suffice for this purpose if the premouse in question has a top extender. We stress that now we talk about premice, i.e. expansions of bookkeeping premice.

Lemma 8.1.2. *There is a Q-sentence φ such that*

$$\langle J^E_\alpha, \emptyset\rangle \textit{ is a premouse } \longleftrightarrow \langle J^E_\alpha, \emptyset\rangle \models \varphi.$$

Proof. In this case the premouse in question agrees with the corresponding bookkeeping premouse, so it suffices to verify a) – f) of the above definition. We already know that a) is a Q-statement; b) and c) are easily seen to be Π_1 (we apply the above Fact for the latter). To see that i) and iii) in d) are Π_1 and ii) is Σ_1 recall first that if M' is as in d), then $\nu \in \mathrm{wfc}(M')$. Łos' Theorem guarantees that any Σ_0 statement over M' can be expressed by a first-order formula over $\langle J^E_\nu, E_\nu\rangle$. Then i) can be expressed as

$$(\forall\alpha)(\forall\xi < \alpha)(\forall x \subset \kappa)(\langle\alpha, x, \xi\rangle \in E \longleftrightarrow \langle[\alpha, \mathrm{id}], [\kappa, f_x], [\xi, \mathrm{id}]\rangle \in^{N'} E^{N'})$$

where $f_x(\eta) = x \cap \eta$ for all $\eta < \kappa$, ii) as $(\exists\alpha, f)\psi(\alpha, f)$ where $\psi(\alpha, f)$ is the conjunction of the formulae

$$(\forall\xi < \nu)([\xi, \mathrm{id}] \in^{M'} [\alpha, f])$$

and

$$(\forall\beta, g \in J^E_\nu)\big[(\forall\xi < \nu)([\xi, \mathrm{id}] \in^{M'} [\beta, g]) \longrightarrow [\beta, g] \notin^{M'} [\alpha, f])\big],$$

and ii) as

$$(\forall\alpha, f)(\psi(\alpha, f) \longrightarrow E^{M'}_{[\alpha, f]} = \emptyset).$$

The clause e) can be verified similarly as in the measures of order 0 case. Recall that the statement $\bar{M} \models \varphi(a_1, \ldots, a_\ell)$ is uniformly $\Pi_1(J_\alpha^E)$, where $\bar{M} \in J_\alpha^E$ is arbitrary structure and φ a formula in the language for this structure. This was proved in Lemma 4.1.2. Recall also that there is a fixed recursive sequence of sentences such that for any premouse N we have: N is sound iff $(\forall n)(N \models \varphi_n)$. We would like to express that for every ν, $\widehat{M||\nu}$ is sound provided it is well founded. If $E_\nu = \varnothing$ then $M||\nu$ agrees with its expansion, and so is well founded. Otherwise we replace $\widehat{M||\nu}$ by the associated term model and require this model to satisfy the above list of sentences. It is clear that if $\widehat{M||\nu}$ exists, i.e. if the above term model is well-founded, then $\widehat{M||\nu}$ is sound. So e) can be expressed as

$$(\forall \nu)(\forall n)(\mathbb{Q}_\nu \models \varphi_n),$$

where

$$\mathbb{Q}_\nu = \begin{cases} S_\nu^E & \text{if } E_\nu = \varnothing \\ \langle Q, \tilde{E}_\nu, \tilde{F}_\nu \rangle & \text{otherwise,} \end{cases}$$

and

$$\begin{aligned} \langle Q, \tilde{E}_\nu \rangle &= \mathbb{D}(J_\tau^E, E_\nu) \\ \tilde{F}_\nu &= \{\langle [\kappa, f_x], [0, c_x] \rangle;\ x \in \mathcal{P}(\kappa) \cap J_\tau^E\} \quad (\text{here } \kappa = \operatorname{cr}(E_\nu)). \end{aligned}$$

This statement is easily seen to be Π_1.

Finally, f) is $\Pi_1(J_\alpha^E)$, as it is equivalent to

$$(\forall \nu, \kappa)\big(J_\nu^E \models [(E_\nu \neq \varnothing \;\&\; \kappa = \operatorname{cr}(E_\nu)) \longrightarrow (\forall \gamma)(\kappa^+ \leq \gamma \longrightarrow E_\gamma \neq \varnothing)]\big).$$

□(Lemma 8.1.2)

Corollary 8.1.3. *The premousehood of $M = \langle J_\alpha^E, \varnothing \rangle$ is preserved downwards by Σ_0-maps and upwards by cofinal Σ_0-maps and Σ_2-maps.* □(Corollary 8.1.3)

Given a bookkeeping premouse $M = \langle J_\alpha^E, E_\alpha \rangle$, let κ be the critical point of E_α. By Lemma 8.1.1 and d) of the definition of premouse, $\alpha = \mathrm{o}(\kappa)$ in $\mathrm{Ult}(J_\tau^E, E_\alpha)$, where $\tau = \kappa^{+M}$ (notice that this ultrapower is an initial segment of $\mathrm{Ult}(J_\alpha^E, E_\alpha)$). So α is represented in the ultrapower as

$$[\kappa, f] \quad \text{where} \quad f(\xi) = \mathrm{o}^M(\xi).$$

This representation works even in the case when the corresponding term model is not well-founded. If π is the associated ultrapower map, then $\pi(\kappa) > \alpha$, as otherwise $\mathbb{D}(M, E_\alpha)$ would satisfy $(\exists \mu < \pi(\kappa))(\mathrm{o}(\mu) = \pi(\kappa))$ and the same would then hold in M by the elementarity of π, contradicting Lemma 8.1.1.

Note also that if M is as above then $\widehat{M}$ is a coherent structure. To see this it suffices to show that F is whole and $\widehat{M} = \langle \mathrm{Ult}(J_\tau^E, F), F \rangle$. The latter actually follows immediately from the former. F is whole as for any $\beta < \pi(\kappa)$ and $f : \kappa \to \kappa$ from J_τ^E, the pair $[\beta, f]$ is equivalent to $[\pi(f)(\beta), \mathrm{id}]$ modulo F. Now if π' is the ultrapower map associated with the ultrapower by F, the canonical map $k : \pi(f)(\beta) \mapsto \pi'(f)(\beta)$ is easily seen to be the identity on $\pi(\kappa) + 1$, which implies that $k = \mathrm{id}$, $\mathrm{Ult}(J_\tau^E, F) = \mathrm{Ult}(J_\tau^E, E_\alpha)$ and $\pi' = \pi$.

We now turn to the internal characterisation of premice with top extenders. Recall that if M' is a premouse with the top extender F which is the expansion of some bookkeeping premouse M of height α, then $E_\alpha^M = F|\alpha$.

Lemma 8.1.4. *Let $M = \langle J_\nu^E, F \rangle$ where $F \neq \varnothing$. Then M is a premouse (that is, the expansion of a bookkeeping premouse) iff the following conditions hold.*

i) *$\langle J_\nu^E, \varnothing \rangle$ is a premouse.*

ii) *$\langle J_\nu^E, F \rangle$ is a coherent structure.*

iii) *F is weakly amenable w.r.t. J_ν^E, in other words, $\mathrm{dom}(F) = \mathcal{P}(\kappa) \cap J_\nu^E$, where $\kappa = \mathrm{cr}(F)$.*

iv) *If F is at κ, λ and $\alpha = \mathrm{o}(\kappa)$ in J_ν^E, then α is primitive recursively closed and*

$$(\forall \xi < \lambda)(\exists g : \kappa \to \kappa)(\exists \beta < \alpha)([\xi, \mathrm{id}] =_F [\beta, g]).$$

In other words, all generators of F are less than α.

Proof. That any premouse satisfies i) – iv) follows from the above remarks. To see the converse we observe that M is the expansion of $\langle J_\alpha^E, F|\alpha \rangle$ and that this structure is a bookkeeping premouse. That $\mathrm{Ult}(J_\tau^E, F) = \mathrm{Ult}(J_\tau^E, F|\alpha)$ and the associated ultrapower maps agree follows from iv), as this guarantees that the canonical embedding from the latter ultrapower to the former defined by $[\xi, f]_{F|\alpha} \mapsto [\xi, f]_F$ for $\xi < \alpha$ is the identity map. This also yields clauses d) and f) on bookkeeping premice. The remaining clauses follow automatically. □(Lemma 8.1.4)

Lemma 8.1.5. *There is a Π_2-sentence φ such that for every transitive rudimentarily closed structure,*

$$M \text{ is a premouse} \longleftrightarrow M \models \varphi.$$

Proof. If $F = \varnothing$ then the premousehood is a Q-condition by Lemma 8.1.2. Otherwise we apply the internal characterisation given by the previous lemma. Clauses i) and ii) are Q-conditions by the above results and iii) and iv) can be expressed in a Π_2-fashion in a straightforward way. □(Lemma 8.1.5)

Corollary 8.1.6. *Premousehood is preserved downwards by Π_1 maps and upwards by Π_2-maps.* □(Corollary 8.1.6)

We stress that clauses iii) and iv) in Lemma 8.1.4 are genuinely Π_2, in other words, they cannot be expressed in a Q-fashion. To demonstrate this we give examples of Σ_0-preserving cofinal maps failing to preserve iii) and iv). Our examples are typical in the sense that they highlight the reason why this is so and also give a hint as to which additional restrictions are to be imposed on Σ_0 cofinal maps in order to make them premousehood preserving ones.

Example. Let $M = \langle J_\alpha^E, E_\alpha \rangle$ be a bookkeeping premouse and $N = \langle J_\nu^E, F \rangle$ its expansion. Assume that κ is not the largest cardinal in J_α^E, where F is at κ, λ. As $\tau = \kappa^{+M}$ indexes a total extender on κ in $\mathrm{Ult}(J_\tau^E, E_\alpha)$, there are unboundedly many indices for total extenders below κ. Consequently, there are unboundedly many ordinals below λ which index total extenders in J_ν^E. Let $\beta > \alpha$ be such an index. Notice that $\mathrm{cr}(E_\beta) > \alpha$. If $N' = \mathrm{Ult}(N, E_\beta)$ and σ is the associated ultrapower map, then σ is obviously Σ_0-preserving and cofinal, but N' is not the expansion of any bookkeeping premouse: If it were, then it would be the expansion of M, as its top extender has critical point κ and N' agrees with N below β.

Example. To produce a counterexample to preservation of weak amenability, it suffices to work with coherent structures instead of premice. Let $M = \langle J_\nu^A, F \rangle$ be a coherent structure such that F is at κ, λ and $\tau = \kappa^{+M}$ is not ω-cofinal. Suppose also that μ is a cardinal in M which is ω-cofinal in the sense of V and lies strictly between τ and λ. Let $\sigma : \bar{M} \to M$ be fully elementary and countable. Then $\bar{M}$ is a coherent structure and σ is not cofinal, as $\sigma''\bar{\tau}$ is not mapped cofinally into τ (as usually, $\bar{x}$ is the σ-preimage of x). By the interpolation lemma, the coarse canonical extension $\tilde{\sigma} : \bar{M} \to \tilde{M}$ of $\sigma \restriction J_{\bar{\mu}}^{\bar{E}}$ exists, and is Σ_0-preserving and cofinal. It follows that $\tilde{M}$ agrees with M below μ, so $\tilde{F}$ cannot be defined on the whole $\mathcal{P}(\kappa)$ in $\tilde{M}$, as $\mathrm{dom}(\tilde{F}) \subset J_{\tilde{\tau}}^E$ where $\tilde{\tau} = \sup(\sigma''\bar{\tau}) < \tau$. The last inequality follows from the fact that $\bar{\tau}$, unlike τ, is ω-cofinal.

As we have already indicated above, those ultrapower maps, which do not fall into one of the previous examples already preserve premousehood:

Lemma 8.1.7. *Let $M = \langle J_\nu^E, F \rangle$ be the expansion of a bookkeeping premouse $\langle J_\alpha^E, E_\alpha \rangle$, where F is an extender at κ, λ.*

a) *If G is a weakly amenable extender w.r.t. M with critical point $\mu \leq \alpha$, then $\mathrm{Ult}(M, G)$ is a premouse.*

b) *If μ is a cardinal in M such that $\mu \leq \alpha$, $\vartheta = \mu^{+M}$ and $\tilde{\sigma} : M \to M'$ is the coarse canonical extension of some $\sigma : J_\vartheta^E \to J_{\vartheta'}^{E'}$ which maps $\tau = \kappa^{+M}$ cofinally into $\sigma(\tau)$, then M' is a premouse.*

Proof. We prove a) and leave the proof of b), which is very much the same, to the reader. Suppose M' is of the form $\langle J_{\nu'}^{E'}, F' \rangle$ and σ is the ultrapower map associated with $\mathrm{Ult}(M, G)$. Notice that the weak amenability of F' follows from the fact that σ maps τ cofinally into $\sigma(\tau)$: As τ is regular in M and G is weakly amenable w.r.t. M, we have $\mu \neq \tau$ and the conclusion follows from the general properties of ultrapowers. It remains to verify iv) of Lemma 8.1.4.

If $\bar{\tau} < \tau = \kappa^{+M}$ is such that $J_{\bar{\tau}}^E$ is a ZFC$^-$ model, then $\bar{F} = F \restriction (\mathcal{P}(\kappa) \cap J_{\bar{\tau}}^E)$ is an element of M and $\mathrm{Ult}(J_{\bar{\tau}}^E, \bar{F})$ is a proper initial segment of J_ν^E. To see this notice that the canonical map $k : \bar{\pi}(f)(\xi) \mapsto \pi(f)(\xi)$, where $\bar{\pi}$ is the ultrapower map associated with $\mathrm{Ult}(J^E_{\bar{\tau}}, \bar{F})$ and π that one associated with $\mathrm{Ult}(J^E_\tau, F)$, is the identity map. Notice also that $\bar{\pi} \in M$.

Now pick an $x \in M'$. Then $x = \sigma(f)(\gamma)$ for some $f : \mu \to M$ from M and $\gamma < \mathrm{length}(G)$. As $f \in J^E_\nu$, there is an $h : \kappa \to J^E_\tau$ in J^E_τ and a $\bar{\gamma}$ such that $f = \pi(h)(\bar{\gamma})$. By the previous paragraph, if we fix $\bar{\tau}$ large enough so that $h \in J^E_{\bar{\tau}}$, then $f = \bar{\pi}(h)(\bar{\gamma})$. By applying σ we obtain $\sigma(f) = \sigma(\pi)(\sigma(h))(\sigma(\bar{\gamma}))$ and $\sigma(\pi)$ is the ultrapower map associated with $\mathrm{Ult}(J^{E'}_{\bar{\tau}'}, \bar{F}')$, where $\bar{\tau}' = \sigma(\bar{\tau})$ and $\bar{F}' = \sigma(\bar{F}) = F' \restriction (\mathcal{P}(\kappa') \cap J^{E'}_{\bar{\tau}'})$. Using the previous paragraph again infer that $\sigma(\pi)(z) = \pi'(z)$ for all appropriate z, where π' is the ultrapower map associated with $\mathrm{Ult}(J^{E'}_{\tau'}, F')$. Thus, if we choose $h' \in J^{E'}_{\bar{\tau}'}$ such that

$$h'(\prec\eta_1, \eta_2\succ) = \sigma(h)(\eta_1)(\eta_2)$$

whenever defined, we obtain $x = \pi'(h')(\prec\sigma(\bar{\gamma}), \gamma\succ)$. We assume $\mu \leq \alpha$, so $\gamma < \sigma(\mu) \leq \alpha' = \sigma(\alpha)$. That $\sigma(\bar{\gamma}) < \alpha'$ is immediate, and the closure properties of α' guarantee that the same applies to $\prec\sigma(\bar{\gamma}), \gamma\succ$. □(Lemma 8.1.7)

Corollary 8.1.8. *If M is a bookkeeping premouse and E^M_ν is a total extender in M, then* $\mathrm{Ult}(\widehat{M}, E^M_\nu)$ *as well as* $\mathrm{Ult}^*(\widehat{M}, E^M_\nu)$ *is a premouse.* □(Corollary 8.1.8)

8.2 Properties of Mice

In this section we describe how to adapt the theory of mice for measures for measures of order 0 to that for non-overlapping extenders. There are only very few differences between both theories, and these differences are mainly technical. Therefore we well keep our exposition brief and leave most of the work to the reader.

The notion of iteration needs a minor revision. The iteration maps are now maps between *premice*, i.e. between the expansions of the bookkeeping premice. Thus clauses d), e) and g) in the present context read:

d) if $E^{M_i}_{\omega\nu_i} = \varnothing$, then $M_{i+1} = M_i||\alpha_i$ and $\pi_{i,i+1} = \mathrm{id} \restriction \widehat{M_i||\alpha_i}$;

e) if $E^{M_i}_{\omega\nu_i} \neq \varnothing$, then $E^{M_i}_{\nu_i}$ is an extender *on* $M_i||\alpha_i$ (i.e. $E^{M_i}_{\nu_i}$ measures all subsets of its critical point which are elements of $M_i||\alpha_i$) and

$$\pi_{i,i+1} : \widehat{M_i||\alpha_i} \xrightarrow[E^{M_i}_{\nu_i}]{*} \widehat{M}_{i+1};$$

g) if λ is a limit ordinal, then

$$\langle \widehat{M}_\lambda, \langle \pi_{i\lambda};\ i < \lambda \rangle \rangle = \lim_{i \leq j < \lambda} (\widehat{M}_i, \pi_{ij}).$$

The entire theory of iterations, copying constructions and the proof of the Dodd–Jensen Lemma are then straightforward generalizations of the corresponding parts of the theory we developed before. The proof of the iterability criterion in Theorem 4.5.1, however, requires a deeper revision. To be more precise, we need a new definition of strong ω-completeness.

Definition. Let M be a premouse and E^M_ν be a total extender in M. We say that E^M_ν is *strongly ω-complete* iff E^M_ν is ω-complete and for every countable $A \subset \widehat{M}$ there is an $H \subset \widehat{M}$ such that

i) $\omega\varrho^{n+1}_M \cup A \subset H$ and H is closed under good $\Sigma^{(n)}_1(\widehat{M})$-functions;

ii) $\omega\varrho^n_M \cap H$ is unbounded in $\omega\varrho^n_M$;

iii) there is a $\prec\succ$-respecting map $\mathfrak{k} : \nu \twoheadrightarrow \kappa$ such that $\mathfrak{k}(\alpha) \in \bigcap(E^M_{\nu,\alpha} \cap H)$ whenever $\alpha \in H \cap \nu$,

where $\kappa = \mathrm{cr}(E^M_\nu)$ and $n \in \omega$ is such that $\omega\varrho^{n+1}_M \leq \kappa < \omega\varrho^n_M$.

With this definition, one can imitate the proof of Theorem 4.5.1, which goes through with minor amendments. Instead of Lemma 4.5.4 we now use:

Lemma 8.2.1. *Let $\mathfrak{I} = \langle N_i, \pi_{ij};\ i \leq j < \theta + 1\rangle$ be a simple normal iteration of N strictly above κ (i.e. $\kappa_0 > \kappa$). Suppose $\omega\varrho^{n+1}_N \leq \kappa < \omega\varrho^n_N$ for an $n \in \omega$ and set $\omega\varrho^{n+1} = \omega\varrho^i_{N_i}$ $(i \leq \theta)$.*

Given a finite $b \subset \widehat{N}_\theta$, there is a finite set $b^\natural \subset \widehat{N}$ with the following properties:

a) *Every $\alpha < \nu_0$ which is $\Sigma^{(n)}_1(\widehat{N}_\theta)$-definable in parameters from $\omega\varrho^{n+1} \cup b$ is $\Sigma^{(n)}_1(\widehat{N})$-definable in parameters from $\omega\varrho^{n+1} \cup b^\natural$.*

b) *For every $x \in \mathcal{P}(\kappa_0) \cap N = \mathcal{P}(\kappa_0) \cap N_\theta$ which is $\Sigma^{(n)}_1(\widehat{N}_\theta)$ definable in parameters from $\omega\varrho^{n+1} \cup b$ there is a $\mu_0 < \omega\varrho^n_N$ with the following property: Given a $\mu_0 \leq \mu < \omega\varrho^n_N$, there is a $\tau_\mu < \kappa_0^{+N}$ such that x is $\Sigma^{(n)}_1(\widehat{N})$-definable in parameters from $\omega\varrho^{n+1} \cup b^\natural \cup \{\tau, \mu\}$, where τ is arbitrary such that $\tau_\mu \leq \tau < \kappa_0^{+N}$.*

The proof of this lemma is very much like that of Lemma 4.5.4, but technically more demanding. The rest of the proof of the generalization of Theorem 4.5.1 is then straightforward. Also, Theorem 4.5.5 and its proof generalize without any substantial revisions. The proof of Theorem 4.5.6 requires a slight revision. Here, we perform the iteration as in the proof of Theorem 4.5.6 to guarantee that all extenders on the M^*-sequence with the critical point smaller than $\mathrm{cr}(E^{M^*}_{\mathrm{top}})$ are ω-complete. Then $E^{M^*}_{\mathrm{top}}$ is ω-complete by Lemma 2.5.5. We then use these facts to show that also all extenders in M^* with indices between $\tau = \mathrm{cr}(E^{M^*}_{\mathrm{top}})^{+M^*}$ and $\mathrm{ht}(M^*)$ are ω-complete. We leave the details to the reader.

The next crucial point of the theory is the proof of the solidity theorem. Again, the proof, as we set it up for measures of order 0, generalizes in a straightforward way with a single exception, namely the proof of coiterability of M with $W = W^{\nu}_{M}$. Assuming that $\omega\varrho^{n+1}_{M} \leq \nu < \omega\varrho^{n}_{M}$, we know that $\omega\varrho^{n+1}_{W} \leq \nu$ and that W is normally iterable above its $(n+1)$-st projectum. This alone, however, need not help if $\mathrm{o}^{W}(\kappa) > \nu$ for some $\kappa < \omega\varrho^{n+1}_{W}$, as we cannot a priori exclude that there is a disagreement between M and W at some $\bar{\nu} \in [\nu, \mathrm{o}^{W}(\kappa))$. However, there really is no such disagreement, as follows from the following condensation-like lemma.

Lemma 8.2.2. *Let $\bar{M}$ and M be Σ_0-iterable premice, let $\kappa \in \bar{M}$ and let $\gamma = \mathrm{o}^{\bar{M}}(\kappa)$. Granting that*

$$\sigma : \widehat{\bar{M}} \to \widehat{M}$$

is a Σ_0-preserving embedding such that $\kappa < \mathrm{cr}(\sigma)$, the premice $\bar{M}$ and M agree below γ, i.e. $E^{\bar{M}} \restriction \gamma = E^{M} \restriction \gamma$.

Proof. (Sketch.) Let $\nu = \mathrm{cr}(\sigma)$ and $\tau = \kappa^{+\bar{M}}$. Notice that we can assume w.l.o.g. that σ is cardinal preserving, as this can fail only if $E^{\bar{M}}_{\mathrm{top}} = \varnothing$ and σ is non-cofinal, in which case we can replace M by the cofinal image of $\bar{M}$ under σ. Let μ be the first point of disagreement between $E^{\bar{M}}$ and E^{M}. Then both $E^{\bar{M}}_{\mu}$ and E^{M}_{μ} are extenders, as $\mu < \mathrm{o}^{\bar{M}}(\kappa) \leq \mathrm{o}^{M}(\kappa)$. We distinguish two cases.

If $\tau < \nu$, then $\tau = \kappa^{+M}$ and, letting $\mu' = \sigma(\mu)$, both $E^{M}_{\mu'}$ and $E^{\bar{M}}_{\mu}$ are extenders on M. Moreover, by the existence of σ, the premouse $\widehat{M}$ is extendible by $E^{\bar{M}}_{\mu}$ and, letting $\widehat{M}' = \mathrm{Ult}(\widehat{M}, E^{M}_{\mu'})$ and $\widehat{M}^{*} = \mathrm{Ult}(\widehat{M}, E^{\bar{M}}_{\mu})$, we can define a Σ_0-preserving embedding $\sigma^{*} : \widehat{M}^{*} \to \widehat{M}'$ by

$$\sigma^{*}(\pi^{*}(f)(\alpha)) = \pi(f)(\sigma(\alpha))$$

where $\pi : \widehat{M} \to \widehat{M}'$ and $\pi^{*} : \widehat{M} \to \widehat{M}^{*}$ are the associated ultrapower maps. The existence of σ^{*} guarantees the Σ_0-iterability of M^{*}, and using the Dodd–Jensen Lemma and the fact that $\pi = \sigma^{*} \circ \pi^{*}$, we can show that M' coiterates with M^{*} to a common simple iterate, say Q. Thus, all maps in the following diagram are total: One more application of the Dodd–Jensen lemma yields (we leave the details to the reader):

$$\tilde{\pi} \circ \pi = \tilde{\pi}^{*} \circ \pi^{*}.$$

Notice that $E^{M^{*}}_{\mu} = \varnothing$ by the coherency condition for $E^{\bar{M}}_{\mu}$, so the iteration from M^{*} to Q is above μ. Thus, for $x \subset \kappa$ and $\alpha < \mu$ we have

$$x \in E^{\bar{M}}_{\mu,\alpha} \longleftrightarrow \alpha \in \pi^{*}(x) \cap \mu = \tilde{\pi}^{*} \circ \pi^{*}(x) \cap \mu = \tilde{\pi} \circ \pi(x) \cap \mu$$

Now if $\mu' > \mu$, the iteration from M' to Q begins with the ultrapower by $E^{M'}_{\mu} = E^{M}_{\mu}$, so $\tilde{\pi} \circ \pi(x) \cap \mu = \tilde{\pi}(x) \cap \mu$ and $\alpha \in \tilde{\pi}(x)$ iff $x \in E^{M}_{\mu,\alpha}$. This proves that $E^{\bar{M}}_{\mu} = E^{M}_{\mu}$ if $\mu' > \mu$. The simpler case where $\mu' = \mu$ is left to the reader.

If $\tau = \nu$, we can proceed exactly as in the proof of Claim 3 from the proof of the solidity theorem for measures of order 0. This time $E^{\bar{M}}_{\mu}$ is an extender on $M||\zeta$ where ζ is maximal such that $\kappa^{+M\|\zeta} = \nu$. Then $\widehat{M\|\zeta}$ is $*$-extendible by $E_\mu^{\bar{M}}$ and the corresponding ultrapower $\widehat{M}^*$ can be embedded into the iterable premouse $\pi'(\widehat{M\|\zeta})$, where π' is the ultrapower map associated with $\mathrm{Ult}(M, E_{\mu'}^M)$. The embedding in question σ^* is defined as above for $f \in \Gamma(\kappa, \widehat{M\|\zeta})$ and $\alpha < \mu$ and is $\Sigma_0^{(n)}$-preserving for n such that $\omega\varrho_{M\|\zeta}^{n+1} \leq \kappa < \omega\varrho_{M\|\zeta}^n$, which guarantees that M^* is normally iterable above κ. Notice that $E_\mu^{M^*} = \varnothing$, so the M^*-side of the coiteration of M against M^* is above μ, which means that these two premice are $*$-coiterable. As in the case of measures of order 0 we show that M and M^* coiterate simply above κ to some Q and that $\pi^* \circ \sigma^* = \pi$, where π and π^* is the iteration map on the M-side and the M^*-side, respectively, and $\sigma^* : M \to M^*$ is the ultrapower map by $E_\mu^{\bar{M}}$. From this equality we infer as before that $E_\mu^{\bar{M}}$ is equal to the first extender used on the M-side of the coiteration, which is nothing other than E_ν^M. □(Lemma 8.2.2)

Notice that the proof of the previous lemma can be incorporated in the proof of the solidity theorem. Although the second part of the proof uses the fact that the initial segments of M are solid, this will not make any harm to the proof of the solidity theorem, as we already know that M is presolid.

The next major theorem regarding mice is the condensation lemma, whose proof does not need any substantial revision. The formulation, however, does:

Theorem 8.2.3. *Let $\bar{M}$ be a premouse, let M be a mouse and let*

$$\sigma : \widehat{\bar{M}} \underset{\Sigma_0^{(n)}}{\longrightarrow} \widehat{M}$$

be such that $\omega\varrho_{\bar{M}}^{n+1} \leq \nu = \mathrm{cr}(\sigma)$. Then $\bar{M}$ is a mouse. Moreover, if $\bar{M}$ is sound above ν, then one of the following holds:

a) *$\bar{M} = \mathrm{core}_\nu(M)$ and σ is the corresponding core map. In this case either*

   i) *M is a simple iterate of $\bar{M}$ above ν and σ is the associated iteration map, or else*

   ii) *there is a $\beta \geq \nu$ such that $\beta = \mathrm{o}^{\bar{M}}(\kappa)$ for some $\kappa < \nu$ and $\mathrm{cr}(E_\beta^M) \neq \varnothing$; then M' is a simple iterate of $\bar{M}$ above ν, where $\widehat{M}' = \mathrm{Ult}^*(\widehat{M}, E_\beta^M)$.*

b) *$\bar{M}$ is a proper initial segment of M.*

c) *$\bar{M}$ is a proper initial segment of M', where $\widehat{M}' = \mathrm{Ult}^*(\widehat{M\|\zeta}, E_\beta^M)$, the ordinal β is as in* aii) *above and ζ is maximal possible, i.e. maximal such that E_β^M measures all subsets of κ in $M\|\zeta$.*

d) *$\bar{M} = M'$ where M' and ζ are as in* c) *and $\zeta < \kappa^{+M}$. Moreover, E_β^M has a single generator, namely κ.*

□(Theorem 8.2.3)

We stress that neither of the clauses of the condensation lemma is vacuous if the mice in question are sufficiently rich; more precisely, extenders with two generators (i.e. extenders which cannot be replaced by measures) already yield examples of mice embeddings $\sigma : \bar{M} \to M$ such that any of a) – d) can hold.

We now turn to the revision of the theory of bicephali. A *prebicephalus* is a structure $M = \langle J_\alpha^E, F^0, F^1\rangle$ such that both F^0, F^1 are extenders of length α and each $M^h = \langle J_\alpha^E, F^h\rangle$ is an expandible premouse ($h = 0, 1$). Notice that $\operatorname{cr}(F^0) = \operatorname{cr}(F^1)$, as the extenders in our premice cannot overlap. An iteration of M would consist of a pair of iterations of M^0, M^1. At every step of such an iteration we have to verify that we obtain a pair of premice corresponding to a prebicephalus. This is a priori not clear, as the extenders are applied to the *expansions* of the bookkeeping premice in the course of iterations. Hence, $\operatorname{Ult}(\widehat{M}^0, G)$ and $\operatorname{Ult}(\widehat{M}^1, G)$ might turn out to be premice whose top extenders are of different lengths; such a pair of premice cannot correspond to any prebicephalus. For this reason, we have to restrict the class of prebicephali of our interest; it will turn out that the restriction we make is vacuous if we deal with *iterable* prebicephali.

Definition. A prebicephalus $M = \langle J_\alpha^E, F^0, F^1\rangle$ is *correct* $\alpha^{+\widehat{M}^0} = \alpha^{+\widehat{M}^1} \overset{\text{def}}{=} \delta_M$ and $E^{\widehat{M}^0} \restriction \delta_M = E^{\widehat{M}^1} \restriction \delta_M$.

The correctness of M says that the expansions $\widehat{M}^0$ and $\widehat{M}^1$ contain the same subsets of α, which means that, given any suitable extender G, the orders of $\kappa = \operatorname{cr}(F^0) = \operatorname{cr}(F^1)$ in $\operatorname{Ult}(\widehat{M}^0, G)$ and in $\operatorname{Ult}(\widehat{M}^1, G)$ will be the same. More formally,

Lemma 8.2.4. *Let $M = \langle J_\alpha^E, F^0, F^1\rangle$ be a correct prebicephalus and let G be a weakly amenable extender w.r.t. M such that $\operatorname{cr}(G) \le \alpha$. Let*

$$\pi^0 : \widehat{M}^0 \underset{G}{\longrightarrow} \widehat{M}_1^0 \quad \textit{and} \quad \pi^1 : \widehat{M}^1 \underset{G}{\longrightarrow} \widehat{M}_1^1.$$

Letting $\delta = \delta_M$ and $E = E^{\widehat{M}^0} \restriction \delta = E^{\widehat{M}^1} \restriction \delta$, the following holds:

a) *The pair of premice M_1^0, M_1^1 determine a correct prebicephalus, namely to $M_1 = \langle J_{\delta_1}^{E^1}, E_{\text{top}}^{M_1^0}, E_{\text{top}}^{M_1^1}\rangle$.*

b) $\pi^0 \restriction J_\delta^E = \pi^1 \restriction J_\delta^E$. □(Lemma 8.2.4)

That iterable prebicephali are correct follows from the following simple criterion:

Lemma 8.2.5. *Let $M = \langle J_\alpha^E, F^0, F^1\rangle$ be a prebicephalus. Assume that N^0, N^1 are Σ_0-mice where*

$$\tilde{\pi}^0 : \widehat{M}^0 \underset{F^1}{\longrightarrow} \widehat{N}^0 \quad \textit{and} \quad \tilde{\pi}^1 : \widehat{M}^1 \underset{F^0}{\longrightarrow} \widehat{N}^1.$$

Then M is correct.

Proof. Let $\delta^h = \alpha^{+\widehat{M}^h}$ for $h = 0, 1$. By coherency, $\delta^h = \alpha^{+N^{1-h}}$. Let N^0_θ, N^1_θ be the result of the Σ_0-coiteration of N^0 against N^1 and let κ_0 be the first critical point in this coiteration. Clearly, $\kappa_0 > \alpha$. Assume w.l.o.g. the N^1-side. Then κ_0 is inaccessible in N^1; hence $\delta_0 < \kappa_0$. If $\delta_1 \leq \kappa_0$ then $\delta_0 = \delta_1$ since N^0, N^1 coincide up to κ_0.

We now show that κ_0 is always at least δ_1. Suppose for a contradiction that $\delta_1 > \kappa_0$. Since κ_0 is not a cardinal in N^0, we truncate N^0 at an ordinal $\omega\eta < \delta_1$. By coherency, N^0, M^1_1 coincide up to $\tilde{\pi}^1(\kappa)$, where κ is the common critical point of F^0, F^1 and $\widehat{M}^1_1 = \text{Ult}(\widehat{M}^1, F^1)$. Hence, $N^0||\eta = M^1_1||\eta$. Now consider the pair of iterations

$$M^1 \underset{F^1}{\longrightarrow} M^1_1,\ M^1_1||\eta \longrightarrow \cdots \longrightarrow N^0_\theta$$

$$M^1 \underset{F^0}{\longrightarrow} N^1 \longrightarrow \cdots \longrightarrow N^1_\theta$$

where the right side of the above diagram comes from the above coiteration of N^0 with N^1. The former is a non-simple Σ_0-iteration from M^1 to N^0_θ, hence N^1_θ is a non-simple iterate of M^1 since it is a segment of N^0_θ. The latter is not a genuine iteration because it uses an outer extender at the first step; nevertheless it yields a Σ_0-preserving map from $\widehat{M}^1$ to $\widehat{N}^1_\theta$. This contradicts the Dodd–Jensen Lemma for Σ_0-iterations. □(Lemma 8.2.5)

Corollary 8.2.6. *Let $M = \langle J^E_\alpha, F^0, F^1\rangle$ be a prebicephalus such that $M||\gamma$ is a mouse whenever $\gamma < \alpha$ and both F^0, F^1 are ω-complete. Then M is correct.*

Bicephalus iterations can be defined as before, but in clauses f) – h) we have to consider expansions instead of bookkeeping premice. Every bicephalus iteration is represented by a pair of premice "iterations" or M^0 and M^1 which are not iterations in the sense of our definition, as they might use external extenders. E.g., the extender F^h_i can be applied to $\widehat{M}^{1-h}_i$. We denote the associated premice iterations by $\mathfrak{I}^0, \mathfrak{I}^1$. Clause g) in the definition of the bicephalus iteration now reads

g) If $F = E^{M_i,h_i}_{\nu_i} \neq \varnothing$ and $\eta_{i'} = \text{ht}(M_i')$ for all $i' \leq i$, then

$$\pi^k_{i,i+1} : \widehat{M}^k_i \underset{F}{\longrightarrow} \widehat{M}^k_{i+1}$$

for $k = 0, 1$.

Clause h) has to be amended in the same way. We can then formulate the bicephalus lemma, whose proof goes through as before. The following is the generalization of the criterion Lemma 4.6.4 for non-overlapping extenders.

Lemma 8.2.7. *Let $M = \langle J^E_\alpha, F^0, F^1\rangle$ be a prebicephalus such that*

a) *$M||\gamma$ is a mouse whenever $\gamma < \alpha$*

b) *F^0 and F^1 are ω-complete.*

Then M is a bicephalus.

From this we obtain the analogue to Corollary 4.6.5 which says that if M satisfies the assumption of the previous lemma, then $F^0 = F^1$. The proof of Lemma 8.2.7 splits into two steps.

Lemma 8.2.8. *Suppose that M is a bicephalus, $\bar{M}$ is a correct prebicephalus and that $\sigma^h : \widehat{\bar{M}}^h \rightarrow \widehat{M}^h$ is a Σ_0-preserving map ($h = 0, 1$) satisfying $\sigma^0 \upharpoonright \bar{\alpha} = \sigma^1 \upharpoonright \bar{\alpha}$ where $\bar{\alpha} = \mathrm{ht}(\bar{M})$. Then $\bar{M}$ is a bicephalus.*

Proof. We show that any simple iteration $\bar{\mathfrak{I}}$ of $\bar{M}$ can be copied onto an iteration $\mathfrak{I}$ of M in the sense that each $\bar{\mathfrak{I}}^h$ is copied onto $\mathfrak{I}^h$ via σ^h; this will obviously prove the lemma. It clearly suffices to show that the copying construction does not break down at successor steps. The main point is to prove is that we can continue the copying construction after any use of an external extender in $\bar{\mathfrak{I}}^h$. Let $\bar{F}_i^h$ and F_i^h be the top extender of $\bar{M}_i^h$ and M_i^h, respectively and let $\bar{\kappa}_i^* = \mathrm{cr}(\bar{F}_i^h)$ for $h = 0, 1$ and $i < \mathrm{length}(\bar{\mathfrak{I}})$. By induction in i we show:

$$\sigma_i^0 \upharpoonright J_{\bar{\delta}_i}^{\bar{E}_i} = \sigma_i^1 \upharpoonright J_{\bar{\delta}_i}^{\bar{E}_i} \tag{8.2.1}$$

$$\langle \sigma_i^h, \sigma_i^{1-h} \upharpoonright \bar{\alpha}_i \rangle : \langle \widehat{\bar{M}}_i^h, \bar{F}_i^{1-h} \rangle \longrightarrow \langle \widehat{M}_i^h, F_i^{1-h} \rangle \tag{8.2.2}$$

where $\bar{\delta}_i = \delta_{\bar{M}_i}$, $\bar{E}_i = E^{\widehat{\bar{M}}_i^0} \upharpoonright \bar{\delta}_i = E^{\widehat{\bar{M}}_i^1} \upharpoonright \bar{\delta}_i$ and similarly for δ_i, E_i.

The latter is an immediate consequence of the former: If $x \in \mathcal{P}(\bar{\kappa}_i^*) \cap \bar{M}_i$ and $\gamma < \bar{\alpha}_i$ (recall that $\bar{\alpha}_i = \mathrm{ht}(\bar{M}_i)$), then

$$x \in \bar{F}_{i,\gamma}^{1-h} \longleftrightarrow \sigma_i^{1-h}(x) \in F_{i,\sigma^{1-h}(\gamma)}^{1-h} \longleftrightarrow \sigma_i^h(x) \in F_{i,\sigma^{1-h}(\gamma)}^{1-h}.$$

The verification of (8.2.1) is trivial at limit stages. Now suppose it holds at i; we shall prove that it also holds at $i+1$. By (8.2.2), the objects σ_{i+1}^h and M_{i+1}^h do exist (recall that we do not have to appeal to (8.2.2) if an *internal* extender is used at the i-th step) and

$$\sigma_{i+1}^h(\bar{\pi}_{i,i+1}^h(f)(\gamma)) = \pi_{i,i+1}^h \circ \sigma_i^h(f)(\sigma_i^{1-h}(\gamma)).$$

By the induction hypothesis, (8.2.1) and the by fact that $\bar{\pi}_{i,i+1}^h$, $\pi_{i,i+1}^h$ does not depend on h we immediately obtain (8.2.1) for $i+1$. □(Lemma 8.2.8)

Lemma 8.2.9. *Let $M = \langle J_\alpha^E, F^0, F^1 \rangle$ be a prebicephalus whose all total extenders are ω-complete. Then M is a bicephalus.*

Proof. Notice first that M is correct; this follows easily from Corollary 8.2.6. Let us prove that M is iterable. Following the usual strategy, suppose for a contradiction that $\mathfrak{I}$ is an iteration of M which cannot be continued. As in the proof of Theorem 4.5.2 we can assume w.l.o.g. that $\mathfrak{I}$ is simple. By the Löwenheim–Skolem argument we construct $\bar{M}$, $\bar{\mathfrak{I}}$ and a map σ such that $\bar{\mathfrak{I}}$ is an iteration of $\bar{M}$ which cannot be continued and

$$\sigma_0^h = \sigma \upharpoonright \widehat{\bar{M}}^h : \widehat{\bar{M}}^h \rightarrow \widehat{M}^h$$

is Σ_0-preserving. We now use the ω-completeness of the total extenders in M to construct a system of Σ_0-embeddings

$$\sigma_i^h : \widehat{M}_i^h \longrightarrow \widehat{M}^h$$

which commutes with the iteration maps $\bar{\pi}_{ij}^h$. As an immediate consequence we obtain that $\bar{\mathfrak{I}}$ can be continued after all, and the desired contradiction will follow immediately.

Let $\langle\langle \bar{\nu}_i, \bar{h}_i\rangle\,;\, i < \text{length}(\bar{\mathfrak{I}})\rangle$ be the iteration indices of $\bar{\mathfrak{I}}$. By induction on i we verify:

$$\sigma_i^0 \restriction J_{\bar{\delta}_i}^{\bar{E}_i} = \sigma_i^1 \restriction J_{\bar{\delta}_i}^{\bar{E}_i} \tag{8.2.3}$$

$$\langle \sigma_i^h, \sigma_i^{1-h} \restriction \bar{\alpha}_i\rangle : \langle \widehat{M}_i^h, \bar{F}_i^{1-h}\rangle \longrightarrow \langle \widehat{M}^h, F^{1-h}\rangle. \tag{8.2.4}$$

The embeddings σ_0^h were defined above, for $i > 0$ we define σ_i^h as follows. Granting that σ_i^h is defined, (8.2.4) follows from (8.2.3) by the argument for the implication (8.2.1) $\longrightarrow$ (8.2.2) from the proof of the previous lemma. Set

$$\nu_i = \sigma_i^h(\bar{\nu}_i) \quad \text{and} \quad \kappa_i = \text{cr}(E_{\nu_i}^{M,h_i}).$$

Pick a map $\mathfrak{k}_i : \nu_i \to \kappa_i$ which fixes X, A where

$$X = (\sigma_i^h)''(\mathcal{P}(\bar{\kappa}_i) \cap \bar{M}_i^h) \quad \text{and} \quad A = (\sigma_i^h)''\bar{\nu}_i$$

(Note: Neither of these definition depends on h.) Such a $\mathfrak{k}_i$ exists, as $E_{\nu_i}^{M,h_i}$ is ω-complete. Now define σ_{i+1}^h in the usual way, i.e.

$$\sigma_{i+1}^h(\bar{\pi}_{i,i+1}^h(f)(\gamma)) = \sigma_i^h(f)(\mathfrak{k}_i \circ \sigma_i^{1-h}(\gamma))$$

for $\gamma < \bar{\nu}_i$. By the induction hypothesis, the right side of the above definition does not depend on h if we restrict ourselves to $f \in J_{\bar{\delta}_i}^{\bar{E}_i}$. This completes the construction at successor steps; the induction at limit steps is again trivial. □(Lemma 8.2.9)

Proof of Lemma 8.2.7. We first observe that M can be iterated by internal extenders. To see this, it obviously suffices to prove that both M^0, M^1 are Σ_0-mice, which is an immediate consequence of Theorem 4.5.6 adapted to the present context. We then proceed as in the proof of that theorem and perform the same normal Σ_0-iteration in order to obtain a prebicephalus M' whose all total extenders with critical points less than $\text{cr}(E_{\text{top}}^{M',h})$ are ω-complete. By Lemma 2.5.5, both top extenders of M' are ω-complete; a straightforward computation which uses the facts about the ω-completeness of the extenders in M' proved so far then yields that all total extenders of M' are ω-complete (we leave this computation to the reader). Lemma 8.2.9 then guarantees that M' is a bicephalus. Let $\pi^h : \widehat{M}^h \to \widehat{M}'^h$ be the pair of iteration maps from the above iteration ($h = 0, 1$). This pair of maps clearly satisfies the assumptions of Lemma 8.2.8, so M is a bicephalus. □(Lemma 8.2.7)

We now turn to the upward extensions of embeddings techniques, a topic that we briefly touched already at the end of the previous section. As we have already mentioned there, it need not be the case that the pseudoultrapower of a premouse by an arbitrary embedding is again a premouse. Recall that the property of being a premouse is a Π_2-property, and the complexity cannot be lowered if the premouse in question has a top extender. Given a Σ_0-preserving cofinal embedding

$$\sigma : J^E_\tau \to J^{\tilde{E}}_{\tilde{\tau}}$$

where J^E_τ is a structure with the largest cardinal κ and a premouse M end-extending J^E_τ such that $\mu = \mathrm{cr}(E^M_{\mathrm{top}}) < \kappa$, it can happen that σ does not map the ordinal $\vartheta = \kappa^{+M}$ cofinally into $\sigma(\vartheta)$. (Recall: τ is a cardinal in M.) If $\omega\varrho^1_M \le \kappa$ or the pseudoultrapower in question is coarse, the associated extension $\tilde{\sigma} : \widehat{M} \to \widehat{M}'$ will be Σ_0-preserving and cofinal, which means that the top extender $\tilde{F}$ of $\tilde{M}$ will be defined precisely on $\mathcal{P}(\tilde{\kappa}) \cap J^{\tilde{E}}_{\tilde{\vartheta}}$, where $\tilde{\kappa} = \sigma(\kappa)$ and $\tilde{\vartheta} = \sup(\sigma''\vartheta) < \vartheta$. Hence $\tilde{F}$, although being the top extender of $\tilde{M}$, is not defined on the entire $\mathcal{P}(\tilde{\kappa}) \cap \tilde{M}$. In other words, $\tilde{F}$ is not weakly amenable w.r.t. $\tilde{M}$, so $\tilde{M}$ fails to be a premouse. There are various ways of dealing with this issue. The easiest one is to avoid the above described situation, which is possible in most constructions. Some arguments, like the proof of the full weak covering lemma for $\boldsymbol{K}$ or that of the combinatorial principle $\Box_\theta$ in $\boldsymbol{L}[E]$ do not allow this, in which case we replace $\tilde{M}$ by M^* where

$$\widehat{M}^* = \mathrm{Ult}^*(\widehat{M||\eta}, \tilde{F})$$

for η maximal with the property that $\tilde{\vartheta}$ is a cardinal in $M||\eta$. Then M^* turns out to be a premouse with fine structural properties which allow to use it as a substitute for $\tilde{M}$. The above discussion, of course, also applies to the case where J^E_τ does not have a largest cardinal. We note that structures M whose top extender is not total are called *protomice* of *fragments*.

By b) of Lemma 8.1.7, the situation described in the previous paragraph is the only one which yields $\tilde{M}$ that is not a premouse. In all other cases we can ask under which assumptions is the premouse $\tilde{M}$ iterable. Here we can simply quote the theorems on upwards extensions that we proved before. Let us state the most important ones, the generalizations of Lemmata 5.6.5 and 5.6.6. In the following lemmata, $Q = J^E_\tau$ and $\tilde{Q} = J^{\tilde{E}}_{\tilde{\tau}}$.

Lemma 8.2.10. *Let* $\mathrm{cf}(\tau) > \omega$, $n = k(M)$, *let both* M *and* $\tilde{Q}$ *be iterable and let one of the following holds:*

- τ *is limit in* M *and* M *is sound above* τ;
- $\tau = \kappa^{+M}$ *and* M *is sound above* κ.

Then the canonical extension $\tilde{\sigma} : \widehat{M} \to \widehat{\tilde{M}}$ *of* σ *exists. Moreover,* $\tilde{M}$ *is normally iterable above* $\tilde{\tau}$ *in the former case and above* $\tilde{\kappa} = \tilde{\sigma}(\kappa)$ *in the latter case.*

Lemma 8.2.11. *Let* $\mathrm{cf}(\tau) > \omega$, *let* Q *have a largest cardinal* κ *and let both* M *and* $\tilde{Q}$ *be* 0*-iterable. Then the canonical* 0*-extension* $\tilde{\sigma} : \widehat{M} \to \widehat{\tilde{M}}$ *of* σ *exists and* $\tilde{M}$ *is* 0*-iterable.*

The proofs of the above lemmata closely follow those of Lemmata 5.6.5 and 5.6.6. In the case of Lemma 8.2.10, we have to apply the new condensation lemma to the embedding $\sigma_1^* : \widehat{\bar{M}}^* \to \widehat{M}$. (See the proof of Lemma 5.6.5.) In the case of Lemma 8.2.11, we have to argue that in the coiteration of M with $\bar{M}^*$, either some M_i is truncated at a point $\eta < \pi_{0,i}(\tau)$ (here $\pi_{0,i} : \widehat{M} \to \widehat{M}_i$ is the iteration map on the M-side) or else $\bar{M}_\vartheta^*$ is a *proper* initial segment of M_ϑ. (See the proof of Lemma 5.6.5.) Assuming that there is no truncation as above, we observe that $\pi_{0,\vartheta}(\tau)$ is defined and of cofinality larger than ω. On the other hand, by examining the $\bar{M}^*$-side of the coiteration, we infer that every successor cardinal in $\widehat{\bar{M}}_\vartheta^*$ is ω-cofinal. First of all, since $\bar{M}^*$ is a coarse pseudoultrapower of a countable premouse, it is easy to see that every successor cardinal in $\widehat{\bar{M}}^*$ larger than κ is ω-cofinal and that the same also applies to $\mathrm{ht}(\widehat{\bar{M}}^*)$. By the existence of σ_1^*, the iteration on the $\bar{M}^*$-side is simple and above $\bar{\tau}^*$ (the latter follows from Lemma 8.2.2 if $\bar{\tau}^*$ is overlapped by an extender). By induction along the iteration on the $\bar{M}^*$-side we then prove that any successor cardinal in $\widehat{\bar{M}}_\vartheta^*$ as well as $\mathrm{ht}(\widehat{\bar{M}}_\vartheta^*)$ is ω-cofinal. Thus, $\bar{M}_\vartheta^*$ must be a proper initial segment of M_ϑ, as otherwise $\pi_{0,\vartheta}(\tau)$, being a non-ω-cofinal cardinal successor of M_ϑ, cannot be an element of $\widehat{\bar{M}}_\vartheta^*$.

Using the upward extensions of embeddings techniques we prove a remarkable lemma, which will be useful in proving weak covering for $\boldsymbol{K}^c$ and in the construction of $\boldsymbol{K}$. Although it is possible to avoid its use in most arguments, the lemma it is an interesting deep fact about ω-completeness on its own. We stress that it is crucial that we deal with non-overlapping extenders and that there does not seem to be a direct generalization of this lemma to higher levels. Recall that in the case of measures of order 0 we proved the following fact: If M is a premouse such that both $\kappa = \mathrm{cr}(E_{\mathrm{top}}^M)$ as well as $\tau = \kappa^{+M}$ are of cofinality larger than ω, then E_{top}^M is ω-complete. In the case of non-overlapping extenders we have:

Lemma 8.2.12. *Let* $M = \langle J_\nu^E, F\rangle$ *be a* Σ_0*-mouse such that both* $\kappa = \mathrm{cr}(F)$ *as well as* $\tau = \kappa^{+M}$ *are of cofinality larger than* ω. *Then* F *is* ω*-complete.*

Proof. Suppose for a contradiction that this fails and fix countable sets $A \subset \nu$ and $X \subset \mathcal{P}(\kappa)$ witnessing this (see the section on extendibility on Chapter 2 for the notation). Thus, A is closed under Gödel's pairing function and there is no mapping $\mathfrak{k} : A \to \kappa$ satisfying $\mathfrak{k}(\alpha) \in x$ whenever $x \in X \cap F_\alpha$. Using the Löwenheim–Skolem argument, construct a countable $\bar{M}$ and an elementary map $\sigma : \widehat{\bar{M}} \to \widehat{M}$ such that $A \cup X \subset \mathrm{rng}(\sigma)$. Letting $\bar{\tau}$ be the preimage of τ under σ and $\tilde{\tau} = \sup(\sigma''\bar{\tau})$, we have $\tilde{\tau} < \tau$, as $\tilde{\tau}$ is ω-cofinal and τ not. By the interpolation lemma (Lemma 3.6.10), there is a structure P and Σ_0-preserving embeddings $\tilde{\sigma} : \widehat{\bar{M}} \to P$ and $\sigma' : P \to \widehat{M}$ such that $\sigma = \sigma' \circ \tilde{\sigma}$. The structure P is the coarse pseudoultrapower of $\widehat{\bar{M}}$ by $\sigma \restriction J_{\bar{\tau}}^{\bar{E}}$ where $\bar{E} = E^{\bar{M}}$ and $\tilde{\sigma}$ is the associated canonical extension of $\sigma \restriction J_{\bar{\tau}}^{\bar{E}}$ to $\widehat{\bar{M}}$. Thus, $\tilde{\sigma}$ is cofinal. Since

$\bar{\kappa} = \sigma^{-1}(\kappa)$ is the critical point of $E_{\text{top}}^{\tilde{M}}$, we can apply b) of Lemma 8.1.7 and conclude that P is a premouse, say P is the expansion of a bookkeeping premouse $\tilde{M}$. And since A and X are contained in the range of σ', the extender $\tilde{F} = E_{\text{top}}^{\tilde{M}}$ fails to be ω-complete.

By Lemma 8.2.2, $\tilde{M}$ is a proper initials segment of M. In fact, $\tilde{M}$ is a proper initial segment of $M||\tau$, as $\tilde{\tau} = \kappa^{+\tilde{M}}$. This means that there is a surjective map $f_0 : \kappa \to \tilde{M}$ in M. Since κ is not ω-cofinal, $A \cup X \subset \text{rng}(f_0 \restriction \beta)$ for some $\beta < \kappa$. Let g be $\text{Col}(\omega, \beta)$-generic over V. Letting $N = \text{Ult}(J_\tau^E, F)$, the structure N is a ZFC$^-$-model and g is $\text{Col}(\omega, \beta)$-generic over N. Thus, there is a surjective map $f_1 : \omega \to \beta$ in $N[g]$, so $A \cup X \subset \text{rng}(f)$ where $f = f_0 \circ f_1$. Notice also that $f \in N$, as both f_0 and f_1 are elements of N. Set

- $\tilde{A}$ = the closure of $\tilde{\nu} \cap \text{rng}(f)$ under Gödel's pairing function;
- $\tilde{X} = \mathcal{P}(\kappa) \cap \tilde{M} \cap \text{rng}(f)$.

Then $(\tilde{A}, \tilde{X})$ is a pair of countable sets in $N[g]$ witnessing that $\tilde{F}$ is not ω-complete, as $A \subset \tilde{A}$ and $X \subset \tilde{X}$. Fix an enumeration $\langle \alpha_i;\ i \in \omega \rangle$ of $\tilde{A}$ in $N[g]$. Consider the set of all pairs of the form $\langle h, n \rangle$ satisfying

- $h : \{\alpha_0, \ldots, \alpha_{n-1}\} \to \kappa$ is an $\prec\ \succ$-respecting function;
- $h(\alpha) \in x$ whenever $x \in \tilde{X} \cap \tilde{F}_\alpha$

and the relation $\mathcal{R}$ on all such pairs defined by

$$\langle h', n' \rangle \mathcal{R} \langle h, n \rangle \longleftrightarrow (n' > n \;\&\; h' \supset h).$$

By the construction of $\tilde{A}$ and $\tilde{X}$, the extender $\tilde{F}$ is ω-complete iff is $\mathcal{R}$ is ill-founded. So $\mathcal{R}$ is well-founded, as we know that $\tilde{F}$ fails to be ω-complete. Notice that $\mathcal{R} \in N[g]$, as $\mathcal{R}$ is locally definable from objects that are elements of $N[g]$. Thus, $\mathcal{R}$ is well-founded in the sense of $N[g]$.

Now let $N' = \text{Ult}(N, F)$ and let $\pi : N \to N'$ be the associated ultrapower map. Since $\text{Col}(\omega, \beta)$ is a small forcing in N, more precisely, as $|\text{Col}(\omega, \beta)|^N = |\beta|^N < \kappa = \text{cr}(F)$, the embedding π naturally extends to the embedding

$$\pi^* : N[g] \to N'[g].$$

Then $\mathcal{R}^* = \pi(\mathcal{R})$ is well-founded in the sense of $N'[g]$, and therefore is well-founded in $V[g]$. We now derive a contradiction by showing that $\mathcal{R}^*$ must be ill-founded in $V[g]$. The relation $\mathcal{R}^*$ is defined in $N'[g]$ from $\pi^*(\tilde{A})$, $\pi^*(\tilde{X})$ and $\pi^*(\tilde{F})$ in the same manner as $\mathcal{R}$ is defined in $N[g]$ from $\tilde{A}$, $\tilde{X}$ and $\tilde{F}$. By the fact that both $\tilde{A}$ and $\tilde{X}$ are countable in $N[g]$ we have $\pi^*(\tilde{A}) = (\pi^*)''\tilde{A}$ and $\pi^*(\tilde{X}) = (\pi^*)''\tilde{X}$. Consequently, given an $\alpha^* \in \pi(\tilde{A})$ and $x^* \in \pi^*(\tilde{X})$, there are $\alpha \in \tilde{A}$ and $x \in \tilde{X}$ such that $(\alpha^*, x^*) = \pi^*(\alpha, x)$, so we have:

$$x^* \in \pi^*(\tilde{F})_{\alpha^*} \longleftrightarrow x \in \tilde{F}_\alpha \longleftrightarrow x \in F_{\sigma'(\alpha)} \longleftrightarrow \sigma'(\alpha) \in \pi^*(x) = x^*$$

This proves that $\sigma' \circ (\pi^*)^{-1} \restriction \pi^*(\tilde{A})$ is a map witnessing the ill-foundedness of $\mathcal{R}^*$, the desired contradiction. □(Lemma 8.2.12)

8.3 Extender Models up to One Strong Cardinal

The general theory of extender models for non-overlapping extenders is a direct generalization of that for measures of order 0. Practically all of the lemmata on weasel coiterations and the canonical ordering of weasels, as stated before, remain valid. Let us mention one new fact – the existence of extender models with strong cardinals. That the extender sequence E witnesses that κ is strong in W can be equivalently expressed as $o^W(\kappa) = \infty$. If we deal with such weasels, some of the proofs on weasel coiterations require a revision (e.g. the proof of Lemma 6.1.7), as we obviously cannot replace W by some of its initial segments anymore. In these cases we use the argument from the proof of the comparison lemma. It is also easy to see that if $o^W(\kappa) = \infty = o^{W'}$, then the coiteration of W with W' terminates below ∞.

The notion of universality is the same as before. It follows immediately from the general properties of iterations that if $o^W(\kappa) = \infty$ for some κ, then W is universal. Moreover, Notice, if $\langle \nu_i;\ i < \infty\rangle$ is an enumeration of all indices ν such that E^W_ν is a total extender on κ, we can form the iteration of length ∞ by applying the extender indexed by $\pi_{0,i}(\nu_i)$ at each step i. At the end we obtain a non-simple iterate W' of W, and it is not hard to see, using the Dodd–Jensen lemma, that W' wins the coiteration against every mouse, i.e. is weakly universal. Also, the sequence of all critical points κ_i of this iteration is normal and unbounded, and $\tau_i = \kappa_i^{+W_i}$ has small cofinality. For this reason, once we deal with models beyond one strong cardinals, the definition of universality must be modified in order to exclude the above described situation. Following Steel [MiS94], the new definition of universality of W requires that if M is any premouse or a weasel, then both

- M-side of the coiteration of W against M is simple;
- each iteration map $\pi^M_{0,i}$ is defined on the entire M.

As we have already mentioned, in the present case, our old definition of universality will suffice. Lemmata on non-universal weasels must be now considerably weakened. In particular, Lemma 6.3.3 has to be reformulated as follows:

Lemma 8.3.1. *Let M be a bookkeeping premouse, $\nu \leq \mathrm{ht}(M)$ and let $\widehat{M}' = \mathrm{Ult}(\widehat{M}, E^M_\nu)$. Then for every $\tau' > \nu$ which is a successor cardinal in $\widehat{M}'$ there is a successor cardinal τ in $\widehat{M}$ and a cofinal map $\Xi : \tau \to \tau'$.* □(Lemma 8.3.1)

This, however does not suffice for the proof of a sufficiently general version of Corollary 6.3.4. We now show that this corollary is false if we deal with extenders which have many generators. For the sake of simplicity, we will work in an extender model W. Suppose there is a κ with $o(\kappa) = \kappa^{++}$; denote this ordinal by β (cardinal functions are computed in W). Set $M = W||\beta$. The extender E^M_β has cofinally many generators below β, so let ν be the κ-th of them. Clearly, $\nu < \beta$. Now if

$$\pi : \widehat{M} \xrightarrow[E^M_\beta]{*} \widehat{M}^*,$$

(note that the $*$-ultrapower agrees with the Σ_0-ultrapower up to $\beta = \omega\varrho_M^\omega$) then every ordinal ζ of the form $\pi(f)(\gamma)$ (here $f \in {}^\kappa\kappa \cap M, \gamma < \nu$) which is less than β is in fact less that ν. Suppose this fails; then $\zeta > \nu$ since ν is a generator, so if $g \in M$ is a function with domain κ such that each $g(\xi)$ is a surjective mapping of $\mathrm{card}(f(\xi))$ onto $f(\xi)$, then $\pi(g)(\gamma)$ maps κ^{+M} onto ζ. Hence, there is an $\eta < \kappa^{+M}$ such that $\pi(g)(\gamma)(\eta) = \nu$. This contradicts the fact that ν is a generator, as $\prec\gamma, \eta\succ < \nu$.

An easy condensation argument together with Lemma 8.2.2 yields that $E_\beta^M|\nu = E_\nu^M$. Recall that if we form the ultrapower

$$\pi' : \widehat{M} \xrightarrow[E_\nu^M]{*} \widehat{M}',$$

then there is a Σ^*-preserving map $\sigma : \widehat{M}' \to \widehat{M}^*$ such that $\pi = \sigma \circ \pi'$. It follows from the computation in the previous paragraph that $\sigma \restriction \nu = \mathrm{id}$ and $\sigma(\nu) = \beta$. Notice that $\beta = \kappa^{++M^*}$. Hence, $\nu = \kappa^{++M'}$. Since $E_\nu^M = E_\beta^M|\nu$, the extender E_ν^M has κ many generators which are cofinal in ν, and this sequence of generators is definable inside a proper initial segment of M. It follows that $\mathrm{cf}^M(\nu) = \kappa$ and it is clear that no successor cardinal of $\widehat{M}$ can be mapped cofinally into κ: For successors $\leq \beta$ this follows from the fact that κ is a cardinal in W; for successors greater than β we can apply a slight modification of Lemma 8.3.1 (which follows by the same proof) and infer that each such successor has cofinality κ^+.

The above discussion enables us to construct a non-universal weasel whose successor cardinals attain arbitrarily large cofinalities. Let $\Im$ be the iteration of M of length ∞ by repeating the images of E_ν^M. Then $\nu_i, \kappa_i = \pi_{0,i}(\nu, \kappa)$ and the iteration is below the ultimate projecta $\omega\varrho_{M_i}^\omega = \pi_{0,i}(\beta)$. Set $W' = \bigcup\{M_i||\kappa_i ; i < \infty\}$. Then W' obviously cannot be universal. Note that $\mathrm{cf}(\nu_i) = \mathrm{cf}(\kappa_i)$ since the cofinality of ν_i as computed in M_i is equal to κ_i. Similarly as above we observe that $\nu_i = \kappa_i^{++M_{i+1}}$. (More precisely: $\kappa_i^{+M_i} = \kappa_i^{+M_{i+1}}$ is the largest cardinal in $J_{\nu_i}^{E_{i+1}}$; on the other hand, ν_i is the critical point of σ_i which is constructed at the i-th step similarly as σ above, therefore it must be a cardinal in M_{i+1}. In fact, if we choose M with $\mathrm{ht}(M) > \beta$, then this additional argument is not necessary.) Now, if θ is regular then $\nu_\theta = \kappa_\theta^{++W'}$ and $\mathrm{cf}(\nu_\theta) = \theta$.

We now revise the K^c-construction. Recall the notation from the section on K^c for measures of order 0. The construction is the same with the single exception that in the present case we deal with extenders instead of measures. Thus, the definition of $N_{\alpha+1}$ now reads:

$$N_{\alpha+1} = \begin{cases} \langle J_{\xi}^{\bar{E}^\alpha}, F\rangle & \text{if } \bar{E}^\alpha_{\omega\xi} = \varnothing \text{ and } F \text{ is the unique } \omega\text{-complete} \\ & \text{extender such that } \langle J_{\xi}^{\bar{E}^\alpha}, F\rangle \text{ is a premouse;} \\ \langle J_{\xi+1}^{\bar{E}^\alpha}, \varnothing\rangle & \text{otherwise.} \end{cases}$$

The definition of N_α for limit α as well as that of K^c is the same as before. Lemmata 6.4.1 and 6.4.2 go through with some minor amendments. Clause a) of Lemma 6.4.1 follows from the generalization of Theorem 4.5.1 to non-overlapping extenders, which we discussed at the beginning of the previous section. Regarding b), in order to prove that all total extenders in $N_{\alpha+1}$ are ω-complete, we also have to take care of those extenders which overlap the ultimate projectum of M_α. More precisely, if $\tau = \kappa^{+M_\alpha} \leq \omega\varrho^{\omega}_{M_\alpha} < \mathrm{o}^{M_\alpha}(\kappa)$, we want to show that $E^{M_\alpha}_\nu$ is ω-complete whenever $\tau \leq \nu \leq \mathrm{o}^{M_\alpha}$. Notice that the critical point of the core map $\sigma_\alpha : \widehat{M}_\alpha \to \widehat{N}_\alpha$ is at least τ, so Lemma 8.2.2 guarantees that the extender sequence of M_α agrees with that of N_α below $\mathrm{o}^{M_\alpha}(\kappa)$. By the induction hypothesis, $E^{M_\alpha}_\nu = E^{N_\alpha}_\nu$ must be ω-complete. The reformulation of Theorem 6.4.3 reads:

Theorem 8.3.2. *Let* $E = E^{K^c}$.

a) *If* E_τ *is a total measure in* K^c, *then* E_τ *is* ω*-complete.*

b) *If* F *is an* ω*-complete extender* on K^c *(thus,* F *measures all sets in* $\mathcal{P}(\mathrm{cr}(F)) \cap K^c$*, but we do* not *assume that* $F \in K^c$*) and* ν *is such that* $\langle J^E_\nu, F\rangle$ *is a premouse, then* $F = E_\nu$.

Proof. To see a), pick an α such that $\nu < \tau_{\alpha\infty}$ (see the definition of K^c in Chapter 6 for the notation). Thus, $E_\nu = E^{N_\alpha}_\nu$ and this extender is clearly total in N_α. By the discussion preceding this theorem, all total extenders in N_α are ω-complete. In particular, this applies to E_ν.

Regarding b), we will find a stage α of the K^c-construction such that F is added on top of N_α and

$$E^{N_\xi}_\nu = F \text{ for all } \xi \geq \alpha. \tag{8.3.1}$$

Define a decreasing sequence of ordinals $\langle\alpha_i\rangle_i$ as follows.

- $\alpha_0 = \infty$.

Granting that α_i is defined and $\mathrm{ht}(N_{\alpha_i}) > \nu$,

- α_{i+1} is the least ordinal ξ such that $\nu < \tau_{\xi\alpha_i}$ if α_i is a limit ordinal;
- $\alpha_{i+1} = \alpha_i - 1$ if α_i is a successor ordinal.

By induction on i we prove that $E^{N_{\alpha_i}}_\nu = E^\eta_\nu$ for all η such that $\alpha_{i+1} \leq \eta \leq \alpha_i$. This is clear if α_i is a limit (∞ is, of course, a limit), so it suffices to show that $E^{N_{\alpha_i}}_\nu = E^{N_{\alpha_{i+1}}}_\nu$ for successor α_i. To simplify the notation, we write ξ for α_{i+1}, so $\alpha_i = \xi + 1$. Let $N_{\alpha_i} = \langle J^{E^{\alpha_i}}_{\zeta^{\alpha_i}}, E^{\alpha_i}_{\omega\zeta^{\alpha_i}}\rangle$. We now consider two cases. If $E^{\alpha_i}_{\omega\zeta^{\alpha_i}} \neq \varnothing$, then $N_\xi = M_\xi = \langle J^{E^{\alpha_i}}_{\zeta^{\alpha_i}}, \varnothing\rangle$ and the conclusion follows immediately. Otherwise ζ^{α_i} is of the form $\zeta + 1$ and $M_\xi = \langle J^{E^{\alpha_i}}_\zeta, E^{\alpha_i}_{\omega\zeta}\rangle$. Letting $\sigma_\xi : \widehat{M}_\xi \to \widehat{N}_\xi$ be the associated core map, we observe that $\mathrm{cr}(\sigma_\xi) \geq \omega\varrho^{\omega}_{M_\xi} \geq \tau$ where $\tau = \kappa^{+K^c} = \kappa^{+M_\xi}$ for $\kappa = \mathrm{cr}(F)$ (recall that we assume that F an extender *on* K^c, i.e. that it measures all subsets of κ in K^c). By Lemma 8.2.2, E^{M_ξ} and E^{N_ξ} agree below $\gamma = \mathrm{o}^{M_\xi}(\kappa)$. We prove that $\mathrm{o}^{N_\xi} = \gamma$, which will yield our desired conclusion $E^{N_\xi}_\nu = E^{M_\xi}_\nu$. For this we use the condensation lemma (Theorem 8.2.3). Since σ_ξ is the core map, either ai) or aii) holds; we show that aii) is false. Indeed, if aii) were the case, then M_ξ and N'_ξ would have the same ultimate projectum where $\widehat{N}'_\xi = \mathrm{Ult}^*(\widehat{N}_\xi, E^{N_\xi}_{\nu_\xi})$ and $\nu_\xi = \mathrm{cr}(\sigma_\xi)$. But we know that $\omega\varrho^\omega_{M_\xi} \leq \nu$, while $\omega\varrho^\omega_{N'_\xi} \geq \pi_\xi(\tau) > \nu$ where $\pi_\xi : \widehat{N}_\xi \to \widehat{N}'_\xi$ is the associated ultrapower map; the inequality "$\geq$" follows from the fact that $\omega\varrho^\omega_{N_\xi} \geq \tau$.

Since α_i constitute a decreasing sequence of ordinals, there is an integer $n > 0$ such that α_{n+1} is not defined. Set $\alpha = \alpha_n$; we show that α meets the requirements specified above. By the definition of α and (8.3.1), $\mathrm{ht}(\alpha) = \nu$ and $E^{N_\alpha}_\nu = E_\nu$. Now $E^{N_\alpha}_\nu = F$, as otherwise the $\boldsymbol{K}^c$-construction would force $E^{N_{\alpha+1}}_\nu = F$, so we would have $E^{N_\alpha}_\nu \neq E^{N_{\alpha+1}}_\nu$, a contradiction to (8.3.1). Consequently, $E_\nu = E^{N_\alpha}_\nu = F$.

□(Theorem 8.3.2)

The notion of A-fullness generalizes to the non-overlapping extenders context in the straightforward way; as before we introduce the A-full $\boldsymbol{K}^c$ and prove that it is universal. The proof of this fact in the measures of order 0 case (see Theorem 6.4.4) was set up so that it can be repeated for A-full weasels in the present sense without any deeper revisions. The analogue to Lemma 6.4.6 now reads:

Lemma 8.3.3. *Let κ be a strong cardinal in $\boldsymbol{V}$. Then κ is strong in $\boldsymbol{K}^c$.*

Proof. (Sketch.) Assuming that the lemma fails, let $\gamma = \mathrm{o}^{\boldsymbol{K}^c}(\kappa) < \infty$. Using the fact that κ is strong in $\boldsymbol{V}$, choose an extender F with critical point κ of sufficient strength so that $M = \mathrm{Ult}(\boldsymbol{V}, F)$ contains $E^{\boldsymbol{K}^c} \restriction \gamma$. This will guarantee that $(\boldsymbol{K}^c)^M = \pi(\boldsymbol{K}^c)$ agrees with $\boldsymbol{K}^c$ below γ. It is then easy to verify that $\langle J^E_\gamma, (F|\gamma) \restriction (\mathcal{P}(\kappa) \cap J^E_\gamma)\rangle$ is a premouse (here $E = E^{\boldsymbol{K}^c}$), so $E_\gamma \neq \varnothing$ by b) of Theorem 8.3.2. Contradiction.

□(Lemma 8.3.3)

The theory of s-mice now translates into the theory of so-called *p-mice* and $0^\ddagger$ translates into $0^\P$, a mouse called 0-*pistol*. A p-premouse is a structure of the form $\langle J^E_\tau, U\rangle$ such that J^E_τ has a largest cardinal κ, U is a weakly amenable measure on J^E_τ and $\mathrm{o}^{J^E_\tau}(\mu) = \tau$ for some $\mu < \kappa$. p-premice can be iterated in the usual way and iterable p-premice are called p-mice. The theory of p-mice is the same as that of s-mice and $0^\P$ is related to p-mice in the same way as $0^\ddagger$ is related to s-mice, i.e. $0^\P$ is the common core of all p-mice. Notice also that, letting M be this common core, $\omega\varrho^1_M = \omega\varrho^\omega_M = \omega$ and $p_M = \varnothing$. Also, by iterating the top measure of M out of the universe we obtain the minimal extender model W for one strong cardinal. A little bit more work yields a an elementary embedding of W or a remarkable set of indiscernibles for W. (Regarding e.g. the former, let $\pi : M \to M_1$ be the ultrapower map by U; the desired embedding is then obtained by iterating the top measures of M and M_1 out of the universe and simultaneous extending of π to the resulting model.) Thus, $0^\P$ represents the *first* mouse which contains two overlapping extenders and corresponds to the sharp for an inner model with a strong cardinal. As in the case of $0^\ddagger$, we can prove that $0^\P$ is a Π_1-singleton (consequently, a Δ^1_3-singleton) and that it cannot be generically added to $\boldsymbol{V}$ by a set forcing.

The statement of the weak covering lemma for K^c does not require any amendment; this applies to both Theorem 6.6.1 and Theorem 6.6.2. The proof of the latter theorem again requires the frequent extensions of embeddings lemma, which we will state below. We shall now focus on the former theorem, as this suffices for the construction of K. The proof of Theorem 6.6.1 was set up so that there is only one place which requires some additional discussion, and this place occurs at the very end of the proof. We briefly recall the situation. Assuming that κ is an ω-closed cardinal in V, we aim to show that $\tau = \kappa^{+K^c}$ is the real κ^+. At the end of the above mentioned proof of Theorem 6.6.1 we constructed a canonical extension $\tilde{\sigma} : M \to N$ of $\sigma \restriction \bar{K}$ to M where M was an iterate of K. More precisely, M was the last model on the K-side of the coiteration of K against $\bar{K}$. This time the domain of $\tilde{\sigma}$ might be a premouse $\widehat{M}$ such that $\mathrm{o}^M(\bar{\mu}) \geq \bar{\tau} = \mathrm{ht}(\bar{K})$ for some $\bar{\mu} < \bar{\kappa} = \sigma^{-1}(\kappa)$, so we have to take care of two issues. First of all, if $\mathrm{cr}(E^M_{\mathrm{top}}) = \bar{\mu}$, we have to ensure that the pseudoultrapower of $\widehat{M}$ by $\sigma \restriction \bar{K}$ is a premouse, i.e. that its top extender is weakly amenable. In order to do this, we construct the map σ at the beginning so that $\mu^{+K^c} \subset \mathrm{rng}(\sigma)$ where $\mu = \sigma(\bar{\mu})$. Simply, if $\mathrm{o}^{K^c}(\mu) \geq \tau$ (there is at most one such μ), we include all ordinals less than μ^{+K^c} in X (see the beginning of that proof). This will guarantee that $\bar{\mu} = \mu$ and that $\bar{K}$ and K have the same subsets of μ. Clearly, X will be of size smaller than κ, as κ is ω-closed.

At this point we know that $\tilde{\sigma} : \widehat{M} \to \widehat{N}$ where N is a premouse. The proof of Theorem 6.6.1 also shows that $\omega\varrho^{n+1}_N \leq \kappa < \omega\varrho^n_N$ and that N is iterable above $\omega\varrho^{n+1}_N$. Our intention is to compare N with K^c. Now if $\mathrm{o}^N(\mu) > \tau$ for some $\mu < \kappa$, we cannot a priori rule out that the first extender used on the N-side has critical point μ, in which case the above guaranteed amount of iterability need not suffice to carry out the comparison. However, we can rule this out with a bit of effort. What we prove is the following:

$$E^{K^c} \restriction \gamma = E^N \restriction \gamma \text{ where } \gamma = \mathrm{o}^N(\mu). \tag{8.3.2}$$

To see this, it is enough to show that each extender E^N_ν with critical point μ is ω-complete. Since N agrees with K^c below τ, we can assume w.l.o.g. that $\nu \geq \tau$. Let $A \subset \nu$ and $X \subset \mathcal{P}(\mu) \cap N$ be countable; we find a mapping $\mathfrak{k} : A \to \kappa$ witnessing that E^N_ν fixes $A \cup X$, which will guarantee the ω-completeness of F. The elementary hull Y of $A \cup X \cup \mu^{+K^c}$ in $\widehat{N||\nu}$ is of size less than κ, so by collapsing we obtain an elementary embedding $\sigma^* : \widehat{N}^* \to \widehat{N||\nu}$ with critical point larger than μ^{+K^c}. By Lemma 8.2.2, N^* is an initial segment of $N||\nu$, so N^*, being of size smaller than κ, must be an initial segment of $N||\kappa = K^c||\kappa$. (Notice that the proof of Theorem 6.6.1 guarantees the Σ_0-iterability of N, which is an amount of iterability sufficient for the application of Lemma 8.2.2.) Thus, all extenders on the N^*-sequence with the critical point μ are ω-complete. Let $\mathfrak{k}^* : A^* \to \kappa$ witness that $E^{N^*}_{\mathrm{top}}$ fixes $A^* \cup X^*$ where A^* and X^* is the preimage of A and X under σ^*, respectively. Given an $x \in X$ and $\alpha \in A$ and letting $\sigma^*(\alpha^*) = \alpha$, we have:

$$x \in E^N_{\nu,\alpha} \longrightarrow x \in E^{N^*}_{\mathrm{top},\alpha^*} \longrightarrow \mathfrak{k}^*(\alpha^*) \in x.$$

Thus, it suffices to set $\mathfrak{k} = \mathfrak{k}^* \circ (\sigma^*)^{-1}$. This completes the proof of the ω-completeness of E_ν^N.

Arguments of the similar sort can be used for the proof of the stronger version of weak covering for $\boldsymbol{K}^c$ (Theorem 6.6.2). Alternatively, one can fix α (see the statement of that theorem for the notation) so that it is not overlapped by any extender; then $\mu > \alpha$ where μ is as in the above discussion and the weak covering lemma will hold at μ by the induction hypothesis. The ω-completeness of E_ν^N is then guaranteed by Lemma 8.2.12. We can then either proceed as above to ensure that N is a premouse or else use the ultrapower $\text{Ult}(N||\eta, E_{\text{top}}^N)$ from the discussion preceding Lemma 8.2.10 if E_{top}^N is not total on N. We then have to argue that this ultrapower is sufficiently iterable so that it can be compared with $\boldsymbol{K}^c$.

Using the weak covering theorem we also obtain the generalization of Lemma 6.6.6 under the assumption that there is no weasel W containing some κ with $\text{o}^W(\kappa) = \infty$. (Note: This is equivalent to saying that there are no strong cardinals in W, as follows from the fact that $W = (\boldsymbol{K}^c)^W$.) Thus, any universal weasel correctly computes cardinal successor on a stationary proper class. If we strengthen our definition of universality in the sense of the discussion preceding Lemma 8.3.1, we can drop the above smallness condition; notice also that if there is a weasel with a strong cardinal, then only weasels containing strong cardinals are universal in the sense of this stronger definition,

Equipped with the weak covering lemma, we can proceed to the theory of $\boldsymbol{K}$. This theory, as it was set up in Chapter 7, can almost literally be translated into the present context, so no substantial revisions are necessary. In particular, both the inductive and Steel's definition of $\boldsymbol{K}$ can be formulated exactly as before, and the construction from Section 7.3 goes through. Let us, however, state the revised definition of a W-correct extender. Given a universal weasel W and an extender F *on* W (i.e. F measures all subsets of $\text{cr}(F)$ in W), F is W-correct iff

- either $\langle J_\nu^{E^W}, U\rangle$ is a premouse or $\langle J_\nu^{E^W}, E_\nu^W, U\rangle$ is an p-premouse;
- $\text{Ult}(W, U)$ exists.

The proofs of Lemmata 7.3.1 – 7.3.3 go through as before; we note that the second sentence in b) of Lemma 7.3.3 this time reads: $\langle J_\nu^E, F\rangle$ is a premouse if $\text{o}^W(\kappa) < \alpha$ for all $\kappa < \alpha$ and $\langle J_\nu^E, E_\nu, F\rangle$ is a p-premouse otherwise. Regarding Lemmata 7.3.4 – 7.3.6, the only deeper revision that has to be made is in the proof of the Claim from the proof of Lemma 7.3.6. There we construct a W-correct measure U and argue that it is ω-complete. In the present context we have a W-correct extender F of length ν and we want to prove that F is ω-complete. For this we use Lemma 8.2.12. The only think we have to verify is that $\langle J_\nu^E, F\rangle$ is Σ_0-iterable (although a closer look at the proof of that lemma shows that far less is necessary), which can be done by a condensation argument using the fact that $\text{cr}(F)^{+W}$ is not ω-cofinal. (Assuming that Σ_0-iterability fails for $M = \langle J_\nu^E, F\rangle$, it will also fail for some countable $\bar{M}$ whose expansion is embeddable into that of M; using the interpolation lemma find an $\tilde{M}$ together with embeddings

$$\tilde{\sigma} : \widehat{M} \to \widehat{\tilde{M}} \quad \text{and} \quad \sigma' : \widehat{\tilde{M}} \to \widehat{M}$$

such that $\text{cr}(F) < \text{cr}(\sigma') = \text{cr}(F)^{+\tilde{M}} < \text{cr}(F)^{+M}$. The existence of $\tilde{\sigma}$ guarantees the failure of Σ_0-iterability for $\tilde{M}$. Now $E^{\tilde{M}}$ and E^M agree below ν by Lemma 8.2.2. To see that they also agree at $\text{ht}(\tilde{M})$, imitate the second part of the proof of that lemma; notice that for this we merely need that $J_\nu^E||\zeta$ is and its image under the ultrapower map by F are iterable where $\zeta < \text{cr}(F)^{+M}$ is arbitrary, and that this follows from the fact that $\omega_1 \subset J^E\nu$.) We stress that the use of Lemma 8.2.12 in the proof of Lemma 7.3.6 can be avoided and replaced by an argument based on the pressing down lemma. This is important in the constructions of $\boldsymbol{K}$ for many strong cardinals (see [Sch1]) from ω-complete extenders, as Lemma 8.2.12 fails at those levels. Lemmata 7.3.7 – 7.3.9 go through as before, so we have

Lemma 8.3.4 ($\neg 0^\P$). *Let F be an extender of length ν that is $\boldsymbol{K}$-correct. Then $F = E_\nu^{\boldsymbol{K}}$.*
□(Lemma 8.3.4)

From this lemma and its the variant for models $\boldsymbol{K}[J_\gamma^E]$ we obtain the rigidity lemma and the generalizations of Theorems 7.4.8 and 7.4.9 and Corollary 7.4.10, which do not need any reformulation. The same also applies to the results on the absoluteness of $\boldsymbol{K}$ relative to inner models and on the generic absoluteness, i.e. to Theorems 7.4.11 and 7.4.12.

The last topic we have to revise is the weak covering lemma for $\boldsymbol{K}$. Again, the lemma holds as stated before, but we have to make some amendments to the proof. More precisely, what needs a reformulation is the frequent extensions of embeddings lemma.

Theorem 8.3.5 (Frequent Extensions of Embeddings Lemma, fine structural version). *Consider a linear system of structures and maps with the following properties:*

a) *$\alpha > \omega_1$ is a regular cardinal;*

b) *$Q = J_\tau^E$ is where $E = E^W$ for some universal weasel W and τ is a cardinal in W (thus, Q is a strong mouse), $\kappa \geq \alpha$ is the largest cardinal in Q and $\text{cf}(\mu^{+Q}) > \omega$ whenever $\alpha \leq \mu < \kappa$;*

c) *$\langle \alpha_\xi;\ \xi \leq \alpha \rangle$ is a normal sequence satisfying $\alpha_\alpha = \alpha$;*

d) *$Q_\xi = J_{\tau_\xi}^{E^\xi}$ for $\xi \leq \alpha$ is a system of premice such that $Q_\alpha = Q$ and $\text{card}(Q_\xi) < \alpha$ for $\xi < \alpha$; moreover, κ_ξ is the largest cardinal in Q_ξ;*

e) *$\sigma_{\xi\xi'} : Q_\xi \to Q_{\xi'}$ is a commutative system of Σ_0-preserving maps such that $\sigma_{\xi\xi'}(\kappa_\xi) = \kappa_{\xi'}$ and, for limit $\lambda \leq \alpha$, $\langle Q_\lambda, \sigma_{\xi\lambda} \rangle_\xi$ is the direct limit of the diagram $\langle Q_\xi, \sigma_{\xi\xi'} \rangle_{\xi \leq \xi' < \lambda}$, that is, $Q_\lambda = \bigcup\{\text{rng}(\sigma_{\xi\lambda});\ \xi < \lambda\}$;*

f) *$\text{cf}(\mu^{+Q_\xi}) > \omega$ whenever $\alpha_\xi \leq \mu < \kappa_\xi$ and $\text{cf}(\xi) > \omega$.*

Setting $\tilde{\tau}_\xi = \sup(\sigma''_{\xi\alpha}\tau_\xi)$ and $\tilde{Q}_\xi = J_{\tilde{\tau}_\xi}^E$, we have

$$\sigma_{\xi\alpha} : Q_\xi \underset{\Sigma_0}{\longrightarrow} \tilde{Q}_\xi \quad \textit{cofinally}.$$

Fix a stationary set $S \subset \alpha$ of ordinals of uncountable cofinality and to each $\xi \in S$ assign a mouse M_ξ such that

A) *M_ξ is a lengthening of Q_ξ, i.e. $E^\xi = E^{M_\xi} \upharpoonright \tau_\xi$;*

B) *τ_ξ is the cardinal successor of κ_ξ in M_ξ;*

C) *One of the following holds:*

  1) $\omega\varrho^{\omega}_{M_\xi} > \kappa_\xi$;

  2) $\omega\varrho^{\omega}_{M_\xi} \leq \kappa_\xi$ *and M_ξ is sound above κ_ξ.*

Under the above assumptions, for all but nonstationarily many ξ we have:

i) *The canonical extension $\tilde{\sigma}_\xi : \widehat{M}_\xi \to \widehat{\tilde{M}}_\xi$ exists;*

ii) *granting that $E^{\tilde{M}_\xi}_{\text{top}}$ is a total extender in $\tilde{M}_\xi$, the structure $\tilde{M}_\xi$ is a premouse that is fully $*$-iterable.*

iii) *if $E^{\tilde{M}_\xi}_{\text{top}}$ is* not *a total extender in $\tilde{M}_\xi$, then, letting $\widehat{N}_\xi = \widehat{\mathrm{Ult}^*(\tilde{M}_\xi \| \eta_\xi, E^{\tilde{M}_\xi}_{\text{top}})}$ where η_ξ is maximal such that $E^{\tilde{M}_\xi}_{\text{top}}$ measures all subsets of $\mathrm{cr}(E^{\tilde{M}_\xi}_{\text{top}})$ in $\tilde{M}_\xi \| \eta_\xi$, the structure N_ξ is a premouse that is fully $*$-iterable.*

The proof the above theorem closely follows that of Theorem 7.5.2. In the present case, we have strengthened the assumptions, as we now cannot prove the iterability of $\tilde{M}_\xi$ above κ in the general case. This is caused by the fact that τ might be overlapped by an extender, i.e. $\gamma_\xi = \mathrm{o}^{\tilde{M}_\xi}(\mu) \geq \tau$ for some $\mu < \kappa$. If C2) holds, the general methods for proving iterability would yield mere iterability of $\tilde{M}_\xi$ above γ_ξ, which need not be enough for the comparison of $\tilde{M}_\xi$ with $\boldsymbol{K}$. The same applies to the iterability of the premice $\bar{M}_\xi$ from the proof of Theorem 7.5.2. There we want to compare a pair of premice $\bar{M}_\xi, \bar{M}_{\xi'}$ and a closer look at the situation reveals that this requires mere normal iterability of these premice above κ_ξ if κ_ξ is not overlapped by an extender, and normal iterability above μ_ξ if $\mu_\xi < \kappa_\xi$ and $\mathrm{o}(\mu_\xi) \geq \tau_\xi$ in the premice in question. In the former case we do not need to take any extra steps. To prove the requisite amount of iterability of $\bar{M}_\xi$ in the latter case, we use the strongness of Q together with Lemma 8.2.2 if $\mu_\xi < \alpha_\xi$ (this will guarantee that a sufficiently long initial segment of $\bar{M}_\xi$ is strong) and a condensation argument together with f) if $\alpha_\xi \leq \mu_\xi < \kappa_\xi$.

Let us give more details regarding this condensation argument. Using the interpolation lemma we construct an embedding $\sigma_N : \widehat{N} \to \widehat{\bar{M}}_\xi$ with the critical point $\bar{\vartheta}$ that is sent by σ_N to $\mu_\xi^{+\bar{M}_\xi}$; this is possible, as $\mathrm{cf}(\mu_\xi^{+\bar{M}_\xi})$ is uncountable. By the standard methods, we can arrange the failure of iterability for N above μ_ξ. By Lemma 8.2.2, N agrees with $\bar{M}_\xi$ below $\mathrm{o}^N(\mu)$, so letting ζ be maximal such that $\bar{\vartheta}$ is a cardinal in $\bar{M}_\xi \| \zeta$, the coiteration of N with $\bar{M}_\xi \| \zeta$ is above $\bar{\gamma} = \mathrm{o}^N(\mu)$. We observe that the map σ_N is sufficiently preserving to guarantee the iterability of N above $\tilde{\gamma}$, which suffices for the coiterability of N with $\bar{M}_\xi||\zeta$. If $\omega\varrho^{n+1}_{\bar{M}_\xi} > \mu_\xi^{+\bar{M}_\xi}$ where n is such that $\omega\varrho^{n+1}_{\bar{M}_\xi} \leq \kappa_\xi < \omega\varrho^{n}_{\bar{M}_\xi}$, then the interpolation map σ_N is $\Sigma_1^{(n)}$-preserving, so it guarantees the iterability of N above $\tilde{\gamma}$. A similar argument can be applied if $\omega\varrho^{n}_{\bar{M}_\xi} < \mathrm{o}^{\bar{M}_\xi}(\mu_\xi)$. In the remaining case we have $\omega\varrho^{n+1}_{\bar{M}_\xi} \leq \mu_\xi$ and $\mathrm{o}^{\bar{M}_\xi}(\mu_\xi) \leq \omega\varrho^{n}_{\bar{M}_\xi}$; notice that $\bar{M}_\xi$ is then fully iterable above the n-th projectum, as we can normally iterate it to a premouse whose all total extenders above the n-th projectum are ω-complete. This guarantees that $\mathrm{core}^n(\bar{M}_\xi)$ exists; we then notice that $\bar{M}_\xi$ can be replaced by this core. This will guarantee that the interpolation map σ_N (now constructed from $\mathrm{core}^n(\bar{M}_\xi)$) is $\Sigma_0^{(n)}$-preserving, which is just enough to ensure the iterability of N above $\tilde{\gamma}$. The just established amount of iterability for N guarantees the coiterability of N with $\bar{M}_\xi||\zeta$, as $\bar{M}_\xi||\zeta$, being a *proper* initial segment of $\bar{M}_\xi$, is fully iterable. It is then easy to see that $\bar{M}_\xi||\zeta$ will win the coiteration, so the iteration map on the N-side π_N will embed $\widehat{N}$ into an iterate of $\bar{M}_\xi||\zeta$ (more precisely, into an expansion thereof), which is iterable. This would guarantee the full iterability for N, a contradiction.

The rest of the proof of the frequent extension of embeddings lemma can be carried out without any essential revision up to the point where we establish the normal iterability of $\bar{M}_\xi$ above $\omega\varrho^{n+1}_{\bar{M}_\xi}$; here n is such that $\omega\varrho^{n+1}_{M_\xi} \leq \kappa_\xi < \omega\varrho^{n}_{M_\xi}$. In order to guarantee the full iterability of $\bar{M}_\xi$, we show that $\bar{M}_\xi||\gamma_\xi$ is strong. This is easy to see if $E = E^{\boldsymbol{K}}$. The above proof namely shows that $\bar{M}_\xi$ is Σ_0-iterable, so in particular, Q is extendible by $E_\nu^{\bar{M}_\xi}$ whenever $\mu^{+\bar{M}_\xi} \leq \nu < \gamma_\xi$. But we know that there is some $\mu' \geq \mu$ such that $\vartheta = \mathrm{cf}((\mu')^{+Q})$ is not ω-cofinal. By Lemma 8.2.12, the canonical extension of the ultrapower map $\pi_\nu : J_\vartheta^E \to \mathrm{Ult}(J_\vartheta^E, E_\nu^{\bar{M}_\xi})$ to $\boldsymbol{K}$ exists, which means that $E_\nu^{\bar{M}_\xi}$ is $\boldsymbol{K}$-correct and, consequently, that $E_\nu^{\bar{M}_\xi} = E_\nu$. In the case of $\boldsymbol{K}^c$ we can argue very much like in the above revision of the proof of weak covering for $\boldsymbol{K}^c$ that $E_\nu^{\bar{M}_\xi}$ is ω-complete (here we use Lemmata 8.2.2 and 8.2.12). However, one can also prove a general lemma:

Lemma 8.3.6. *Let J_τ^E be strong where τ is a cardinal in some universal weasel such that $E^W \restriction \tau = E \restriction \tau$ and let J_β^E be a lengthening of J_τ^E. (In this lemma, we do not assume that J_τ^E has a largest cardinal.) Suppose that there is a regular α in $\boldsymbol{V}$ such that $\mathrm{cf}(\lambda^{+J_\tau^E}) > \omega$ for all $\lambda \geq \alpha$ in J_τ^E. Suppose further that $\mathrm{o}^{J_\beta^E}(\mu) = \beta$ for some μ such that $\mu^{+J_\tau^E} < \tau$. Then J_β^E is strong.*

Proof. (Sketch.) We verify the two clauses from the definition of strongness. Assuming that M is a premouse satisfying $E^M \restriction \beta = E$ which is normally iterable beyond $\beta+1$, we have to show that M is fully iterable and that $\mathrm{core}(M)$ agrees with J_β^E below β. Regarding the full iterability of M, we can proceed as in the above proof of the iterability of $\bar{M}_\xi$ above μ_ξ. Regarding the initial segment of the core, notice that this follows from the condensation lemma, once we have ruled out clause aii). By the argument from the proof of Theorem 8.3.2, aii) can occur only if $\tilde{M}_\xi$ projects to μ. Now let W be a universal weasel from the statement of this lemma; then $\tilde{M}_\xi$ coiterates with W above μ and W wins the coiteration, which means that

$$\Sigma^*(\widehat{\tilde{M}}_\xi) \cap \mathcal{P}(\mu) \subset W \subset J_\tau^E \subset \widehat{\tilde{M}}_\xi .$$

Contradiction. □(Lemma 8.3.6)

In the above revision of the proof of the frequent extensions of embeddings lemma we assumed that $\tilde{M}_\xi$ is a premouse. If this were not the case, the whole construction would contain one more additional step – namely the transfer from $\tilde{M}_\xi$ to the ultrapower $\mathrm{Ult}^*(\tilde{M}_\xi || \zeta, E_{\mathrm{top}}^{\tilde{M}_\xi})$ described above. The same also applies to the structures $\bar{M}_\xi$. What has to be done is to establish the relationship between the fine structure of $\tilde{M}_\xi$ and the associated ultrapower. The arguments described above will become somewhat more technical, but no substantially new ideas are needed here. □(Theorem 8.3.5)

Finally we state the coarse version of the frequent extensions of embeddings lemma, whose proof is just a mix of the techniques described above and those from the proof of Theorem 7.5.3

Theorem 8.3.7 (Frequent Extensions of Embeddings Lemma, coarse version). *Consider J_τ^E, α, κ and the system of structures Q_ξ and maps $\sigma_{\xi\xi'}$ satisfying* a) – f) *from the previous theorem. Define $\tilde{Q}_\xi$ as before, fix a stationary set $S \subset \alpha$ of ordinals of uncountable cofinality and to each $\xi \in S$ assign a Σ_0-mouse M_ξ satisfying* A) *and* B) *of the previous theorem.*

Then for all but nonstationarily many ξ we have:

i) *The 0-extension $\tilde{\sigma}_\xi : \widehat{M}_\xi \to \widehat{\tilde{M}}_\xi$ of σ_ξ exists.*

ii) *Granting that $\tilde{M}_\xi$ is a premouse, it is Σ_0-iterable.* □(Theorem 8.3.7)

The proof of the weak covering theorem for K requires only some minor revisions, which are clear from our above discussion of weak covering for K^c and the proof of the frequent extensions of embeddings lemma. At the end of the proof we obtain a liftup $\tilde{M}_\xi$ which is iterable above $\omega\varrho_{\tilde{M}_\xi}^{n+1}$ where n is such that $\omega\varrho_{\tilde{M}_\xi}^{n+1} \leq \kappa < \omega\varrho_{\tilde{M}_\xi}^{n}$. We now need to make sure that the coiteration of $\tilde{M}_\xi$ with K is above κ on the $\tilde{M}_\xi$-side. If κ is not overlapped by any extender in $\tilde{M}_\xi$, we can then proceed as in the measures of order 0 case. Now suppose that $\gamma_\xi = \mathrm{o}^{\tilde{M}_\xi}(\mu) \geq \tau$ for some $\mu < \kappa$. By the Lemma 8.3.6 (or, by the discussion in the paragraph preceding this lemma) we know that $E_\nu^{\tilde{M}_\xi} = E_\nu^K$ whenever $\tau \leq \nu < \gamma_\xi$, so the coiteration on the $\tilde{M}_\xi$-side is in fact above γ_ξ. This completes the sketch of the proof of the weak covering theorem.

Notes

The core model theory for non-overlapping extenders was first developed by Jensen [Je1]; another approach is taken by Koepke [Ko89], where a different fine structure is developed. Our presentation differs from both and is substantially simplified by using coherent structures as expansions. Fragments were first considered by Friedman and Wylie in [Wy90]; they were later used in covering arguments and in the analysis of the structure of extender models in [Mi1] and [SZ1].

Chapter 9
Overlapping Extenders

In this section we develop the theory of mice with overlapping extenders which can be used to build extender models beyond one strong cardinal. The models constructed at present reflect large cardinal principles with consistency strength hovering at the level of many Woodin cardinals, yet well below that of a measurable Woodin cardinal. However, we shall focus primarily on the properties of iterable premice, as the model constructions exceed the framework of this book.

We chose to use short extenders with dense indexing in the case of non-overlapping extenders. In the present case, we shall go to the other extreme – following Jensen, we shall use long extenders, which require very sparse indexing. There are many alternatives to this approach; the Mitchell–Steel indexing, for instance, uses much shorter extenders than we do. In fact, we shall use the longest possible extenders, while they use the shortest possible. Each approach has its own advantages and disadvantages. With our indexing, the theory of premice becomes much more elegant than with any shorter indexing; on the other hand, the price for this is the complexity of iterability proofs.

The notion of iteration requires a reformulation when dealing with overlapping extenders, as linear iterations no longer suffice for the comparison process. We shall replace them with iteration trees with iteration maps along their branches. With the advent of iteration trees a new phenomenon arises: at limit steps of an iteration we have to choose a cofinal well-founded branch through the tree, that is, a cofinal branch such that the direct limit along it is well-founded. An assignment picking such branches is called an *iteration strategy* and iterability is now synonymous to the existence of an iteration strategy. Thus, iterability is no longer a Π_1-notion as it was below one strong cardinal. As it might be expected, constructing iteration strategies is a deep problem, which turns out to be connected to smallness assumptions. Roughly speaking, in the absence of Woodin cardinals, all iteration trees have exactly one cofinal well-founded branch, which gives rise to a so-called uniqueness strategy. It is worth noting that all constructions of iteration strategies known at present build on simulating the uniqueness by putting additional requirements on well-founded branches.

As we have already mentioned above, we shall focus on fine structure of iterable premice. We first introduce the notions of premice and iterability, develop the copying construction and prove a type of Dodd–Jensen lemma. This will allow us to establish solidity and prove the condensation lemma. Finally, we give a criterion for uniqueness of well-founded branches.

9.1 Premice and Iteration Trees

The extender sequences associated with our new premice guarantee that any premouse is a coherent structure. Such extenders are whole, i.e. their lengths are the images of their critical points under the associated ultrapower maps. In this sense they are *long*. The comparison process in our earlier settings was based on the fact that the critical points of extenders in the coiteration do not overlap those extenders used earlier. It is not hard to see that this will fail if we attempt to coiterate the present premice linearly. This deficiency can be remedied by applying each extender arising in the comparison process to the earliest model to which it can possibly be applied. This gives rise to a tree structure called an *iteration tree* and it follows that the above requirement on the critical points holds along the branches, which suffices for the proof of the comparison lemma. As the ultrapowers we are now building use external extenders, an additional argument is needed to see that iteration maps have all requisite preservation properties.

Definition. A *potential premouse* is an acceptable structure $M = \langle J_\nu^E, E_{\omega\nu}\rangle$ satisfying:

a) E is a set of triples $\langle \bar\nu, x, y\rangle$ for $\bar\nu \le \nu$ such that, setting

$$E_{\omega\bar\nu} = \{\langle x, y\rangle;\ \langle \bar\nu, x, y\rangle \in E\},$$

the structure $M||\bar\nu = \langle J_{\bar\nu}^E, E_{\omega\bar\nu}\rangle$ is coherent whenever $E_{\omega\bar\nu} \ne \varnothing$.

b) For every $\bar\nu \le \nu$, if $E_{\omega\bar\nu} \ne \varnothing$, then $E_{\omega\bar\nu}$ is weakly amenable w.r.t. $M||\bar\nu$.

c) $M||\bar\nu$ is sound for every $\bar\nu < \nu$.

If $E_{\omega\bar\nu} \ne \varnothing$, then by a) and b) of the above definition, $J_{\bar\nu}^E = \mathrm{Ult}(J_{\bar\tau}^E, E_{\omega\bar\nu})$ where $\bar\tau$ is the cardinal successor of $\mathrm{cr}(E_{\omega\bar\nu})$ in $J_{\bar\nu}^E$. It follows that $J_{\bar\nu}^E$ is a ZFC$^-$ model, so $\omega\bar\nu = \bar\nu$. Furthermore, if $\zeta \ge \bar\tau$ is such that $\bar\tau$ is still a cardinal in J_ζ^E, then $\mathbb{D}(J_\zeta^E, E_{\omega\bar\nu})$ is defined and the regularity of $\bar\tau$ in $M||\bar\nu$ guarantees that $\bar\nu$ is the image of $\bar\tau$ under the ultrapower map and lies in the well-founded core of $\mathbb{D}$. It follows that $E_{\bar\nu}^{\mathbb{D}} = \varnothing$ as otherwise $\langle J_{\bar\nu}^E, E_{\bar\nu}^{\mathbb{D}}\rangle$ would be a coherent structure giving rise to a cofinal map from the cardinal successor of $\mathrm{cr}(E_{\bar\nu}^{\mathbb{D}})$ in $J_{\bar\nu}^E$ to $\bar\nu$ definable (in fact, Δ_1-definable) over $\langle J_{\bar\nu}^E, E_{\bar\nu}^{\mathbb{D}}\rangle$. Such a map would collapse $\bar\nu$ inside $\mathbb{D}$, which is impossible by our above observation. We have thus proved:

Lemma 9.1.1. *Let M be a premouse and let E_ν^M be a total extender in M. If $M' = \mathrm{Ult}(M, E_\nu^M)$, then $E^{M'} \restriction \nu = E^M \restriction \nu$ and $E_\nu^{M'} = \varnothing$.* □(Lemma 9.1.1)

In other words, the coherency condition, which we had to postulate explicitly in our earlier definitions of premice, follows now automatically from a) and b). It is also not hard to see that being a potential premouse is a Π_2-condition. As one of the requirements on potential premousehood is the weak amenability of the top extender, the example from the section on premice for non-overlapping extenders shows that potential premousehood is not preserved upwards under cofinal Σ_0-maps.

Lemma 9.1.2. *There is a Π_2-sentence φ such that for every transitive rudimentarily closed structure $M = \langle U, E, F\rangle$ we have:*

$$M \text{ is a potential premouse} \quad \longleftrightarrow \quad M \models \varphi.$$

Consequently, potential premousehood is preserved downwards by Σ_1-maps and upwards by Σ_2-maps.

Proof. This is very much like the proof of Lemmata 8.1.2 and 8.1.5. Clause a) can be expressed in a Q-fashion, b) in Π_2-fashion and c) in Π_1-fashion. The details are left to the reader. □(Lemma 9.1.2)

To ensure that the comparison of a pair of premice terminates, we will need an appropriate initial segment condition. Here we shall face a more delicate issue than it would seem at first glance, as this time the initial segment condition will not, in general, be preserved downwards by Σ_1-maps.

Definition. Let F be a whole extender over M at κ, λ which is weakly amenable w.r.t. M. We say that the ordinal $\bar{\lambda} < \lambda$ is a *cutpoint* of F iff for all $f : \kappa \to \kappa$ in M and $\alpha < \bar{\lambda}$ we have $\pi(f)(\alpha) < \bar{\lambda}$ where π is the associated ultrapower map, that is,

$$\prec\alpha, \bar{\lambda}\succ \in F(\{\prec\eta, \eta'\succ \in \kappa;\ f(\eta) < \eta'\}).$$

In other words, $\bar{\lambda}$ is a cutpoint of F iff $F|\bar{\lambda}$ is a whole extender over M. It follows easily that the set of all cutpoints of F is closed in λ.

If $\langle J_\nu^A, F\rangle$ is a coherent structure, $\pi : J_\tau^A \to J_\nu^A$ is the ultrapower map determined by F, $\bar{\lambda} \leq \lambda$ and $\bar{\pi} : J_\tau^A \to J_{\bar{\nu}}^{\bar{A}}$ the ultrapower map determined by $F|\bar{\lambda}$, then the canonical map $k : J_{\bar{\nu}}^{\bar{A}} \to J_\nu^A$ defined by $\bar{\pi}(f)(\alpha) \mapsto \pi(f)(\alpha)$ for $f : \kappa \to \kappa$ and $\alpha < \bar{\lambda}$ is Σ_0 cofinal and sends $\bar{\pi}(\kappa)$ to λ. If $\bar{\lambda}$ is a cutpoint of F, then $\mathrm{cr}(k) = \bar{\lambda} = \bar{\pi}(\kappa)$. Consequently, $k(\bar{\lambda}) = \lambda$ and $\bar{A} \cap J_{\bar{\lambda}}^{\bar{A}} = A \cap J_{\bar{\lambda}}^{A}$. The converse also holds: If $\mathrm{cr}(k) = \bar{\lambda}$, then $\bar{\lambda}$ is a cutpoint of F. Any cutpoint $\bar{\lambda}$ of F is a limit (in fact inaccessible) cardinal in $J_{\bar{\nu}}^{\bar{A}}$, and, by the preservation properties of k, also a limit cardinal in J_ν^A. Moreover, since $J_{\bar{\lambda}}^{\bar{A}}$ is the $V_{\bar{\lambda}}$ of $J_{\bar{\nu}}^{\bar{A}}$, it is also the $V_{\bar{\lambda}}$ of J_ν^A.

As $\bar{F} = F|\bar{\lambda}$ is whole, $\langle J_{\bar{\nu}}^{\bar{A}}, \bar{F}\rangle$ is also a coherent structure. We show that k is Σ_0-preserving as a map from $\langle J_{\bar{\nu}}^{\bar{A}}, \bar{F}\rangle$ to $\langle J_\nu^A, F\rangle$, and so Σ_0 cofinal in this sense. Clearly, it suffices to prove that $k(\bar{F} \cap a) = F \cap k(a)$ for every $a \in J_{\bar{\nu}}^{\bar{A}}$. Notice that if $\zeta < \tau$ is such that κ is the largest cardinal in S_ζ^A and $f : \kappa \to \mathcal{P}(\kappa) \cap S_\zeta^A$ is an enumeration of the power set of κ in S_ζ^A, then $\pi(f) \restriction \kappa$ is the enumeration of $\pi''(\mathcal{P}(\kappa) \cap S_\zeta^A) = \pi''\mathcal{P}(\kappa) \cap \pi(S_\zeta^A)$ in $\langle J_\nu^A, F\rangle$. So

$$F \cap \pi(S_\zeta^A) = \{\langle \pi(f)(\xi) \cap \kappa, \pi(f)(\xi)\rangle;\ \xi < \kappa\}$$

and an analogous formula holds for $\bar{F} \cap \bar{\pi}(S_\zeta^A)$ and $\bar{\pi}(f)$. As $k \circ \bar{\pi}(f) = \pi(f)$, we have $k(\bar{F} \cap \bar{\pi}(S_\zeta^A)) = F \cap \pi(S_\zeta^A)$. The conclusion then follows from the fact that both π and $\bar{\pi}$ are cofinal.

Definition. The *initial segment condition*, briefly ISC, for a potential premouse $M = \langle J_\nu^E, E_{\omega\nu}\rangle$ is the following statement:

$$\text{If } \bar{\nu} \leq \nu,\ E_{\bar{\nu}} \neq \varnothing \text{ and } \lambda^* \text{ is a cutpoint of } E_{\bar{\nu}}, \text{ then } E_{\bar{\nu}}|\lambda^* \in M||\bar{\nu}.$$

The initial segment condition is a requirement parallel to soundness. It ensures that the extenders put on the sequence do not contain any redundant information. To illustrate this, let $\langle J_\nu, F\rangle$ be a premouse representing $0^\#$ and $\langle J_{\nu'}, F'\rangle = \mathrm{Ult}(\langle J_\nu, F\rangle, F)$. Then F' is an extender with critical point $\lambda = \mathrm{length}(F)$. Now if $\langle J_{\nu^*}, F^*\rangle = \mathrm{Ult}(\langle J_\nu, F\rangle, F')$, then λ is a cutpoint of F^* and $F^*|\lambda = F$, but F^* and F contain essentially the same information, namely, they both correspond to $0^\#$. So F^* is an example of an extender that does not satisfy the initial segment condition. It cannot be on a premouse sequence, as we choose the shorter extender F to represent $0^\#$.

Quoting the above remarks, if $\bar{\lambda}$ is a cutpoint of E_ν, then $\bar{\lambda}$ is a limit cardinal in M. The initial segment condition guarantees that $\bar{F} = E_\nu|\bar{\lambda}$ is an element of M, so $\mathrm{Ult}(J_\nu^E, \bar{F})$ exists, as this ultrapower is internal. Using the canonical map k from above we infer that $V_{\bar{\pi}(\kappa)} = V_{\bar{\lambda}}$ of J_ν^E is contained in $\mathrm{Ult}(J_\nu^E, \bar{F})$, which means that $\bar{F}$ is a superstrong extender in J_ν^E. We have thus shown:

Corollary 9.1.3. *If $\bar{\lambda}$ is a cutpoint of E^M_{top}, then $E^M_{\mathrm{top}}|\bar{\lambda}$ is a superstrong extender in M.*

□(Corollary 9.1.3)

Thus, the existence of cutpoints in the presence of the initial segment condition is a very strong large cardinal assumption and the mice used for the construction of extender models at the level of Woodin cardinals will not, of course, have any cutpoints.

We shall later see that, for any cutpoint $\bar{\lambda}$, the restriction $E_\nu^M|\bar{\lambda}$ is actually on the sequence E^M if M is a sufficiently iterable premouse. This will follow from the condensation lemma (or, alternatively, from a comparison argument). However, for the purpose of constructing extender models it is more convenient to adopt the above weaker formulation of the initial segment condition, as this simplifies the verification of premousehood considerably.

Recall that a potential premouse is *active* if it has a top extender and *passive* otherwise. A natural attempt at a definition of premouse would be to restrict the class of potential premice to those which satisfy the initial segment condition. Unfortunately, this definition does not seem to guarantee the preservation of the initial segment condition downwards under Σ_1-maps when we deal with active premice. Since ultrapower maps do preserve the initial segment condition upwards, the above deficiency can be remedied with a slight revision of the intended notion of premouse (and, consequently, of the notion of embedding between a pair of premice). This, in turn, requires a classification of active premice according to the distribution of cutpoints of the top extender.

In the following, it will be convenient to denote the length of an extender F associated with a coherent structure by $\lambda(F)$; so $\lambda(F) = F(\mathrm{cr}(F))$.

Definition. Given an active potential premouse $M = \langle J_\nu^E, F\rangle$, let C_M denote the set of all cutpoints of F. We say that M is of type

- A iff $C_M = \varnothing$;
- B iff C_M is nonempty, but bounded in $\lambda(F)$;
- C iff C_M is unbounded in $\lambda(F)$.

If M is of type B, then C_M, being closed, has a largest element, which we denote by λ^*_M. Let γ_M be the height of the potential premouse with top extender $F|\lambda^*_M$.

We are now ready to give a rigorous definition of premouse. This definition will agree with our earlier attempt in the case of type A and type C potential premice; in the case of type B potential premice we add an additional predicate naming the restriction of the top extender to the largest cutpoint.

Definition. We say that the structure $M = \langle J^E_\nu, F, e\rangle$ is a *premouse* iff $M^- = \langle J^E_\nu, F\rangle$ is a potential premouse satisfying the initial segment condition and $e = \varnothing$ unless M^- is an active type B potential premouse, in which case $e = \{F|\lambda^*_{M^-}\}$.

The above definition implicitly modifies the language of premice, which is augmented by the additional predicate symbol. The only real difference, however, is limited to the case of type B premice. The presence of the predicate e guarantees that any embedding with the target structure M will have $F|\lambda^*_{M^-}$ in its range, which suffices for the downward preservation of the initial segment condition. Notice also that since the language of type B premice is different from that of potential premice, all fine structural parameters (like projecta, standard parameters and standard codes) are computed relative to the additional predicate e. For this reason, in the future we shall consider mainly embeddings of premice of the same type and call such embeddings *premice embeddings.* A typical example of such embeddings are iteration maps; actually, although the following lemma and corollary make it clear that iteration maps are automatically of this kind, even if we do not include e as a part of the language. There is one important reason why we also have to consider embeddings which fail to be premice embeddings. Starting from type C premice, collapsing constructions often yield premice of type A or B, so the associated uncollapsing embeddings then fail to be premice embeddings. Since this phenomenon seems to be unavoidable, one has to be prepared to deal with such situations. Lemma 9.1.7 gives some information on the downward preservation of premousehood in this case.

Convention. In order to simplify the notation, we shall systematically use e as a suppressed parameter, unless the particular situation requires us to do otherwise. Thus, when talking about premice, we shall usually refer to structures of the form $\langle J^E_\nu, F\rangle$, but keep in mind that there is the additional predicate if the premouse in question is of type B.

That all extenders on E^M that are elements of M satisfy the initial segment condition, can be easily expressed in a Π_1-fashion. Thus, the premousehood of passive premice is preserved under embeddings with a low preservation degree.

Lemma 9.1.4. *There is a Q-sentence φ such that for any transitive rudimentarily closed structure $M = \langle U, E\rangle$ we have:*

$$M \text{ is a passive premouse} \quad \longleftrightarrow \quad M \models \varphi.$$

Consequently, passive premousehood is preserved downwards by Σ_0 maps and upwards by cofinal Σ_0 maps and Σ_2 maps. □(Lemma 9.1.4)

If $\bar\lambda$ is a cutpoint of $E_{\bar\nu}$ and $\bar M$ is the coherent structure associated with $E_{\bar\nu}|\bar\lambda$, then $C_{\bar M} = C_M \cap \bar\lambda$. It follows that $\bar M$ is a potential premouse satisfying the initial segment condition, as it agrees with M below $\bar\lambda$. (That the initial segment condition holds for all $E^{\bar M}_\alpha \in \bar M$ follows from the previous lemma.) Thus, $\bar M$ will become a premouse after adding the predicate e.

We now turn to preservation lemmata for active premice. The following notation will often prove useful: If F is the top extender of a coherent structure M and π is the ultrapower map determined by F, we shall write $F(f)(\alpha)$ instead of $\pi(f)(\alpha)$. This will not do any harm, as in most cases $\pi(f)(\alpha)$ is an ordinal less than $\lambda(F)$ or a subset of such an ordinal, in which case $f : \kappa \to \kappa$ or $f : \kappa \to \mathcal{P}(\kappa)$. Such an f can be recursively encoded in a subset a_f of κ, so we can think of $F(f)$ as of $F(a_f)$. Another way of justifying this is that any $\pi(f)$ can be expressed in the form $\bar\pi(f)$ where $\bar\pi = \pi \restriction J^E_{\bar\tau}$ for some $\bar\tau < \kappa^{+M}$. Such a $\bar\pi$ is the ultrapower map associated with $\mathrm{Ult}(J^E_{\bar\tau}, \bar F)$, where $\bar F = F \restriction (\mathcal{P}(\kappa) \cap J^E_{\bar\tau})$, and so both $\bar F$ and $\bar\pi$ are elements of M.

Lemma 9.1.5. *Let $M = \langle J^E_\nu, F\rangle$ be an active premouse and let $\sigma : M \to M'$ be such that one of the following holds:*

a) *M is of type A or B and σ is Σ_2-preserving;*

b) *M is of type C and $\sigma(\omega\varrho^1_M) = \omega\varrho^1_{M'}$;*

c) *σ is Σ_0 cofinal, and maps λ cofinally to its image, where F is at κ, λ.*

Here, we consider σ to be an embedding between two ***potential premice****, i.e. we do* ***not*** *consider the predicate e as the part of the language when talking about the preservation properties of σ.*

Then M' is a premouse of the same type as M.

Proof. By the previous lemmata, it suffices to verify the initial segment condition for the top extender F' of M', as all other requirements for premousehood of M' are met. Let F' be at κ', λ'. Notice first that the statement $F(f)(\alpha) < \bar\lambda$ is Δ_1 and the same holds with $=$ or $>$ in place of $<$. Next, we observe that being a cutpoint is a Π_1 condition. Indeed, $\bar\lambda$ is a cutpoint of F iff

$$(\forall f)(\forall\alpha)[(f : \kappa \to \kappa \;\&\; \alpha < \bar\lambda) \longrightarrow F(f)(\alpha) < \bar\lambda].$$

It follows that the statement "No $\bar\lambda < \lambda$ is a cutpoint of F" is Π_2.

Suppose M is of type A, i.e. F has no cutpoints. By the previous paragraph, this can be expressed in a Π_2 fashion, so it is preserved upwards if the requirement a) on σ holds. It follows that F' does not have any cutpoints either and so M is a type A premouse. Now assume σ satisfies c). Pick $\alpha' < \lambda'$ and set $\alpha = (\sigma^{-1})''\alpha'$. Then $\alpha < \lambda$, as σ maps λ cofinally to λ'. But α is not a cutpoint of F, so there is some $\bar{\alpha} < \alpha$ and some $f : \kappa \to \kappa$ such that $F(f)(\bar{\alpha}) \geq \alpha$. Applying σ to this inequality yields $F(f')(\bar{\alpha}') \geq \sigma(\alpha) \geq \alpha'$, where $f' = \sigma(f)$ and $\bar{\alpha}' = \sigma(\bar{\alpha}) < \alpha'$. This shows that α' is not a cutpoint of F' and, as α' was chosen arbitrarily, that M' is of type A.

Now let M be a type B premouse. Using the arguments from the previous paragraphs it follows easily that $\sigma(\lambda^*_M) = \lambda^*_{M'}$. That $e = F|\lambda^*_M$ can be expressed in a Π_1 fashion, namely as the conjunction of the following two statements:

- e is a map and $(\forall x)(x \in \mathrm{dom}(e) \longleftrightarrow x \subset \kappa)$;
- $(\forall x)(\forall y)(\langle x, y\rangle \in F \longrightarrow \langle x, y \cap \lambda^*_M\rangle \in e)$.

As σ is Π_1-preserving, $\sigma(e) = F'|\lambda^*_{M'}$.

Finally let M be of type C. If b) holds, it suffices to show that the statements:

- There are unboundedly many cutpoints below λ.
- There are unboundedly many ordinals ζ below λ such that $F|\zeta \in M$.

are $Q^{(1)}$. To see this, it suffices to prove that $\lambda = \omega\varrho^1_M$. Clearly, $\omega\varrho^1_M \leq \lambda$, as any ordinal larger than λ is of the form $F(f)(\alpha)$ for some $f : \kappa \to \kappa$ and $\alpha < \lambda$, so that $h_M(\lambda) = M$. On the other hand, if $A(x, y)$ is $\Sigma_1(M)$, $\beta < \lambda$ and $p \in M$, then by the previous observation, $p \in h_M(\gamma)$ for some $\gamma < \lambda$. Pick a cutpoint $\bar{\lambda}$ of F which is larger than $\max\{\beta, \gamma\}$ and form the coherent structure $\bar{M}$ determined by $F|\bar{\lambda}$. Then $\bar{M} \in M$ by the initial segment condition. The canonical map $k : \bar{M} \to M$ defined by $[\xi, f]_{F|\bar{\lambda}} \mapsto [\xi, f]_F$ is Σ_0 cofinal, has critical point $\bar{\lambda}$ and p in its range. If $k(\bar{p}) = p$ and $\bar{A}$ is the $\Sigma_1(\bar{M})$ relation with the same definition as A, then $\bar{A} \in M$, and $\{\alpha < \beta;\ A(\alpha, p)\} = \{\alpha < \beta;\ \bar{A}(\alpha, \bar{p})\}$, so this set is also in M. It follows that $\omega\varrho^1_M$ cannot be strictly less than λ, so the equality holds.

If M is type C and c) holds, then $\sigma''C_M$ is cofinal in λ' by our requirements on σ and $\sigma''C_M \subset C_{M'}$ by the above results. It follows that M' is of type C. Now by the argument used for type B premice, $F'|\sigma(\bar{\lambda}) = \sigma(F|\bar{\lambda}) \in M'$ whenever $\bar{\lambda} \in M$, which verifies the initial segment condition for F'. □(Lemma 9.1.5)

Notice that the above proof also shows that if M is a type C premouse, then $p_M \subset \lambda$, i.e. $p^{(1)}_M = \varnothing$ and M is 1-sound.

The statement of the previous lemma might seem a bit technical, nevertheless, most of the maps arising in natural constructions will satisfy its assumptions. In particular, this applies to ultrapower maps used in iterations.

Corollary 9.1.6. *Let M be an active premouse and G a weakly amenable extender w.r.t. M with critical point $\mu < \lambda(E^M_{\mathrm{top}})$. Then $\mathrm{Ult}^k(M, G)$ is a premouse of the same type as M for all $k \leq \omega$.*

Proof. Notice that $\mathrm{Ult}^k(M, G)$ either agrees with the coarse ultrapower, or else $k > 0$ and $\omega\varrho^1_M > \mu$. In the former case, the associated ultrapower map σ satisfies the assumptions in clause c) of the previous lemma, so the conclusion holds. In the latter case, σ is Σ_2 preserving, which takes care of type A and B premice. Finally suppose that M is a type C premouse. Then either $k > 1$ and $\omega\varrho^2_M > \mu$, in which case σ is $\Sigma^{(1)}_2$-preserving, or else $k = 1$ or $\omega\varrho^2_M \leq \mu$ and σ is $\Sigma^{(1)}_0$-preserving. As M is 1-sound by the above observation, σ is in fact $\Sigma^{(1)}_0$ cofinal, and so $Q^{(1)}$-preserving. This completes the proof for type C premice. □(Corollary 9.1.6)

Lemma 9.1.7. *Given an active premouse $M = \langle J^E_\nu, F\rangle$, consider $\sigma : \bar{M} \to M$ as an embedding of potential premice such that $e^M = E^M_{\mathrm{top}}|\lambda^*_M \in \mathrm{rng}(\sigma)$ whenever M is of type B. Assume further that σ satisfies one of the following:*

a) *M is of type A or B and σ is Σ_1-preserving;*

b) *M is of type C and σ is $\Sigma^{(1)}_1$-preserving;*

c) *both M and $\bar{M}$ are of type C and σ is $\Sigma^{(1)}_0$-preserving.*

Then $\bar{M}$ is a premouse. Moreover, a) *guarantees that $\bar{M}$ is of the same type as M. The same is true for* b), *granted that $\mathrm{rng}(\sigma)$ is the $\Sigma^{(1)}_1$-hull of some subset of M. In all these cases, σ is a premouse embedding. However, this need not be the case in general.*

Proof. The proof is of the same spirit as that of the previous lemma and corollary, so we will leave it to the reader. Regarding c), however, let us note that if $\lambda = \sigma(\bar{\lambda})$ for some cutpoint $\bar{\lambda}$ of $\bar{M}$, then the acceptability of M guarantees that $E^M_{\mathrm{top}}|\lambda \in J^E_\mu$ where $\mu = \lambda^{+M}$ and $E = E^M$; this fact can then be expressed in a $\Sigma^{(1)}_0$-fashion as

$$(\exists f^1 \in J^E_\mu)(\forall x^0, y^0)(\langle x^0, y^0\rangle \in E^M_{\mathrm{top}} \longleftrightarrow \langle x^0, y^0 \cap \lambda\rangle \in f^1).$$

We again stress that in b) it does *not* follow that $\bar{M}$ must be of type C, unless $\mathrm{rng}(\sigma)$ is the $\Sigma^{(1)}_1$-hull of some subset of M. The reason why it is true in this special case is that, by Lemma 1.11.1 we have $\omega\varrho^1_{\bar{M}} = \lambda_{\bar{M}}$, so the statement "I am of type C" can be uniformly expressed in a $Q^{(1)}$-fashion over both M and $\bar{M}$. □(Lemma 9.1.7)

Our next goal is to develop a method for comparing premice. As we have already mentioned in the introduction, this requires a more sophisticated notion of iteration than the one we used below one strong cardinal. These new iterations will carry a tree structure describing where the iteration maps live. We shall begin with the notion of *generalized iteration.* Although we shall never use this notion in its full generality, it will serve as a base for our terminology.

Definition. A tree $T = \langle\theta, \leq_T\rangle$ on an ordinal θ is an *iteration tree* iff

a) 0 is the root of T and each successor ordinal $\alpha < \theta$ has an immediate T-predecessor $T(\alpha) < \alpha$;

b) if $\alpha < \theta$ is a limit ordinal, then $\alpha = \sup\{\xi < \alpha;\ \xi <_T \alpha\}$.

It is easy to see that $\xi <_T \zeta$ implies $\xi < \zeta$ and that $\{\xi < \alpha;\ \xi <_T \alpha\}$ is closed in α for all $\alpha < \theta$ and c.u.b. in α if α is limit.

Definition. A *generalized iteration* of a premouse M of length $\theta \leq \infty$ with *iterates* M_i, *indices* ν_i and η_i, tree T and *iteration maps* π_{ij} is the 5-tuple

$$\mathfrak{I} = \langle\langle M_i;\ i < \theta\rangle, \langle \nu_i;\ i \in B\rangle, \langle \eta_i;\ i+1 < \theta\rangle, \langle \pi_{ij};\ i \leq_T j < \theta\rangle, T\rangle$$

satisfying:

a) T is an iteration tree.

b) Each M_i is a premouse and $M_0 = M$.

c) $\langle \pi_{ij};\ i \leq_T j\rangle$ is a commutative system of partial maps, where $\pi_{ij} : M_i \to M_j$.

d) Setting $\xi_i = T(i+1)$, we have $\eta_i \leq \mathrm{ht}(M_{\xi_i})$ and for every $j < \theta$ there are only finitely many i such that $\xi_i <_T j$ and $\eta_i < \mathrm{ht}(M_{\xi_i})$.

e) If $i \notin B$ and $i+1 < \theta$, then $i = T(i+1)$, $M_{i+1} = M_i||\eta_i$ and $\pi_{i,i+1} = \mathrm{id}$.

f) If $i \in B$, then $i+1 < \theta$ and $E_{\nu_i}^{M_i} \neq \varnothing$. Setting $\kappa_i = \mathrm{cr}(E_{\nu_i}^{M_i})$ and $\tau_i = \kappa_i^{+M_i||\nu_i}$, we have $\tau_i = \kappa_i^{+M_{\xi_i}||\eta_i}$, $E^{M_i} \restriction \tau_i = E^{M_{\xi_i}} \restriction \tau_i$ and

$$\pi_{\xi_i,i+1} : M_{\xi_i}||\eta_i \xrightarrow[E_{\nu_i}^{M_i}]{*} M_{i+1}$$

g) If $i < \theta$ is limit, then $\langle M_i, \pi_{hi};\ h <_T i\rangle$ is the direct limit of the diagram $\langle M_h, \pi_{\bar{h}h};\ \bar{h} \leq_T h <_T i\rangle$.

Given an iteration $\mathfrak{I}$, we denote the objects from the above definition related to $\mathfrak{I}$ by $M_i^{\mathfrak{I}}, \nu_i^{\mathfrak{I}}, \eta_i^{\mathfrak{I}}, \pi_i^{\mathfrak{I}}, B^{\mathfrak{I}}, T^{\mathfrak{I}} \ldots$ We also set

$$\begin{aligned}
\kappa_i^{\mathfrak{I}} &= \mathrm{cr}(E_{\nu_i^{\mathfrak{I}}}^{M_i^{\mathfrak{I}}}),\\
\lambda_i^{\mathfrak{I}} &= \lambda(E_{\nu_i^{\mathfrak{I}}}^{M_i^{\mathfrak{I}}}),\\
\tau_i^{\mathfrak{I}} &= (\kappa_i^{\mathfrak{I}})^{+M_i^{\mathfrak{I}}||\nu_i^{\mathfrak{I}}} \text{ and}\\
D^{\mathfrak{I}} &= \{i+1;\ \eta_i^{\mathfrak{I}} < \mathrm{ht}(M_{\xi_i^{\mathfrak{I}}}^{\mathfrak{I}})\}.
\end{aligned}$$

Of course, in order to simplify the notation, we shall omit subscripts and superscripts whenever possible. We shall often write M_i^* instead of $M_{\xi_i}||\eta_i$.

The elements of $D^{\mathfrak{I}}$ are called *truncation points*. Clause d) is equivalent to saying that $D^{\mathfrak{I}} \cap i$ is finite. For limit i, this implies that from certain i_0 on, all maps along the branch $\{h;\ h <_T i\}$ are total, so the direct limit in g) exists. Notice also that clause f) guarantees that

$$E^{M_i}_{\nu_i} \text{ is weakly amenable w.r.t. } M_{\xi_i}||\eta_i. \tag{9.1.1}$$

It is common to use the term "iteration tree" when referring to the whole iteration $\Im$ rather than $T^{\Im}$ and we shall occasionally do so, as the meaning will be clear from the context. An iteration $\Im$ is *direct* iff $i \in B$ for *all* i such that $i+1 < \theta$, in other words, when an extender is applied at each stage in $\Im$. We shall say that $\Im$ is *padded* in order to stress that $\Im$ need not be direct. The iterations which are relevant for us are those which arise in the comparison process, which motivates the following definition.

Definition. An iteration $\Im$ is *normal* iff

a) $\nu_h < \nu_i$ whenever $h, i \in B$ and $h < i$;

b) $\xi_i =$ the least $\xi \in B$ such that $\kappa_i < \lambda_\xi$;

c) $\eta_i =$ the maximal $\eta \leq \mathrm{ht}(M_{\xi_i})$ such that $\tau_i = \kappa_i^{+M_{\xi_i}||\eta}$.

For a normal iteration $\Im$, the inductive application of the coherency condition yields:

- $E^{M_i} \restriction \nu_i = E^{M_j} \restriction \nu_i$ whenever $i \in B$ and $i \leq j$.
- If $i \in B$ and $i < j$, then ν_i is a cardinal in M_j, but not necessarily in M_i.

This means that ν_i is a successor cardinal in M_j for $i < j$ and, by acceptability, λ_i, κ_i and τ_i are cardinals in M_j. As λ_j is the largest cardinal in $M_j||\nu_j$ and is a limit cardinal there, we have $\lambda_i < \lambda_j$. It also follows that η_i is defined and

$$\mathcal{P}(\kappa_i) \cap M_i||\nu_i = \mathcal{P}(\kappa_i) \cap M_{\xi_i}||\nu_{\xi_i} = \mathcal{P}(\kappa_i) \cap M_{\xi_i}||\eta_i,$$

so $E^{M_i}_{\nu_i}$ can be applied to $M_{\xi_i}||\eta$ for all η such that $\tau_i \leq \eta \leq \eta_i$. Notice that η_i determines the longest initial segment of M_{ξ_i} to which $E^{M_i}_{\nu_i}$ can be applied. The normality of $\Im$ also implies:

$$\xi_i <_T \xi_j \longrightarrow \kappa_i < \lambda_{\xi_i} \leq \kappa_j, \tag{9.1.2}$$

that is, the critical points *along the branches* strictly increase and the extenders do not overlap, as $\xi_j \geq i+1$. This is the point which links the present notion of normal iteration with that in the linear case and which is crucial for the comparison process.

In order to carry out an iteration, we face a new problem: at limit stages we have to choose a branch through the iteration tree built so far in such a way that the direct limit along this branch is well-founded. This leads to the notion of *iteration strategy.*

Definition. An iteration strategy is a partial function $\mathbb{S}$ on iterations of limit length such that if $b = \mathbb{S}(\Im)$, then b is a *cofinal well-founded branch* through $\Im$, that is,

a) b is a branch through $T^{\Im}$ cofinal in length$(\Im)$;

b) $D^{\Im} \cap b$ is finite;

c) the direct limit along b, i.e. $\lim\langle M_i, \pi_{ij};\ i, j \in b \ \&\ i \leq_T j\rangle$ is well-founded.

An iteration $\mathfrak{I}$ is an $\mathbb{S}$-iteration iff $\{h;\ h <_T i\} = \mathbb{S}(\mathfrak{I} \upharpoonright i)$ for every limit $i < \text{length}(\mathfrak{I})$.

Again, we will be primarily interested in *normal iteration strategies,* i.e. iteration strategies which are needed for the comparison process.

Definition. An iteration strategy $\mathbb{S}$ is a *normal iteration strategy* iff any normal $\mathbb{S}$-iteration $\mathfrak{I}$ can be extended by $\mathbb{S}$, that is:

a) If $\mathfrak{I}$ is of limit length, then $\mathbb{S}(\mathfrak{I})$ exists.

b) If $\mathfrak{I}$ is of successor length k and $\nu \geq \sup\{\nu_h + 1;\ h < k\}$ is such that $E_\nu^{M_k} \neq \varnothing$, then $\mathfrak{I}$ has a normal extension $\mathfrak{I}'$ of length $k+1$ with $\nu_k^{\mathfrak{I}'} = \nu$. In other words, setting $\nu_k = \nu$, the ultrapower $\text{Ult}^*(M_{\xi_k} || \eta_k, E_{\nu_k}^{M_k})$ exists, where ξ_k and ν_k are determined by the requirements on normal iterations.

A premouse M is *θ-normally iterable* iff it has a normal iteration strategy of length θ. (That is, a normal iteration strategy which gives rise to a normal iteration of length θ.) M is *normally iterable* iff it is ∞-normally iterable.

Definition. The iteration strategy $\mathbb{S}$ is the *uniqueness strategy* iff $\mathbb{S}(\mathfrak{I})$, if defined, is the unique cofinal well-founded branch through $\mathfrak{I}$. A premouse M is *θ-uniquely normally iterable* iff M is θ-normally iterable by the uniqueness strategy.

The notions of normal iteration and normal iterability *above,* resp. *beyond* a point are straightforward generalizations of the corresponding notions for linear iterations. The same applies to coiterations, which we nevertheless introduce formally.

Definition. Given two premice M and N, a *coiteration* of M, N of length $\theta + 1$ is a pair of normal iterations

$$\mathfrak{I}^M = \langle\langle M_i;\ i < \theta + 1\rangle, \langle \nu_i;\ i \in B^M\rangle, \dots, T^M\rangle$$
$$\mathfrak{I}^N = \langle\langle N_i;\ i < \theta + 1\rangle, \langle \nu_i;\ i \in B^N\rangle, \dots, T^N\rangle$$

satisfying:

a) $M_0 = M$ and $N_0 = N$.

b) If $i < \theta$, then $\nu_i \simeq$ the least ν such that $E_\nu^{M_i} \neq E_\nu^{N_i}$.

c) $B^M \cup B^N = \theta$.

We say that the coiteration $\langle\mathfrak{I}^M, \mathfrak{I}^N\rangle$ of M, N *terminates* iff ν_θ is undefined.

Clearly, if M and N are normally iterable via some iteration strategies $\mathbb{S}$ and $\mathbb{S}'$, then there is a unique coiteration of M and N such that $\mathfrak{I}$ is an $\mathbb{S}$-iteration and $\mathfrak{I}'$ an $\mathbb{S}'$-iteration. Notice also that, concerning iterations, everything discussed so far goes through for *potential premice,* that is, none of the notions introduced required to refer to the initial segment condition. However, the initial segment condition guarantees the following useful property of normal iterations. Let us say that two extenders $\bar{F}$ and F are *compatible* iff $\bar{F} = F|\bar{\lambda}$ for some $\bar{\lambda} \leq \lambda(F)$ or vice versa. Then:

$$\text{No two distinct extenders used in a normal iteration of a premouse are compatible.} \tag{9.1.3}$$

Indeed, if $E_{\nu_i}^{M_i}$ were compatible with $E_{\nu_j}^{M_j}$, then, assuming w.l.o.g. that $i < j$, we would have $E_{\nu_i}^{M_i} = E_{\nu_j}^{M_j}|\lambda_i$ and λ_i would be a cutpoint of $E_{\nu_j}^{M_j}$. By the initial segment condition, $E_{\nu_i}^{M_i} \in M_j$, so $E_{\nu_i}^{M_i}$ would give rise to a map which is an element of M_j and which maps τ_i cofinally into ν_i. As ν_i is a successor cardinal in M_j, this is impossible. For coiterations of premice we obtain:

$$\text{If } E_{\nu_i}^{M_i} \text{ and } E_{\nu_j}^{N_j} \text{ are extenders used in the coiteration of premice } M \text{ and } N\text{, then } E_{\nu_i}^{M_i} \text{ and } E_{\nu_j}^{N_j} \text{ are incompatible.} \tag{9.1.4}$$

For $i = j$ this follows from the fact that the extenders $E_{\nu_i}^{M_i}$ and $E_{\nu_i}^{N_i}$ arise at the i-th stage in the coiteration and, therefore, are different; for $i \neq j$ this follows by the above argument.

Lemma 9.1.8. *If $\langle \Im^M, \Im^N \rangle$ is a coiteration of premice M and N, then its length is strictly less than $\theta = \max\{\mathrm{card}(M)^+, \mathrm{card}(N)^+\}$.*

Proof. We shall closely follow the proof of Lemma 4.4.1. We suppose that $\langle \Im^M, \Im^N \rangle$ is a coiteration of M with N of length θ, pick a sufficiently large regular τ and construct an elementary map $\sigma : H \to H_\tau$ of size less than θ such that the above coiteration is in the range of σ, and both H and $\mathrm{rng}(\sigma) \cap \theta$ are transitive. Setting $\bar{\theta} = \mathrm{rng}(\sigma) \cap \theta$, we have $\bar{\theta} = \mathrm{cr}(\sigma)$ and $\sigma(\bar{\theta}) = \theta$. Let also $\sigma(\bar{\Im}) = \Im^M$ and $\sigma(\bar{\Im}') = \Im^N$. As before we show that:

- $\bar{\Im} = \Im^M \restriction \bar{\theta}$ and $\bar{\Im}' = \Im^N \restriction \bar{\theta}$;
- $\sigma \restriction M_{\bar{\theta}} = \pi_{\bar{\theta}\theta}^M$ and $\sigma \restriction N_{\bar{\theta}} = \pi_{\bar{\theta}\theta}^N$, where $\pi_{\bar{\theta}\theta}^M : M_{\bar{\theta}} \to M_\theta$ and $\pi_{\bar{\theta}\theta}^N : N_{\bar{\theta}} \to N_\theta$ are the corresponding iteration maps, and these maps are total.

It follows that for some $i, j < \theta$ we have:

- $\bar{\theta} = T^M(i+1)$ and $\bar{\theta} = \kappa_i^M$;
- $\bar{\theta} = T^N(j+1)$ and $\bar{\theta} = \kappa_j^N$.

Applying (9.1.2), we see that $\mathrm{cr}(\pi_{i+1,\theta}^M) \geq \lambda_i^M$, so given an ordinal $\alpha < \lambda_i^M$,

$$\alpha \in E_{\nu_i}^{M_i}(x) \longleftrightarrow \alpha \in \pi_{\bar{\theta},i+1}^M(x) \longleftrightarrow \alpha \in \pi_{i+i,\theta} \circ \pi_{\bar{\theta},i+1}^M(x) = \pi_{\bar{\theta}\theta}^M(x) = \sigma(x).$$

The same computation for $E_{\nu_j}^{N_j}$ then yields that $E_{\nu_i}^{M_i}$ and $E_{\nu_j}^{N_j}$ are compatible, contradicting (9.1.4). □(Lemma 9.1.8)

Two important applications of the comparison process are the proofs of solidity and condensation for premice. In these proofs, it was essential that one side of the coiteration was *above* the critical point of the canonical witness map and the condensation embedding, respectively. The reason why it was possible to guarantee this was the fact that we dealt with measures or, more generally, *non-overlapping* extenders. In the present context, however, it is easy to find a coiteration of two premice where critical points drop below any reasonably chosen point. Therefore, in order to be able to generalize the results on solidity and condensation to mice in our present sense, we introduce a new technique. The idea is simple: whenever the critical point of the extender we are applying drops below our chosen point, we apply this extender to a premouse in the iteration which is carefully-chosen, but different from the one given by the rules of normal iterations.

Definition. A *phalanx* is a triple $\langle M_{-1}, M_0, \lambda\rangle$ where

a) M_{-1} and M_0 are premice;

b) $\lambda \in M_{-1} \cap M_0$ is a cardinal in M_0;

c) $E^{M_{-1}} \upharpoonright \lambda = E^{M_0} \upharpoonright \lambda$;

d) if $\bar\lambda < \lambda$, then $\bar\lambda$ is a cardinal in M_0 iff $\bar\lambda$ is a cardinal in M_{-1}.

We included clause d) only for convenience, as the proofs of general theorems become simpler and all phalanxes arising in our applications will automatically satisfy d). Otherwise, the whole theory also goes through with the more liberal definition omitting d).

Normal iterations of a phalanx can be viewed as normal iterations of M_0 (that is, any such iteration starts by choosing an extender from E^{M_0}) with one extra property: whenever $\kappa_i = \mathrm{cr}(E^{M_i}_{\nu_i})$ drops below λ, we apply $E^{M_i}_{\nu_i}$ to M_{-1}. What we obtain is a tree structure with two roots, -1 and 0. More formally:

Definition. $\mathfrak{I}$ is a *normal iteration of a phalanx* $\langle M_{-1}, M_0, \lambda\rangle$ iff, setting $\lambda_{-1} = \lambda$, $\mathfrak{I}$ obeys the rules of normal iterations for premice with the only exception that $\lambda_{i+1} = \lambda_i$ is allowed if $-1 = T(i+1)$.

Thus, it might happen that $\lambda_0 = \lambda_{-1}$, as we are completely free to choose $\lambda_{-1} = \lambda$. Moreover, if $E^{M_{-1}}_\lambda \neq \varnothing$ and κ_i is the cardinal predecessor of λ in M_{-1}, then $\eta_i = \lambda$ and $M_{i+1} = \mathrm{Ult}(M_{-1}||\lambda, E^{M_i}_{\nu_i})$. The structure M_{i+1} is of height ν_i, as $\lambda = (\kappa_i)^{+M_i||\nu_i}$, M_{i+1} agrees with M_i below ν_i and its top extender fails to satisfy the initial segment condition, as it is the "stretch" of $E^{M_{-1}}_\lambda$ by $E^{M_i}_{\nu_i}$. More precisely, $E^{M_{i+1}}_{\mathrm{top}} = E^{M_i}_{\nu_i} \circ E^{M_{-1}}_\lambda$, as can easily be seen. Thus, $E^{M_{i+1}}_{\nu_i}|\kappa_i = E^{M_{-1}}_\lambda$, but $E^{M_{i+1}}_\lambda = \varnothing$, as λ is the cardinal successor of κ_i in M_h for all $h > 0$. In order to keep the normality of the iteration, $E^{M_{i+1}}_{\nu_i}$ is the only possible extender which can be applied at the $(i+1)$-st stage of the iteration, so $\nu_{i+1} = \nu_i$ and $\lambda_{i+1} = \lambda_i$. Notice that $\kappa_{i+1} = \mathrm{cr}(E^{M_{i+1}}_{\nu_{i+1}}) = \mathrm{cr}(E^{M_{-1}}_\lambda) < \lambda$, so $\xi_{i+1} = -1$ and $-1 = T(i+2)$. By d) in the definition of phalanx, $\lambda(E_{\text{top}}^{M^*_{i+1}}) \geq \kappa_i > \kappa_{i+1}$, so $M_{i+2} = \text{Ult}^*(M^*_{i+1}, E_{\nu_{i+1}}^{M_{i+1}})$ is a premouse. Notice also that for $i > -1$, the situation $\lambda_{i+1} = \lambda_i$ can only occur if $E_\lambda^{M_{-1}}$ is a superstrong extender in M_{-1}. Thus, such a situation will not arise if we only consider "smaller" models, i.e. models below a superstrong cardinal. Extender models we are able to construct today range at the level of a Woodin limit of Woodin cardinals, which is well below this bound.

Definition. Let $\Im$ be a normal iteration of a phalanx $\langle M_{-1}, M_0, \lambda\rangle$ and $0 < i+1 < \text{length}(\Im)$. We say that $i+1$ is an *anomaly* iff $T(i+1) = -1$, κ_i is the largest cardinal less than λ in M_{-1} and $\eta_i < \text{ht}(M_{-1})$. If, moreover, $\eta_i = \lambda$ and $E_\lambda^{M_{-1}} \neq \varnothing$, we call $i+1$ a *strong anomaly.*

Anomalies in the weaker sense will be discussed in connection with copying constructions in the next section. Clearly, at most one anomaly can occur in a branch. Moreover, if $i+1$ is a strong anomaly, then $\{-1, i+1\}$ is a maximal branch in T, i.e. there is no j such that $i+1 <_T j$. To summarize the above discussion, we obtain the following analogues to (9.1.3).

Each M_i in a phalanx iteration is a premouse unless $i = h+1$ is a strong anomaly, in which case $E_{\text{top}}^{M_i} = E_{\nu_h}^{M_h} \circ E_\lambda^{M_{-1}}$ where M_h is a premouse, $-1 = T(i) = T(i+1)$ and $\{-1, i\}$ is a maximal branch in T, i.e. there is no j such that $i <_T j$. (9.1.5)

If $i \neq j$ then $E_{\nu_i}^{M_i}$ and $E_{\nu_j}^{M_j}$ are incompatible. (9.1.6)

To see the latter, it suffices to check the case where both $E_{\nu_i}^{M_i}$ and $E_{\nu_j}^{M_j}$ are stretches of $E_\lambda^{M_{-1}}$, as the other cases either follow as in the premouse case or else are trivial. So suppose $E_{\nu_i}^{M_i} = E_{\nu_k}^{M_k} \circ E_\lambda^{M_{-1}}$ and $E_{\nu_j}^{M_j} = E_{\nu_\ell}^{M_\ell} \circ E_\lambda^{M_{-1}}$ where $E_{\nu_k}^{M_k}$ and $E_{\nu_\ell}^{M_\ell}$ are premice extenders. This means that $E_{\nu_k}^{M_k}$ and $E_{\nu_\ell}^{M_\ell}$ are incompatible, so pick an $a \subset \kappa_k = \kappa_\ell$ such that $E_{\nu_k}^{M_k}(a)$ and $E_{\nu_\ell}^{M_\ell}(a)$ are incompatible, in the sense that $E_{\nu_k}^{M_k}(a) \neq E_{\nu_\ell}^{M_\ell}(a) \cap \lambda_k$ and vice versa. As $a \in M_{-1}||\lambda$, setting $\mu = \text{cr}(E_\lambda^{M_{-1}})$, we can find a $\vartheta < \mu^{+M_{-1}||\lambda}$ such that $a \in \pi(M_{-1}||\vartheta)$ where π is the ultrapower map associated with $E_\lambda^{M_{-1}}$. Let $x \subset \mu \times \mu$ encode an enumeration of $\mathcal{P}(\mu) \cap M_{-1}||\vartheta$. Then, for some $\zeta < \kappa_k$, we have $a = (E_\lambda^{M_{-1}}(x))_\zeta$ together with

$$(E_{\nu_i}^{M_i}(x))_\zeta = (E_{\nu_k}^{M_k} \circ E_\lambda^{M_{-1}}(x))_\zeta = E_{\nu_k}^{M_k}((E_\lambda^{M_{-1}}(x))_\zeta) = E_{\nu_k}^{M_k}(a)$$

and

$$(E_{\nu_j}^{M_j}(x))_\zeta = (E_{\nu_\ell}^{M_\ell} \circ E_\lambda^{M_{-1}}(x))_\zeta = E_{\nu_\ell}^{M_\ell}((E_\lambda^{M_{-1}}(x))_\zeta) = E_{\nu_\ell}^{M_\ell}(a),$$

which means that $E_{\nu_i}^{M_i}(x)$ and $E_{\nu_j}^{M_j}(x)$ are incompatible.

Definition. A *coiteration* of a phalanx $\langle M_{-1}, M_0, \lambda\rangle$ with a premouse N is a pair of iterations $\langle \Im^M, \Im^N\rangle$ such that $\Im^M$ is a normal iteration of $\langle M_{-1}, M_0, \lambda\rangle$, $\Im^N$ is a normal iteration of N, i.e. $N_0 = N$ and for $i \geq 0$, both $\Im^M$ and $\Im^N$ obey the same rules as in the premice coiteration.

An analogue of (9.1.4) for phalanx-premouse coiterations is:

$$\text{In the coiteration of a phalanx } \langle M_{-1}, M_0, \lambda\rangle \text{ with a premouse } N, \text{ all pairs of extenders } E^{M_i}_{\nu_i}, E^{N_j}_{\nu_j} \text{ are incompatible with the only possible exception occurring when } i = h+1 \text{ is a strong anomaly, } j = 0 \text{ and } \nu_0 = \lambda. \tag{9.1.7}$$

Clearly, if $E^{M_i}_{\nu_i}$ and $E^{N_j}_{\nu_j}$ are compatible, then this is possible only if $E^{M_i}_{\nu_i}$ fails to satisfy the initial segment condition and $E^{N_j}_{\nu_j} = E^{M_i}_{\nu_i}|\lambda_j$, as $E^{N_j}_{\nu_j}$ is always a premouse extender. Thus, i is a strong anomaly, say $i = h+1$. As $E^{M_i}_{\nu_i}|\kappa_h$ is not on the sequence E^{M_i}, we have $\lambda_j \leq \kappa_h$. This can only happen if $j = 0$, so $\lambda_0 = \lambda$.

The above discussion gives us all the requisite facts to prove the comparison lemma for phalanx-premouse coiterations.

Lemma 9.1.9. *If $\langle \Im^M, \Im^N\rangle$ is a coiteration of a phalanx $\langle M_{-1}, M_0, \lambda\rangle$ with a premouse N, then its length is strictly less than*

$$\theta = \max\{\mathrm{card}(M_{-1})^+, \mathrm{card}(M_0)^+, \mathrm{card}(N)^+\}.$$

Proof. We proceed as in the proof of Lemma 9.1.9. Assuming that the coiteration has length at least θ, we find a $\bar\theta < \theta$ such that the iteration map on the N-side $\pi^N_{\bar\theta\theta}$ and that on the phalanx side $\pi^M_{\bar\theta\theta}$ have the same critical point $\bar\theta$ and

$$\pi^N_{\bar\theta\theta}(x) = \pi^M_{\bar\theta\theta}(x) \text{ whenever } x \subset \bar\theta \text{ is an element of both } M_{\bar\theta} \text{ and } N_{\bar\theta}.$$

Letting i and j be such that $T^{\Im^M}(i+1) = \bar\theta = T^{\Im^N}(j+1)$, we infer that $E^{N_j}_{\nu_j}$ and $E^{M_i}_{\nu_i}$ is a pair of compatible extenders. By (9.1.7), this can happen only if i is a strong anomaly and $j = 0$. But $\bar\theta$ is definitely different from -1, so i cannot be a strong anomaly and $E^{N_j}_{\nu_j}$ and $E^{M_i}_{\nu_i}$ cannot be compatible, a contradiction. □(Lemma 9.1.9)

In the following we prove another crucial property of coiterations, namely, that one side does not contain any truncations on the *main* branch, i.e. on the branch $\{i;\ i < \theta\}$ where M_θ is the final model in the iteration. (Let us note that the notion of main branch can be defined for generalized iterations in the same manner.) In the theory of small extender models, the proof of this fact was based on the observation that none of the premice which arise in a normal iteration after the first truncation point are sound. The failure of soundness after a truncation was a consequence of the preservation properties of the iteration maps, more precisely, the preservation of the standard parameters as well as the projecta up to the critical points. There are several ways of guaranteeing the above preservation properties of the iteration maps in the current context of non-linear iterations. One, particularly useful, builds on the fact that all extenders used in a normal iteration are *close* to the premice they are applied to. This can be considered a generalization of the situation in the linear case where those extenders were elements of the respective premice and therefore trivially close to them.

Lemma 9.1.10. *Let $\mathfrak{I}$ be a normal iteration of a premouse M and $i+1 < \text{length}(\mathfrak{I})$. Then $E^{M_i}_{\nu_i}$ is close to M^*_i.*

Proof. It suffices to verify that $E^{M_i}_{\nu_i}$ is Σ_1-amenable w.r.t. M^*_i, as the weak amenability follows immediately. Assume the contrary. If i is the least counterexample to the lemma, then $\xi_i < i$ and, consequently, $E^{M_i}_{\nu_i}$ is the top extender of M_i, as $\lambda_{\xi_i} > \kappa_i$ is a limit cardinal in M_i. For $j+1 < \text{length}(\mathfrak{I})$ set

- $\hat{\kappa}_j = \text{cr}(E^{M_j}_{\text{top}})$ and $\hat{\tau}_j = \hat{\kappa}_j^{+M_j}$;
- $\hat{\xi}_j =$ the least ξ such that $\hat{\kappa}_j < \lambda_\xi$;
- $\hat{\eta}_j =$ the maximal η such that $\mathcal{P}(\hat{\kappa}_j) \cap M_j = \mathcal{P}(\hat{\kappa}_j) \cap (M_{\hat{\xi}_j} || \hat{\eta}_j)$;
- $\hat{M}_j = M_{\hat{\xi}_j} || \hat{\eta}_j$.

Let ι be the least j with the property:

$$\text{There is a } \Sigma_1(M_j) \text{ subset of } \hat{\tau}_j \text{ which is not } \Sigma_1(\hat{M}_j). \tag{9.1.8}$$

Clearly, i satisfies (9.1.8), so $\iota \leq i$. It is also clear that $\hat{\xi}_\iota < \iota$. We next show that ι must be a successor ordinal. For, if ι were a limit, we could fix a $\xi <_T \iota$ larger than $\hat{\xi}_\iota$ such that there is no truncation point on the branch between ξ and ι, and a $b \subset \hat{\tau}_\iota$ together with a parameter $p \in M_\iota$ satisfying

- b is $\Sigma_1(M_\iota)$ in p, b witnesses (9.1.8) for ι and $p \in \text{rng}(\pi_{\xi\iota})$.

Choosing an $\bar{\iota}$ such that $\xi <_T \bar{\iota} <_T \iota$, we have $\hat{\kappa}_\iota < \lambda_{\bar{\iota}}$ and $\text{cr}(\pi_{\bar{\iota}\iota}) \geq \lambda_\xi > \hat{\kappa}_\iota$. As $\pi_{\bar{\iota}\iota}$ is Σ_1-preserving (this follows from the general properties of fine ultrapowers), the above clause holds with $\bar{\iota}$ in place of ξ. It follows that

- $\hat{\kappa}_{\bar{\iota}} = \hat{\kappa}_\iota$ and $\hat{\tau}_{\bar{\iota}} = \hat{\tau}_\iota$;
- $\hat{\xi}_{\bar{\iota}} = \hat{\xi}_\iota$ and $\hat{M}_{\bar{\iota}} = \hat{M}_\iota$;
- if $\pi_{\bar{\iota}\iota}(\bar{p}) = p$, then b is $\Sigma_1(M_{\bar{\iota}})$ in $\bar{p}$ by the same definition as in M_ι.

This means that $\bar{\iota}$ also satisfies the condition (9.1.8), which contradicts the minimality of ι.

So $\iota = h+1$ for some h and $M_\iota = \text{Ult}^*(M^*_h, E^{M_h}_{\nu_h})$. As in the limit case, fix a $b \subset \hat{\tau}_\iota$ witnessing (9.1.8) and a parameter p such that b is $\Sigma_1(M_\iota)$ in p. Recall that i is the least index with the property that $E^{M_i}_{\nu_i}$ fails to be close to M^*_i and $\iota \leq i$. This means that $E^{M_h}_{\nu_h}$ is close to M^*_h, so $\omega\varrho^{n+1}_{M^*_h} = \omega\varrho^{n+1}_{M_\iota}$ and $\omega\varrho^n_{M_\iota} > \pi_{h\iota}(\kappa_h) = \lambda_h$ where n is such that $\omega\varrho^{n+1}_{M^*_h} \leq \kappa_h < \omega\varrho^n_{M^*_h}$. We observe that $n = 1$. Notice first that $\hat{\kappa}_\iota < \lambda_h$, as $\hat{\xi}_\iota \leq h$. But $\hat{\kappa}_\iota$, being the critical point of the top extender of M_ι, is in the range of $\pi_{h\iota}$, so it must be less than κ_h. And, as κ_h is a limit cardinal in both M^*_h and M_ι, we have $\hat{\tau}_\iota < \kappa_h$. Now if $\omega\varrho^1_{M_\iota} > \hat{\tau}_\iota$, we would have $b \in M_\iota$ and, consequently, $b \in \hat{M}_\iota$, as $\hat{M}_\iota$ and M_ι agree up to $\lambda_{\hat{\xi}_\iota}$ and $\lambda_{\hat{\xi}_\iota}$ is a cardinal in M_ι. This would contradict our choice of b. Thus, $\omega\varrho^1 \leq \hat{\tau}_\iota$ and $n = 1$. By Lemma 3.3.6, there is a Σ_1-formula $\varphi^*(u, v)$ and a parameter $p^* \in M^*_h$ such that for every $\xi < \kappa_h$,

$$\xi \in b \longleftrightarrow M^*_h \models \varphi^*(\xi, p^*). \tag{9.1.9}$$

We claim that $M^*_h = M_{\xi_h}$. Suppose that M^*_h is a *proper* initial segment of M_{ξ_h}. We derive a contradiction to our choice of b. This is immediate if $\hat{\xi}_\iota < \xi_h$ or else $\hat{\xi}_\iota = \xi_h$ and M^*_h is a *proper* initial segment of $\hat{M}_\iota$, as in both cases we obtain $b \in \hat{M}_\iota$. It suffices to apply (9.1.9) and, in the latter case also the fact that $\lambda_{\hat{\xi}_\iota} \in M^*_h$ is a cardinal in M_{ξ_h}. It remains to discuss the case where $\hat{\xi}_\iota = \xi_h$ and $M^*_h = \hat{M}_\iota$ (notice that M^*_h is always an initial segment of $\hat{M}_\iota$ if $\hat{\xi}_\iota = \xi_h$). The desired contradiction then follows directly from (9.1.9) which in the present case asserts that b is $\mathbf{\Sigma}_1(\hat{M}_\iota)$.

It is now easy to see that, setting $k = \xi_h$,

- $\hat{\kappa}_k = \hat{\kappa}_\iota$ and $\hat{\tau}_k = \hat{\tau}_\iota$;
- $\hat{\xi}_k = \hat{\xi}_\iota$ and $\hat{M}_k = \hat{M}_\iota$.

Applying these equalities to (9.1.9), we obtain that $b \subset \hat{\tau}_k$ is $\mathbf{\Sigma}_1(M_k)$, but not $\mathbf{\Sigma}_1(\hat{M}_k)$. But ι was chosen to be minimal with this property, a contradiction. □(Lemma 9.1.10)

The same proof yields the corresponding lemma on normal phalanx iterations. However, we have to restrict the class of phalanxes in order to get the version of the lemma useful for the later applications.

Definition. The phalanx $\langle M_{-1}, M_0, \lambda\rangle$ is *good* iff $\mathcal{P}(\lambda) \cap M_0 \subset M_{-1}$.

We shall later see that all phalanxes which arise in condensation arguments are good, which will enable us to prove results on solidity and condensation.

Lemma 9.1.11. *Let $\Im$ be a normal iteration of a good phalanx $\langle M_{-1}, M_0, \lambda\rangle$. If $i+1 <$ length($\Im$) is not an anomaly, then $E^{M_i}_{\nu_i}$ is close to M^*_i. If $i+1$ is an anomaly, then $E^{M_i}_{\nu_i,\alpha} \in M_{-1}$ whenever $\alpha < \lambda_i$.*

Proof. Again, it suffices to show that, given an arbitrary i, either $i+1$ is not an anomaly and $E^{M_i}_{\nu_i}$ is Σ_1-amenable w.r.t. M^*_i or else $i+1$ is an anomaly and $E^{M_i}_{\nu_i,\alpha} \in M_{-1}$ for every $\alpha < \lambda_i$. Suppose i is a counterexample to this. Let ι be the least index j with the property that $E^{M_j}_{\nu_j}$ is an extender. Then $\iota \geq 0$ and $\lambda_\iota > \lambda$, as λ is a cardinal in M_0. Notice also that $M_\iota = M$. This enables us to imitate the argument from the proof of the previous lemma and show that $E^{M_i}_{\nu_i}$ is the top extender of M_i. For j less than the length of the coiteration define $\hat{\kappa}_j$, $\hat{\tau}_j$, $\hat{\xi}_j$ and $\hat{M}_j$ as before. Say that j is a *quasianomaly* iff $\hat{\xi}_j = -1$, $\hat{\kappa}_j$ is the cardinal predecessor of λ in M_{-1} and λ is not a cardinal in M_{-1}. It follows immediately that i satisfies one of the following clauses:

- $i+1$ is not a quasianomaly and there is a set $b \in \mathcal{P}(\hat{\tau}_i) \cap \Sigma_1(M_i)$ which is not $\Sigma_1(\hat{M}_i)$;

- $i+1$ is a quasianomaly and there is a set $b \in \mathcal{P}(\hat{\tau}_i) \cap \Sigma_1(M_i)$ which is not in M_{-1}.

Now consider i^* minimal with this property; we can adapt the rest of the proof of the previous lemma to the current situation and derive a contradiction. The details are left to the reader. We just note that, if $i^* = h+1$ is an anomaly, then the argument from the proof of Lemma 9.1.10 which proves the implication $b \in M_j \longrightarrow b \in \hat{M}_j$ cannot be used here to show $b \in M_{i^*} \longrightarrow b \in M_{-1}$, as our induction hypothesis is not enough to guarantee this. Instead, we observe that $\hat{\tau}_{i^*} < \kappa_h$ where κ_h is a cardinal in all models of the iteration and apply acceptability. □(Lemma 9.1.11)

Corollary 9.1.12. *Let $\Im$ be a normal premouse iteration and let the iteration map $\pi_{\xi_i,j} : M_i^* \to M_j$ be total. Then M_i^* and M_j have the same projecta which are less or equal than κ_i, and $\pi_{\xi_i,j}$ is Σ^*-preserving.*

If $\Im$ is a normal iteration of a good phalanx, the same is true. If $i+1$ is an anomaly, then $\omega\varrho^{\omega}_{M_i^*} = \kappa_i$ and, letting n be the least such that $\omega\varrho^{n+1}_{M_i^*} = \kappa_i$, we have $\omega\varrho^{\omega}_{M_j} = \omega\varrho^{n+1}_{M_j} = \kappa_i$. Moreover, $\pi_{-1,i+1}$ is both $\Sigma_0^{(n)}$ cofinal, and Σ^*-preserving.

Proof. Apply the two previous lemmata and the fact that fine ultrapowers by close extenders preserve projecta below the critical points and give rise to Σ^*-preserving ultrapower embeddings (see the corresponding section in the chapter on fine ultrapowers). For phalanx iterations, the same argument goes through unless $i+1$ is not anomalous.

Let us now discuss the case where $i+1$ is an anomaly. Notice first that, by the definition of phalanx, κ_i is a cardinal in M_{-1}, so M_i^* cannot project strictly below κ_i. Let n be as in the statement of the lemma. Using the fact that $E^{M_i}_{\nu_i}$ is weakly amenable w.r.t. M_i^*, we infer that $\omega\varrho^{n+1}_{M_{i+1}} \leq \kappa_i$. As $E^{M_i}_{\nu_i,\alpha} \in M_{-1}$ for all $\alpha < \lambda_i$, a slight variation of the argument from the proof of Lemma 3.2.7 shows that any $\mathbf{\Sigma}_1^{(n)}(M_{i+1})$ subset of κ_i is an element of M_{-1}. As κ_i is a cardinal in M_{-1}, any such *bounded* subset of κ_i is in $M_{-1}||\kappa_i$, and therefore in M_{i+1}. So $\omega\varrho^{n+1}_{M_{i+1}}$ cannot be strictly less than κ_i, which yields the equality $\omega\varrho^{n+1}_{M_{i+1}} = \kappa_i$. That $\pi_{-1,i+1}$ is $\Sigma_0^{(n)}$ cofinal follows from the soundness of M_i^*. Finally, by induction on $m \geq n$ we show that if $A(z_1, \ldots, z_\ell)$ is $\mathbf{\Sigma}_1^{(m)}(M_{i+1})$, then $A \cap (M_{i+1}||\kappa_i) \in M_{-1}$. This follows from the fact that all M_j agree below κ_i and $M_i^* \in M_{-1}$, so any relation obtained by substituting parameters from M_i^* and quantifying over some initial segment M_i^* must also be in M_{-1}. As above, this implies that M_{i+1} cannot project strictly below κ_i, so $\omega\varrho^{\omega}_{M_{i+1}} = \kappa_i$. Knowing that $\pi_{-1,i+1}$

is $\Sigma_1^{(n)}$-preserving, an easy induction shows that $\pi_{-1,i+1}$ is actually Σ^*-preserving. □(Corollary 9.1.12)

By Lemmata 9.1.10 and 9.1.11, if $i + 1$ is a truncation point, then M_j is never sound, provided that $i + 1 \leq_T j$. This follows as in the proof of Lemma 4.4.2. Granting that $\Im$ is a premouse iteration or else that $i + 1$ is not an anomaly, M_i^* and M_j have the same subsets of κ_i and, as we have already pointed out, also the same projecta below κ_i (here we assume w.l.o.g. that there is to truncation point $i' + 1$ such that $i + 1 <_T i' + 1 <_T j$). That M_j is not sound follows from the fact that $\pi_{\xi_i j}$ preserves the standard parameter above κ_i, which, in turn, is a consequence of the solidity of M_i^*. If $i + 1$ *is* an anomaly, then the second part of the previous corollary guarantees that all premice on the branch from $i + 1$ to j have the same $(n + 1)$-st projectum, which is sufficient to carry out the above argument and infer that M_j is not sound. We thus have the following generalization of Corollary 4.4.3:

Lemma 9.1.13. *Assume that M and N are premice and one of the following holds:*

- *M and N coiterate to some M' and N';*
- *M and N agree up to λ, the phalanx $\langle N, M, \lambda\rangle$ is good and coiterates with N to some M' and N'.*

If both sides of the coiteration involve truncations on their main branches, then $M' = N'$. □(Lemma 9.1.13)

For presolid premice we obtain that at least one side of the coiteration is *simple,* i.e. it does not involve a truncation on its main branch.

Lemma 9.1.14. *Let M and N be premice satisfying the assumptions of the previous lemma. Then at least one side of the coiteration of N with M, resp. $\langle M, N, \lambda\rangle$ is simple.*

Proof. We follow the proof of Lemma 5.1.4. Let $i + 1$ and $j + 1$ be the last truncation point on the M-side and N-side, respectively. Also, let T and U be the respective iteration trees. Set $\xi = T(i + 1)$ and $\zeta = U(j + 1)$. Also, let π be the iteration map from M_i^* to the last premouse on the M-side and let π' be that from N_j^* to the last premouse on the N-side. Using the fact that the projecta and the standard parameter are preserved along branches with no truncations we infer that $M_i^* = \mathrm{core}(M')$ and π is the core map. Similarly, $N_i^* = \mathrm{core}(N')$ and π' is the core map. By the previous lemma, $N_i^* = M_i^*$ and $\pi' = \pi$. This allows us to conclude that $E_{\nu_i}^{M_i}$ and $E_{\nu_j}^{N_j}$ are compatible, which contradicts (9.1.6).

If we replace M by the phalanx $\langle N, M, \lambda\rangle$, then the same argument goes through and we have that $E_{\nu_i}^{M_i}$ and $E_{\nu_j}^{N_j}$ are compatible. By (9.1.7), $i = h + 1$ must be a strong anomaly, $j = 0$, $\nu_0 = \lambda$ and $\lambda_0 = \kappa_h$. Letting k be such that $k + 1$ is on the main branch of U and $1 = U(k + 1)$, we have $\kappa_k = \lambda_0$, as otherwise κ_k would be larger than λ_0, which would contradict the fact that $\pi'(\kappa_0) = \pi(\kappa_0) \geq E_{\nu_i}^{M_i}(\kappa_0) > \lambda_0$. It follows that, given an $a \in \mathrm{rng}(E_\lambda^N)$, the set $E_{\nu_k}^{N_k}(a)$ is equal to $E_{\nu_h}^{M_h}(a) \cap \lambda_k$ or, vice versa, $E_{\nu_h}^{M_h}(a) = E_{\nu_k}^{N_k}(a) \cap \lambda_h$. By the proof of (9.1.6), the extenders $E_{\nu_k}^{N_k}$ and $E_{\nu_h}^{M_h}$ are compatible, which contradicts (9.1.7). □(Lemma 9.1.14)

The fact that one side of a coiteration is simple is still not enough to obtain a meaningful outcome of a comparison of N with a good phalanx $\langle N, M, \lambda\rangle$. The reason is simple: It might happen that the premouse N is on the main branch of the phalanx side of the coiteration. In order to exclude this, we need some form of the Dodd–Jensen Lemma, which will be discussed in the next section.

Let us finally note that the notion of phalanx can be generalized in the sense that one allows more premice in a phalanx. There are many arguments which seem to require the use of such phalanxes; however, phalanxes of length 2, i.e. those in the sense of our definition, entirely suffice for the purposes of this chapter.

9.2 Copying and the Dodd–Jensen Property

The first part of this section is devoted to copying of nonlinear iterations. Unlike in the linear case, here we need some extra verification to show that the copying construction can be carried out and that the copying maps have the required preservation properties. Having developed the basic facts about copying, we can prove the Dodd–Jensen Lemma. However, the full Dodd–Jensen Lemma holds only for a relatively small class of iterable premice. In order to develop general fine structure theory for premice arising in the constructions of extender models, it is sufficient to consider a weak version of the Dodd–Jensen lemma for countable premice due to Neeman and Steel.

First of all, we define the notion of copy of an iteration. Here, we adopt a fairly general approach which will later enable us to unify our terminology. Recall that a map σ between two structures is *cardinal preserving* iff σ-images of cardinals of the domain structure are cardinals in the target structure.

Definition. Let $\bar{M}$ and M be premice and $\sigma : \bar{M} \to M$ be Σ_0-preserving. Given a generalized iteration $\bar{\Im}$ of $\bar{M}$ of length θ, we say that a generalized iteration $\Im = \sigma\bar{\Im}$ of M is a *copy* of $\bar{\Im}$ by σ with *copying* maps $\langle \sigma_i;\ i < \theta\rangle$ iff the following hold:

a) $\text{length}(\Im) = \theta$, $B^{\Im} = B^{\bar{\Im}}$ and $T^{\Im} = T^{\bar{\Im}}$.

b) Each $\sigma_i : \bar{M}_i \to M_i$ is both Σ_0-preserving and cardinal preserving, $\sigma_0 = \sigma$ and $\sigma_i \circ \bar{\pi}_{hi} = \pi_{hi} \circ \sigma_h$ whenever $h \leq_T i$.

c) If $i = h + 1$ and $h \notin B$, then $\eta_h = \sigma_h(\bar{\eta}_h)$ and $\sigma_i = \sigma_h \restriction (\bar{M}_h||\bar{\eta}_h)$. Here, $\sigma(\mathbf{On} \cap \bar{M}_h) = \mathbf{On} \cap M_h$.

d) If $i = h + 1$ and $h \in B$, the following is satisfied:

i) $\sigma_i \restriction \bar{\lambda}_h = \sigma_h \restriction \bar{\lambda}_h$;

ii) $\sigma_{\xi_h}(\bar{\eta}_h) = \eta_h$;

iii) letting $\sigma_h^* = \sigma_{\xi_h} \restriction \bar{M}_h^*$, the map $\sigma_h^* : \bar{M}_h^* \to M_h^*$ is $\Sigma_0^{(n)}$-preserving whenever $\omega\varrho^n_{\bar{M}_h^*} > \bar{\kappa}_h$;

iv) $\langle \sigma^*, \sigma_h \restriction \bar{\lambda}_h \rangle : \langle \bar{M}_h^*, E_{\bar{\nu}_h}^{\bar{M}_h} \rangle \longrightarrow \langle M_h^*, E_{\nu_h}^{M_h} \rangle$;

v) given an $f \in \Gamma(\bar{\kappa}_h, M_h^*)$ and an $\alpha < \bar{\lambda}_h$,

$$\sigma_i(\bar{\pi}_{\xi_h,i}(f)(\alpha)) = \pi_{\xi_h,i} \circ \sigma_h^*(f)(\sigma_h(\alpha)).$$

e) If i is limit, then $\sigma_i(x) = \pi_{hi} \circ \sigma_h(x_h)$ whenever $h <_T i$ and $x_h \in \bar{M}_h$ is such that $\bar{\pi}_{hi}(x_h) = x$.

The copying maps are thus defined in the only possible way to make the whole diagram commutative and keep enough agreement between them. As in the linear case, these maps are uniquely determined, provided that they exist. If $\bar{\mathfrak{I}}$ is normal, then

$$\sigma_j \restriction (\bar{\nu}_i + 1) = \sigma_i \restriction (\bar{\nu}_i + 1) \text{ whenever } i \leq j, \tag{9.2.1}$$

as follows by a simple induction on j.

Lemma 9.2.1. *Let $\sigma : \bar{M} \to M$ be Σ^*-preserving, $\bar{\mathfrak{I}}$ be a normal iteration on $\bar{M}$ and $\mathfrak{I} = \sigma\bar{\mathfrak{I}}$. Then for all $i <$ length$(\bar{\mathfrak{I}})$ we have*

$$\sigma_i : \bar{M}_i \underset{\Sigma_0^{(n)}}{\longrightarrow} M_i$$

whenever $\omega\varrho^n_{\bar{M}_i} \geq \bar{\lambda}'_i = \sup\{\bar{\lambda}_h;\ h < i\}$.

Proof. We can w.l.o.g. assume that $\bar{\mathfrak{I}}$ is direct and proceed by induction on i. The case $i = 0$ is trivial. Suppose i is a successor ordinal, say $i = h + 1$. Then $\bar{\kappa}_h \geq \bar{\lambda}'_{\xi_h}$ by the normality of $\bar{\mathfrak{I}}$. The induction hypothesis guarantees that σ_h^* is $\Sigma_0^{(n)}$-preserving where $n = n_h$ is such that $\omega\varrho^{n+1}_{\bar{M}_h^*} \leq \bar{\kappa}_h < \omega\varrho^n_{\bar{M}_h^*}$ and $\kappa_h = \sigma_h^*(\bar{\kappa}_h) < \omega\varrho^n_{M_h^*}$. By Lemma 3.4.4, the map σ_i is $\Sigma_0^{(n)}$-preserving and by the weak amenability of $E_{\bar{\nu}_h}^{\bar{M}_h}$ w.r.t. $\bar{M}_h^*$ and properties of fine ultrapowers we have $\omega\varrho^{n+1}_{M_i} \leq \kappa_h < \lambda_h < \omega\varrho^n_{M_i}$. This verifies the induction step at i.

If i is limit, notice that the sequence $\langle n_h;\ h+1 <_T i \rangle$ is non-decreasing, so we can pick an h sufficiently large such that n_h (see the previous paragraph) remains fixed, say $n_h = n$, along the interval $[h+1, i)_T = \{j;\ h+1 \leq j < i\}$. By a simple induction argument we then obtain $\omega\varrho^{n+1}_{\bar{M}_i} < \bar{\lambda}'_i \leq \omega\varrho^n_{\bar{M}_i}$. Furthermore, for $j \in [h+1, i)$, the map σ_j is $\Sigma_0^{(n)}$-preserving, so σ_i is $\Sigma_0^{(n)}$-preserving as well by the properties of direct limits. This verifies the induction step for limit i. □(Lemma 9.2.1)

A minor modification of the above proof yields a version of the copying lemma for embeddings with lower preservation degree.

Lemma 9.2.2. *Let $\sigma : \bar{M} \to M$ be $\Sigma_0^{(n)}$-preserving and $\omega\varrho_{\bar{M}}^{n+1} < \omega\varrho_{\bar{M}}^{n}$. Then the conclusion of the previous lemma holds for normal iterations above $\omega\varrho_{\bar{M}}^{n+1}$.*

Proof. In order to determine the preservation degree of σ_{h+1}, we need to know that σ_h^* is $\Sigma_0^{(m)}$-preserving whenever $\bar{\kappa}_h < \omega\varrho_{\bar{M}_h^*}^{m}$. This follows by our assumption that $\bar{\mathfrak{I}}$ is above $\omega\varrho_{\bar{M}}^{n+1}$ if $T(h+1) = 0$ and $h+1$ is not a truncation point, and by the argument in the proof of the previous lemma otherwise. The rest of the proof is the same as before. □(Lemma 9.2.2)

We shall later state a variant of the copy lemma for phalanx iterations, which will prove to be much more useful than Lemma 9.2.2. However, the proof of Lemma 9.2.2 lemma can serve as a motivation here, as most of the copying constructions which involve phalanx iterations start from maps with relatively low preservation degree.

Definition. Let $\sigma : \bar{M} \to M$ where $\bar{M}$ and M are premice. Given an iteration strategy $\mathbb{S}$ on iterations of M, we define an iteration strategy $\sigma\mathbb{S}$ on the iterations of $\bar{M}$ by

$$\sigma\mathbb{S}(\bar{\mathfrak{I}}) \simeq \mathbb{S}(\sigma\bar{\mathfrak{I}}).$$

So far, we have only described some properties of copying maps, granting that the copies of the iterations in question exist. The following lemmata show that these copies really do exist, provided that M has an iteration strategy.

Lemma 9.2.3. *Let $\sigma : \bar{M} \to M$ be Σ^*-preserving and suppose that M has a normal iteration strategy $\mathbb{S}$. Let $\bar{\mathfrak{I}}$ be a normal $\sigma\mathbb{S}$-iteration of $\bar{M}$. Then $\mathfrak{I} = \sigma\bar{\mathfrak{I}}$ exists and is a normal $\mathbb{S}$-iteration of M.*

Proof. We proceed by the induction on the length of iterations. Again, we consider w.l.o.g. only iterations which are direct. The case where $i = 1$ or i is a limit are trivial, so let us assume that i is a successor ordinal, say $i = h+1$. Set $\mathfrak{I}' \simeq \sigma(\bar{\mathfrak{I}} \restriction h)$. Suppose first that h is a limit ordinal. By the induction hypothesis, $\mathfrak{I}'$ exists and is an $\mathbb{S}$-iteration of M. By the definition of $\sigma\mathbb{S}$ we have $\sigma\mathbb{S}(\bar{\mathfrak{I}} \restriction h) = \mathbb{S}(\mathfrak{I}')$; denote this branch by b. Let $\mathfrak{I}$ be the extension of $\mathfrak{I}'$ determined by the branch b, i.e. the final model in $\mathfrak{I}$ is M_h, where $\langle M_h, \pi_{kh}\rangle$ is the direct limit along b in $\mathfrak{I}'$ and $k <_T h$ for every $k \in b$. Then $\mathfrak{I}$ is an $\mathbb{S}$-iteration of length $h+1$ and the formula

$$\bar{\pi}_{kh}(y) \mapsto \pi_{kh} \circ \sigma_k(y)$$

defines the embedding $\sigma_h : \bar{M}_h \to M_h$. It is clear that $\mathfrak{I} = \sigma\bar{\mathfrak{I}}$.

Now suppose that h is a successor ordinal, say $h = k+1$. Let $\xi = T^{\bar{\mathfrak{I}}}(h)$. This means that $\bar{\lambda}_j \leq \bar{\kappa}_k < \bar{\lambda}_\xi$ whenever $j < \xi$. Now notice that $\sigma_k(\bar{\lambda}_j) = \lambda_j$ for all $j < k$; this follows from the fact that $\sigma_j(\bar{\lambda}_j) = \lambda_j$ and (9.2.1). As a consequence we have that $\lambda_j \leq \kappa_k < \lambda_\xi$, which means that if $\mathfrak{I}$ is a normal iteration of M of length $h+1$ extending $\mathfrak{I}'$, then $\xi = T^{\mathfrak{I}}(h)$, i.e. $T^{\mathfrak{I}} = T^{\bar{\mathfrak{I}}}$. Hence $M_h = \mathrm{Ult}^*(M_\xi || \eta_\xi, E_{\nu_k}^{M_k})$ where M_h is the last premouse in $\mathfrak{I}$. If h is a truncation point in $\bar{\mathfrak{I}}$, it is easy to see that $\eta_k = \sigma_\xi(\bar{\eta}_k)$.

Otherwise $\bar{\tau}_k$ is a cardinal in $\bar{M}_\xi$, so $\tau_k = \sigma_\xi(\bar{\tau}_k)$, as σ_ξ is cardinal preserving. Once we have proved

$$\langle \sigma_k^*, \sigma_k \restriction \bar{\lambda}_k \rangle : \langle \bar{M}_k^*, E_{\bar{\nu}_k}^{\bar{M}_k} \rangle \longrightarrow \langle M_k^*, E_{\nu_k}^{M_k} \rangle, \tag{9.2.2}$$

we can use Lemma 3.4.4 to define σ_h, as the assumptions are satisfied by Lemma 9.2.1. It will then follow that σ_h is cardinal preserving as well: If $n > 0$, then σ_h is in fact Σ_1-preserving. If $n = 0$, notice that the ultrapower map $\bar{\pi}_{\xi h}$ is cofinal, so σ_h maps every $\bar{M}_h$-cardinal of the form $\bar{\pi}_{\xi h}(\mu)$ to $\pi_{\xi h} \circ \sigma_\xi^*(\mu)$, and the latter is certainly a cardinal in M_h, as σ_ξ^* is cardinal preserving by the induction hypothesis.

It remains to prove (9.2.2). Here we use the fact that, by (9.2.1), there is enough agreement between the maps σ_k and σ_ξ so that $\sigma_k(a) = \sigma_\xi(a)$ whenever $a \in \mathcal{P}(\bar{\kappa}_k) \cap \bar{M}_k$. Then $\alpha \in E_{\bar{\nu}_k}^{\bar{M}_k}$ iff $\sigma_\xi(\alpha) = \sigma_k(\alpha) \in E_{\nu_k}^{M_k}$. □(Lemma 9.2.3)

Let us note that the requirement on the copying maps to be cardinal preserving is needed to guarantee that $\eta_h = \sigma_{\xi_h}(\bar{\eta}_h)$. This follows automatically if σ_{ξ_h} is Σ_1-preserving, however, we are unable to prove this if $\xi_h = \ell + 1$ and $\omega\varrho^1_{\bar{M}_\ell^*} \leq \bar{\kappa}_\ell$. However, the fact that the copying map is cardinal preserving is preserved in the course of the induction.

Lemma 9.2.4. *Let $\sigma : \bar{M} \to M$ be Σ^*-preserving where $\bar{M}$ and M are premice. Given a normal iteration strategy $\mathbb{S}$ for M, the derived strategy $\sigma\mathbb{S}$ is a normal iteration strategy for $\bar{M}$.*

Proof. We have to verify that any normal $\sigma\mathbb{S}$-iteration $\bar{\mathfrak{I}}$ of $\bar{M}$ has a proper extension. Let $\mathfrak{I} = \sigma\bar{\mathfrak{I}}$ and let $i = \text{length}(\bar{\mathfrak{I}}) = \text{length}(\mathfrak{I})$. If i is a limit, then both $\mathbb{S}$ and $\sigma\mathbb{S}$ determine the same branch through $T = T^{\bar{\mathfrak{I}}} = T^{\mathfrak{I}}$ and as in the proof of the previous lemma we conclude that the direct limit along b in $\bar{\mathfrak{I}}$ can be embedded into the direct limit along b in $\mathfrak{I}$. The latter, however, is well-founded, as $\mathbb{S}$ is a normal iteration strategy, so the former is well-founded as well. If i is a successor ordinal, say $i = h+1$, we again proceed as in the above proof and show that $\bar{\xi}_h = \xi_h$, $M_h^* = \sigma_\xi(\bar{M}_h^*)$ where $\xi = \bar{\xi}_h$ and (9.2.2) holds with h in place of k. By Lemma 3.4.4, the term model $\mathbb{D}^*(\bar{M}_h^*, E_{\bar{\nu}_h}^{\bar{M}_h})$ can be embedded into $\mathbb{D}^*(M_h^*, E_{\nu_h}^{M_h})$. As in the limit case we then infer that this ultrapower must be well-founded. □(Lemma 9.2.4)

A similar conclusion can be obtained for iterations above a projectum for σ with weaker preservation properties. Here we use Lemma 9.2.2 instead of Lemma 9.2.1.

Let us also make the following observation. Say that M has the *normal uniqueness property* iff every normal iteration of M of limit length has at most one cofinal well-founded branch. It follows easily from the previous lemma that if $\sigma : \bar{M} \to M$ is Σ^*-preserving and $\bar{M}$ has the normal uniqueness property, then $\sigma\mathbb{S}$ is the normal uniqueness strategy for $\bar{M}$ whenever $\mathbb{S}$ is a normal iteration strategy for M.

We now state a deeper result on the copying maps. We have seen in the previous section that the iteration maps are always Σ^*-preserving. The same conclusion follows for the copying maps and, in fact, the proofs of these results are of the same flavour.

Lemma 9.2.5. *Let $\sigma : \bar{M} \to M$ be Σ^*-preserving where $\bar{M}$ and M are premice. Let $\bar{\Im}$ be a normal iteration of $\bar{M}$ and $\Im = \sigma\bar{\Im}$ with copying maps σ_i. Then each $\sigma_i : \bar{M}_i \to M_i$ is Σ^*-preserving.*

Proof. The conclusion will follow immediately from Lemma 3.4.5, once we have shown

$$\langle \sigma_i^*, \sigma_i \restriction \bar{\lambda}_i \rangle : \langle \bar{M}_i^*, E_{\bar{\nu}_i}^{\bar{M}_i} \rangle \stackrel{*}{\longrightarrow} \langle M_i^*, E_{\nu_i}^{M_i} \rangle. \tag{9.2.3}$$

The proof of (9.2.3) is very much like that of Lemma 9.1.10. If i is a counterexample to (9.2.3), then $\xi_i < i$ and both $E_{\bar{\nu}_i}^{\bar{M}_i}$ and $E_{\nu_i}^{M_i}$ are the top extenders of $\bar{M}_i$ and M_i, respectively. To see the former, notice that if $\bar{\alpha} < \bar{\lambda}_i$ and $\alpha = \sigma_i(\bar{\alpha})$, then

$$x \in E_{\bar{\nu}_i,\bar{\alpha}}^{\bar{M}_i} \quad\longleftrightarrow\quad (\exists y)(\langle x, y\rangle \in E_{\bar{\nu}_i}^{\bar{M}_i} \;\&\; \bar{\alpha} \in y)$$

and the relation $x \in E_{\nu_i,\alpha}^{M_i}$ is defined in the same manner. It is then easy to see that the former definition is $\Sigma_1(\bar{M}_i^*)$ in $\bar{\alpha}$ and the latter $\Sigma_1(M_i^*)$ in α, which rules out the possibility $\xi_i = i$. Now if the two above extenders were not the top extenders, by acceptability we would have $E_{\bar{\nu}_i,\bar{\alpha}}^{\bar{M}_i} \in \bar{M}_i||\bar{\lambda}_{\xi_i}$ and $E_{\nu_i,\alpha}^{M_i} \in M_i||\lambda_{\xi_i}$, and so $\sigma_i(E_{\bar{\nu}_i,\bar{\alpha}}^{\bar{M}_i}) = E_{\nu_i,\alpha}^{M_i}$. The agreement between σ_{ξ_i} and σ_i stated in (9.2.1) then guarantees that also $\sigma_i^*(E_{\bar{\nu}_i,\bar{\alpha}}^{\bar{M}_i}) = E_{\nu_i,\alpha}^{M_i}$, contradicting the fact that i is a counterexample to (9.2.3).

Define $\hat{\kappa}_i$, $\hat{\tau}_i$, $\hat{M}_i$ and $\hat{\xi}_i$ as in the proof of Lemma 9.1.10 and let $\hat{\bar{\kappa}}_i$, $\hat{\bar{\tau}}_i$, $\hat{\bar{M}}_i$ and $\hat{\bar{\xi}}_i$ be defined in the same manner for $\bar{\Im}$. Notice that $\hat{\bar{\xi}}_i = \hat{\xi}_i$. By the failure of (9.2.3), the following statement holds for index i:

There are $\bar{a} \subset \hat{\bar{\tau}}_i$ and $a \subset \hat{\tau}_i$ such that:

- $\bar{a}$ is $\Sigma_1(\bar{M}_i)$ in some $\bar{p}$ and a is $\Sigma_1(M_i)$ in $p = \sigma_i(\bar{p})$ by the same definition.
- There is no $\bar{r} \in \hat{\bar{M}}_i$ such that $\bar{a}$ is $\Sigma_1(\hat{\bar{M}}_i)$ in $\bar{r}$ and a is $\Sigma_1(\hat{M}_i)$ in $r = \sigma_i^*(\bar{r})$ by the same defintion.

(9.2.4)

From now on suppose that i is minimal with this property. The argument for $\xi_i < i$ from the previous paragraph can be easily modified to show that $\hat{\xi}_i < i$. Also, i must be a successor ordinal, as otherwise we could find a $j + 1 <_T i$ such that $\hat{\bar{\kappa}}_i < \bar{\lambda}_j$ and $\bar{p} = \bar{\pi}_{j+1,i}(\bar{p}_{j+1})$. Then $\hat{\kappa}_i < \lambda_j$ and $p = \pi_{j+1,i}(p_{j+1})$ where $p_{j+1} = \sigma_{j+1}(\bar{p}_{j+1})$. It follows that $\hat{\bar{\xi}}_{j+1} = \hat{\xi}_{j+1} = \xi$, $\bar{a}$ is $\Sigma_1(\bar{M}_{j+1})$ in $\bar{p}_{j+1}$ and a is $\Sigma_1(M_{j+1})$ in p_{j+1} by the same definition. This means that (9.2.4) would hold for $j + 1$ in place of i, which would contradict the minimality of i.

Let $i = h + 1$. As in the proof of Lemma 9.1.10 we infer that $\hat{\bar{\kappa}}_i < \bar{\lambda}_h$, so in fact $\hat{\bar{\kappa}}_i < \bar{\kappa}_h$ and so $\hat{\kappa}_i < \kappa_h$. Consequently, $\hat{\xi}_i \leq \xi_h$. By the minimality of i, all $i' < i$ satisfy (9.2.3), which means that σ_ι is Σ^*-preserving whenever $\iota \leq i$. Thus, σ_ι^* is also Σ^*-preserving. This in particular applied to $\iota = h, i$. We now observe that $\omega\varrho^1_{\bar{M}^*_h} \leq \bar{\kappa}_h$. Suppose this fails. By the preservation properties of σ^*_h we obtain $\omega\varrho^1_{M^*_h} > \kappa_h$, and by the properties of fine ultrapowers we would have $\omega\varrho^1_{\bar{M}_i} > \bar{\lambda}_h > \hat{\bar{\tau}}_h$ and $\omega\varrho^1_{M_i} > \lambda_h > \hat{\tau}_h$. It follows that $\bar{a} \in \bar{M}_i$ and $a \in M_i$. Now $\bar{a}$ is characterized as follows:

$$y = \bar{a} \longleftrightarrow y \subset \hat{\bar{\tau}}_i \ \& \ (\forall \eta^1 < \hat{\bar{\tau}}_i)(\eta^1 \in y \longleftrightarrow \bar{A}(\eta^1, \bar{p})),$$

where $\bar{A}$ is $\Sigma_1(\bar{M}_i)$. Notice that a is characterized in the same manner over M_i in $\hat{\tau}_i$ and p, where A is $\Sigma_1(M_i)$ by the same definition as $\bar{A}$. The above statement is a $\Sigma_0^{(1)}$-statement. Notice that σ_i is $\Sigma_0^{(1)}$-preserving (in fact, the induction hypothesis guarantees that σ_i is Σ^*-preserving), which yields $\sigma_i(\bar{a}) = a$. As $\xi < i$, $\bar{\lambda}_\xi$ and λ_ξ are limit cardinals in $\bar{M}_i$ and M_i respectively, and by acceptability we have $\bar{a} \in \bar{M}_i||\bar{\lambda}_\xi = \bar{M}_\xi||\bar{\lambda}_\xi$ and $a \in M_i||\lambda_\xi = M_\xi||\lambda_\xi$. The agreement between σ_ξ and σ_i then yields $\hat{\sigma}_i(\bar{a}) = a$, which contradicts (9.2.4).

We thus have $\omega\varrho^1_{\bar{M}^*_h} \leq \bar{\kappa}_h$ and, as σ^*_h is Σ^*-preserving, also $\omega\varrho^1_{M^*_h} \leq \kappa_h$. To see the latter, notice that the formula $(\forall\eta^1)(\eta^1 \leq \bar{\kappa}_h)$ is preserved under σ^*_h. This means that both $\mathrm{Ult}^*(\bar{M}^*_h, E^{\bar{M}_h}_{\bar{\nu}_h})$ and $\mathrm{Ult}^*(M^*_h, E^{M_h}_{\nu_h})$ are in fact coarse ultrapowers. Then $\bar{p} = \bar{\pi}_{\xi_h,i}(\bar{f})(\bar{\alpha})$ for some $\bar{f} \in \bar{M}^*_h$ and $\bar{\alpha} < \bar{\lambda}_h$ and $p = \pi_{\xi_h,i}(f)(\alpha)$ where $f = \sigma^*_h(\bar{f})$ and $\alpha = \sigma_h(\bar{\alpha})$. By Lemmata 9.1.10 and 3.2.9, $\bar{a}$ is $\Sigma_1(\bar{M}^*_h)$ in $\bar{f}$ and some $\bar{r}$, where $E^{\bar{M}_h}_{\bar{\nu}_h,\bar{\alpha}}$ is $\Sigma_1(\bar{M}^*_h)$ in $\bar{r}$. Moreover, the Σ_1-definition of $\bar{a}$ over $\bar{M}^*_h$ is uniform and depends merely on the Σ_1-definitions of $\bar{A}$ over $\bar{M}_i$ and $E^{\bar{M}_h}_{\bar{\nu}_h,\bar{\alpha}}$ over $\bar{M}^*_h$ from $\bar{r}$. Also, the corresponding conclusion can be made for a, M^*_h and $E^{M_h}_{\nu_h,\alpha}$, where $E^{M_h}_{\nu_h,\alpha}$ is Σ_1-definable over M^*_h in some r. As (9.2.3) holds for h in place of i, the Σ_1-definitions of $E^{\bar{M}_h}_{\bar{\nu}_h,\bar{\alpha}}$ and $E^{M_h}_{\nu_h,\alpha}$ are the same and $r = \sigma^*_h(\bar{r})$. It follows that $\bar{a}$ is $\Sigma_1(\bar{M}^*_h)$ in $\bar{f}, \bar{r}$ and a is $\Sigma_1(M^*_h)$ in f, r by the same definition.

We know that $\xi \leq \xi_h$. Suppose that they are equal. As $\hat{\bar{\kappa}}_i < \bar{\kappa}_h$, we have $\hat{\bar{\eta}}_i \geq \bar{\eta}_h$ and, similarly, $\hat{\eta}_i \geq \eta_h$. If $\hat{\bar{\eta}}_i = \bar{\eta}_h$, then $\hat{\eta}_i = \eta_h$, so $\hat{\bar{M}}_i = \bar{M}^*_h$ and $\hat{M}_i = M^*_h$. By the conclusion of the previous paragraph, (9.2.4) would fail, a contradiction to our choice of i. If $\hat{\bar{\eta}}_i < \bar{\eta}_h$, then $\hat{\eta}_i < \eta_h$ and $\hat{\sigma}_i(\bar{M}^*_h) = \sigma_{\xi_h}(\bar{M}^*_h) = M^*_h$. By the conclusion of the previous paragraph, $\bar{a} \in \hat{\bar{M}}_i$, $a \in \hat{M}_i$ and $\hat{\sigma}_i(\bar{a}) = a$, a contradiction again. It follows that $\xi < \xi_h$. We show that $\bar{M}^*_h = \bar{M}_{\xi_h}$. Indeed, if $\bar{M}^*_h \in \bar{M}_{\xi_h}$, then $M^*_h \in M_{\xi_h}$, so again by the conclusion of the previous paragraph we obtain $\bar{a} \in \bar{M}_{\xi_h}$, $a \in M_{\xi_h}$ and $\sigma_{\xi_h}(\bar{a}) = a$. But $\bar{\lambda}_\xi$ and λ_ξ are limit cardinals in $\bar{M}_{\xi_h}$ and M_{ξ_h}, respectively, so by acceptability and the agreement between σ_ξ and σ_{ξ_h} we have $\hat{\sigma}_i(\bar{a}) = \sigma_\xi(\bar{a}) = a$, which again contradicts our choice of i.

As a consequence of the previous paragraph we have that $\bar{M}^*_h = \bar{M}_{\xi_h}$ and $M^*_h = M_{\xi_h}$. Also notice that $\hat{\bar{\kappa}}_{\xi_h} = \hat{\bar{\kappa}}_i$ and $\hat{\kappa}_{\xi_h} = \hat{\kappa}_i$. Thus, $\xi_{\xi_h} = \xi$. By our earlier computation, $\bar{a}$ is $\Sigma_1(\bar{M}_{\xi_h})$ in $\bar{f}, \bar{r}$ and a is $\Sigma_1(M_{\xi_h})$ in f, r by the same definition. This means that (9.2.4) is true for $\xi_h \leq h < i$, which contradicts the minimality of i. It follows that (9.2.4) must fail for all i, and so (9.2.3) is true after all. □(Lemma 9.2.5)

In the previous section we proved lemmata about phalanx coiterations under the assumption that such coiterations can be carried out. This assumes sufficient amount of iterability of the phalanxes in question. We have seen that iterability of premice can be guaranteed by embedding the premice in question into iterable premice. The same strategy also works for phalanxes.

Definition. We say that $\sigma : M \to N$ is an *embedding* of a phalanx $\langle N, M, \lambda \rangle$ into the premouse N iff $\sigma \restriction \lambda = \text{id}$ and σ is both cardinal preserving and $\Sigma_0^{(n)}$-preserving whenever $\omega \varrho_M^n > \lambda$.

Notice that if a phalanx $\langle N, M, \lambda \rangle$ can be embedded into N, then $\langle N, M, \lambda \rangle$ is automatically a good phalanx. Let us also note that we should have defined a phalanx embedding as a pair of maps $\langle \text{id}, \sigma \rangle$ where $\text{id} : N \to N$, however, we can consider this identity embedding as a suppressed parameter.

Definition. Given an embedding σ of $\langle N, M, \lambda \rangle$ into N and a normal iteration $\bar{\mathfrak{I}}$ of $\langle N, M, \lambda \rangle$, we define a copy $\mathfrak{I} = \sigma \bar{\mathfrak{I}}$ of $\bar{\mathfrak{I}}$ by σ with copying maps

- $\sigma_i : M_i \to N_i$ for $i \geq 0$ and
- $\sigma_{-1} = \text{id} : M_{-1} \to N_0$ (recall that $M_{-1} = N_0 = N$)

as follows.

a) $T = T^{\mathfrak{I}}$ is the same as $\bar{T} = T^{\bar{\mathfrak{I}}}$ with the only exception that the two roots of $\bar{T}$ are glued together. More formally:

$$0 <_T i \longleftrightarrow (0 <_{\bar{T}} i \vee -1 <_{\bar{T}} i);$$
$$h <_T i \longleftrightarrow h <_{\bar{T}} i \quad \text{for } h > 0.$$

b) $\sigma_{-1} = \text{id}$ and $\sigma_0 = \sigma$. Moreover, $\sigma_i \circ \bar{\pi}_{ih} = \pi_{ih} \circ \sigma_h$ whenever $h <_{\bar{T}} i$.

c) $B^{\mathfrak{I}} = B^{\bar{\mathfrak{I}}} \stackrel{\text{def}}{=} B$ and $\sigma_i(\bar{\nu}_i) = \nu_i$ for $i \in B$. Thus, $\sigma_i(\bar{\kappa}_i, \bar{\tau}_i) = (\kappa_i, \tau_i)$. Also, $\bar{\xi}_i = \xi_i$ unless $\bar{\xi}_i = -1$, in which case $\xi_i = 0$.

d) If $i \notin B$, then $\sigma_{i+1} = \sigma_i \restriction M_{i+1}$.

e) If $i \in B$ and $i+1$ is not an anomaly, then setting $\sigma_i^* = \sigma_{\bar{\xi}_i} \restriction M_i^* : M_i^* \to N_i^*$, we have

$$\langle \sigma_i^*, \sigma_i \restriction \bar{\lambda}_i \rangle : \langle M_i^*, E_{\bar{\nu}_i}^{M_i} \rangle \longrightarrow \langle N_i^*, E_{\nu_i}^{N_i} \rangle$$

and $\sigma_{i+1} : M_{i+1} \to N_{i+1}$ is defined by

$$\sigma_{i+1}(\bar{\pi}_{\bar{\xi}_i, i+1}(f)(\alpha)) = \pi_{\xi_i, i+1} \circ \sigma_i^*(f)(\sigma_i(\alpha)).$$

where $f \in \Gamma(\bar{\kappa}_i, M_i^*)$ and $\alpha < \bar{\lambda}_i$.

f) If $i + 1$ is an anomaly, then

$$\sigma_{i+1} : M_i^* \to \pi_{0,i+1}(M_i^*)$$

is defined by

$$\sigma_{i+1}(\bar{\pi}_{-1,i+1}(f)(\alpha)) = \pi_{0,i+1}(f)(\sigma_i(\alpha))$$

where $f \in \Gamma(\bar{\kappa}_i, M_i^*)$ and $\alpha < \bar{\lambda}_i$.

g) If i is limit, then σ_i is defined in the usual way, i.e. by the formula

$$\bar{\pi}_{hi}(z) \mapsto \pi_{hi} \circ \sigma_h(z).$$

The argument for (9.2.1) goes through also in the present context and we obtain the same amount of agreement for $i \geq 0$, provided that $i + 1$ is not an anomaly. By the agreement between M and N we also have $\sigma_i \restriction \lambda = \mathrm{id}$ for all $i \geq 0$. For anomalous $i+1$ we have a weaker form of agreement $\sigma_j \restriction \bar{\nu}_i = \sigma_i \restriction \bar{\nu}_i$, as $\sigma_{i+1}(\bar{\nu}_i)$ is the cardinal successor of λ_i in a proper initial segment of N_{i+1} that is shorter than ν_i, as we shall see below.

If $i + 1$ as an anomaly, then $i + 1$ is a truncation point in $\bar{\mathfrak{I}}$, but not in $\mathfrak{I}$. The premouse M_{i+1} is embedded into $\pi_{0,i+1}(M_i^*)$ where M_i^* is the longest initial segment of N which does not collapse λ. Thus, $\pi_{0,i+1}(M_i^*)$ is the longest initial segment of M_{i+1} which does not collapse $\pi_{0,i+1}(\lambda)$. It is clear that $\pi_{0,i+1}(M_i^*)$ projects to λ_i. Now if $i + 1 = T(j + 1)$, then $\kappa_j \geq \lambda_i$, so $\eta_j \leq \mathrm{ht}(\pi_{0,i+1}(M_i^*))$. It follows that $\langle \sigma_j^*, \sigma_j \restriction \bar{\lambda}_j \rangle : \langle M_j^*, E_{\bar{\nu}_j}^{M_j} \rangle \longrightarrow \langle N_j^*, E_{\nu_j}^{N_j} \rangle$ and σ_j^* is $\Sigma_0^{(n)}$-preserving whenever $\bar{\kappa}_j < \omega\varrho_{M_j^*}^n$, so σ_{j+1} is defined as in e).

The fact that $\eta_i \leq \mathrm{ht}(\pi_{0,i+1}(M_i^*))$ enables us to repeat the argument from the proof of Lemma 9.1.1 and establish:

Lemma 9.2.6. *Let σ be an embedding of the phalanx $\langle N, M, \lambda \rangle$ into N, let $\bar{\mathfrak{I}}$ be a normal iteration of this phalanx and let $\mathfrak{I} = \sigma\bar{\mathfrak{I}}$. Then for each $i \geq 0$, the copying map σ_i is cardinal preserving and*

$$\sigma_i : M_i \underset{\Sigma_0^{(n)}}{\longrightarrow} N_i$$

whenever $\omega\varrho_{M_i}^n \geq \bar{\lambda}_i' = \sup\{\bar{\lambda}_h;\ -1 \leq h < i\}$. □(Lemma 9.2.6)

Given an iteration strategy $\mathbb{S}$ for M, we define the derived iteration strategy $\sigma\mathbb{S}$ for $\langle N, M, \lambda \rangle$ in the obvious way, i.e. we would like to let $\sigma\mathbb{S}(\bar{\mathfrak{I}})$ to be equal to $\mathbb{S}(\mathfrak{I})$. This is not literally possible, as $\bar{T}$ has two roots, so $\sigma\mathbb{S}$ is defined by

$\sigma\mathbb{S}(\bar{\mathfrak{I}}) \simeq$ the unique branch $\bar{b}$ through $\bar{T}$ such that $\bar{b} - \{0, 1\} = \mathbb{S}(\mathfrak{I}) - \{0, 1\}$.

As before we then show:

Lemma 9.2.7. *Let σ be an embedding of a phalanx $\langle N, M, \lambda\rangle$ into N, let $\mathbb{S}$ be a normal iteration strategy for N and $\bar{\mathbb{S}} = \sigma\mathbb{S}$. Then $\bar{\mathbb{S}}$ is a normal iteration strategy for $\langle N, M, \lambda\rangle$ and if $\bar{\mathfrak{I}}$ is a normal $\bar{\mathbb{S}}$-iteration of $\langle N, M, \lambda\rangle$, then $\sigma\bar{\mathfrak{I}}$ is a normal $\mathbb{S}$-iteration of N.*

□(Lemma 9.2.7)

We now establish stronger preservation properties for the copying maps, similar to those in Lemma 9.1.5. This time we have to formulate the lemma more carefully, as the embedding $\sigma : M \to N$ is not fully Σ^*-preserving and the situation might be complicated by the presence of anomalies.

Lemma 9.2.8. *Let σ be an embedding of the phalanx $\langle N, M, \lambda\rangle$ into the premouse N, let $\bar{\mathfrak{I}}$ be a normal iteration of $\langle N, M, \lambda\rangle$ and let $\mathfrak{I} = \sigma\bar{\mathfrak{I}}$ with the copying maps $\langle\sigma_i;\ i < \text{length}(\bar{\mathfrak{I}})\rangle$. Then* (9.2.3) *holds whenever $i+1$ is not an anomaly.*

Proof. The proof closely follows that of Lemma 9.2.5 and Lemma 9.1.11. We sketch the crucial point and leave the details to the reader. Again, any witness i to the failure of the lemma instances (9.2.4) and is such that $i+1$ is not anomalous. It follows that there is a minimal i satisfying (9.2.4) which is not a quasianomaly and such that, if i is an anomaly, then (9.2.4) holds with $\pi_{0i}(M^*_{i-1})$ in place of i. As before we show that such an i must be a successor ordinal, say $i = h+1$.

If i is not an anomaly, we can argue exactly as in the proof of Lemma 9.2.5, as this time we have (9.2.3) with h in place of i. However, if $0 <_{\bar{T}} i$ and there is no truncation point $i' \leq_{\bar{T}} i$, we have to refine the part of the argument where we show that $\bar{a}$ is $\Sigma_1(M^*_h)$ in $\bar{f}, \bar{r}$ and a is $\Sigma_1(N^*_h)$ in f, r by the same definition. The reason for this is that σ is not necessarily Σ^*-preserving (indeed, in most typical applications it will not be), so by our assumption on i, we cannot guarantee that σ^*_h has preservation properties which will ensure that $\omega\varrho^1_{N^*_h} \leq \kappa_h$ if we know that $\omega\varrho^1_{M^*_h} \leq \bar{\kappa}_h$. (Notice that $M^*_h = M_{\xi_h}$ and $N^*_h = N_{\xi_h}$ in this case.) To overcome this obstacle, we argue as follows. Form the coarse ultrapower $\tilde{N}_i = \text{Ult}(N^*_h, E^{N_h}_{\nu_h})$ with the associated ultrapower map $\tilde{\pi}_i$. Define the embeddings

- $\bar{\sigma}_i : M_i \to \tilde{N}_i$ by $\bar{\sigma}_i(\bar{\pi}_{\xi_h,i}(f)(\alpha)) = \tilde{\pi}_i \circ \sigma^*_h(f)(\sigma_h(\alpha))$ for $f \in M^*_h$ and $\alpha < \bar{\lambda}_h$;
- $\tilde{\sigma}_i : \tilde{N}_i \to N_i$ by $\tilde{\sigma}_i(\tilde{\pi}_i(f)(\alpha)) = \pi_{\xi_h,i}(f)(\alpha)$ for $f \in N^*_h$ and $\alpha < \lambda_h$.

It is a routine to verify that both $\bar{\sigma}_i$ and $\tilde{\sigma}_i$ are Σ_0-preserving and that $\sigma_i = \tilde{\sigma}_i \circ \bar{\sigma}_i$. Moreover, notice that both $\tilde{N}_i$ and N_i have top extenders with critical point $\hat{\kappa}_i$ and that $\hat{\tau}_i < \kappa_h < \lambda_h$. As $\tilde{\sigma} \restriction \lambda_h = \text{id}$, the embedding $\tilde{\sigma}_i$ maps $\tilde{N}_i$ cofinally into N_i, and so is Σ_1-preserving. Now suppose $\bar{a}, \bar{p}, \bar{f}, \bar{\alpha}$ and a, p, f, α are as in the proof of Lemma 9.2.5; then α is in the range of $\tilde{\sigma}_i$ and, letting $\tilde{p}$ be such that $p = \tilde{\sigma}_i(\tilde{p})$, we have $\tilde{p} = \tilde{\pi}_i(f)(\alpha)$. Moreover, as $\tilde{\sigma}_i$ is Σ_1-preserving, a is $\Sigma_1(\tilde{N}_i)$ in $\tilde{\pi}_i(f)(\alpha)$ and $\bar{a}$ is $\Sigma_1(M_i)$ in $\bar{\pi}_{\xi_h,i}(\bar{f})(\bar{\alpha})$ by the same definition. Now we can proceed as in the proof of Lemma 9.2.5 and apply the argument from the proof of Lemma 3.2.7 (or, in fact, that from the proof of Lemma 2.3.3) to get that $\bar{a}$ is $\Sigma_1(M^*_h)$ in $\bar{f}, \bar{r}$ and a is $\Sigma_1(N^*_h)$ in f, r by the same definition. (See the proof of Lemma 9.2.5 for the definition of $\bar{r}$ and r.)

If i is an anomaly, then $\bar{\kappa}_h$ is a cardinal in all models of $\bar{\mathfrak{I}}$ and $\hat{\bar{\tau}}_i < \bar{\kappa}_h < \lambda$. Consequently, $\sigma_i \upharpoonright \hat{\bar{\tau}}_i = \mathrm{id}$. This implies that σ_i is cofinal, as the top extenders of both M_i and $\pi_{0i}(M_h^*)$ have the same critical point $\hat{\bar{\kappa}}_i$. It follows that σ_i is always Σ_1-preserving, so $\bar{b} = b$. By Lemma 9.1.11, $E^{M_h}_{\bar{\nu}_h,\alpha} \in N$ for all $\alpha < \bar{\lambda}_h$; the argument from the proof of Lemma 3.3.6 together with the fact that $M_h^* \in N$ then yields that $b \in N$. The desired contradiction then follows immediately. □(Lemma 9.2.8)

As an immediate consequence we have:

Lemma 9.2.9. *Under the assumptions of Lemma* 9.2.8, *the following holds:*

a) If $-1 <_{\bar{T}} i$ *and there is no anomaly* $h \leq_{\bar{T}} i$, *then* $\sigma_i : M_i \to N_i$ *is* Σ^**-preserving.*

b) If there is a truncation point $h \leq_{\bar{T}} i$ *which is not an anomaly, then* σ_i *is* Σ^**-preserving.*

Proof. For a), recall that $\sigma_{-1} : N \to N$ is the identity map, and therefore Σ^*-preserving. For b), recall that if $h+1$ is the first truncation point $\leq_{\bar{T}} i$ which is not an anomaly, then $\sigma_h^* : M_h^* \to N_h^*$ is Σ^*-preserving. In both cases the conclusion follows by the inductive application of the previous lemma and Lemma 3.4.5. □(Lemma 9.1.9)

If $\sigma : M \to N$ is strongly $\Sigma_h^{(n+1)}$-preserving ($h = 0, 1$) in the sense of the definition preceding Lemma 3.4.8 where n satisfies $\omega\varrho_M^{n+1} \leq \lambda < \omega\varrho_M^n$, then the application of that lemma yields

$$\begin{array}{l}\sigma_i : M_i \to N_i \text{ is strongly } \Sigma_h^{(n+1)}\text{-preserving whenever} \\ n \text{ is as above, } 0 \leq_{\bar{T}} i \text{ and there is no truncation point} \\ h \leq_{\bar{T}} i. \text{ Moreover, } \omega\varrho_M^{n+1} \leq \lambda < \omega\varrho_M^n.\end{array} \tag{9.2.5}$$

We stated this fact merely to illustrate that one can obtain better preservation properties of the copying maps; however, we will not need this in the applications.

We now turn to the Dodd–Jensen properties. This ingredient yields the canonical ordering of mice, as we have seen in the linear case. Moreover, the Dodd–Jensen argument is also needed to see that if $\sigma : M \to N$ is sufficiently preserving, then the M-side of the coiteration is simple, which in turn is one of the vital points in the proof of the solidity theorem and condensation lemma. One version of the Dodd–Jensen property is the Dodd–Jensen Lemma, whose proof is based on the fact that $\sigma\mathfrak{I}$ always exist for σ as above and an arbitrary iteration $\mathfrak{I}$ of M. This is true in the linear case; in fact, this is also true of iterations which pick unique branches, but it might fail in the general case. Since we would like to establish fine structural properties of sufficiently iterable premice without any limitations on the iterations which arise in the comparison process, we use a modification of the Dodd–Jensen argument, which yields a weak form of the Dodd–Jensen Lemma sufficient to carry out the comparison arguments.

In order to establish any kind of the Dodd–Jensen Lemma, we need a stronger notion of iterability than that of normal iterability. This was already clear in the linear case, as the proof of the Dodd–Jensen Lemma required the ability to form linear concatenations of iterations. A closer examination of the whole comparison theory would reveal that linear concatenations of length $\omega+1$ of normal iterations would suffice for this purpose. This turns out to generalize to non-linear iterations.

Definition. Let M be a premouse and θ, μ be ordinals. A generalized iteration $\mathfrak{I}$ is a *(θ, μ)-iteration* iff $\mathfrak{I}$ has length θ and it is a linear concatenation of less than μ normal iterations in the following sense: There is a $\bar{\mu} < \mu$ and a normal sequence $\langle \theta_i;\ i \leq \bar{\mu}\rangle$ such that $\theta_0 = 0, \theta_{\bar{\mu}} = \theta$ and for all $i < \bar{\mu}$ we have that $\mathfrak{I}_i = \mathfrak{I} \restriction [\theta_i, \theta_{i+1}]$ is a normal iteration of some initial segment $M^+_{\theta_i}$ of M_{θ_i}. (This involves a slight abuse of notation; to make the last part more formal we have to require $\mathfrak{I}'_i$ to be normal where $\mathfrak{I}'_i$ is of length $\theta_{i+1} + 1$, $\mathfrak{I}'_i \restriction [\theta_i, \theta_{i+1}] = \mathfrak{I}_i$, $h <_{T'_i} \theta_i$ and $M_h^{\mathfrak{I}'_i} = M^+_{\theta_i}$ for the associated iteration tree T'_i and any $h \leq \theta_i$.)

$\mathfrak{I}$ is a $(< \theta, \mu)$-iteration of M iff $\mathfrak{I}$ is a $(\bar{\theta}, \mu)$-iteration of M for some $\bar{\theta} < \theta$.

Given two generalized iterations $\mathfrak{I}_1$ and $\mathfrak{I}_2$ such that the last premouse of $\mathfrak{I}_1$ is the root of $\mathfrak{I}_2$, we define the *concatenation* $\mathfrak{I}_1{}^\frown\mathfrak{I}_2$ in the obvious way, i.e. as the unique iteration $\mathfrak{I}$ satisfying:

- Letting θ_1+1 and θ_2 be the length of $\mathfrak{I}_1$ and $\mathfrak{I}_2$ and T, T_1, T_2 be the corresponding iteration tree, we set $i <_T j$ iff one of the following holds:
  - $i, j < \theta_1 + 1$ and $i <_{T_1} j$;
  - $i <_{T_1} \theta_1$ and $j = \theta_1 + j'$ for some $j' < \theta_2$;
  - $i = \theta_1 + i'$ and $j = \theta_2 + j'$ for some i', j' such that $i' <_{T_2} j'$.
- $\mathfrak{I} \restriction (\theta+1) = \mathfrak{I}_1$ and $\mathfrak{I} \restriction [\theta_1+1, \theta_1+1+\theta_2] = \mathfrak{I}_2$ (this again involves a slight abuse of notation similar to that in the above definition).

Given a sufficiently preserving embedding $\sigma : M \to N$ and a (θ, ν)-iteration $\mathfrak{I}$ of M, the copy $\sigma\mathfrak{I}$ of $\mathfrak{I}$ is obtained as the concatenation of the copies of the normal iterations which determine $\mathfrak{I}$.

Definition. A (θ, μ)-iteration strategy for M is an iteration strategy $\mathbb{S}$ for M such that any $(< \theta, \mu)$-iteration $\mathfrak{I}$ of M can be extended by $\mathbb{S}$. More precisely:

- If $\mathfrak{I}$ is of limit length, then $\mathbb{S}(\mathfrak{I})$ is defined.
- If $\mathfrak{I}$ is of successor length and the last normal component of $\mathfrak{I}$ is $\mathfrak{I}_k$, then $\mathfrak{I}_k$ can be continued as a normal iteration.

A premouse M is (θ, μ)-iterable iff M has a (θ, μ)-iteration strategy.

If σ is as above and $\mathbb{S}$ is a (θ, μ)-iteration strategy for N, the derived iteration strategy $\sigma\mathbb{S}$ for M is defined as in the case of normal iterations, i.e. $\sigma\mathbb{S}(\mathfrak{I}) = \mathbb{S}(\sigma\mathfrak{I})$. As in the case of normal iterations we have: if $\sigma : M \to N$ is Σ^*-preserving, $\mathbb{S}$ is a

(θ, μ)-iteration strategy for N and a (θ, μ)-iteration $\bar{\mathfrak{I}}$ of M is a $\sigma\mathbb{S}$-iteration, then $\sigma\bar{\mathfrak{I}}$ exists and is an $\mathbb{S}$-iteration of N.

We now state the Dodd–Jensen Lemma. However, we can prove it only for premice which are (θ, μ)-uniquely iterable for suitable θ and μ, in other words, which are iterable by the (θ, μ)-uniqueness strategy.

Theorem 9.2.10 (Dodd–Jensen Lemma). *Let θ be a regular cardinal and suppose that M is (θ, ω_1)-uniquely iterable. Let $\mathbb{S}$ be the corresponding iteration strategy. If M' is an $\mathbb{S}$-iterate of M and there is a Σ^*-preserving embedding $\sigma : M \to M'$, then, letting $\mathfrak{I}$ be the corresponding $(< \theta, \omega_1)$-iteration, we have:*

a) *There is no truncation point on the main branch of $\mathfrak{I}$.*

b) *If $\pi : M \to M'$ is the associated iteration map, then $\pi(\xi) \leq \sigma(\xi)$ for all $\xi \in \mathbf{On} \cap M$.*

It follows that M' cannot be both a simple and a non-simple iterate of M and that iteration maps are uniquely determined, i.e. if $\mathfrak{I}'$ is another $(< \theta, \mu)$-iteration of M with the last model M' and $\pi' : M \to M'$ is the associated iteration map, then $\pi' = \pi$.

Proof. The proof is basically a repetition of that of the Dodd–Jensen Lemma for linear iterations. It is based on the fact that if $\sigma : M \to N$ is Σ^*-preserving, $\mathbb{S}$ is the uniqueness strategy for M and N has an iteration strategy $\mathbb{S}'$, then $\mathbb{S} = \sigma\mathbb{S}'$. Indeed, if we know that $\mathbb{S} \restriction \lambda = \sigma\mathbb{S}' \restriction \lambda$ and that $\mathfrak{I}$ is an $\mathbb{S}$-iteration of M of length λ, then $\mathfrak{I}' = \sigma\mathfrak{I}$ exists and is an $\mathbb{S}'$-iteration of N of length λ. Granting that λ is limit, set $b = \mathbb{S}'(\mathfrak{I}')$. Then b is a cofinal well-founded branch in $\mathfrak{I}$, as the direct limit along b in $\mathfrak{I}$ can be embedded into the direct limit along b in $\mathfrak{I}'$. But then $b = \mathbb{S}(\mathfrak{I})$, as $\mathbb{S}$ is the uniqueness strategy. The details of the proof are left to the reader. Notice that the above argument also proves that $\mathbb{S}'$ is the uniqueness strategy for N. □(Theorem 9.2.10)

A straightforward modification of the methods which we will develop in the next section yields solidity for premice of size θ which are uniquely $(\theta + 1, \mu)$-iterable. As a consequence we have that if we coiterate a pair of such premice, at least one side of the coiteration does not involve a truncation on the main branch. This allows us to introduce the canonical ordering of such premice exactly as we did it in the case of linear iterability, i.e. we set

$$M <^* N \longleftrightarrow \text{the } N\text{-side of the coiteration of } M \text{ against } N \text{ is non-simple;}$$

$$M \sim^* N \longleftrightarrow M \text{ and } N \text{ coiterate simply to a common premouse.}$$

Here, "non-simple" means that the iteration has a truncation point on its main branch, while "simple" means the opposite. As before we prove that $<^*$ is a well-founded quasi-ordering on the above class of premice and $\sim^*$ a congruence relation on $<^*$.

As we have already mentioned, for a general proof of the solidity theorem and condensation lemma we need a weaker version of the Dodd–Jensen Lemma, as we are unable to prove the existence of uniqueness strategy for arbitrary iterable premice. However, for countable premice we are able to find iteration strategies which are in some sense minimal and which will suffice for the argument. The following notion of weak iterability then proves to be useful, as it allows to establish all requisite fine structural properties of premice and all premice which arise in K^c-constructions prove to be weakly iterable in this sense.

Definition. A premouse M is *weakly iterable* iff any countable $\bar{M}$ which is elementarily embeddable into M has a $(\omega_1 + 1, \omega_1)$-iteration strategy.

Indeed, solidity is a first-order property of M. Hence, if $M \in H_\theta$ for some large θ and $\sigma : \bar{H} \to H_\theta$ is elementary for some countable $\bar{H}$ and such that $\sigma(\bar{M}) = M$, then $\bar{M}$ has an $(\omega_1 + 1, \omega_1)$-iteration strategy. Now, we can turn this strategy into a minimal strategy which can be used to prove that $\bar{M}$ is solid, so $M = \sigma(\bar{M})$ must be solid as well. A similar argument shows that if M is weakly iterable, then the condensation lemma holds of M, as it can be proved for any countable $\bar{M}$ which is elementarily embeddable into M.

Definition. Let M be a countable premouse and let $e = \langle e_i;\ i \in \omega\rangle$ be an enumeration of M. We say that an (θ, ω_1)-iteration strategy $\mathbb{S}$ for M is *e-minimal* iff for every $(< \theta, \omega_1)$-iteration $\mathfrak{I}$ of M which is an $\mathbb{S}$-iteration we have: If M' is the last model of $\mathfrak{I}$ and $\sigma : M \to M'$ is Σ^*-preserving, then:

a) There is no truncation on the main branch of $\mathfrak{I}$.

b) $\pi(e) \leq_{\text{lex}} \sigma(e)$, that is, $\pi(e_i) < \sigma(e_i)$ where i is the least such that $\pi(e_i) \neq \sigma(e_i)$.

As in the case of the Dodd–Jensen Lemma, it follows that if $\mathbb{S}$ is a minimal iteration strategy for M and M' is an iterate of M, then M' cannot be both a simple and a non-simple iterate of M. Also, iteration maps arising in $\mathbb{S}$-iterations are minimal: If $\mathfrak{I}$ and $\mathfrak{I}'$ are $\mathbb{S}$-iterations of M with the last premouse M' and no truncations on their main branches, then $\pi' = \pi$ where π and π' are the associated iteration maps.

Theorem 9.2.11 (Neeman–Steel Lemma). *If M is a countable premouse which has an (ω_1+1, ω_1)-iteration strategy, then M has an e-minimal (ω_1+1, ω_1)-iteration strategy for any enumeration e of M.*

Proof. Fix an $(\omega_1 + 1, \omega_1)$-iteration strategy $\mathbb{S}$ for M and an enumeration e of M. We find an $\mathbb{S}$-iteration $\mathfrak{I}^*$ of M with the last premouse M^* together with a Σ^*-preserving embedding $\sigma^* : M \to M^*$ such that the following holds:

> Given an iteration $\mathfrak{I}'$ of M^* with the last model M' such that $\mathfrak{I}^{*\frown}\mathfrak{I}'$ is an $\mathbb{S}$-iteration and M is Σ^*-embeddable into M',
>
> a) there is no truncation point on the main branch of $\mathfrak{I}'$;
>
> b) if $\pi' : M^* \to M'$ is the iteration map associated with $\mathfrak{I}'$ and $\sigma : M \to M'$ a Σ^*-preserving embedding, then
>
> $$\pi' \circ \sigma^*(e) \leq_{\text{lex}} \sigma(e).$$

Granting that we have such $\mathfrak{I}^*$, M^* and σ^*, define an iteration strategy $\mathbb{S}^*$ for M by

$$\mathbb{S}^*(\mathfrak{I}) = \text{the branch through } T^{\mathfrak{I}} \text{ which is an end-tail of } \mathbb{S}(\mathfrak{I}^{*\frown}\sigma^*\mathfrak{I}).$$

$\mathbb{S}^*$ is clearly an (ω_1+1, ω_1)-iteration strategy for M. Notice that, by definition, if $\mathfrak{I}$ is $(<\omega_1+1, \omega_1)$-iteration of M which is an $\mathbb{S}^*$-iteration, then $\sigma^*\mathfrak{I}$ exists and $\mathfrak{I}^{*\frown}\sigma^*\mathfrak{I}$ is an $\mathbb{S}$-iteration of M^*. Let us show that $\mathbb{S}^*$ is e-minimal. Given an $\mathbb{S}^*$-iteration $\mathfrak{I}$ of M with the last premouse N and a Σ^*-preserving embedding $\sigma : M \to N$, let N^* be the last premouse of $\sigma^*\mathfrak{I}$. The corresponding copying map $\sigma' : M^* \to N^*$ is obviously Σ^*-preserving, hence so is the embedding $\sigma' \circ \sigma : M \to N$. As $\mathfrak{I}^{*\frown}\sigma^*\mathfrak{I}$ is an $\mathbb{S}$-iteration, a) guarantees that there is no truncation point on the main branch of $\sigma^*\mathfrak{I}$. Consequently, there is no truncation point on the main branch of $\mathfrak{I}$. Thus, both iteration maps $\pi : M \to N$ and $\pi^* : M^* \to N^*$ are total and $\pi^* \circ \sigma^* = \sigma' \circ \pi$. From this identity and b) we obtain

$$\sigma' \circ \pi(e) = \pi^* \circ \sigma^*(e) \leq_{\text{lex}} \sigma' \circ \sigma(e),$$

which immediately yields $\pi(e) \leq_{\text{lex}} \sigma(e)$.

It remains to find $\mathfrak{I}^*$, M^* and σ^*. Given two $(<\omega_1, \omega_1)$-iterations $\mathfrak{I}_1$ and $\mathfrak{I}_2$ of M with last premice N_1 and N_2 respectively, we set

$$\mathfrak{I}_2 \, R \, \mathfrak{I}_1 \quad \longleftrightarrow \quad \begin{array}{l} \mathfrak{I}_2 = \mathfrak{I}_1{}^\frown\mathfrak{I}_{12} \text{ for some iteration } \mathfrak{I}_{12} \text{ of } N_1 \\ \text{with a truncation on its main branch.} \end{array}$$

The relation R is clearly well-founded, as any infinite R-decreasing chain $\langle \mathfrak{I}_k;\ k \in \omega \rangle$ would provide us with the $(<\omega_1, \omega_1)$-iteration $\mathfrak{I}_{12}{}^\frown\mathfrak{I}_{23}{}^\frown\mathfrak{I}_{34}{}^\frown\ldots$ which has infinitely many truncation points on its main branch. Let $\mathfrak{K}$ be the class of all $(<\omega_1, \omega_1)$-iterations $\mathfrak{I}$ of M with the property that M is Σ^*-embeddable into the last model of $\mathfrak{I}$. As R is well-founded, we can pick an R-minimal element $\mathfrak{I}_0$ of $\mathfrak{K}$; let M_0 be its last premouse and $\sigma_0 : M \to M_0$ a Σ^*-preserving embedding. By the R-minimality of $\mathfrak{I}_0$ we have:

> If $\mathfrak{I}'$ is an iteration of M_0 with last premouse M' such that $\mathfrak{I}_0{}^\frown\mathfrak{I}'$ is an $\mathbb{S}$-iteration and M is Σ^*-embeddable into M', then there is no truncation on the main branch of $\mathfrak{I}'$. (9.2.6)

We inductively define a sequence of iterations $\mathfrak{I}_k$ for $k > 0$ such that each $\mathfrak{I}_k$ is an iteration of M_{k-1} with the last premouse M_k. Moreover, $\mathfrak{I}_0{}^\frown\mathfrak{I}_1{}^\frown\ldots{}^\frown\mathfrak{I}_k$ is an $\mathbb{S}$-iteration of M for every k. Denote the iteration map associated with $\mathfrak{I}_k$ by $\pi_{k-1,k}$; then $\pi_{ij} : M_i \to M_j$ is the corresponding composition of these maps. Together with $\mathfrak{I}_k$ we choose a Σ^*-preserving embedding $\sigma_k : M \to M_k$ such that

i) $\pi_{jk} \circ \sigma_j(e_i) = \sigma_k(e_i)$ whenever $i < j \leq k$;

ii) if $\mathfrak{I}'$ is an iteration of M_{k+1} with the last premouse M' such that $\mathfrak{I}_0{}^\frown\mathfrak{I}_1{}^\frown\ldots{}^\frown\mathfrak{I}_{k+1}{}^\frown\mathfrak{I}'$ is an $\mathbb{S}$-iteration and $\sigma' : M \to M'$ is a Σ^*-preserving embedding such that $\pi' \circ \sigma_{k+1}(e_i) = \sigma'(e_i)$ for all $i < k$ where π' is the iteration map associated with $\mathfrak{I}'$, then $\pi' \circ \sigma_{k+1}(e_k) \leq \sigma'(e_k)$.

Assume we have already constructed $\mathfrak{I}_1, \ldots, \mathfrak{I}_k$ and $\sigma_1, \ldots, \sigma_k$. Let $\mathfrak{K}_k$ be the set of all pairs $\langle \mathfrak{I}', \sigma' \rangle$ such that

- $\mathfrak{I}'$ is an iteration of M_k such that $\mathfrak{I}_0 \frown \mathfrak{I}_1 \frown \ldots \frown \mathfrak{I}_k \frown \mathfrak{I}'$ is an $\mathbb{S}$-iteration;
- M is Σ^*-embeddable into the last premouse M' of $\mathfrak{I}'$;
- $\sigma' : M \to M'$ is a Σ^*-preserving embedding such that $\pi' \circ \sigma_k(e_i) = \sigma'(e_i)$ for all $i < k$ where π' is the iteration map associated with $\mathfrak{I}'$.

Notice that π' as above is always total; this follows from (9.2.6) where we substitute $\mathfrak{I}_0 \frown \mathfrak{I}_1 \frown \ldots \frown \mathfrak{I}_k \frown \mathfrak{I}'$ for $\mathfrak{I}'$. Furthermore, $\mathfrak{K}_k$ is nonempty, as $\langle \mathfrak{I}_k^{\mathrm{triv}}, \sigma_k \rangle$ is certainly in $\mathfrak{K}_k$, where $\mathfrak{I}_k^{\mathrm{triv}}$ is the iteration of M_k of length 0. On $\mathfrak{K}_k$ we define a relation R_k by

$$\langle \mathfrak{I}'', \sigma'' \rangle \, R_k \langle \mathfrak{I}', \sigma' \rangle \longleftrightarrow \begin{array}{l} - \ \mathfrak{I}'' = \mathfrak{I}' \frown \mathfrak{I}^+ \text{ for some iteration } \mathfrak{I}^+ \text{ of } M'; \\ - \ \sigma''(e_k) < \pi^+ \circ \sigma'(e_k) \text{ where } \pi^+ : M' \to M'' \\ \quad \text{is the iteration map associated with } \mathfrak{I}^+. \end{array}$$

The relation R_k is well-founded, as any infinite R_k-descending chain

$$\langle \mathfrak{I}^{(1)}, \sigma(1) \rangle, \langle \mathfrak{I}^{(2)}, \sigma^{(2)} \rangle, \langle \mathfrak{I}^{(3)}, \sigma^{(3)} \rangle \ldots$$

would generate a $(< \omega_1, \omega_1)$-iteration $\mathfrak{I}^{(12)} \frown \mathfrak{I}^{(23)} \frown \ldots$ with ill-founded direct limit. The situation is depicted in the following diagram, where each $\pi^{(i,i+1)}$ is the iteration map associated with $\mathfrak{I}^{(i,i+1)}$.

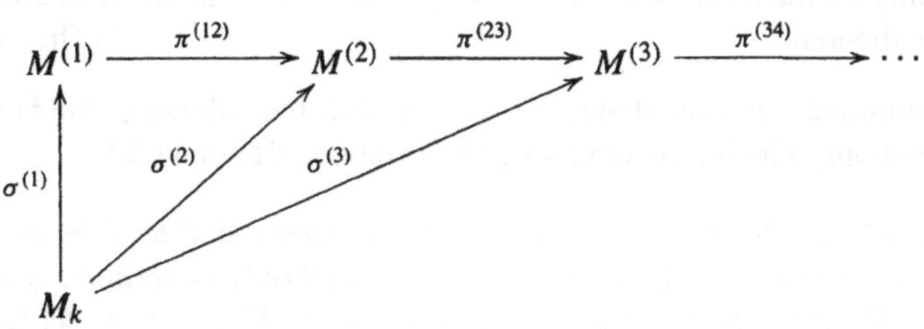

The threads

$$t_i = \langle \pi^{(ij)} \circ \sigma^{(i)}(e_k);\ i \le j \rangle$$

would constitute an infinite descending sequence in the above direct limit, as follows from the fact that $\sigma^{(i+1)}(e_k) < \pi^{(i,i+1)} \circ \sigma^{(i)}(e_k)$.

Having verified the well-foundedness of R_k, we pick an R_k-minimal element $\langle \mathfrak{I}_{k+1}, \sigma_{k+1} \rangle$ of $\mathfrak{K}_k$, set M_{k+1} to be the last premouse of $\mathfrak{I}_{k+1}$ and $\pi_{k,k+1}$ the associated iteration map. Let us verify i) and ii) for $k+1$. To see i), notice that

$$\pi_{j,k+1} \circ \sigma_j(e_i) = \pi_{k,k+1} \circ \pi_{jk} \circ \sigma_j(e_i) = \pi_{k,k+1} \circ \sigma_k(e_i) = \sigma_{k+1}(e_i)$$

where the second equality follows by the induction hypotheses and the last one by the fact that $\langle \mathfrak{I}_{k+1}, \sigma_{k+1} \rangle \in \mathfrak{K}_k$. Clause ii) follows by the R_k-minimality of $\langle \mathfrak{I}_{k+1}, \sigma_{k+1} \rangle$: If $\mathfrak{I}'$, M', π' and σ' are as in ii), then $\langle \mathfrak{I}_{k+1} {}^\frown \mathfrak{I}', \sigma' \rangle$ is in $\mathfrak{K}_k$, so necessarily $\pi' \circ \sigma_{k+1}(e_k) \leq \sigma'(e_k)$.

Let $\langle M^*, \pi_k;\ k \in \omega \rangle$ be the direct limit of the diagram $\langle M_k, \pi_{jk};\ j \leq k < \omega \rangle$. Furthermore, let $\mathfrak{I}^*$ be the concatenation of all iterations $\mathfrak{I}_k$ extended by the above direct limit; so the last premouse of $\mathfrak{I}^*$ is M^*. Similarly, let $\mathfrak{I}^*_k$ be the concatenation of all iterations $\mathfrak{I}_j$ for $j > k$ extended by the direct limit; clearly $\mathfrak{I}_0 {}^\frown \mathfrak{I}_1 {}^\frown \ldots {}^\frown \mathfrak{I}_k {}^\frown \mathfrak{I}^*_k = \mathfrak{I}^*$. Define the embedding $\sigma^* : M \to M^*$ by

$$\sigma^*(e_i) = \pi_k \circ \sigma_k(e_i) \quad \text{where } k > i.$$

Then σ^* is well-defined, as follows from i). We verify that $\mathfrak{I}^*$, M^* and σ^* satisfy a) and b). Pick any iteration $\mathfrak{I}'$ of M^* such that $\mathfrak{I}^* {}^\frown \mathfrak{I}'$ is an $\mathbb{S}$-iteration. Let M' be the last premouse of $\mathfrak{I}'$. Suppose M is Σ^*-embeddable into M'. Setting $\mathfrak{I}'_0 = \mathfrak{I}^*_0 {}^\frown \mathfrak{I}'$, we have that $\mathfrak{I}_0 {}^\frown \mathfrak{I}'_0$ is an $\mathbb{S}$-iteration of M_0 with the last premouse M', so (9.2.6) guarantees that there is no truncation point on the main branch of $\mathfrak{I}'_0$. Consequently, there is no truncation on the main branch of $\mathfrak{I}'$.

Towards the verification of b), denote the iteration map associated with $\mathfrak{I}'$ by π'. Assuming that $\pi' \circ \sigma(e_k) \neq \sigma'(e_k)$ for some k, pick k to be minimal with this property. Notice that $\mathfrak{I}'_{k+1} = \mathfrak{I}^*_{k+1} {}^\frown \mathfrak{I}'$ is an iteration of M_{k+1} such that $\mathfrak{I}_0 {}^\frown \mathfrak{I}_1 {}^\frown \ldots {}^\frown \mathfrak{I}_{k+1} {}^\frown \mathfrak{I}'_{k+1}$ is an $\mathbb{S}$-iteration and $\pi' \circ \pi_{k+1}$ is the associated iteration map. Moreover, for each $i < k$ we have $(\pi' \circ \pi_{k+1}) \circ \sigma_{k+1}(e_i) = \pi' \circ \sigma^*(e_i) = \sigma'(e_i)$, so

$$\pi' \circ \sigma^*(e_k) = (\pi' \circ \pi_{k+1}) \circ \sigma_{k+1}(e_k) \leq \sigma'(e_k).$$

The inequality on the right side follows by ii). This verifies b) and thus completes the proof of the theorem. □(Theorem 9.2.11)

Using minimal iteration strategies, we establish the following fundamental fact about coiterations, which is an obvious generalization of Lemma 5.1.6.

Lemma 9.2.12. *Let N be a countable presolid premouse and let $\mathbb{S}$ be an e-minimal $(\omega_1 + 1, \omega_1)$-iteration strategy for N where e is some fixed enumeration of N. Given a phalanx $\langle N, M, \lambda \rangle$ together with an embedding σ of $\langle N, M, \lambda \rangle$ into N, let $\langle \bar{\mathfrak{I}}, \mathfrak{I} \rangle$ be the coiteration of $\langle N, M, \lambda \rangle$ with N which uses the iteration strategy $\mathbb{S}$ on the N-side and $\sigma\mathbb{S}$ on the phalanx side. By Lemma 9.1.8, this coiteration terminates after countably many steps. Let N_θ and M_θ be the last premouse on the N-side and the phalanx side, respectively. Then:*

a) *$0 \leq_{\bar{T}} \theta$ where $\bar{T} = T^{\bar{\mathfrak{I}}}$.*

b) *M_θ is an initial segment of N_θ.*

c) *There is no truncation point on the main branch of $\bar{T}$.*

Proof. We first fix the notation. Recall that $\mathfrak{I}$ and $\bar{\mathfrak{I}}$ is the N-side and the phalanx side of the coiteration, respectively. Consistently with the statement of the lemma, we denote the premice on the N-side of the coiteration by N_i and those on the phalanx side by M_i. Furthermore, let N_i' be the premice of $\sigma\bar{\mathfrak{I}}$ and let $\sigma_i : M_i \to N_i'$ be the copying maps. Finally, let π_{ij}, $\bar{\pi}_{ij}$ and π_{ij}' be the iteration maps of $\mathfrak{I}$, $\bar{\mathfrak{I}}$ and $\sigma\bar{\mathfrak{I}}$, respectively.

We first verify a). Suppose for a contradiction that $-1 \leq_{\bar{T}} \theta$. We shall discuss several cases. If N_θ is a proper initial segment of M_θ, then there is no truncation on the main branch of $\mathfrak{I}$, as N_θ is sound (see the remarks above Lemma 9.1.13). Thus, $\pi_{0,\theta}$ is total. Clearly, $\sigma_\theta \restriction N_\theta : N_\theta \to \sigma(N_\theta)$ is fully elementary, so $\sigma_\theta \circ \pi_{0,\theta} : N \to \sigma_\theta(N_\theta)$ is Σ^*-preserving. But $\sigma(N_\theta)$, being a proper initial segment of N_θ', is a non-simple $\mathbb{S}$-iterate of N, which contradicts the e-minimality of $\mathbb{S}$ by the previous theorem. Notice that this argument also goes through in the case where $0 \leq_{\bar{T}} \theta$, which proves b).

From now on we assume that M_θ is an initial segment of N_θ. If there is a non-anomalous truncation point $i+1 <_{\bar{T}} \theta$, then Lemma 9.1.14 guarantees that there is no truncation on the main branch of $\mathfrak{I}$, so $\pi_{0,\theta} : N \to N_\theta$ is total. Moreover, N_θ is an initial segment of M_θ, as M_θ fails to be sound (again, see the remarks above Lemma 9.1.13), so $N_\theta = M_\theta$. By Lemma 9.2.9, $\sigma_\theta : M_\theta \to N_\theta'$ is Σ^*-preserving, so the same holds of $\sigma_\theta \circ \pi_{0,\theta} : N \to N_\theta'$. But N_θ' is a non-simple $\mathbb{S}$-iterate of N, as the copying construction guarantees that $i+1$ is a truncation point of $\sigma\bar{\mathfrak{I}}$. This yields a contradiction to the e-minimality of $\mathbb{S}$ as before.

Next assume that the only truncation point $i+1 \leq_{\bar{T}} \theta$ is anomalous. As above we infer that $\pi_{0,\theta} : N \to N_\theta$ is total and that $N_\theta = M_\theta$. Let n be the least such that $\omega\varrho^{n+1}_{M_i^*} = \kappa_i$. By Corollary 9.1.12, κ_i is both the $(n+1)$-st and the ultimate projectum of each premouse M_j on the main branch of $\bar{\mathfrak{I}}$ such that $i+1 \leq_{\bar{T}} j$. This in particular applies to $M_\theta = N_\theta$. It follows that $\operatorname{cr}(\pi_{0,\theta}) \geq \kappa_i$, as otherwise $\omega\varrho^{n+1}_{N_\theta}$ would be larger than κ_i. Thus, κ_i is both the $(n+1)$-st and the ultimate projectum of N and of all premice on the main branch of $\mathfrak{I}$. Let A be a $\mathbf{\Sigma}_1^{(n)}(N)$-relation such that $A \cap \kappa_i \notin N$. An iterative application of Lemma 3.3.1 yields that there is a $\mathbf{\Sigma}_1^{(n)}(M_{i+1})$-relation A' such that $A' \cap \kappa_i = A \cap \kappa_i$. By the argument from the proof of Corollary 9.1.12, $A' \cap \kappa_i \in N$, which contradicts our choice of A.

It remains to discuss the case where there is no truncation point on the main branch of $\bar{\mathfrak{I}}$. Notice that neither can M_θ be a proper initial segment of N_θ, nor can there be a truncation on the main branch of $\mathfrak{I}$: Otherwise M_θ would be a non-simple iterate of N, and at the same time, $\bar{\pi}_{-1,\theta} : N \to M_\theta$ would be a Σ^*-preserving embedding, which would again contradict the previous theorem. It follows that $M_\theta = N_\theta$ and that $\pi_{0,\theta}$ is total. As $\pi_{0,\theta}$ is an $\mathbb{S}$-iteration map,

$$\pi_{0,\theta}(e) \leq_{\text{lex}} \bar{\pi}_{-1,\theta}(e). \tag{9.2.7}$$

By a) of Lemma 9.2.9, the map $\sigma_\theta : M_\theta \to N_\theta'$ is Σ^*-preserving, so the composition $\sigma_\theta \circ \pi_{0,\theta} : N \to N_\theta'$ is Σ^*-preserving as well. As $\sigma\bar{\mathfrak{I}}$ is an $\mathbb{S}$-iteration with the last premouse N_θ', there is no truncation on the main branch of $\sigma\bar{\mathfrak{I}}$. Hence $\pi_{0,\theta}'$ is total and

$$\sigma_\theta \circ \bar{\pi}_{-1,\theta}(e) = \pi_{0,\theta}' \circ \operatorname{id}(e) = \pi_{0,\theta}'(e) \leq_{\text{lex}} \sigma_\theta \circ \pi_{0,\theta}(e),$$

where the first equality follows from the commutativity of the copying diagram and the inequality from the Neeman–Steel Lemma. By eliminating σ_θ from the resulting inequality we obtain $\bar{\pi}_{-1,\theta}(e) \leq_{\text{lex}} \pi_{0,\theta}(e)$, which together with (9.2.7) yields $\bar{\pi}_{-1,\theta} = \pi_{0,\theta}$. From this, using the standard computation and the fact that the extenders applied along branches do not overlap, we infer that $E^{M_i}_{\nu_i}$ is compatible with $E^{N_j}_{\nu_j}$, where $E^{M_i}_{\nu_i}$ and $E^{N_j}_{\nu_j}$ is the first extender applied on the main branch of $\bar{\Im}$ and $\Im$, respectively. By (9.1.7), this means that i is a strong anomaly and $j = 0$. Let h and k be such that $i = h + 1$, $k + 1 <_T \theta$ and $1 = T(k + 1)$. We can now proceed as in the proof of Lemma 9.1.14 (see last paragraph of this proof) and obtain a contradiction to (9.1.7) by showing that $E^{M_h}_{\nu_h}$ and $E^{N_k}_{\nu_k}$ are compatible. This completes the proof of a) of the lemma.

As b) was proved at the beginning of the proof of a), we can proceed directly to the proof of c), which is very much like that of b). If $i + 1$ is a truncation point on the main branch of $\bar{\Im}$, then it is also a truncation point on the main branch of $\sigma\bar{\Im}$, so N'_θ is a non-simple iterate of N. We show that N is Σ^*-embeddable into N'_θ, and thus obtain a contradiction to the Neeman–Steel Lemma. We already know that M_θ is an initial segment of N_θ. Now M_θ cannot be a proper initial segment of N_θ, as it fails to be sound because of the truncation on the main branch of $\bar{\Im}$. Hence $M_\theta = N_\theta$. By Lemma 9.1.14, there is no truncation point on the main branch of $\Im$, so $\pi_{0,\theta}$ is total. Also, σ_θ is Σ^*-preserving by Lemma 9.2.9, so $\sigma_\theta \circ \pi_{0,\theta} : N \to N'_\theta$ is Σ^*-preserving as well. This completes the proof of the lemma. □(Lemma 9.2.12)

Corollary 9.2.13. *Assume that M and N are as in the previous lemma. Then M has an $(\omega_1 + 1, \omega_1)$-iteration strategy.*

Proof. We will stick to the notation of the previous lemma. By the conclusions of that lemma, $\bar{\pi}_{0,\theta} : M \to M_\theta$ is total, and therefore Σ^*-preserving. Also, M_θ is an $\mathbb{S}$-iterate of N, as is witnessed by $\Im$ followed by an additional truncation of N_θ, if necessary. Letting $\Im^*$ be this iteration from N to M_θ, we have that M_θ has an $(\omega_1 + 1, \omega_1)$-iteration strategy $\mathbb{S}'$ assigning to each iteration $\Im'$ of M_θ the branch b determined by $\mathbb{S}(\Im^{*\frown}\Im')$. Then $\bar{\pi}_{0,1}\mathbb{S}'$ is an $(\omega_1 + 1, \omega_1)$-iteration strategy for M. □(Corollary 9.2.13)

We are now able to establish the following generalization of Lemma 5.1.7, which yields, among other things, the universality of good parameters for weakly iterable presolid premice. Let us first state a fact about coiterations, which is used in the proof of the next lemma and which will be also useful elsewhere.

In the coiteration of N with M or a phalanx $\langle N, M, \lambda\rangle$, if $E^{N_i}_{\nu_i} \neq \varnothing$, then ν_0 is a cardinal in N_{i+1} and thus in N_j for any $j > i$. (9.2.8)

This is clear if $E^{N_0}_{\nu_0} \neq \varnothing$. Otherwise $E^{M_0}_{\nu_0} \neq \varnothing$, so ν_0 is a cardinal in M_i whenever $i > 0$. Thus, ν_0 is a cardinal in $N_{i+1}||\nu_i = M_{i+1}||\nu_i$. Now if $E^{N_i}_{\nu_i} \neq \varnothing$, then ν_i is a cardinal in N_{i+1}, hence by acceptability, ν_0 is a cardinal in N_{i+1} as well.

Lemma 9.2.14. *Let M be a countable presolid premouse and let $\mathbb{S}$ be an e-minimal (ω_1+1, ω_1)-iteration strategy where e is a fixed enumeration of M. Given a parameter $p \in P_M^{n+1}$, let $\bar{M}$ be the transitive collapse of the hull $\tilde{h}_M^{n+1}(\omega\varrho_M^{n+1} \cup \{p\})$ and let $\sigma : \bar{M} \to M$ be the associated uncollapsing map. Then σ is an embedding of the phalanx $\langle M, \bar{M}, \omega\varrho_M^{n+1}\rangle$ into M, the coiteration of M with the phalanx $\langle M, \bar{M}, \omega\varrho_M^{n+1}\rangle$ terminates after $\theta < \omega_1$ many steps and, letting $\mathfrak{I}$ and $\bar{\mathfrak{I}}$ be the iteration on the M-side and the phalanx side respectively, we have:*

a) $0 \leq_{\bar{T}} \theta$;

b) $\bar{M}_\theta = M_\theta$;

c) *neither of $\mathfrak{I}$ and $\bar{\mathfrak{I}}$ involves a truncation on its main branch;*

d) *$\bar{M}$ and M have the same $(n+1)$-st projectum $\omega\varrho^{n+1}$ and the critical points of the extenders applied to the models on the main branches of $\mathfrak{I}$ and $\bar{\mathfrak{I}}$ are at least $\omega\varrho^{n+1}$.*

Here, T is the iteration tree and M_θ the last premouse of $\mathfrak{I}$, and $\bar{T}$ and $\bar{M}_\theta$ denote the corresponding objects for $\bar{\mathfrak{I}}$.

Moreover, σ is Σ^*-preserving and $\bar{M}$ is (ω_1+1, ω_1)-iterable.

As an immediate consequence of the above we obtain c) – i) *of Lemma* 5.1.7. *In particular, p is universal and can be lengthened to a parameter $p' \in P_M^*$.*

Proof. By construction and the proof of Lemma 1.11.2, $\omega\varrho_{\bar{M}}^{n+1} = \omega\varrho_M^{n+1}$; denote this ordinal by $\omega\varrho^{n+1}$. The same lemma also yields that σ is Σ^*-preserving. Lemma 9.2.12 guarantees that a) holds, $\bar{M}_\theta$ is an initial segment of M_θ and that there is no truncation on the main branch of the phalanx side. By a), the fact that the critical points of the ultrapowers along the main branch of the phalanx side are above $\omega\varrho^{n+1}$, Lemma 9.1.11 and Lemma 3.3.1, all premice on the main branch of the phalanx side have the same $(n+1)$-st projectum $\omega\varrho^{n+1}$, and also the same collection of $\mathbf{\Sigma}_1^{(n)}$-definable subsets of $\omega\varrho^{n+1}$. Let A be $\Sigma_1^{(n)}(M)$ in p such that $a = A \cap \omega\varrho^{n+1} \notin M$ and let $\bar{A}$ be $\Sigma_1^{(n)}(\bar{M})$ in $\bar{p}$ by the same definition, where $\sigma(\bar{p}) = p$. The map σ is $\Sigma_1^{(n)}$-preserving, so $\bar{A} \cap \omega\varrho^{n+1} = a$. By the above mentioned properties of coiterations, there is an $A' \in \mathbf{\Sigma}_1^{(n)}(\bar{M}_\theta)$ such that $A' \cap \omega\varrho^{n+1} = \bar{A} \cap \omega\varrho^{n+1} = a$.

Now if $\bar{M}_\theta$ were a proper initial segment of M_θ, we would have $a \in M_\theta$. If $M_\theta = M$, this would yield an immediate contradiction to the choice of A. If $M_\theta \neq M$, then $E_{\nu_i}^{M_i} \neq \varnothing$ for some i, so ν_0 is a cardinal in M_θ by (9.2.8). But $\nu_0 \geq \omega\varrho^{n+1}$, as $\bar{M}$ agrees with M below this point. Moreover, $\nu_0 > \omega\varrho^{n+1}$, as $\omega\varrho^{n+1}$ is a cardinal in both $\bar{M}$ and M, while ν_0 fails to be, as it indexes an extender in at least one of them (here, we assume w.l.o.g. that $\nu_0 < \mathrm{ht}(\bar{M})$). Thus a, being a bounded subset of ν_0, must be an element of $M_\theta || \nu_0$. Consequently, $a \in M$, as M_θ agrees with M below ν_0. This yields a contradiction again, and thus completes the proof of b).

The proof of c) is of the same spirit. Assuming that there is a truncation on the main branch of $\mathfrak{I}$, let $i+1$ be the last one and $k = T(i+1)$. Recall that $\nu_i > \omega\varrho^{n+1}$

for all i, so $\lambda_i \geq \omega\varrho^{n+1}$. If $\omega\varrho_{M_i^*}^{n+1} > \kappa_i$, then $\omega\varrho_{M_\theta}^{n+1} > \pi_{k,\theta}(\kappa_i) = \lambda_i \geq \omega\varrho^{n+1}$, a contradiction. Thus, $\omega\varrho_{M_i^*}^{n+1} \leq \kappa_i$. By Lemmata 9.1.10 and 3.3.1, all premice M_j on the main branch of $\mathfrak{I}$ with $i+1 \leq_T j$ have the same $(n+1)$-st projectum, namely $\omega\varrho_{M_i^*}^{n+1}$. This yields $\omega\varrho_{M_i^*}^{n+1} = \omega\varrho_{M_\theta}^{n+1} = \omega\varrho^{n+1}$ and, consequently, $\omega\varrho^{n+1} \leq \kappa_i$. By the same lemmata, $a = A'' \cap \omega\varrho^{n+1}$ for some $A'' \in \Sigma_1^{(n)}(M_i^*)$. It follows that $a \in M_k$. We show that a is actually an element of M, which contradicts the choice of A. If $M_k = M$, this is immediate. If $M_k \neq M$, then there is an $h < k$ such that $E_{\nu_h}^{M_h} \neq \varnothing$, so (9.2.8) guarantees that ν_0 is a cardinal in M_k. By acceptability, $a \in M_k||\nu_0 = M||\nu_0$. This completes the proof of c).

Notice also that, knowing c), the first part of the proof of c) also proves d). The rest of the lemma is trivial or follows by the arguments in the proof of Lemma 5.1.7.

□(Lemma 9.2.14)

The next corollary follows by considering countable elementary substructures.

Corollary 9.2.15. *Let M be weakly iterable presolid premouse and let n, p, $\bar{M}$ and σ be as in the previous lemma. Then $\bar{M}$ and M have the same $(n+1)$-st projectum, $\bar{M}$ is weakly iterable and clauses* c) – i) *of Lemma* 5.1.7 *hold. In particular, p is universal and can be lengthened to a parameter $p' \in P_M^*$ and σ is Σ^*-preserving.* □(Corollary 9.2.15)

Recall that, by the previous lemma and corollary, $p_{M,n} = p_M - \omega\varrho_M^n$ and core(M) exists and is sound whenever M is a weakly iterable premouse.

9.3 Solidity and Condensation

The technical means for comparing sufficiently iterable premice which we developed in the previous section enables us to prove the solidity theorem and condensation lemma. The proofs of these theorems are of the same spirit as those we presented for linear iterations; however, the present level of generality makes them technically more demanding. In particular, new kinds of difficulties arise once we deal with superstrong extenders. One way of dealing with this problem is to work with k-iterations, which allows us to avoid such difficulties (see [Je97]). However, this makes the whole theory even less transparent, so we instead stick to $*$-iterations and employ extra arguments to take care of the above mentioned pathologies.

Theorem 9.3.1 (Solidity Theorem). *Every weakly iterable premouse is solid.*

Proof. Fix a weakly iterable premouse M. Our intention is to prove that M is solid. We shall closely follow the proof of Theorem 5.2.1. In order to make the argument go through, we impose the following restrictions on M:

A) M is countable and (ω_1+1, ω_1)-iterable.

B) If $\bar{M}$ is a countable $(\omega_1 + 1, \omega_1)$-iterable premouse such that either

- $\mathrm{ht}(\bar{M}) < \mathrm{ht}(M)$ or else
- $\mathrm{ht}(\bar{M}) = \mathrm{ht}(M)$ and $p_{\bar{M}} <^* p_M$,

then $\bar{M}$ is solid.

C) M is sound.

Let us verify that these restrictions will not affect the generality of the result. First of all, by a Löwenheim–Skolem argument, we can construct an elementary embedding $\sigma : \bar{M} \to M$ where $\bar{M}$ is countable. As M is weakly iterable, $\bar{M}$ is (ω_1+1, ω_1)-iterable. Now it suffices to show that $\bar{M}$ is solid, as solidity, being an internal first order property of M, is preserved under σ. This justifies A). The second restriction can be considered an induction hypothesis, where the induction is on the height of countable premice and the well-ordering $<^*$ of standard parameters. Regarding C), any M satisfying A) and B) satisfies the assumptions of Corollary 9.2.15, so by the remarks after this corollary, core(M) exists and is sound. Now the associated core map is Σ^*-preserving, maps the standard parameter of core(M) onto the standard parameter of M and preserves generalized witnesses, so the solidity of core(M) readily yields the solidity of M.

Fix an enumeration $e = \langle e_i;\ i \in \omega \rangle$ of M and an e-minimal $(\omega_1 + 1, \omega_1)$-iteration strategy $\mathbb{S}$ for M. Such a strategy exists by the Neeman–Steel Lemma. Let $\nu \in p_M$ and suppose we have already proved that $W_M^{\nu'} \in M$ for all $\nu' \in p_M$ which are larger than ν. Our task now is to show that $W = W_M^{\nu} \in M$ as well. Let n be such that $\omega\varrho_M^{n+1} \leq \nu < \omega\varrho_M^n$. We now list some properties of the canonical witness map $\sigma : W \to M$, which follow exactly as in the proof of Theorem 5.2.1:

- $\sigma \upharpoonright \nu = \mathrm{id}$ and $\sigma(\nu) > \nu$. Thus, ν is a regular cardinal in W.
- σ is $\Sigma_1^{(n)}$-preserving.
- $q \in \mathrm{rng}(\sigma)$ where $q = p_M - (\nu + 1)$. We set $\bar{q} = \sigma^{-1}(q)$.

We next verify that W is a premouse. This follows immediately from a) of Lemma 9.1.7 if M is passive or active of type A or B, as σ is Σ_1-preserving. Now suppose M is active of type C. Then $\omega\varrho_M^1 = \lambda_M$ and $p_M - \lambda_M = \varnothing$, so $n > 0$. This means that σ is in fact $\Sigma_1^{(1)}$-preserving, so the conclusion then follows from b) of the same lemma.

It follows easily that $\langle M, W, \nu \rangle$ is a phalanx and that σ is an embedding of this phalanx into M. Let $\langle \mathfrak{I}, \bar{\mathfrak{I}} \rangle$ be the coiteration of M against $\langle M, W, \nu \rangle$, where $\mathfrak{I}$ is an $\mathbb{S}$-iteration and $\bar{\mathfrak{I}}$ an $\sigma\mathbb{S}$-iteration. This coiteration terminates at some countable θ. Denote the iteration tree, premice and the iteration maps associated with $\bar{\mathfrak{I}}$ by $\bar{T}$, W_h and $\bar{\pi}_{hk}$, and those associated with $\mathfrak{I}$ by T, M_h and π_{hk}, respectively. By Lemma 9.2.12 we have:

- $0 \leq_{\bar{T}} \theta$.
- W_θ is an initial segment of M_θ.

- There is no truncation on the main branch of $\bar{\Im}$.

As in the proof of Theorem 5.2.1, $\omega\varrho_W^{n+1} \leq \nu$ and W is $\mathbf{\Sigma}_1^{(n)}$-encodable into a subset a of ν. Our task is to show that $a \in M$, as this yields $W \in M$.

For the rest of the proof, let i be such that M_{i+1} is the result of the first ultrapower on the main branch of $\Im$. More formally, i is the unique ordinal satisfying

$$i + 1 \leq_T \theta \quad \& \quad h \notin B^{\Im} \text{ whenever } h <_T i + 1. \tag{9.3.1}$$

Clearly, $M_h = M$ for all $h <_T i+1$. Furthermore, let $j+1$ be the last truncation point on the main branch of $\Im$, if there is any. We define

$$\xi^* \simeq \begin{cases} \theta \text{ if } W_\theta \in M_\theta, \text{ in which case we set } M^*_{\xi^*} = W_\theta; \\ \xi_j \text{ if } W_\theta = M_\theta \text{ and } j \text{ is defined, and set } M^*_{\xi^*} = M^*_j. \end{cases} \tag{9.3.2}$$

As W agrees with M below ν, each iteration index ν_h is larger than or equal to ν. Moreover, $\nu_h = \nu$ can only happen if $h = 0$, in which case $E_{\nu_h}^{M_h} = E_\nu^M$ is a superstrong extender in M. To see this, notice that if $\lambda = \lambda(E_\nu^M)$, then from the agreement between W and M and the fact that ν is a cardinal in W we infer that λ is a cardinal in W, so the preservation properties of σ guarantee that $\lambda = \sigma(\lambda)$ is a cardinal in M. From now on, we reserve the symbols κ, τ and λ to denote the relevant ordinals for such E_ν^M, that is:

$$\kappa = \operatorname{cr}(E_\nu^M), \quad \lambda = \lambda(E_\nu^M) \quad \text{and} \quad \tau = \kappa^{+M}. \tag{9.3.3}$$

The crucial point of the proof is the observation that ξ^* is always defined. This observation is based on the fact that $\nu \in p_M$. The intuition behind (9.3.4) is that removing an element from the standard parameter causes an essential loss of information. Let us first state the obvious reformulation of the fact that ξ^* is defined:

$$\begin{array}{l}\text{Either there is a truncation on the main branch of } \Im \\ \text{or else } W_\theta \text{ is a } \textit{proper} \text{ initial segment of } M_\theta.\end{array} \tag{9.3.4}$$

Before proceeding further with the proof, we derive a useful consequence of the failure of (9.3.4).

Claim 1. Suppose that (9.3.4) fails, so there are no truncations along the main branch on either side of the coiteration and $W_\theta = M_\theta$. Then the following equalities hold (compare with the equalities which arise in the proof of Theorem 5.2.1):

- $\omega\varrho_W^{n+1} = \omega\varrho_{M_\theta}^{n+1} = \omega\varrho_{M_{i+1}}^{n+1} \stackrel{\text{def}}{=} \omega\varrho^{n+1}$;
- $\mathcal{P}(\omega\varrho^{n+1}) \cap W = \mathcal{P}(\omega\varrho^{n+1}) \cap M_\theta = \mathcal{P}(\omega\varrho^{n+1}) \cap M_{i+1}$;
- $\mathcal{P}(\omega\varrho^{n+1}) \cap \mathbf{\Sigma}_1^{(n)}(W) = \mathcal{P}(\omega\varrho^{n+1}) \cap \mathbf{\Sigma}_1^{(n)}(M_\theta) = \mathcal{P}(\omega\varrho^{n+1}) \cap \mathbf{\Sigma}_1^{(n)}(M_{i+1})$.

Proof. The equalities on the left hand side follow from the fact that there are no truncations along the main branch of $\bar{\mathfrak{I}}$ and that the iteration is above $\nu \geq \omega\varrho_W^{n+1}$ on this branch. By the failure of (9.3.4), the equalities on the right hand side will follow once we have shown that the branch $[i+1, \theta)_T$ is above $\omega\varrho_{M_{i+1}}^{n+1}$. But this is certainly true, as either $M_{i+1} = M_\theta$, in which case there is nothing to prove, or else $i+1 = T(k+1)$ for some k, which clearly must be larger than i. Thus, $\omega\varrho_{M_{i+1}}^{n+1}$ cannot be larger than κ_k, because this would mean that $\omega\varrho_{M_\theta}^{n+1} > \lambda_k > \nu_i \geq \nu$, which contradicts the above computations. □(Claim 1)

Claim 2. The ordinal ξ^* is always defined, i.e. (9.3.4) holds.

Proof. Towards a contradiction, suppose (9.3.4) holds. We shall consider two cases. If $\nu_i > \nu$ or else $\nu_i = \nu$ and $\omega\varrho_M^{n+1} > \tau$, we infer that $\kappa_i \geq \omega\varrho_M^{n+1}$. This will guarantee that M and M_{i+1} have the same $(n+1)$-st projectum $\omega\varrho^{n+1}$ and the same power sets of $\omega\varrho^{n+1}$, so from the above equalities we obtain:

$$\omega\varrho_W^{n+1} = \omega\varrho_M^{n+1} = \omega\varrho^{n+1} \tag{9.3.5}$$

$$\mathcal{P}(\omega\varrho^{n+1}) \cap W = \mathcal{P}(\omega\varrho^{n+1}) \cap M \tag{9.3.6}$$

The rest of the proof of (9.3.4) in this case is a literal repetition of that of Claim 2 in the proof of Theorem 5.2.1. We merely mention that it uses (9.3.5) and (9.3.6) to find a set $b \subset \omega\varrho_M^{n+1}$ and an $r \subset \nu$ such that b is $\Sigma_1^{(n)}(M)$ in $q \cup r$, but not an element of M. However, the existence of such a set contradicts the fact that $q \cup r <^* p_M - \omega\varrho_M^{n+1} = p_{M,n+1}$ (see the end of the previous section).

To see the inequality $\omega\varrho_M^{n+1} \leq \kappa_i$, assume the contrary, i.e. $\kappa_i < \omega\varrho_M^{n+1}$. Then $\tau_i \leq \omega\varrho_M^{n+1}$. By applying the iteration map along the main branch of $\mathfrak{I}$ we obtain $\omega\varrho_{M_\theta}^{n+1} \geq \pi_{0,\theta}(\tau_i) \geq \nu_i$. If $\nu_i > \omega\varrho_W^{n+1}$, this yields an immediate contradiction. Otherwise we still have $\omega\varrho_M^{n+1} > \tau$, so $\omega\varrho_{M_{i+1}}^{n+1} > \nu \geq \omega\varrho_W^{n+1}$, a contradiction again. This completes the proof of (9.3.4) in the first case.

In the remaining case $\nu_i = \nu$ and $\tau = \omega\varrho_M^{n+1}$, so $i = 0$ and $E_{\nu_i}^{M_i} = E_\nu^M$ is a superstrong extender in M. We first show:

$$\omega\varrho_W^{n+1} = \omega\varrho_{M_1}^{n+1} = \nu. \tag{9.3.7}$$

By our earlier computations, it suffices to verify the equality on the right side. By the properties of fine ultrapowers, $\omega\varrho_{M_1}^{n+1} \geq \pi_{0,1}(\tau) = \nu$, so our task is to verify the converse. But this follows from the fact that $p_M \in R_M^{n+1}$, which implies that $M_1 = \tilde{h}_{M_1}^{n+1}(\nu \cup \{\pi_{0,1}(p_M)\})$.

As $\bar{\pi}_{0,\theta} : W \to W_\theta$ is Σ^*-preserving and W_θ is an $\mathbb{S}$-iterate of M, the premouse W has an (ω_1+1, ω_1)-iteration strategy $\mathbb{S}'$ defined by

$$\mathbb{S}'(\mathfrak{I}') = \text{the branch through } T^{\mathfrak{I}'} \text{ determined by } \mathbb{S}(\mathfrak{I}^\frown\mathfrak{I}').$$

We now use the assumption B) to show that W is solid. This follows immediately if the height of W is strictly less than that of M, so suppose that $\mathrm{ht}(W) = \mathrm{ht}(M)$. Notice that $\bar{q} \in R_W^{n+1}$, so $p_W - \nu = p_{W,n+1} \leq^* \bar{q} \leq^* q$ where the equality on the left follows from Lemma 9.2.14. If any of the inequalities above is strict, we have $p_W <^* p_M$. Otherwise $p_W - \nu = q <^* q \cup \{\nu\} \leq^* p_M$, which yields $p_W <^* p_M$ again. This proves that W is solid.

We next show that actually

$$p_W - \nu = \bar{q}. \tag{9.3.8}$$

Let $B(u, v)$ be $\Sigma_1^{(n)}(M)$ such that $b = B(-, p_M) \cap \tau \notin M$ and let B_1 be $\Sigma_1^{(n)}(M)$ by the same definition. Notice that $b \cap \zeta \in M$ whenever $\zeta < \tau$, as $\tau = \omega\varrho_M^{n+1}$. Fixing any such ζ, the statement

$$(\forall\eta^{n+1} < \zeta)[\eta^{n+1} \in b \cap \zeta \longleftrightarrow A(\eta^{n+1}, p_M)]$$

is $\Sigma_0^{(n+1)}$, and therefore is preserved upward under $\pi_{0,1}$. It follows that $b_1 \cap \pi_{0,1}(\zeta) = \pi_{0,1}(b \cap \zeta)$ where $b_1 = A_1(-, \pi_{0,1}(p_M)) \cap \nu$. As $\pi_{0,1}$ maps τ cofinally into ζ, we can reconstruct b inside M, provided that $b_1 \in M$. Indeed,

$$b = \bigcup\{b^* \in J_\tau^E;\ \pi^*(b^*) = b_1 \cap \zeta' \text{ for some } \zeta' < \nu\}$$

where $\pi^* = \pi_{0,1} \restriction J_\tau^E$. That b is contained in the set on the right follows from the above observation that $\pi^*(b^*)$ is an initial segment of b_1 whenever $b^* \in J_\tau^E$ is an initial segment of b. On the other hand, if $\pi^*(b^*)$ is an initial segment of b_1, pick an initial segment b' of b such that $b^* \subset \eta$ for some $\eta \in b'$. This is possible, as b is unbounded in τ. Then $\pi^*(b^*)$ is an initial segment of $\pi^*(b')$, as both are initial segments of b_1 and $\pi^*(b')$ is longer. Consequently, b^* is an initial segment of b' and therefore is an initial segment of b. This proves the above equality. Now $\pi^* \in J_\mu^E$ where $\mu = \lambda^{+M} = \sigma(\nu)$, as π^* is the ultrapower map associated with $\mathrm{Ult}(J_\tau^E, E_\nu^M)$ and both J_μ^E as well as E_ν^M are in J_μ^E. Now if $b_1 \in M$, then $b_1 \in J_\mu^E$, so the whole reconstruction of b takes place in $J_\mu^E \subset M$.

In the following we show that the assumption $p_W - \nu <^* \bar{q}$ implies $b_1 \in M$. Thus, by the previous paragraph, $p_W - \nu <^* \bar{q}$ would actually yield that $b \in M$, which is false by our choice of B. This proves (9.3.8).

We know that b_1 is a $\mathbf{\Sigma}_1^{(n)}(M_1)$ subset of ν, so by Claim 1 and (9.3.7), b_1 is $\mathbf{\Sigma}_1^{(n)}(W)$. By Lemma 9.2.14, $\bar{W} = \mathrm{core}_W^{n+1}$ exists, is sound above $\omega\varrho_W^{n+1} = \omega\varrho^{n+1}$ and b_1 is $\mathbf{\Sigma}_1^{(n)}(\bar{W})$. Consequently, $b_1 = \bar{A}(-, (p_{\bar{W}} - \omega\varrho^{n+1}) \cup r) \cap \nu$ for some $\bar{A}$ which is $\Sigma_1^{(n)}(\bar{W})$ and some finite $r \subset \omega\varrho^{n+1}$. Due to (9.3.7) and the fact that the associated core map $\bar{\sigma} : \bar{W} \to W$ is Σ^*-preserving we finally obtain $b_1 = A(-, (p_W - \nu) \cup r) \cap \nu$ where A is $\Sigma_1^{(n)}(W)$ by the same definition as $\bar{A}$. Now $\sigma : W \to M$ is $\Sigma_1^{(n)}$-preserving and $\sigma \restriction \nu = \mathrm{id}$, so letting A' be the $\Sigma_1^{(n)}(M)$ relation by the same definition as A, we have $b_1 = A'(-, \sigma(p_W - \nu) \cup r) \cap \nu$. But if $p_W - \nu <^* \bar{q}$, then $\sigma(p_W - \nu) <^* q$. Let $\alpha \in q - \sigma(p_W - \nu)$ be such that $q - (\alpha + 1) = \sigma(p_W - \nu) - (\alpha + 1)$, let k be such that $\omega\varrho_M^{k+1} \leq \alpha < \omega\varrho_M^k$ and let $\sigma_\alpha : W_M^\alpha \to M$ be the canonical witness map.

If $k = n$, then σ_α is $\Sigma_1^{(n)}$-preserving, $\sigma_\alpha \restriction \alpha = \mathrm{id}$ and $\sigma(p_W - \nu) \cup r$ is in the range of σ_α. Now if A_α is $\Sigma_1^{(n)}(W_M^\alpha)$ by the same definition as A' and q_α is the σ_α-preimage of $\sigma(p_W - \nu) \cup r$, we have $A_\alpha(-, q_\alpha) \cap \nu = A'(-, \sigma(p_W - \nu) \cup r) \cap \nu = b_1$. But, by our induction hypothesis, we already know that $W_M^\alpha \in M$, so b_1, being definable over W_M^α, is also an element of M.

If $k < n$, then

$$b_1 = A'(-, \sigma(p_W - \nu) \cup r) \cap \nu = A''(-, s') \cap \nu$$

for some A'' which is $\Sigma_1^{(n-(k+1))}(M^{k+1,s})$ where $s = \sigma(p_W - \nu) - \omega\varrho_M^{k+1}$ and $s' = (\sigma(p_W - \nu) \cup r) \cap \omega\varrho_M^{k+1}$. Now $A_M^{k+1,s}$ is $\Sigma_1^{(k)}(M)$ in s, say $A_M^{k+1,s} = A^{k+1}(-, s) \cap H_M^{k+1}$ where A^{k+1} is $\Sigma_1^{(k)}(M)$. Furthermore, the canonical witness map σ_α is $\Sigma_1^{(k)}$-preserving and $s \cup s' = \sigma(p_W - \nu) \cup r$ is in the range of σ_α, so $A_M^{k+1,s} = A_\alpha(-, \sigma_\alpha^{-1}(s)) \cap H_M^{k+1}$ where this time A_α is $\Sigma_1^{(k)}(W_M^\alpha)$ by the same definition as A^{k+1}. (Notice that $H_M^{k+1} \subset W_M^\alpha$.) Again, we apply the induction hypothesis $W_M^\alpha \in M$ to infer that $A_M^{k+1,s} \in M$, and then the above observation that b_1 is definable from $A_M^{k+1,s}$ to conclude that $b_1 \in M$. This completes the proof of the fact that $b_1 \in M$ from the assumption $p_W - \nu <^* \bar{q}$.

□(9.3.8)

The solidity of W and (9.3.4) enable us to prove that $\pi_{0,\theta}(q) = \bar{\pi}_{0,\theta}(\bar{q})$. By the fine structural properties of iteration maps, $\bar{\pi}_{0,\theta}(p_W) \in P_{W_\theta}^*$. Now as W is solid, each element β of p_W has a generalized witness w.r.t. W and p_W in W whose image under $\bar{\pi}_{0,\theta}$ is a generalized witness for $\bar{\pi}_{0,\theta}(\beta)$ w.r.t. W_θ and $\bar{\pi}_{0,\theta}(p_W)$ in W_θ. So $\bar{\pi}_{0,\theta}(p_W) = p_{W_\theta}$ and $\bar{\pi}_{0,\theta}(\bar{q})$ is a top segment of p_{W_θ}, as $\bar{q}$ is a top segment of p_W. Similarly, $\pi_{0,\theta}(p_M) \in P_{M_\theta}^*$ and each element of $\pi_{0,\theta}(q)$ has a generalized witness w.r.t. M_θ and $\pi_{0,\theta}(p_M)$ in M_θ; the latter follows from our induction hypothesis that each element of q has a generalized witness w.r.t. M and p_M in M. Thus, $\pi_{0,\theta}(q)$ is a top segment of $p_{M_\theta} = p_{W_\theta}$ (recall that $M_\theta = W_\theta$ by the failure of (9.3.4)). Since q and $\bar{q}$ are of the same size, $\pi_{0,\theta}(q)$ and $\bar{\pi}_{0,\theta}(\bar{q})$ must be of the same size as well, so $\pi_{0,\theta}(q) = \bar{\pi}_{0,\theta}(\bar{q})$.

Now pick a relation $\bar{A}(u, v)$ which is $\Sigma_1^{(n)}(W)$ and such that $\bar{A}(-, \bar{q}) \cap \omega\varrho^{n+1} \notin W$. This is possible by (9.3.7) and (9.3.8). We know that the iteration along the main branch of $\bar{\Im}$ is above ν and, by the proof of Claim 1, the iteration $\Im$ is above ν along $[i+1, \theta)_T$. Thus, letting $q_1 = \pi_{0,1}(q)$ and A_1 be the $\Sigma_1^{(n)}(M_1)$-relation by the same definition as $\bar{A}$, the result of the previous paragraph guarantees that $A_1(-, q_1) \cap \omega\varrho^{n+1} = \bar{A}(-, \bar{q}) \cap \omega\varrho^{n+1}$. Using Claim 1 we then conclude that $A_1(-, q_1) \cap \omega\varrho^{n+1} \notin M_1$. Now let A be the $\Sigma_1^{(n)}(M)$-relation by the same definition as A_1. We show that $b = A(-, q) \cap \omega\varrho_M^{n+1} \notin M$. Indeed, if $b \in M$, we would have

$$(\forall \xi^{n+1})[\xi^{n+1} \in b \longleftrightarrow A(\xi^{n+1}, q)].$$

This statement, being $\Sigma_1^{(n+1)}$, is preserved upward under $\pi_{0,1}$, as $\pi_{0,1}$ is Σ^*-preserving. So $\pi_{0,1}(b) = \{\xi < \nu;\ A_1(\xi, q_1)\}$, which contradicts the above proved fact that the set on the right is not an element of M_1.

Recapitulating the above argument, the set A we have found is $\Sigma_1^{(n)}(M)$ and such that $A(-,q) \cap \omega\varrho_M^{n+1} \notin M$. However, this contradicts the fact that q is a *proper* top segment of p_M and thus completes the proof of Claim 2 in the case where $\nu_i = \nu$ and $\omega\varrho_M^{n+1} = \tau$. □(Claim 2)

Recall the notation and setting from the beginning of the proof of the solidity theorem. In particular, a is a $\mathbf{\Sigma}_1^{(n)}(W)$ subset of ν encoding W. Also, i, j, ξ^* and $M^*_{\xi^*}$ were defined in (9.3.1), (9.3.2) and the remarks following these formulae.

Claim 3. The set a is $\mathbf{\Sigma}_1^{(n)}(M^*_{\xi^*})$. Consequently, $a \in M_{\xi^*}$.

Proof. If $\xi^* = \theta$ then there is nothing to prove, as $M^*_{\xi^*} = W_\theta$ and a is obviously $\Sigma_1^{(n)}(W_\theta)$. So suppose $\xi^* < \theta$, in which case $W_\theta = M_\theta$ and $\xi^* = \xi_j$.

We recall some basic facts about the coiteration. We know that the main branch of $\bar{\mathfrak{I}}$ is above ν and without truncation points. As there are no truncation points along $(\xi_j, \theta)_T$, we can repeat the proof of Claim 1 and infer that all premice on these branches together with M^*_j have the same $(n+1)$-st projectum $\omega\varrho^{n+1} \leq \nu$ and that the iteration along $[\xi_j, \theta)_T$ is above $\omega\varrho^{n+1}$, but not above the n-th projectum. An iterative application of Lemma 3.3.6 yields

$$\mathcal{P}(\kappa_k) \cap \mathbf{\Sigma}_1^{(n)}(M^*_k) = \mathcal{P}(\kappa_k) \cap \mathbf{\Sigma}_1^{(n)}(M_\theta) = \mathcal{P}(\kappa_k) \cap \mathbf{\Sigma}_1^{(n)}(W)$$

whenever $\xi_j \leq_T k \leq_T \theta$ and $\kappa_k \geq \nu$, which immediately yields $a \in \mathbf{\Sigma}_1^{(n)}(M^*_j)$ for j such that $\kappa_j \geq \nu$.

From now on assume that $\kappa_j < \nu$. Recall that all premice of $\mathfrak{I}$ agree below ν, as $\nu_0 \geq \nu$. Furthermore, the witness map $\sigma : W \to M$ is cardinal preserving and $\mathrm{cr}(\sigma) = \nu$. This means that any cardinal in J^E_ν is in fact a cardinal in M and, consequently, in any premouse of the iteration $\mathfrak{I}$. Thus, if $j+1$ is a truncation point and $\kappa_j < \nu$, then κ_j must be the largest cardinal in J^E_ν, as otherwise $\mathcal{P}(\kappa_j) \cap M_{\xi_j}$ would be contained in J^E_ν and no truncation would be needed. It follows that the iteration $\mathfrak{I}$ is above $\lambda_j > \nu$ along $[j+1, \theta)_T$, so the argument from the previous paragraph yields $a \in \mathbf{\Sigma}_1^{(n)}(M_{j+1})$.

We now show that a is $\mathbf{\Sigma}_1^{(n)}(M^*_j)$. Notice that $\nu = \tau_i$, as M_i agrees with W_i below ν_i and ν is the cardinal successor of τ_i in W_i (however, this is not relevant for the proof). Fix a relation $A(u, v)$ from $\Sigma_1^{(n)}(M_1)$ and a parameter $p \in M_{j+1}$ such that $\xi \in a \longleftrightarrow A(\xi, p)$. We will now imitate the corresponding part of the proof of Theorem 5.2.1, namely the argument from the second paragraph below the statement of Claim 2. That argument makes use of the sequence of functions $\langle f^*_\xi;\ \xi < \nu\rangle \in \Sigma_1(M^*_j)$ introduced in the proof of Lemma 4.1.2; recall that $\pi(f^*_\xi)(\kappa_j) = \xi$ for any $\xi < \nu$ where $\pi = \pi_{\xi_j, j+1}$. By the facts about our coiteration summarized above, $\omega\varrho^{n+1}_{M^*_j} \leq \kappa_j < \omega\varrho^n_{M^*_j}$. Letting $\bar{A}$ be a $\Sigma_0^{(n)}(M_{j+1})$-relation such that $A(x, y)$ is of the form $(\exists z^n)\bar{A}(z^n, x, y)$, the relation "$\xi \in a$" can be expressed as

$$(\exists \zeta < \omega\varrho^m_{M^*_j})[M_{j+1} \models (\exists z^n \in \pi(S^E_\zeta)\bar{A}(z^n, \pi(f^*_\xi)(\kappa_j), \pi(f)(\alpha))]$$

where $p = \pi(f)(\alpha)$ for some $f \in \Gamma(\kappa_j, M_j^*)$ and $\alpha < \lambda_j$. Letting $\bar{B}$ be the $\Sigma_0^{(n)}(M)$-relation with the same definition as $\bar{A}$, the above formula can be equivalently rewritten as

$$(\exists\zeta < \omega\varrho_{M_j^*}^m)(u(\zeta,\xi) \in E_{\nu_j, \prec\kappa_j,\alpha\succ}^{M_j})$$

where

$$u(\zeta,\xi) = \{\prec\eta,\eta'\succ \in \kappa_j;\ M_j^* \models (\exists z^n \in S_\zeta^E)\bar{B}(z^n, f_\xi^*(\eta), f(\eta'))\}.$$

The collection $\{u(\zeta,\xi);\ \zeta < \omega\varrho_{M_j^*}^n\ \&\ \xi < \nu\}$ is obviously $\mathbf{\Sigma}_0^{(n)}(M)$. Furthermore, Lemma 9.1.10 guarantees that the ultrafilter $E_{\nu_j,\prec\kappa_j,\alpha\succ}^{M_j}$ is $\mathbf{\Sigma}_1(M_j^*)$, so a can be defined inside M. It is then straightforward to verify that a is $\mathbf{\Sigma}_1^{(n)}(M_j^*)$. □(Claim 3)

The proof now splits into three cases.

Case 1. It is *not* the case that both E_ν^M is a superstrong extender in M and $\omega\varrho_M^m = \tau$ for some $m > n$, where $\tau = \kappa^{+M}$ and $\kappa = \mathrm{cr}(E_\nu^M)$.

If $M_{\xi^*} = M$, then the desired conclusion $a \in M$ follows immediately from Claim 3. From now on assume $M_{\xi^*} \neq M$, which means that $E_{\nu_h}^{M_h}$ is an extender for some $h < \xi^*$. By (9.2.8), ν_0 is a cardinal in M_ξ. If $\nu_0 > \nu$, then $a \in M_{\xi^*}||\nu_0$ by acceptability. Consequently, $a \in M$, as M agrees with M_{ξ^*} below ν_0.

From now on assume that $\nu_0 = \nu$. By (9.3.3), $E_{\nu_0}^{M_0} = E_\nu^M$ is a superstrong extender in M. We first observe that $a \in M_1$. This is clear if $E_{\nu_h}^{M_h} = \varnothing$ whenever $1 \leq h < \xi^*$, in which case $M_{\xi^*} = M_1$. If $E_{\nu_h}^{M_h}$ is an extender for some such h, then ν_h is a cardinal in M_{ξ^*}, so $a \in M_1$ by the argument from the previous paragraph.

Knowing that $a \in M_1$, we show that a is actually in M. The crucial point here is the fact that $E_\nu^M \in M$. Let m be such that $\omega\varrho_M^{m+1} \leq \kappa < \omega\varrho_M^m$. Then $\omega\varrho_M^m > \tau$, as we assume that no projectum of M is equal to τ. It follows that $\omega\varrho_{M_1}^m > \nu = \pi(\tau)$ where $\pi = \pi_{0,1}$. As $\omega\varrho_{M_1}^m$ is a cardinal in M_1 and $a \subset \nu$, we have $a \in H_{M_1}^m$, so $a = \pi(f)(\alpha)$ for some $f \in M$ and $\alpha < \lambda$. Fix a sequence of functions $g = \langle g_\zeta;\ \kappa \leq \zeta < \tau\rangle \in M$ such that each g_ζ is a bijection between κ and ζ. (Notice that there is a canonical sequence of such bijections uniformly definable over J_τ^E.) Then each $\xi < \nu$ can be expressed in the form $\pi(g_\zeta)(\beta) = E_\nu^M(g_\zeta)(\beta)$ for some ζ such that $\pi(\zeta) > \xi$ and some $\beta < \lambda$. This enables us to express the membership relation "$\xi \in a$" as

$$(\exists\zeta < \tau)(\exists\beta < \lambda)[\xi = E_\nu^M(g_\zeta)(\beta)\ \&\ \pi(g_\zeta)(\beta) \in \pi(f)(\alpha)].$$

Using Łos' Theorem, this can be rewritten in the form

$$(\exists\zeta < \tau)(\exists\beta < \lambda)[\xi = E_\nu^M(g_\zeta)(\beta)\ \&\ \prec\beta,\alpha\succ \in E_\nu^M(u_\zeta)]$$

where $u_\zeta = \{\prec\eta,\eta'\succ \in \kappa;\ g_\zeta(\eta) \in f(\eta')\}$. This statement can be expressed in a $\mathbf{\Sigma}_0$ fashion over $M||\nu^{+M} \in M$, as the parameters τ, λ, E_ν^M and $\langle u_\zeta\rangle_\zeta$ are in $M||\nu^{+M}$. Consequently, $a \in M$. □(Case 1)

Before proceeding to the discussion of the remaining cases, we shall introduce an auxiliary notion of *ζ-witness*, which is a weakening of the notion of solidity witness.

Definition. Let N be an acceptable structure and let $p \subset N$ be a finite sequence of ordinals. Given an $\alpha \in p$ and a $\zeta \leq \alpha$, let k be such that $\omega\varrho_N^{k+1} \leq \alpha < \omega\varrho_N^k$. The transitive collapse $W_N^{\nu,p}(\zeta)$ of $\tilde{h}_N^{k+1}(\zeta \cup \{p-(\alpha+1)\})$ is called the *standard ζ-witness* for α w.r.t. N and p. If $p = p_N$, we write briefly $W^\nu(\zeta)$. A pair $\langle Q, r\rangle$ is called a *generalized ζ-witness* for α w.r.t. N and p iff

a) Q is an acceptable structure and $\zeta \subset Q$;

b) $r \subset Q$ is a finite set of ordinals;

c) given a $\Sigma_1^{(n)}$-formula $\varphi(v_0, \dots, v_\ell)$ and an ℓ-tuple of ordinals $\xi_1, \dots, \xi_\ell < \zeta$, we have

$$N \models \varphi(p-(\alpha+1), \xi_1, \dots, \xi_\ell) \longleftrightarrow Q \models \varphi(r, \xi_1, \dots, \xi_\ell)$$

It is clear that the standard α-witness for α w.r.t. N and p is nothing other than the standard witness for α w.r.t. N and p in our usual sense. Similarly, any generalized α-witness for α w.r.t. N and p is a generalized witness for α w.r.t. N and p in our usual sense. Notice that if $W_N^{\alpha,p}(\zeta) \in N$ and $\bar{\zeta} < \zeta$, then $W_N^{\alpha,p}(\bar{\zeta}) \in N$ as well. The proof of Lemma 1.12.3 can be easily modified to yield:

$W_N^{\alpha,p}(\zeta) \in N$ iff there is a generalized ζ-witness for α w.r.t. M and p in N. (9.3.9)

If there is a generalized ζ-witness for α w.r.t. M and p in N and $\bar{\zeta} < \zeta$, then there is also a generalized $\bar{\zeta}$-witness for α w.r.t. M and p in N. (9.3.10)

The other arguments concerning witnesses can be correspondingly modified as well; we will list merely those results which are relevant for the rest of the proof of the solidity theorem. In the following, n has the same meaning as in the above definition.

The property of being a generalized ζ-witness for α w.r.t. N and p is a $\Pi_1^{(n)}$-statement over N. (9.3.11)

If $q = p_{N,n+1} - (\alpha+1)$, $N = \tilde{h}_N^{n+1}(\alpha \cup \{p, \alpha\})$ and $W_N^{\alpha,q}(\zeta) \in N$ for all $\zeta < \alpha$, then $q \cup \{\alpha\} = p_{N,n+1} - \alpha$. (9.3.12)

If N is a weakly iterable presolid premouse, then (9.3.12) holds with p_N in place of $p_{N,n+1}$, as follows from Corollary 9.2.15. (9.3.13)

Let us now return back to the proof of the solidity theorem. As we have seen, the standard argument for proving solidity does not seem to go through if the restrictions of Case 1 fail. However, that argument still proves the following weaker version of solidity for M:

If $\nu \in p_M$ then $W_M^\nu(\zeta) \in M$ whenever $\zeta < \nu$. (9.3.14)

Indeed, as $W^{\nu}_M(\zeta)$ is encodable in a $\Sigma_1^{(n)}(W)$ subset a_ζ of $\zeta < \nu$, the proof of Claim 3 guarantees that a_ζ is $\Sigma_1^{(n)}(M^*_{\xi^*})$, so $a_\zeta \in M_{\xi^*}$. As before, it suffices to show that $a_\zeta \in M$, as $W^{\nu}_M(\zeta)$ can be reconstructed from a_ζ inside M. Following the proof of Case 1, either $M_{\xi^*} = M$, in which case there is nothing to prove, or else ν_0 is a cardinal in M_{ξ^*}, so $a_\zeta \in M_{\xi^*}||\nu_0$, as this time a_ζ is a *bounded* subset of ν. By the agreement between M and M_{ξ^*} we then have $a_\zeta \in M$. □(9.3.14)

In the following we will discuss the remaining cases of the proof of the solidity theorem, so we return to the notation set up at the beginning of this proof. From now on we assume that E^M_ν is superstrong in M and that $\omega\varrho^m_M = \tau$ for some $m > n$. Recall also that at this point we know that $W^{\nu'}_M \in M$ for all $\nu' \in p_M$ which are larger than M and that $W^{\nu}_M(\zeta) \in M$ whenever $\zeta < \nu$. Our aim is to prove that $W^{\nu}_M \in M$.

Case 2. E^M_ν is superstrong in M and $\omega\varrho^{n+2}_M \geq \tau$.

Let $\bar{M}$ be the transitive collapse of $\tilde{h}^{n+2}_M(\kappa \cup \{p_M\})$ and let $\bar{\sigma} : \bar{M} \to M$ be the associated uncollapsing map. By construction, the map $\bar{\sigma}$ is $\Sigma_1^{(n+1)}$-preserving and $\tilde{h}^{n+2}_{\bar{M}}(\kappa \cup \{\bar{p}\}) = \bar{M}$ where $\bar{p}$ is the $\bar{\sigma}$-preimage of p_M. It follows that $\omega\varrho^{n+2}_{\bar{M}} \leq \kappa$, as $\bar{M}$ can be encoded into a $\Sigma_1^{(n+1)}(\bar{M})$ subset b of κ. The preservation degree of $\bar{\sigma}$ together with the fact that $\bar{\sigma} \upharpoonright \kappa = \mathrm{id}$ then guarantees that b is a $\Sigma_1^{(n+1)}(M)$ subset of $\kappa < \omega\varrho^{n+2}_M$, so $\bar{M} \in M$. In particular, $\mathrm{ht}(\bar{M}) < \mathrm{ht}(M)$. The above discussion also shows that $\bar{\sigma}$ is an embedding of the phalanx $\langle M, \bar{M}, \kappa\rangle$ into M, hence by Corollary 9.2.13, $\bar{M}$ has an $(\omega_1 + 1, \omega_1)$-iteration strategy. Our assumption B) now guarantees that $\bar{M}$ is solid.

In the following we show that $\bar{p} = p_{\bar{M}}$. By the above discussion, any bounded subset of κ which is definable over $\bar{M}$ is an element of M and therefore an element of $M||\kappa = \bar{M}||\kappa$, as κ is a cardinal in M. So $\bar{M}$ cannot project below κ, which means that $\omega\varrho^{\omega}_{\bar{M}} = \omega\varrho^{n+2}_{\bar{M}} = \kappa$. As an immediate consequence we have that $\bar{p} \in R^*_{\bar{M}}$. To get the desired conclusion, it suffices to prove that each $\nu' \in \bar{p}$ larger than $\bar{\nu} = \bar{\sigma}^{-1}(\nu)$ has a generalized witness w.r.t. $\bar{M}$ and $\bar{p}$ in $\bar{M}$ and that for each $\zeta < \bar{\nu}$, $\bar{\nu}$ has some generalized ζ-witness w.r.t. $\bar{M}$ and $\bar{p}$ in $\bar{M}$. By our assumption and (9.3.14), we know that this holds with ν, p_M and M in place of $\bar{\nu}$, $\bar{p}$ and $\bar{M}$. Thus, we will be done if we manage to show that these properties are preserved downward under $\bar{\sigma}$. The statement "There is a generalized witness for $\bar{\sigma}(\nu')$ w.r.t. M and p_M" is $\Sigma_2^{(k)}$ for k such that $\omega\varrho^{k+1}_{\bar{M}} \leq \nu' < \omega\varrho^k_{\bar{M}}$ and the same is true when we consider generalized $\bar{\sigma}(\zeta)$-witnesses for $\bar{\sigma}(\nu)$, so it is enough to verify that $\bar{\sigma}$ is $\Sigma_2^{(k)}$-preserving for every $k \leq n$. But this follows by the downwards extension of embeddings lemmata (Lemma 1.3.2 and 1.7.3) and from the facts that

$$\bar{\sigma} \upharpoonright H^{n+1}_{\bar{M}} : \bar{M}^{n+1,\bar{p}} \underset{\Sigma_1}{\longrightarrow} M^{n+1}$$

and that M is sound, i.e. that $p_M = \bar{\sigma}(\bar{p}) \in R^*_M$.

Knowing that $\bar{M}$ is solid and $\bar{p} = p_{\bar{M}}$, we can conclude that $W^{\bar{\nu}}_{\bar{M}} \in \bar{M}$. But then, the preservation properties of $\bar{\sigma}$ guarantee that $\sigma(W^{\bar{\nu}}_{\bar{M}})$ is a generalized witness for ν w.r.t. M and p_M. □(Case 2)

Case 3. E_ν^M is superstrong in M and $\omega\varrho_M^{n+2} < \tau = \omega\varrho_M^{n+1}$.

Assume first that $\text{Ult}^*(M, E_\nu^M) = \text{Ult}^n(M, E_\nu^M)$. Notice that $\nu_0 = \nu$ in the coiteration $\langle \Im, \tilde{\Im} \rangle$, i.e. E_ν^M is the first extender used on the M-side of the coiteration. By Claim 3, $a \in M_{\xi^*}$. Now we can proceed exactly as in Case 1: first we show that $a \in M_1 = \text{Ult}^*(M, E_\nu^M)$. By our assumption, however, $M_1 = \text{Ult}^n(M, E_\nu^M)$. Now $a \in H_{M_1}^n$, as $\nu < \omega\varrho_{M_1}^n$, so $a = \pi(f)(\alpha)$ for some $f \in M$ and $\alpha < \lambda$ where π is the associated ultrapower map. The rest of the proof of Case 1 then yields $a \in M$.

From now on assume that $\text{Ult}^*(M, E_\nu^M) \neq \text{Ult}^n(M, E_\nu^M)$. We use a different argument to show that $W_M^\nu \in M$. Let $M' = \text{Ult}^*(M, E_\nu^M)$ and $M^* = \text{Ult}^n(M, E_\nu^M)$. Furthermore, let $\pi : M \to M'$ and $\pi^* : M \to M^*$ be the associated ultrapower maps and $\sigma' : M^* \to M'$ be the canonical embedding defined by $\sigma'(\pi^*(f)(\alpha)) = \pi(f)(\alpha)$ for $f \in \Gamma^n(\kappa, M)$ and $\alpha < \lambda$. Let us summarize some useful facts about M^*, M', σ' and the ultrapower maps.

a) π is Σ^*-preserving and $\Sigma_2^{(k)}$-preserving for every $k \leq n$. Also, π is $\Sigma_0^{(n+1)}$ cofinal and $\omega\varrho_{M'}^{n+1} = \nu = \pi(\tau)$.

b) π^* is $\Sigma_0^{(n)}$ cofinal, maps τ cofinally into $\nu = \pi^*(\tau)$ and ν cofinally into $\pi^*(\nu)$.

c) $\tilde{h}_{M'}^{n+1}(\nu \cup \{\pi(p_M)\}) = M'$.

d) $\tilde{h}_{M^*}^{n+1}(\nu \cup \{p^*\})$ where $p^* = \pi^*(p_M)$. It follows that $\omega\varrho_{M^*}^{n+1} \leq \nu$.

e) If $\alpha \in p_M$ and $\alpha > \nu$, then $\pi^*(\alpha)$ has a generalized witness w.r.t. M^* and p^* in M^*. Furthermore, $\pi^*(\nu)$ has a generalized ζ-witness w.r.t. M^* and p^* for cofinally many $\zeta < \pi^*(\nu)$.

f) $p^* - \pi^*(\nu) = p_{M^*} - \pi^*(\nu)$.

g) $\sigma' \restriction (\nu + 1) = \text{id}$.

h) $\sigma' : M^* \to M'$ is $\Sigma_0^{(n)}$-preserving and *not* cofinal.

Clauses a) and b) follow from the general properties of fine ultrapowers. To see that $\pi(\tau) = \nu$ notice that any good $\mathbf{\Sigma}_1^{(n)}(M)$ function $f : \kappa \to \tau$ must be bounded in τ, as otherwise we could define, in a $\Sigma_1^{(n)}$-fashion, a subset of κ coding the order type τ. This would contradict the fact that $\tau = \omega\varrho_M^{n+1}$. That π^* maps ν cofinally into $\pi^*(\nu)$ follows from the fact that π, E_ν^M and $E_{\pi^*(\nu)}^{M^*}$ map τ, τ and ν cofinally into ν, ν and $\pi^*(\nu)$, respectively. We leave the details to the reader. Clauses c) and e) follow from the general properties of fine ultrapowers and the fact that M is sound, in particular, from the equality $\tilde{h}_M^{n+1}(\tau \cup \{p_M\}) = M$. e) is a consequence of the preservation properties of π^*, our assumption that $W_M^\alpha \in M$ for all $\alpha \in p_M$ larger than ν and our partial result that $W_M^\nu(\zeta) \in M$ for all $\zeta < \nu$, as we know from b) that π^* maps ν cofinally into $\pi^*(\nu)$. From e), (9.3.10), (9.3.12) and (9.3.13) we immediately obtain f). To see g), notice that $\sigma' \restriction (\lambda + 1) = \text{id}$, as we use the same extender in both ultrapowers; this immediately implies that in fact $\sigma \restriction \nu = \text{id}$. The equality $\sigma(\nu) = \nu$ then follows from the latter part of a). Since Łos' theorem for $\Sigma_0^{(n)}$-formulae holds for both ultrapowers, the standard verification yields that σ' is $\Sigma_0^{(n)}$-preserving. If σ' were cofinal, then σ' would be $\Sigma_1^{(n)}$-preserving. By g) and the obvious fact that $\sigma'(p^*) = \pi(p_M)$ we know that $\nu \cup \{\pi(p_M)\} \subset \text{rng}(\sigma')$, so c) would imply that $\text{rng}(\sigma') = M'$. This would mean that $M' = M^*$ and $\sigma' = \text{id}$, contradicting our assumption that $M' \neq M^*$.

Claim 4. M^* is solid.

Proof. Assume that $\bar{M} \in M$ is a premouse satisfying one of the following:

H0 There is a $\Sigma_1^{(n)}$-preserving embedding $\bar{\sigma} : \bar{M} \to \tilde{M}$ for some *proper* initial segment $\tilde{M}$ of M such that $\bar{\sigma} \restriction \omega\varrho_{\bar{M}}^{n+1} = \text{id}$;

H1 $n = 0$, the premice M and $\bar{M}$ are both active and there is a Σ_0-preserving embedding $\bar{\sigma} : \bar{M} \to \tilde{M}$ where $\tilde{M} = \langle J_\delta^E, E_{\text{top}}^M \cap J_\delta^E \rangle$ for some $\delta \in M$ larger than λ_M such that $\bar{\sigma} \restriction \omega\varrho_{\bar{M}}^1 = \text{id}$.

We show that $\bar{M}$ is solid. For $\bar{M}$ as above, $\bar{\sigma}$ is an embedding of the phalanx $\langle M, \bar{M}, \omega\varrho_{\bar{M}}^{n+1} \rangle$ into M. This is clear if H0 holds. Under H1, however, $\bar{\sigma}$ is still an embedding of $\bar{M}$ into M which is both Σ_0-preserving and cardinal preserving. The former is obvious, the latter follows from the fact that $\bar{\sigma}$ maps the largest cardinal $\lambda_{\bar{M}}$ of $\bar{M}$ to the largest cardinal λ_M of M. By Corollary 9.2.13, $\bar{M}$ has an (ω_1+1, ω_1)-iteration strategy. But as $\bar{M} \in M$, the height of $\bar{M}$ is less than that of M. So our assumption B) guarantees that $\bar{M}$ is solid.

Now notice that the statement $(\forall \bar{M})[(\text{H0} \vee \text{H1}) \longrightarrow \bar{M} \text{ is solid}]$ is Π_1 over M, and therefore is preserved upward under ultrapower maps. In particular, granting that M^+ is M' or $\text{Ult}^*(M', E_{\nu'}^{M'})$ for some ν' and that $\bar{M}'$ is such that H0 or H1 holds with M^+ and $\bar{M}'$ in place of M and $\bar{M}$ and with some $\bar{\sigma}' : \bar{M}' \to \tilde{M}$ such that $\bar{\sigma}' \in M^+$, then $\bar{M}'$ is solid. We show that $\bar{M}' = M^*$ meets these conditions, which will prove the claim. Before proceeding further, recall the properties of the embedding $\sigma' : M^* \to M'$.

Suppose first that $n = 0$ and $\omega\delta = \sup[(\sigma')''(\mathbf{On} \cap M^*)]$. Then $\delta < \text{ht}(M')$ by h). If $\bar{M}$ is passive, set $\tilde{M} = J_\delta^{E'}$ where $E' = E^{M'}$. Then $\sigma' : M^* \to \tilde{M}$ is Σ_1-preserving and also satisfies the rest of H0. If $\bar{M}$ is active, then so is M' and, setting $\tilde{M} = \langle J_\delta^{E'}, E_{\text{top}}^{M'} \cap J_\delta^{E'} \rangle$, the map $\sigma' : M^* \to \tilde{M}$ meets the requirements of H1. It remains to prove that M^* and σ' are both in M'. Notice that σ' as a map into $\tilde{M}$ is Σ_1-preserving. From this, d) and g) we infer that $\text{rng}(\sigma') = h_{\tilde{M}}(\nu \cup \{\sigma'(p^*)\}) \in M'$, so M^* is the transitive collapse of $\text{rng}(\sigma')$ and σ' is the corresponding uncollapsing map. The conclusion then follows from the fact that this collapse can be constructed inside M'.

From now on assume that $n > 0$. Let $\omega\tilde{\varrho} = \sup[(\sigma')''\omega\varrho_{M^*}^n]$. By h), $\omega\tilde{\varrho} < \omega\varrho_{M'}^n$. Thus, $\tilde{A} = A_M^n \cap J_{\tilde{\varrho}}^{E'} \in M'$. By f) and general properties of reducts and $\Sigma_0^{(n)}$-embeddings, we have

$$\sigma' \restriction H_{M^*}^n : (M^*)^n \underset{\Sigma_0}{\longrightarrow} \tilde{N} \quad \text{cofinally,} \tag{9.3.15}$$

where $\tilde{N} = \langle J_{\tilde{\varrho}}^{E'}, \tilde{A} \rangle$.

The identity map $\text{id} \restriction \tilde{N} : \tilde{N} \to (M')^n$ is a Σ_0-preserving embedding, so by the downward extension of embeddings lemmata, there is a uniquely determined structure $\tilde{M}$, a parameter $r \in R^*_{\tilde{M}}$ and a $\Sigma_0^{(n)}$-preserving embedding $\sigma'_1 : \tilde{M} \to M'$ such that $\tilde{N} = \tilde{M}^{n,r}$, $\sigma'_1 \restriction \tilde{N} = \text{id}$ and $\sigma'_1(r) = \pi(p_M) \cap \omega\tilde{\varrho}$. Applying the downward extension of embeddings lemmata to (9.3.15), we infer that there are unique $\bar{N}, \bar{r} \in R^n_{\bar{N}}$ and a $\Sigma_0^{(n)}$-preserving map $\sigma'_0 : \bar{N} \to \tilde{M}$ such that $\bar{N}^{n,\bar{r}} = (M^*)^n$ and $\sigma'_0 \restriction H^n_{\bar{N}} = \sigma' \restriction H^n_{M^*}$. Finally notice that $\sigma'_1 \circ \sigma'_0 \restriction H^n_{\bar{N}} : \bar{N}^{n,\bar{r}} \to (M')^n$ is $\Sigma_0^{(n)}$-preserving, agrees with $\sigma' \restriction H^n_{M^*}$ and $\sigma'_1 \circ \sigma'_0(\bar{r}) = \pi(p_M) - \omega\varrho^n_{M'} = \sigma'(p_{M^*} - \omega\varrho^n_{M^*})$, so one more application of the downward extension of embeddings lemmata yields $\bar{N} = M^*$, $\bar{r} = p_{M^*} - \omega\varrho^n_{M^*}$ and $\sigma'_1 \circ \sigma'_0 = \sigma'$. Informally, the structure $\tilde{M}$ together with the maps σ'_0 and σ'_1 constitute an interpolation to the embedding $\sigma' : M^* \to M'$. Notice that σ'_0 is in fact $\Sigma_0^{(n)}$ cofinal, and therefore is $\Sigma_1^{(n)}$-preserving. As we know that $W^\alpha_{M^*} \in M^*$ for every $\alpha \in p_{M^*} - \omega\varrho^n_{M^*}$, the image of each such witness under σ'_0 is a generalized witness for $\sigma'_0(\alpha)$ w.r.t. $\tilde{M}$ and r in $\tilde{M}$. This together with the fact that $r \in R^n_{\tilde{M}}$ implies that $r = p_{\tilde{M}} - \omega\varrho^n_{\tilde{M}}$, $\tilde{M}$ is n-solid and n-sound. As an immediate consequence of this, d) and g) we have that $\text{rng}(\sigma'_0) = \tilde{h}^n_{\tilde{M}}(\nu \cup \{\sigma'_0(p^*)\})$. Thus, M^* is the transitive collapse of $\text{rng}(\sigma'_0)$, and σ'_0 is the corresponding uncollapsing map, so both M^* and σ'_0 are reconstructible from $\tilde{M}$, σ'_0 and ordinals below ν inside any structure with enough closure properties. Notice also that $\tilde{M}$ can be encoded into a set which is rudimentary over $\tilde{M}^{n,r} = \tilde{N} \in M'$, so $\tilde{M} \in M'$, as it can be reconstructed from $\tilde{N}$ inside M'.

In the following we show that $\tilde{M}$ is a proper initial segment of M^+ where M^+ is as above. For this we use a comparison argument. We first observe that $\tilde{M}$ is a premouse. This is clear if M' is of type A or B, as our assumption $n > 0$ implies that the embedding σ'_1 is Σ_1-preserving; the conclusion then follows from a) of Lemma 9.1.7. If M' is of type C, then $n > 1$, so σ'_1 is in fact $\Sigma_1^{(1)}$-preserving and the conclusion follows from b) of the same lemma. It is then easy to verify that σ'_1 is an embedding of the phalanx $\langle M', \tilde{M}, \omega\tilde{\varrho}\rangle$ into M'. As M' is an iterate of M (indeed, it is a single ultrapower of M), it has an $(\omega_1 + 1, \omega_1)$-iteration strategy, and since M' is countable, we can pick such a strategy $\mathbb{S}_{M'}$ which is minimal relative to some fixed enumeration e' of M'. Let $\langle \mathfrak{I}', \tilde{\mathfrak{I}}\rangle$ be the coiteration of M' against $\langle M', \tilde{M}, \omega\tilde{\varrho}\rangle$ using the iteration strategy $\mathbb{S}_{M'}$ on the M'-side and $\sigma'_1\mathbb{S}_{M'}$ on the phalanx side. Let M'_h, π'_{hk} and $\tilde{M}_h, \tilde{\pi}_{hk}$ denote the premice and the iteration maps on the M'-side and the phalanx side, respectively. This coiteration terminates at some countable θ' and Lemma 9.2.12 guarantees that 0 is on the main branch of $\tilde{\mathfrak{I}}$, there is no truncation point on this branch (so the iteration map $\tilde{\pi}_{0,\theta'}$ is total) and that $\tilde{M}_{\theta'}$ is an initial segment of $M'_{\theta'}$. We show that

Either there is a truncation on the main branch of $\mathfrak{I}'$ or else $\tilde{M}_{\theta'}$ is a *proper* initial segment of $M'_{\theta'}$. (9.3.16)

Assume that this is not the case, i.e. that both iterations are without truncations on their main branches and that $\tilde{M}_{\theta'} = M'_{\theta'}$. We know that the main branch of $\tilde{\mathfrak{I}}$ is above

$\omega\tilde{\varrho} = \omega\varrho^n_{\tilde{M}}$, so all premice along this branch have the same n-th projectum $\omega\tilde{\varrho}$. By construction, however, $\omega\varrho^n_{M'} > \omega\tilde{\varrho}$, so

$$\omega\tilde{\varrho} = \omega\varrho^n_{\tilde{M}_{\theta'}} = \omega\varrho^n_{M'_{\theta'}} \geq \omega\varrho^n_{M'} > \omega\tilde{\varrho},$$

which is absurd. This proves (9.3.16).

(9.3.16) enables us to define j' and ξ' in the same manner as we defined j and ξ^* before, i.e. $j'+1$ is the last truncation point on the main branch of $\mathfrak{I}'$ provided that there is any and ξ' is either θ' if $\tilde{M}_{\theta'}$ is a proper initial segment of $M'_{\theta'}$ or else $\xi_{j'}$. Also, let $M^*_{\xi'}$ be either $\tilde{M}_{\theta'}$ if $\xi' = \theta'$ or else the result of the last truncation on the main branch of $\mathfrak{I}'$. We show that $\tilde{M} = M^*_{\xi'}$. This is clear if $\xi' = \theta'$: In this case $M^*_{\xi'} = \tilde{M}_{\theta'}$ is a proper initial segment of $M'_{\theta'}$, so $\tilde{M}_{\theta'} = \tilde{M}$, as otherwise $\tilde{M}_{\theta'}$ would fail to be sound (recall that the iteration along the main branch of $\tilde{\mathfrak{I}}$ is above $\omega\tilde{\varrho} = \omega\varrho^n_{\tilde{M}}$). In the remaining case $\tilde{M}_{\theta'} = M'_{\theta'} \stackrel{\text{def}}{=} Q$ and $\xi' = \xi_{j'}$. We show

$$\kappa_{j'} \geq \omega\tilde{\varrho} = \omega\varrho^n_{M^*_{\xi'}} \tag{9.3.17}$$

$$\pi'_{\xi',\theta'}(p_{M^*_{\xi'}} - \omega\tilde{\varrho}) = p_Q - \omega\tilde{\varrho} = \tilde{\pi}_{0,\theta'}(p_{\tilde{M}} - \omega\tilde{\varrho}). \tag{9.3.18}$$

From this and the facts that $M^*_{\xi'}$ is sound and $\tilde{M}$ is n-sound we infer:

$$\mathrm{rng}(\tilde{\pi}_{0,\theta}) = \tilde{h}^n_Q(\omega\tilde{\varrho} \cup \{p_Q - \omega\tilde{\varrho}\}) = \mathrm{rng}(\pi'_{\xi',\theta'}),$$

so

$$\tilde{M} = \text{the transitive collapse of } \tilde{h}^n_Q(\omega\tilde{\varrho} \cup \{p_Q - \omega\tilde{\varrho}\}) = M^*_{\xi'}.$$

If $M_{\xi'} = M$, then $\tilde{M}$ is obviously a proper initial segment of M. Otherwise there are two possibilities. The first of them is $\nu_0 > \omega\tilde{\varrho}$, in which case $\tilde{M} = M^*_{\xi'}$ is a proper initial segment of $M'_{\xi'}||\nu_0$, as ν_0 is a cardinal in $M'_{\xi'}$ and $\tilde{M}$ projects to $\omega\tilde{\varrho}$. From the amount of agreement between $M'_{\xi'}$ and M' we obtain that $\tilde{M}$ is a proper initial segment of M. The second possibility is $\nu_0 = \omega\tilde{\varrho}$, in which case $E^{M'}_{\omega\tilde{\varrho}}$ is a superstrong extender in M' and $M'_1 = \mathrm{Ult}^*(M', E^{M'}_{\omega\tilde{\varrho}})$. The same argument as above with M'_1 and ν_1 in place of M' and ν_0 then shows that $\tilde{M}$ is a proper initial segment of M'_1.

It remains to prove (9.3.17) and (9.3.18). To see the former, we note that $\nu_{j'} > \omega\tilde{\varrho}$. Otherwise $j' = 0$ and the first extender applied along the main branch of $\mathfrak{I}'$ would be $E^{M'}_{\omega\tilde{\varrho}}$, which can easily be seen to be superstrong in M'. So $1 = j' + 1$ could not be a truncation point. Knowing this, $\kappa_{j'}$ cannot be less than the n-th projectum of $M^*_{\xi'}$, as otherwise $\omega\varrho^n_Q$ would be larger or equal than $\nu_{j'}$ contradicting the fact that it must be equal to $\omega\tilde{\varrho}$. So $\omega\varrho^n_{M^*_{\xi'}} \leq \kappa_{j'}$, which implies (9.3.17). Regarding the latter, both objects on the left and on the right of the equality (9.3.18) are in P^n_Q by the general properties of fine ultrapowers. That both are equal to $p_Q - \omega\tilde{\varrho}$ then follows by the standard argument from the fact that $M^*_{\xi'}$ is solid and $\tilde{M}$ is solid above $\omega\tilde{\varrho}$. □(Claim 4)

Let W^* be the standard witness for $\pi^*(\nu)$ w.r.t. M and p^*, let $q^* = p^* - (\pi^*(\nu)+1)$ and let $\bar{q}^*$ be the preimage of q^* under the associated witness map. By f) and Claim 4, $W^* \in M^*$. We now use this fact to show that $W \in M$. Since W^* is essentially a subset of $\pi^*(\nu)$, both W^* and q^* are elements of $H^{M^*}_{\varepsilon^*}$ where $\varepsilon^* = \pi^*(\varepsilon)$ for $\varepsilon = \nu^{+M}$. Hence there are $f, g \in H^M_\varepsilon$ such that $W^* = \pi^*(f)(\alpha)$ and $q^* = \pi^*(g)(\beta)$ for some $\alpha, \beta < \lambda$. Notice that we can assume w.l.o.g. that $\varepsilon \in M$, as otherwise $W \in M$ for trivial reasons. Now working in M, fix a transitive elementary substructure H of H^M_ε of size less than ε such that $\nu \cup \{\nu, f, g\} \subset H$. Then $H \in H^M_\varepsilon$ and the same also applies to $H' = \mathrm{Ult}(H, E^M_\nu)$ and the associated ultrapower map π'. Set $W' = \pi'(f)(\alpha)$ and $q' = \pi'(g)(\beta)$.

We next show that W can be obtained as a hull over W'. Fix a $\Sigma^{(n)}_1$-formula $\varphi(v_0, v_1, \dots, v_\ell)$. Given $\xi_1, \dots, \xi_\ell < \nu$, the following statements are equivalent:

- $W \models \varphi(q, \xi_1, \dots, \xi_\ell)$
- $M \models \varphi(q, \xi_1, \dots, \xi_\ell)$
- $M^* \models \varphi(q^*, \pi^*(\xi_1), \dots, \pi^*(\xi_\ell))$
- $W^* \models \varphi(\bar{q}^*, \pi^*(\xi_1), \dots, \pi^*(\xi_\ell))$
- $\pi^*(f)(\alpha) \models \varphi(\pi^*(g)(\beta), \pi^*(\xi_1), \dots, \pi^*(\xi_\ell))$
- $\prec\alpha, \beta\succ \in E^M_\nu(\{\prec\eta_1, \eta_2\succ \in \kappa;\ f(\eta_1) \models \varphi(g(\eta_2), \xi_1, \dots, \xi_\ell)\})$
- $\pi'(f)(\alpha) \models \varphi(\pi'(g)(\beta), \pi'(\xi_1), \dots, \pi'(\xi_\ell))$
- $W' \models \varphi(q', \pi'(\xi_1), \dots, \pi'(\xi_\ell))$

The equivalence between the fifth and sixth statement follows from the fact that the fifth statement is Σ_0 in the parameters $\pi^*(f)(\alpha)$, $\pi^*(g)(\beta)$ and $\pi^*(\xi_h)$ over M^*. Now if $\phi(u_1, \dots, u_\ell)$ is a $\Sigma^{(n)}_1$-formula and $x_1, \dots, x_\ell \in W$, fix $\xi_h < \nu$ such that $x_h = \tilde{h}^{n+1}(\xi_h, \bar{q})$. By the above result we then have

$$\begin{aligned} W \models \psi(x_1, \dots, x_\ell) &\longleftrightarrow W \models \psi(\tilde{h}^{n+1}(\xi_1, \bar{q}), \dots, \tilde{h}^{n+1}(\xi_\ell, \bar{q})) \\ &\longleftrightarrow W' \models \psi(\tilde{h}^{n+1}(\pi'(\xi_1), q'), \dots, \tilde{h}^{n+1}(\pi'(\xi_\ell), \bar{q})). \end{aligned}$$

Letting $\psi(u)$ be the formula $(\exists z^0)(z^0 = u)$, we obtain:

$$\tilde{h}^{n+1}_W(\xi, \bar{q}) \text{ is defined} \quad \text{iff} \quad \tilde{h}^{n+1}_{W'}(\pi'(\xi), q') \text{ is defined.} \tag{9.3.19}$$

Consequently, the assignment $\tilde{h}^{n+1}_W(\xi, \bar{q}) \mapsto \tilde{h}^{n+1}_{W'}(\pi'(\xi), q')$ is a well-defined map $\tilde{\pi} : W \to W'$; the rightmost side of the above equivalence can then be rewritten as $W' \models \psi(\tilde{\pi}(x_1), \dots, \tilde{\pi}(x_\ell))$. It follows that $\tilde{\pi}$ is a $\Sigma^{(n)}_1$-preserving embedding of W into W'. Another consequence of (9.3.19) is that W is the transitive collapse of $\tilde{h}^{n+1}_{W'}((\mathrm{rng}(\pi') \cap \nu) \cup \{q'\})$. But we know that both W' and π' are elements of H^M_ε, so $\mathrm{rng}(\pi') \cap \nu$ is also in H^M_ε. Consequently, $W \in H^M_\varepsilon$, too. This completes the proof of Case 3, and thus the proof of the solidity theorem. □(Theorem 9.3.1)

Having established the solidity theorem, we are ready to prove the condensation lemma. The proof is again based on the comparison argument and we will also borrow many arguments from the above proof of the solidity theorem.

Theorem 9.3.2 (Condensation Lemma). *Let M and $\bar{M}$ be premice of the same type and let $\sigma : \bar{M} \to M$ be a premice embedding which is both cardinal preserving and $\Sigma_0^{(n)}$-preserving for n such that $\sigma \restriction \omega\varrho_{\bar{M}}^{n+1} = \mathrm{id}$. Assume further that M is weakly iterable. Then $\bar{M}$ is weakly iterable, and therefore solid. Moreover, granted that $\bar{M}$ is sound above $\nu = \mathrm{cr}(\sigma)$, exactly one of the following holds:*

a) *$\bar{M} = \mathrm{core}_\nu(M)$ and σ is the associated core map.*

b) *$\bar{M}$ is a proper initial segment of M.*

c) *$\bar{M} = \mathrm{Ult}^*(M||\delta, E_\alpha^M)$ where δ is maximal with the property that ν is still a cardinal in $M||\delta$, $\alpha \leq \delta$, $\kappa = \mathrm{cr}(E_\alpha^M)$ is the cardinal predecessor of ν in $M||\delta$ and at the same time the unique generator of E_α^M.*

d) *$\bar{M}$ is a proper initial segment of* $\mathrm{Ult}(M||\nu, E_\nu^M)$.

Before going to the proof of the condensation lemma, let us make the following comments. In d), $\mathrm{Ult}(M||\nu, E_\nu^M)$ can be obviously replaced by $\mathrm{Ult}(J_{\tilde{\tau}}^E, E_\nu^M)$ where $\tilde{\tau} = \mathrm{cr}(E_\nu^M)^{++M}$ and if σ is cardinal preserving, also by $\mathrm{Ult}(M, E_\nu^M)$. Regarding the critical point of σ, we might also allow ν to be the height of $\bar{M}$, in which case ν would not be a critical point in the usual sense. As $\sigma \restriction \bar{M} = \mathrm{id}$ then, the conclusion of the condensation lemma follows immediately. Notice that in both c) and d), $\bar{M}$ is an ultrapower by an internal extender. In c), this extender is essentially a measure. In d), it is easy to see that E_ν^M is superstrong in M. Furthermore, the ultrapower in d) is really *coarse,* so $\bar{M}$ can be considered an initial segment of an internal ultrapower in this case.

Proof of Theorem 9.3.2. The conclusion on weak iterability of $\bar{M}$ follows by an easy Löwenheim–Skolem argument, so we leave it to the reader and rather focus on the second part of the theorem. From now on assume that M is countable and (ω_1+1, ω_1)-iterable. The proof proceeds by induction on the length of M, so we make the following additional assumption:

$$\begin{array}{l}\text{The condensation lemma holds for any } N, \bar{N} \\ \text{and } \bar{\sigma} : \bar{N} \to N \text{ such that } \mathrm{ht}(N) < \mathrm{ht}(M).\end{array} \tag{9.3.20}$$

Fix an (ω_1+1, ω_1)-iteration strategy $\mathbb{S}$ for M which is e-minimal for some enumeration e of M. The assumptions of the theorem clearly guarantee that $\langle M, \bar{M}, \nu\rangle$ is a phalanx and that σ is an embedding of this phalanx into M. Letting $\langle \mathfrak{I}, \bar{\mathfrak{I}}\rangle$ be the coiteration of M against $\langle M, \bar{M}, \nu\rangle$ using the iteration strategy $\mathbb{S}$ on the M-side and $\sigma\mathbb{S}$ on the phalanx side, denote the premice, the iteration maps and the iteration tree on the M-side by M_h, π_{hk} and T, and those on the phalanx side by $\bar{M}_h$, $\bar{\pi}_{hk}$ and $\bar{T}$. The coiteration $\langle \mathfrak{I}, \bar{\mathfrak{I}}\rangle$ terminates at some countable θ and Lemma 9.2.12 guarantees the following:

- 0 is on the main branch of $\mathfrak{I}$.
- There is no truncation on the main branch of $\mathfrak{I}$.

- $\bar{M}_\theta$ is an initial segment of M_θ.

We shall adopt the same notation as in the proof of the solidity theorem: $i+1$ is the result of the first ultrapower on the main branch of $\Im$ and $j+1$ is the last truncation point on this branch, provided that there is any. Furthermore, $\xi^* = \theta$ if $\bar{M}_\theta$ is a *proper* initial segment of M_θ and $\xi^* = \xi_j$ if $\bar{M}_\theta = M_\theta$ and j is defined. In the former case we set $M^* = \bar{M}_\theta$, in the latter case $M^* = M_j^*$. Clearly, M^* is a proper initial segment of M_{ξ^*}. Notice that we can also w.l.o.g. assume that

$$\omega\varrho_{\bar{M}}^{n+1} \leq \nu < \omega\varrho_{\bar{M}}^{n}.$$

If this fails, we can always replace n by that $\bar{n} < n$ which satisfies the above formula. The rest of the proof reduces to the discussion of four cases.

Case 1. There is no truncation on the main branch of $\Im$ and $\bar{M}_\theta = M_\theta$.

We show that a) of the condensation lemma holds. Assume first that $\kappa_i \geq \omega\varrho_M^{n+1}$. Following the proof of Claim 1 of the solidity theorem, we obtain:

(E_0) $\omega\varrho_{\bar{M}}^m = \omega\varrho_M^m \stackrel{\text{def}}{=} \omega\varrho^m$ whenever $m > n$;

(E_1) both sides of the coiteration are above $\omega\varrho^{n+1}$;

(E_2) $\mathcal{P}(\omega\varrho^{n+1}) \cap \bar{M} = \mathcal{P}(\omega\varrho^{n+1}) \cap M$;

(E_3) $\bar{\pi}_{0,\theta}(p_{\bar{M}}) = p_{M'} = \pi_{0,\theta}(p_M)$.

Clause (E_3) follows from the properties of iteration maps, as we know that both M and $\bar{M}$ are solid. The next step is to show:

$$\sigma \text{ is } \Sigma_0^{(n)} \text{ cofinal.} \tag{9.3.21}$$

Assume this fails; then

$$\omega\tilde{\varrho} \stackrel{\text{def}}{=} \sup(\sigma''\omega\varrho_{\bar{M}}^n) < \omega\varrho_M^n.$$

Fix a $\Sigma_1^{(n)}(M)$-relation $A(u,v)$ such that $a = A(-,p_M) \cap \omega\varrho^{n+1} \notin M$. By ($E_0$), ($E_3$) and the preservation properties of the iteration maps, $\bar{A}(-,p_{\bar{M}}) \cap \omega\varrho^{n+1} = a$ where $\bar{A}$ is $\Sigma_1^{(n)}(\bar{M})$ by the same definition as A. Suppose $\bar{A}(u,v) \longleftrightarrow (\exists z^n \bar{B})(z^n,u,v)$ for some $\Sigma_0^{(n)}(\bar{M})$-relation $\bar{B}$. It is then easy to see that

$$\bar{A}(x,y) \longleftrightarrow (\exists z^n \in J_{\tilde{\varrho}}^E) B(z^n, \sigma(x), \sigma(y))$$

where we have $E = E^M$ and B is the $\Sigma_0^{(n)}(M)$-relation by the same definition as $\bar{B}$. In particular, the membership $\xi \in a$ can be expressed in a $\Sigma_0^{(n)}(M)$ fashion as $(\exists z^n \in J_{\tilde{\varrho}}^E) B(z^n, \xi, \sigma(p_{\bar{M}}))$, so $a \in M$. This contradicts our choice of a, and thus proves (9.3.21).

An argument of similar spirit yields in fact the following:

$$\sigma(p_{\bar{M}}^{n+1}) = p_M^{n+1}. \tag{9.3.22}$$

Notice that any $\alpha \in \sigma(p_{\bar{M}}^{n+1})$ has a generalized witness w.r.t. M and $\sigma(p_{\bar{M}})$ in M. This follows from the solidity of $\bar{M}$ and the preservation properties of σ – at this point we know that σ is $\Sigma_1^{(n)}$-preserving. Thus, $\sigma(p_{\bar{M}}^{n+1}) \leq^* p_M^{n+1}$. Assume that this equality is strict; let $\alpha \in p_M^{n+1} - \sigma(p_{\bar{M}}^{n+1})$ be such that $p_M^{n+1} - (\alpha+1) = \sigma(p_{\bar{M}}^{n+1}) - (\alpha+1)$. Then $\nu \cup \sigma(p_{\bar{M}}) \subset \mathrm{rng}(\sigma_\alpha)$ where $\sigma_\alpha : W_M^\alpha \to M$ is the canonical witness map associated with W_M^α. Since this witness map is $\Sigma_1^{(n)}$-preserving, the whole $\mathrm{rng}(\sigma) = \tilde{h}_M^{n+1}(\nu \cup \{\sigma(p_{\bar{M}})\})$ is contained in $\mathrm{rng}(\sigma_\alpha)$ and the preimage of $\mathrm{rng}(\sigma)$ under σ_α is equal to $X = \tilde{h}_W^{n+1}(\nu \cup \{\sigma(p_\alpha)\})$ where $W = W_M^\alpha$ and $p = \sigma_\alpha^{-1} \circ \sigma(p_{\bar{M}})$. It is then easy to see that collapsing this hull X gives rise to a $\Sigma_1^{(n)}$-preserving embedding $\bar{\sigma}_\alpha : \bar{M} \to W$ which is the identity on ν. If we now choose the set $a \subset \omega\varrho^{n+1}$ as in the proof of (9.3.21), we conclude that the membership $\xi \in a$ can be expressed in a $\Sigma_1^{(n)}$-fashion over $W \in M$, so we again arrive at the contradiction $a \in M$. This completes the proof of (9.3.22).

Now (9.3.21) and (9.3.22) imply that $\sigma : \bar{M} \to M$ meets the assumptions of Lemma 1.11.2, so σ is in fact Σ^*-preserving and $\sigma(p_{\bar{M}}) \in P_M^*$. However, as $\bar{M}$ is solid, each element of $\sigma(p_{\bar{M}})$ has a generalized witness w.r.t. M and $\sigma(p_{\bar{M}})$ in M, so in fact $\sigma(p_{\bar{M}}) = p_M$. Consequently, a) of the condensation lemma holds. This completes the proof of Case 1 granting that $\kappa_i \geq \omega\varrho_M^{n+1}$.

In the following we show that κ_i must always be larger than or equal to $\omega\varrho_M^{n+1}$. Suppose for a contradiction that $\kappa_i < \omega\varrho_M^{n+1}$. Since this implies that $\omega\varrho_M^{n+1} \geq \nu_i \geq \nu$, it can happen only when $\nu_i = \nu$, in which case $i = 0$ and $E_{\nu_i}^{M_i} = E_\nu^M$ is a superstrong extender in M. This time the iteration $\mathfrak{I}$ is above $\omega\varrho_{M_1}^{n+1}$ along the branch $[1, \theta)_T$, so we obtain (E_0) – (E_3) with M_1 in place of M. Furthermore, $\omega\varrho_M^{n+1} = \tau_i < \nu = \omega\varrho_{\bar{M}}^{n+1}$. Using the arguments from above and those from the proof of (9.3.8) we again prove (9.3.21) and (9.3.22). Since σ is the identity map on $\omega\varrho_M^{n+1}$, these two facts immediately imply that M and $\bar{M}$ have the same $(n+1)$-st projectum. (Choose a $\Sigma_1^{(n)}(M)$-relation $D(u, v)$ such that $d = D(-, p_M^{n+1}) \cap \tau_i \notin M$. If $\bar{D}$ is $\Sigma_1^{(n)}(\bar{M})$ by the same definition, then $d = \bar{D}(-, p_{\bar{M}}^{n+1}) \cap \tau_i \notin \bar{M}$.) However, we know that the projecta of M and $\bar{M}$ are different, a contradiction.

Now if $B(u, v)$ is $\Sigma_1^{(n)}(M)$ and $r \in M$ then, following the proof of (9.3.8), we infer that $B_1(-, \pi_{0,1}(r)) \cap \nu$ is the union of all sets $\pi_{0,1}(B(-, r) \cap \zeta)$ for $\zeta < \tau$, where B_1 is the $\Sigma_1^{(n)}(M_1)$-relation by the same definition as B. From this we obtain that $B(-, r) \cap \tau$ can be reconstructed from $B_1(-, \pi_{0,1}(r)) \cap \nu$ as the union of all sets x such that $\pi_{0,1}(x)$ is an initial segment of $B_1(-, \pi_{0,1}(r)) \cap \nu$. Knowing that $B_1(-, \pi_{0,1}(r)) \cap \nu$ is an element of M_1, we can conclude that $B(-, r) \cap \tau$ is in M as well.

Now assuming that σ is not cofinal, set $\omega\tilde{\varrho} = \sup(\sigma''\omega\varrho_{\bar{M}}^n)$; then $\omega\tilde{\varrho} < \omega\varrho_M^n$. Pick a $\Sigma_1^{(n)}(M)$-relation $B(u, v)$ such that $B(-, p_M^{n+1}) \cap \tau \notin M$; by the previous paragraph,

$B_1(-, \pi_{0,1}(p_M^{n+1})) \cap \nu \notin M_1$. As before we then show that $B_1(-, \pi_{0,1}(p_M^{n+1})) \cap \nu = \bar{B}(-, p_{\bar{M}}^{n+1}) \cap \nu$ for some $\bar{B}$ which is $\Sigma_1^{(n)}(\bar{M})$, so $B_1(-, \pi_{0,1}(p_M^{n+1})) \cap \nu$ is $\Sigma_0^{(n)}(M)$ and therefore an element of M. Finally $B(-, p_M^{n+1}) \cap \tau$ must be in M by the previous paragraph, a contradiction. This proves (9.3.21). The proof of (9.3.22) is almost the same; this time we obtain that $B_1(-, \pi_{0,1}(p_M^{n+1})) \cap \nu$ is $\Sigma_1^{(n)}(M)$ in $\sigma(p_{\bar{M}}^{n+1})$; so the assumption $\sigma(p_{\bar{M}}^{n+1}) <^* p_M^{n+1}$ would imply $B_1(-, \pi_{0,1}(p_M^{n+1})) \in M$ and yield a contradiction as above. The inequality $\sigma(p_{\bar{M}}^{n+1}) \leq^* p_M^{n+1}$ follows, as before, from the fact that $W_M^\alpha \in M$ for every $\alpha \in p_M^{n+1}$, which in turn is a consequence of the preservation properties of σ and the solidity of $\bar{M}$. This completes the discussion of Case 1. □(Case 1)

From now on we assume that ξ^* is defined, i.e.

$$\begin{array}{l}\text{either there is a truncation on the main branch of } \mathfrak{I} \\ \text{or else } \bar{M}_\theta \text{ is a proper initial segment of } M.\end{array} \tag{9.3.23}$$

Case 2. $\bar{M}$ is *not* sound.

We show that c) of the condensation lemma is the case. Notice first that $\bar{M}_\theta$ fails to be sound as well, so $\bar{M}_\theta = M_\theta$ and ξ_j is defined. Setting $p^* = p_{M_j^*}$, we have $\pi_{\xi_j,\theta}(p^*) = p_{M_\theta} = \bar{\pi}_{0,\theta}(p_{\bar{M}})$. Furthermore, $\bar{M}$, M_θ and M_j^* have the same $(n+1)$-st projectum $\omega\varrho^{n+1} \leq \nu, \kappa_j$. Notice that $\kappa_j \geq \nu$, is not possible, as otherwise the following would be true:

$$\begin{aligned}\operatorname{rng}(\bar{\pi}_{0,\theta}) &= \bar{\pi}''_{0,\theta}\tilde{h}_{\bar{M}}^{n+1}(\nu \cup \{p_{\bar{M}}\}) = \tilde{h}_{M_\theta}^{n+1}(\nu \cup \{p_{M_\theta}\}) \\ &= \pi''_{\xi_j,\theta}\tilde{h}_{M_j^*}^{n+1}(\nu \cup \{p^*\}) = \operatorname{rng}(\pi_{\xi_j,\theta}).\end{aligned}$$

The first and the last equality follows from the soundness of $\bar{M}$ above ν and from the full soundness of M, respectively. Thus, we would obtain the consequence $\bar{M} = M_i^*$ which is obviously false, as $\bar{M}$ is not sound. So $\kappa_j < \nu$, and the only way this can happen is if κ_j is equal to the largest cardinal κ^* in the structure J_ν^E where $E = E^M$. To see this, recall that any J_ν^E-cardinal is a cardinal both in $\bar{M}$ and M, as ν is a cardinal in $\bar{M}$ and σ is cardinal preserving. Furthermore, each of these ordinals remains a cardinal in any M_h, so $j+1$ would fail to be a truncation point if $\kappa_j < \kappa^*$. The following are more or less immediate consequences of the above discussion:

- $\nu_j > \nu$, so $\lambda_j > \nu$ as well and $\operatorname{cr}(\pi_{j+1,\theta}) > \nu$.
- $\omega\varrho^{n+1} \leq \kappa^*$.
- $\bar{M} = \operatorname{Ult}^*(M_j^*, U)$ where $U = E_{\nu_j,\kappa^*}^{M_j}$. So $\bar{M}$ is sound above $\kappa^* + 1$.

We give more details regarding the last assertion. Let $\bar{M}' = \operatorname{Ult}^*(M_j^*, U)$ and let π^0 be the associated ultrapower map. Define $\pi^1 : \bar{M}' \to M_{j+1}$ by $\pi^1(f)(\kappa^*) =$

$\pi_{\xi_j,j+1}(f)(\kappa^*)$ for $f \in \Gamma(\kappa^*, M_j^*)$. Notice that $\omega\varrho_{M_j^*}^{n+1} \leq \kappa^* < \omega\varrho_{M_j^*}^{n}$, so Łos' Theorem for $\Sigma_0^{(n)}$-formulae holds for both $\mathrm{Ult}^*(M_j^*, U)$ as well as $\mathrm{Ult}^*(M_j^*, E_{\nu_j}^{M_j})$ and the standard argument shows that π^1 is $\Sigma_0^{(n)}$ cofinal and that $\pi^1 \restriction (\kappa^* + 1) = \mathrm{id}$. By the general properties of ultrapowers we have $\bar{M}' = \tilde{h}_{\bar{M}'}^{n+1}(\kappa^* \cup \{\kappa^*, \pi^0(p^*)\})$, so $\mathrm{rng}(\pi^1) = \tilde{h}_{M_{j+1}}^{n+1}(\kappa^* \cup \{\kappa^*, p_{M_{j+1}}\})$. Since both $E_{\nu_j}^{M_j}$ and U are weakly amenable w.r.t. M_j^*, all three premice M_j^*, $\bar{M}'$ and M_{j+1} have the same subsets of κ^* and therefore also the same successor of κ^*, namely ν. Now $\mathrm{cr}(\pi^1) \geq \nu$, so the ordinal ν is contained in $\tilde{h}_{M_{j+1}}^{n+1}(\kappa^* \cup \{\kappa^*, p_{M_{j+1}}\})$. Since the critical points of both $\pi_{j+1,\theta}$ and $\bar{\pi}_{0,\theta}$ are at least ν, we conclude that $\bar{M} = \tilde{h}_{\bar{M}}^{n+1}(\kappa^* \cup \{\kappa^*, p_{\bar{M}}\})$, as ν is contained in this hull. It follows that $\bar{M}$ is the transitive collapse of $(\pi_{j+1,\theta}^{-1} \circ \bar{\pi}_{0,\theta})''\tilde{h}_{\bar{M}}^{n+1}(\kappa^* \cup \{\kappa^*, p_{\bar{M}}\}) = \tilde{h}_{M_{j+1}}^{n+1}(\kappa^* \cup \{\kappa^*, p_{M_{j+1}}\})$, so $\bar{M}' = \bar{M}$.

Letting i^* be the least ordinal h such that $E_{\nu_h}^{M_h} \neq \varnothing$, we have $\lambda_{i^*} \geq \kappa^*$. Since M_{i^*} agrees with $\bar{M}_{i^*}$ below ν_{i^*} and ν is a successor cardinal in $\bar{M}_{i^*}$, we conclude that $\lambda_{i^*} > \nu > \kappa^*$ whenever $\nu_{i^*} > \nu$, so $\xi_j = i^*$. Notice also that ν_{i^*} is always larger than ν, as otherwise $i^* = 0$ and $\lambda_{i^*} = \kappa^*$, so $\xi_j > 0$. Consequently, ν is a cardinal in M_{ξ_j}, so $j + 1$ could not be a truncation point. This proves that M_j^* is actually a proper initial segment of M, so we have

$$\bar{M} = \mathrm{Ult}^*(M||\eta_j, U). \tag{9.3.24}$$

In the following we show that U can be replaced by some E_α^M which is generated by U. More precisely, let $J_\alpha^{E^*} = \mathrm{Ult}(J_\nu^E, U)$, let π^* be the associated ultrapower map and let $F^* = \pi^* \restriction (\mathcal{P}(\kappa^*) \cap J_\nu^E)$. We show that $F^* = E_\alpha^M$ for some $\alpha \leq \mathrm{ht}(M||\eta_j)$; the latter will follow from the fact that ν is the cardinal successor of κ^* in $M^* = \langle J_\alpha^{E^*}, F^* \rangle$.

The first step towards the above assertion is the following observation:

$$\text{Either } \nu_j \in M_j \text{ or else there is a truncation point } k + 1 \leq_T j. \tag{9.3.25}$$

Assume for a contradiction that this fails; then $\pi_{0,j}$ is total and $E_{\nu_j}^{M_j}$ is the top extender of M_j. Let M_{j^*+1} be the result of the first ultrapower on the branch $[0, j]_T$. Hence $M_{\xi_{j^*}} = M$ and $M_{j^*+1} = \mathrm{Ult}(M, E_{\nu_{j^*}}^{M_{j^*}})$. Now κ_{j^*} cannot be $\geq \kappa^*$, as we would obtain $\mathcal{P}(\kappa^*) \cap M = \mathcal{P}(\kappa^*) \cap M_j = \mathcal{P}(\kappa^*) \cap J_\nu^E$. This is, however impossible, as we know that the set on the left is properly larger than the one on the right. So $\kappa_{j^*} < \kappa^*$. We also know that $\lambda_{j^*} \geq \kappa^*$; let us show that this inequality is in fact strict. Indeed, the equality $\lambda_{j^*} = \kappa^*$ would yield $\nu_{j^*} = \nu$ and $j^* = 0$. By the discussion above (9.3.24), however, ν_0 must be larger than ν. So $\pi_{0,j}(\kappa_{j^*}) \geq \lambda_{j^*} > \kappa^*$ (it is easy to see that we have an equality on the left). Now κ^*, being the critical point of $E_{\mathrm{top}}^{M_j}$, must be in the range of $\pi_{0,j}$, so say $\kappa^* = \pi_{0,j}(\bar{\kappa})$. It follows that $\bar{\kappa} < \kappa_{j^*}$. But then $\kappa^* = \pi_{0,j}(\bar{\kappa}) = \bar{\kappa}$, as κ_{j^*} is the critical point of $\pi_{0,j}$. Contradiction. This completes the proof of (9.3.25).

Let $M' = M_j||\nu_j$ if $\nu_j < \mathrm{ht}(M_j)$ and $M' = M_k^*$ otherwise, where $k + 1$ is the largest truncation point on the branch $[0, j]_T$. Also, let $k' = j$ if the former holds and

$k' = \xi_k$ otherwise. If $\nu_j = \mathrm{ht}(M_j)$, then k is defined, $\mathrm{cr}(E^{M'}_{\mathrm{top}}) = \kappa^*$ and $E^{M'}_{\mathrm{top},\kappa^*} = U$. These equalities are immediate consequences of the fact that $\mathrm{cr}(\pi_{\xi_k,j}) = \kappa_k > \kappa^*$. Indeed, M' and M_j then have the same power sets of κ^* and given any $x \subset \kappa^*$, we have $\pi_{\xi_k,j}(x) = x$, so

$$x \in U \longleftrightarrow \kappa^* \in E^{M_j}_{\mathrm{top}}(x) \longleftrightarrow \kappa^* \in E^{M'}_{\mathrm{top}}(x).$$

To see that $\kappa_k > \kappa^*$, notice first that $\kappa_k \geq \kappa^*$, as otherwise $k+1$ would fail to be a truncation point (recall that κ^* is a cardinal in each M_h). But κ_k cannot be equal to κ^*, as κ^*, being the critical point of $E^{M_j}_{\mathrm{top}}$ must be in the range of $\pi_{\xi_k,j}$, which would contradict the fact that $\kappa^* = \kappa_k$ is the critical point of $\pi_{\xi_k,j}$.

We are now ready to show that F^* is on the $E^{M_{k'}}$-sequence. For this we use the condensation lemma and (9.3.20). Let $\bar\nu \in M$ be such that $E^M_{\bar\nu} \neq \varnothing$, let $\bar U = E^M_{\bar\nu,\bar\kappa}$ where $\bar\kappa = \mathrm{cr}(E^M_{\bar\nu})$ and let $\bar M^*$ be defined from $\bar U$ in the same manner as M^* was defined from U (M^* was defined below (9.3.24)). If $E^M_{\bar\nu}$ is of type B or C, we replace it by the cutoff at its least cutpoint; this will guarantee that both $M||\bar\nu$ and $\bar M^*$ will be of the same type, namely type A. It follows that $M||\bar\nu$ and $\bar M^*$ have the same successor of $\bar\kappa$, say $\bar\tau$. Furthermore, $\omega\varrho^1_{\bar M^*} \leq \bar\tau$, $p_{\bar M^*} - \nu = \varnothing$ and $\bar M^*$ is sound above $\bar\tau$. Indeed, any element of $\bar M^*$ is of the form $\bar\pi(f)(\bar\kappa) = E^{\bar M^*}_{\mathrm{top}}(f)(\bar\kappa)$ for some $f : \bar\kappa \to J^E_{\bar\tau}$ where $\bar\pi$ is the ultrapower map associated with $\mathrm{Ult}(J^E_{\bar\tau}, \bar U)$, and thus is $\Sigma_1(\bar M^*)$-definable from ordinals less than $\bar\tau$. (Recall that any f as above can be uniformly encoded into a subset of $\bar\kappa$ in a Σ_1-fashion). Łos' Theorem for Σ_0-formulae implies that the formula $E^{\bar M^*}_{\mathrm{top}}(f)(\bar\kappa) \mapsto E^M_{\bar\nu}(f)(\bar\kappa)$ defines a Σ_0-preserving map from $\bar M^*$ into $M||\bar\nu$ which is the identity on $\bar\nu$; call this map $\bar\sigma$. By (9.3.20), the condensation lemma can be applied to $M||\bar\nu$, $\bar M^*$ and $\bar\sigma$. If a) holds, then $\bar M$ is the core of $M||\bar\nu$, so $\bar M = M||\bar\nu$, as $M||\bar\nu$ is sound. If b) holds, then $\bar M^*$ is a proper initial segment of $M||\bar\nu$. Notice that $\mathrm{cr}(\bar\sigma) > \bar\tau$; denote this critical point by $\bar\nu^*$. From this we infer that both c) and d) are false. Indeed, c) would imply that $\bar M^*$ fails to be sound above the cardinal predecessor of $\bar\nu^*$ which is $\geq \bar\tau$ and d) would require that $\bar M^*$ cannot project below $\bar\nu^*$. This means that $\bar M^*$ is an initial segment of $M||\bar\nu$, so the following statement holds in M:

$$(\forall\bar\nu)[E_{\bar\nu} \neq \varnothing \longrightarrow (\exists\alpha \leq \bar\nu)\psi(\alpha, \bar\nu)] \tag{9.3.26}$$

where $\psi(\alpha, \bar\nu)$ is the formula

E_α is a total extender in $M||\bar\nu$ with a single generator $\mathrm{cr}(E_\alpha) = \mathrm{cr}(E_{\bar\nu})$.

It is a routine to verify that the statement (9.3.26) is Π_1, so it is preserved under iteration maps. As Π_1-statements are reflected by transitive substructures, (9.3.26) holds in any iterate of M, no matter whether the iterate in question is simple or non-simple. In particular, (9.3.26) holds in $M_{k'}$. Thus, there is an $\alpha \leq \mathrm{ht}(M')$ such that $E^{M_{k'}}_\alpha$ has a single generator κ^* and $E^{M_{k'}}_{\alpha,\kappa^*} = E^{M'}_{\mathrm{top},\kappa^*} = U$. In other words, $E^{M_{k'}}_\alpha = F^*$.

It remains to show that $M_{k'}||\alpha = M||\alpha$. Since this is clear if $M_{k'} = M$, we can assume w.l.o.g. that $M_{k'} \neq M$. Then $E^{M_h}_{\nu_h} \neq \varnothing$ for some $h < k$, so ν_0 is a cardinal in $M_{k'}$. The discussion above (9.3.24) yields that $\nu_0 > \nu$, so $\alpha < \nu_0$, as $M_{k'}||\alpha$ projects to ν. The conclusion then follows from the agreement up to ν_0 between M and $M_{k'}$. This completes the proof in Case 2. □(Case 2)

From now on we assume that $\bar{M}$ is sound. We show that $\bar{M} = M^*_{\xi^*}$. This is clear if $\bar{M}_\theta$ is a proper initial segment of M_θ (so $\xi^* = \theta$): In this case we have $\bar{M}_\theta = \bar{M}$, as otherwise $\bar{M}_\theta$ would fail to be sound. Now assume $\bar{M}_\theta = M_\theta$, so j is defined. From the general properties of iterations we infer (using the standard arguments) that M^*_j and $\bar{M}$ are both sound, have the same $(n+1)$-st projectum, say $\omega\varrho^{n+1}$, and both $\pi_{\xi_j,\theta} : M^*_j \to M_\theta$ and $\bar{\pi}_{0,\theta} : \bar{M} \to M_\theta$ are Σ^*-preserving and agree with the identity map below $\omega\varrho^{n+1}$. Thus,

$$\bar{M} = \text{the transitive collapse of } \tilde{h}^{n+1}_{M_\theta}(\omega\varrho^{n+1} \cup \{p_{M_\theta}\}) = M^*_j.$$

Case 3. $\nu_0 > \nu$.

We show that b) of the condensation lemma is the case. If $M_{\xi_j} = M$, this is immediate. Otherwise ν_0 is a cardinal in M_{ξ_j}, so the fact that $\bar{M}$ projects to ν guarantees that $\bar{M}$ is in fact a proper initial segment of $M_{\xi_j}||\nu_0$. The rest then follows from the agreement below ν_0 between M and M_{ξ_j}. □(Case 3)

Case 4. $\nu_0 = \nu$.

Then E^M_ν is a superstrong cardinal in M and $M_1 = \text{Ult}^*(M, E^M_\nu)$. Let κ, τ and λ have the same meaning as in the proof of the solidity theorem, i.e. $\kappa = \text{cr}(E^M_\nu)$, $\tau = \kappa^{+M}$ and $\lambda = \lambda(E^M_\nu)$. As $\nu_1 > \nu$, the same argument as in Case 3 proves that $\bar{M}$ is a proper initial segment of M_1, so at this point we have proved a version of the condensation lemma where d) is replaced by the weaker

d*) $\bar{M}$ is a proper initial segment of $\text{Ult}^*(M, E^M_\nu)$.

As in the proof of the solidity theorem we observe that if $\omega\varrho^m_M \neq \tau$ for all $m \in \omega$, then $\text{Ult}^*(M, E^M_\nu)$ and $\text{Ult}(M, E^M_\nu)$ have the same cardinal successor of ν, say $\tilde{\nu}$, and agree below this successor. It is also easy to see that the initial segment of any of these ultrapowers up to $\tilde{\nu}$ is precisely the ultrapower $\text{Ult}(J^E_{\tilde{\tau}}, E^M_\nu)$ where $\tilde{\tau} = \tau^{+M}$. This yields clause d). From now on assume that $\omega\varrho^m_M = \tau$ for some $m > n$. We will discuss several possibilities.

Case 4.1. σ is not cofinal and $n > 0$.

Then $\omega\tilde{\varrho} = \sup(\sigma''\omega\varrho^n_{\bar{M}}) < \omega\varrho^n_M$. We construct a premouse $\tilde{M}$ together with embeddings $\sigma_0 : \bar{M} \to \tilde{M}$ and $\sigma_1 : \tilde{M} \to M$ interpolating σ as we did in the proof of Claim 4 of the proof of the solidity theorem. Thus, $\tilde{M}^{n,p_{\tilde{M}}} = \langle J^E_{\tilde{\varrho}}, \tilde{A}\rangle$ where $\tilde{A} = \bigcup\{\sigma(A^n_{\bar{M}} \cap x);\ x \in H^n_{\bar{M}}\} = A^{n,\sigma(p_{\bar{M}})}_M \cap J^E_{\tilde{\varrho}}$ and $\tilde{M}$ is the unique acceptable structure with this property. The embeddings σ_0 and σ_1 are $\Sigma^{(n)}_0$-preserving; σ_0 is, moreover, cofinal. $\tilde{M}$ is sound and solid above $\omega\tilde{\varrho} = \omega\varrho^n_{\bar{M}}$ and $\sigma_1 \restriction \omega\tilde{\varrho} = \text{id}$. Also, $\tilde{M} \in M$, as $\tilde{M}$ is reconstructible from $\tilde{A} \in M$. It follows that $\text{ht}(\tilde{M}) < \text{ht}(M)$. By the preservation properties of σ_0 and σ_1 and by Lemmata 9.1.5 and 9.1.7, $\tilde{M}$ is a premouse; Corollary 9.2.13, then guarantees that $\tilde{M}$ has an (ω_1+1, ω_1)-iteration strategy. We now apply the condensation lemma to $\bar{M}$, $\tilde{M}$ and σ_0. Notice first that $\nu = \mathrm{cr}(\sigma_0)$ and that $\tilde{M}$ agrees with M below $\sigma(\nu) > \nu$, so $\tilde{M}$ cannot project below $\sigma(\nu)$, as $\sigma(\nu)$ is a cardinal in M. It follows that $\bar{M}$ cannot be the core of $\tilde{M}$ above ν, as the $(n+1)$-st projecta of $\bar{M}$ and $\tilde{M}$ differ. Thus, clause a) of the condensation lemma fails. Clause b) fails too, as $E^{\bar{M}}_\nu = \varnothing$ and $E^{\tilde{M}}_\nu = E^M_\nu$ is a superstrong extender in $\tilde{M}$. Finally, clause c) is false, because we assume that $\bar{M}$ is sound. Thus, $\bar{M}$ must be a proper initial segment of $\mathrm{Ult}(\tilde{M}||\nu, E^{\tilde{M}}_\nu) = \mathrm{Ult}(M||\nu, E^M_\nu)$. This completes the proof of Case 4.1.

Case 4.2. σ is not cofinal and $n = 0$.

If $E^{\bar{M}}_{\mathrm{top}} = \varnothing$, let J^E_α be the cofinal image of $\bar{M}$ under σ. Then $\sigma : \bar{M} \to J^E_\alpha$ is Σ_0 cofinal and we can proceed exactly as in the discussion at the very beginning of the proof of the condensation lemma. From now on assume that $E^{\bar{M}}_{\mathrm{top}} \neq \varnothing$. Here we cannot imitate the argument from Case 4.1, as the intermediate structure $\tilde{M}$ might fail to be a premouse. More precisely, the top extender of $\tilde{M}$ might fail to be weakly amenable w.r.t. $\tilde{M}$. To get around this obstacle, we will adopt the arguments from the proofs of Case 3 and Claim 4 in the proof of the solidity theorem.

We first observe that M can be replaced by $\tilde{M} = \mathrm{core}_\tau(M)$. Assume we are able to prove that in $\tilde{M}$ we have:

$$(\forall N, \bar{N}, \tilde{\sigma}, \tilde{\nu})[\psi(N, \bar{N}, \tilde{\sigma}, \tilde{\nu}) \longrightarrow \psi'(\bar{N}, N, \tilde{\nu})] \tag{9.3.27}$$

where $\psi'(\bar{N}, N, \tilde{\nu})$ says

$$\bar{N} \text{ is a proper initial segment of } \mathrm{Ult}(N||\tilde{\nu}, E^{\tilde{M}}_{\tilde{\nu}})]$$

and $\psi(N, \bar{N}, \tilde{\sigma}, \tilde{\nu})$ is the conjunction of the following statements:

- N is a coherent structure and $\lambda_{\bar{M}} \in M$;
- $(\forall x, y \in N)(\langle x, y\rangle \in E^N_{\mathrm{top}} \longleftrightarrow \langle x, y\rangle \in E^{\bar{M}}_{\mathrm{top}})$;
- $\bar{N}$ is a sound premouse and $\omega\varrho^1_{\bar{M}} \leq \tilde{\nu}$;
- $\tilde{\sigma} : \bar{N} \to N$ is Σ_0-preserving and cofinal and $\mathrm{cr}(\tilde{\sigma}) = \tilde{\nu}$;
- $E^N_{\tilde{\nu}} \neq \varnothing$.

Then the same holds in M, as (9.3.27) is obviously a Π_1-statement and the associated core map $\sigma_\tau : \mathrm{core}_\tau(M) \to M$ is Σ^*-preserving. Let $\bar{M}'$ be the cofinal image of $\bar{M}$ under σ. Granting that $\bar{M}'$, $\bar{M}$ and σ are elements of M, it is clear that M satisfies $\psi(\bar{M}', \bar{M}, \sigma, \nu)$. Thus, we can conclude that $\bar{M}$ is a proper initial segment of $\mathrm{Ult}(\bar{M}'||\nu, E^M_\nu) = \mathrm{Ult}(M||\nu, E^M_\nu)$, which yields clause d) of the condensation lemma. Now $\bar{M}'$ is clearly in M, as $\bar{M}' = \langle J^E_\zeta, E^M_{\mathrm{top}} \cap J^E_\zeta\rangle$ and M is an amenable structure;

here $\omega\zeta = \sup(\sigma''(\mathbf{On} \cap \bar{M}))$. Since $\bar{M}$ is equal to the hull $h_{\bar{M}}(\nu \cup \{p_{\bar{M}}\})$ and the embedding $\sigma : \bar{M} \to \bar{M}'$ is Σ_1-preserving with the critical point ν, we conclude that $\mathrm{rng}(\sigma) = h_{\bar{M}'}(\nu \cup \{\sigma(p_{\bar{M}})\})$, the structure $\bar{M}$ is the transitive collapse of this hull and σ is the associated uncollapsing map. But J^E_β is a ZFC$^-$ model where $\beta = \mathrm{ht}(M)$, so both $\bar{M}$ as well as σ can be reconstructed from $\bar{M}'$, ν and $\sigma(p_{\bar{M}})$ inside M. This completes the reduction of M to $\tilde{M}$.

From now on suppose that M is sound above τ. As in Case 3 in the proof of the solidity theorem we construct the ultrapowers $M' = \mathrm{Ult}^*(M, E^M_\nu)$ and $M^* = \mathrm{Ult}(M, E^M_\nu)$ together with the associated ultrapower embeddings $\pi : M \to M'$ and $\pi^* : M \to M^*$ and the canonical embedding $\sigma' : M^* \to M$. Then a) – h) listed in that proof hold; moreover, e) can be strengthened to hold of any $\alpha \geq \tau$, so f) holds with ν in place of $\pi^*(\nu)$ and M^* is sound above ν. Finally, g) can be strengthened to $\mathrm{cr}(\sigma') = \nu^{+M^*}$, as otherwise M^* and M' would have the same cardinal successor of ν, so $\bar{M}$, being a proper initial segment of M' (this follows from d*), would automatically be a proper initial segment of M^* and we would obtain the conclusion d) for free. We now apply the weaker version of the condensation lemma, which we established above, to M', M^* and σ'. Clause a) is false, as σ' fails to be cofinal. To exclude c) and d*), notice that, letting N be the common part of M^* and M' below $\mathrm{cr}(\sigma')$, the largest cardinal ν of N is not a limit cardinal in N. From this discussion we obtain that M^* must be a proper initial segment of M'.

Recall that both $\bar{M}$ and σ are elements of M, as we have proved above. Set $\bar{M}^* = \pi^*(\bar{M})$ and define the map $\sigma^* : \bar{M}^* \to M^*$ by

$$\sigma^*(\pi^*(f)(\alpha)) = \pi^*(\sigma \circ f)(\alpha)$$

for $f \in {}^\kappa\bar{M} \cap M$ and $\alpha < \lambda$. The expression on the right does make sense: $\sigma \circ f \in M$, as both σ and f are in M. Given a Σ_0-formula $\varphi(v_1, \ldots, v_\ell)$, functions $f_1, \ldots, f_\ell \in {}^\kappa\bar{M} \cap M$ and $\alpha_1, \ldots, \alpha_\ell$, the following are equivalent:

- $\bar{M}^* \models \varphi(\pi^*(f_1)(\alpha_1), \ldots, \pi^*(f_\ell)(\alpha_\ell))$;
- $\prec\alpha_1, \ldots, \alpha_\ell\succ \in E^M_\nu(\{\prec\eta_1, \ldots, \eta_\ell\succ \in \kappa;\ \bar{M} \models \varphi(f_1(\eta_1), \ldots, f_\ell(\eta_\ell))\}) = E^M_\nu(\{\prec\eta_1, \ldots, \eta_\ell\succ \in \kappa;\ M \models \varphi(\sigma \circ f_1(\eta_1), \ldots, \sigma \circ f_\ell(\eta_\ell))\})$;
- $M^* \models \varphi(\pi^*(\sigma \circ f_1)(\alpha_1), \ldots, \pi^*(\sigma \circ f_\ell)(\alpha_\ell))$.

The equality between the two sets in the middle statement follows from the fact that σ is Σ_0-preserving; indeed, this guarantees that $\bar{M} \models \varphi(f_1(\eta_1), \ldots, f_\ell(\eta_\ell))$ iff $M \models \varphi(\sigma \circ f_1(\eta_1), \ldots, \sigma \circ f_\ell(\eta_\ell))$. An immediate consequence of the above equivalence is that σ^* is a well-defined map which is Σ_0-preserving. Moreover, σ^* is defined on the entire $\bar{M}^*$, as any element of $\bar{M}^*$ is of the form $\pi^*(f)(\alpha)$ for some f and α as above. By the preservation properties of π^*, $\bar{M}^*$ is sound and its first projectum is $\leq \nu^* = \pi^*(\nu)$. Also notice that ν^* is the critical point of σ^*: If $\xi < \pi^*(\lambda)$, then $\xi = \pi^*(f)(\alpha)$ for some $f : \kappa \to \lambda$ from M and $\alpha < \lambda$, so $\sigma \circ f = f$ and $\sigma^*(\xi) = \pi^*(\sigma \circ f)(\alpha) = \pi^*(f)(\alpha) = \xi$. From the definition of σ^* we immediately obtain $\sigma^*(\pi^*(\lambda)) = \pi^*(\lambda)$, so σ^* remains the identity map on $\pi^*(\lambda)^{+\bar{M}^*} = \nu^*$. On the other hand, $\sigma^*(\nu^*) > \nu^*$, as ν^* indexes an extender in M^*, but not in $\bar{M}^*$. This completes the proof of the fact that $\nu^* = \mathrm{cr}(\sigma^*)$. Finally we note that the above argument which shows that $\bar{M} \in M$ applied to M^*, $\bar{M}^*$, and σ^* yields that both σ^* and $\bar{M}^*$ are elements of M'; here we use the fact proved above that M^* is a proper initial segment of M' (and thus of $M'||\nu^{+M'}$).

Recall that we know that the condensation lemma holds for any proper initial segment of M, so the following variation on (9.3.27) holds in M:

$$(\forall N, \bar{N}, \tilde{\sigma}, \tilde{\nu})[\tilde{\psi}(N, \bar{N}, \tilde{\sigma}, \tilde{\nu}) \longrightarrow \tilde{\psi}'(\bar{N}, N, \tilde{\nu})] \tag{9.3.28}$$

where $\tilde{\psi}'(\bar{N}, N, \tilde{\nu})$ says

$$\bar{N} \text{ is a proper initial segment of } \mathrm{Ult}(N||\tilde{\nu}, E^N_{\tilde{\nu}})$$

and $\tilde{\psi}(N, \bar{N}, \tilde{\sigma}, \tilde{\nu})$ is the conjunction of the latter three statements from ψ together with

- N is a proper initial segment of M.

As before, (9.3.28) is preserved under π', and we have seen in the previous paragraph that $\tilde{\psi}(M^*, \bar{M}^*, \sigma^*, \nu^*)$ holds true. Thus, $\bar{M}^*$ is a proper initial segment of $\mathrm{Ult}(M^*||\nu^*, E^{M^*}_{\nu^*})$ and this fact can obviously be expressed inside M^* in a Σ_0-fashion in parameters from $\mathrm{rng}(\pi^*)$. Since π^* is Σ_0 cofinal, we conclude that $\bar{M}$ is a proper initial segment of $\mathrm{Ult}(M||\nu, E^M_\nu)$. This completes the proof of Case 4.2.

From now on we assume that σ is $\Sigma_0^{(n)}$ cofinal. We first derive d) of the condensation lemma on the assumption that $\omega\varrho^{n+1}_{\tilde{M}} > \nu$. We find a premouse $\tilde{M}$ together with the maps $\sigma_0 : \bar{M} \to \tilde{M}$ and $\sigma_1 : \tilde{M} \to M$ such that $\sigma_1 \circ \sigma_0 = \sigma$, both σ_0 and σ_1 are $\Sigma_0^{(n)}$ cofinal and $\sigma_1 \restriction \nu = \mathrm{id}$. Here, $\tilde{M}$ is the transitive collapse of $\tilde{h}^{n+1}_M(\nu \cup \{\nu, \sigma(p_{\bar{M}})\})$, the embedding σ_1 is the associated uncollapsing map and $\sigma_0 = \sigma_1^{-1} \circ \sigma$. It is easy to verify that both σ_0 and σ_1 have all the requisite properties and that $\mathrm{cr}(\sigma_0) = \nu$. To see that $\tilde{M}$ is a premouse, we can assume w.l.o.g. that $\nu < \lambda_{\tilde{M}}$, as otherwise the initial segment condition immediately yields that $\tilde{M}$ is an initial segment of M. But then either σ_1 is Σ_1-preserving and both M and $\tilde{M}$ are potential premice of the same type A or B or else $n > 0$ and both M and $\tilde{M}$ are of type C. That $\tilde{M}$ is a premouse then follows from Lemma 9.1.7. By the existence of σ_1 and Corollary 9.2.13, $\tilde{M}$ has an $(\omega_1 + 1, \omega_1)$-iteration strategy. Also, from the definition of the premouse $\tilde{M}$ and the maps σ_h we infer that $\tilde{M} = \tilde{h}^{n+1}_{\tilde{M}}(\nu \cup \{\nu, \sigma_0(p_{\bar{M}})\})$, so $\tilde{M}$ is encodable into a $\Sigma_1^{(n)}(M)$ of ν. Hence $\tilde{M}$, being reconstructible from this set inside M, is an element of M and therefore is of height strictly less than that of M. By (9.3.20), we can apply the condensation lemma to $\tilde{M}$, $\bar{M}$ and σ_0. We again show that clause d) applies, which means that $\bar{M}$ is a proper initial segment of $\mathrm{Ult}(\tilde{M}||\nu, E^{\tilde{M}}_\nu) = \mathrm{Ult}(M||\nu, E^M_\nu)$. To exclude a) we recall that $\omega\varrho^\omega_{\bar{M}} = \nu$, so a) would mean that $\omega\varrho^\omega_{\tilde{M}} = \nu$. This is, however, impossible, as ν indexes an extender on the $\tilde{M}$-sequence, and therefore cannot be an $\tilde{M}$-cardinal. To exclude b), recall that ν cannot index an extender on the $E^{\bar{M}}$-sequence, while $E^M_\nu \neq \varnothing$. Finally, c) fails, as $\bar{M}$ is sound. This completes the discussion in the case where $\omega\varrho^{n+1}_M > \nu$.

We shall now focus on the case where σ is $\Sigma^{(n)}_0$ cofinal and $\omega\varrho^{n+1}_M \leq \nu$. We first observe that $\sigma(p^{n+1}_{\bar{M}}) <^* p^{n+1}_M$. Indeed, by the preservation properties of σ and the fact that $\bar{M}$ is solid, each element of $\sigma(p^{n+1}_{\bar{M}})$ has a generalized witness w.r.t. M and $\sigma(p^{n+1}_{\bar{M}})$ in M, which yields the inequality $\leq^*$. But the equality is not possible, as otherwise $\bar{M} = \mathrm{core}_\nu(M)$, i.e. Lemma 1.11.2 would guarantee that $\omega\varrho^m_{\bar{M}} = \omega\varrho^m_M$ for all $m \geq n+1$. However, we know that $\omega\varrho^\omega_{\bar{M}} = \nu$, while $\omega\varrho^\omega_M \leq \tau < \nu$. So there is an $\alpha \in p^{n+1}_M - \sigma(p_{\bar{M}})$ such that $\sigma(p_{\bar{M}}) - (\alpha+1) = p_M - (\alpha+1)$. Clearly, $\alpha \geq \nu$. We now distinguish two cases.

Case 4.3. $\alpha > \nu$.

By the solidity theorem, $W = W^\alpha_M \in M$, so $\mathrm{ht}(W) < \mathrm{ht}(M)$. Also, letting $\sigma_\alpha : W \to M$ be the canonical witness map, we have $\mathrm{rng}(\sigma) \subset \mathrm{rng}(\sigma_\alpha)$, so letting $\bar{\sigma}_\alpha = \sigma^{-1}_\alpha \circ \sigma$, we have a $\Sigma^{(n)}_1$-preserving embedding of $\bar{M}$ into W with the critical point ν. By the proof of the solidity theorem, W is a premouse and has an (ω_1+1, ω_1)-iteration strategy. All of this guarantees that we can apply the condensation lemma to W, $\bar{M}$ and $\bar{\sigma}_\alpha$. We again show that d) applies, so $\bar{M}$ is a proper initial segment of $\mathrm{Ult}(W||\nu, E^W_\nu) = \mathrm{Ult}(M||\nu, E^M_\nu)$. Clauses b) and c) can be seen to fail exactly as in the discussion preceding Case 4.3. To exclude a), notice that $\omega\varrho^\omega_{\bar{M}} = \nu$, while $\omega\varrho^\omega_W \geq \nu^{+M}$; this inequality follows from acceptability and the facts that W agrees with M up to $\alpha \geq \nu^{+M}$ and $W \in M$. This completes the proof of Case 4.3.

Case 4.4. $\alpha = \nu$

In this case $\bar{M} = W^\nu_M$. We first observe that M can be taken to be sound. Letting $\tilde{M} = \mathrm{core}(M)$ and $\tilde{\nu}$ be the preimage of ν under the associated core map $\tilde{\sigma}$, recall that $\tilde{\sigma}(W^{\tilde{\nu}}_{\tilde{M}})$ is a generalized witness for ν w.r.t. M and p_M in M. By the proof of Lemma 1.12.2, there is a canonical $\Sigma^{(n)}_0$-preserving embedding $\sigma^* : W^\nu_M \to \tilde{\sigma}(W^{\tilde{\nu}}_{\tilde{M}})$ which is the identity on ν and an element of M as well.

Consider a premouse $W^* \in \tilde{M}$ satisfying:

i) W^* is sound and $\omega\varrho^{n+1}_{W^*} \leq \tilde{\nu}$.

ii) There is a $\Sigma^{(n)}_0$-preserving embedding $\sigma^* : W^* \to W^{\tilde{\nu}}_{\tilde{M}}$ such that $\sigma^* \restriction \tilde{\nu} = \mathrm{id}$.

Letting $\tilde{\sigma}_{\tilde{\nu}} : W^{\tilde{\nu}}_{\tilde{M}} \to \tilde{M}$ be the canonical witness map associated with $W^{\tilde{\nu}}_{\tilde{M}}$, the composition $\tilde{\sigma}_{\tilde{\nu}} \circ \sigma^* : W^* \to \tilde{M}$ is a $\Sigma^{(n)}_0$-preserving embedding with the critical point $\tilde{\nu}$. Granting that the condensation lemma holds with $\tilde{M}$ in place of M, we conclude that W^* is a proper initial segment of $\mathrm{Ult}(\tilde{M}||\tilde{\nu}, E^{\tilde{M}}_{\tilde{\nu}})$. To see this, we have to exclude a) – c). Regarding b) and c), we can proceed as before. To see that a) fails, notice that $W^* \in \tilde{M}$, and therefore W^* cannot be $\mathrm{core}_{\tilde{\nu}}(\tilde{M})$.

Summarizing the discussion of the previous paragraph, in $\tilde{M}$ we have "Any premouse W^* satisfying i) and ii) above is a proper initial segment of $\mathrm{Ult}(\tilde{M}||\tilde{\nu}, E_{\tilde{\nu}}^{\tilde{M}})$". This statement, being Π_1, is preserved upward under $\tilde{\sigma}$, so any premouse $W^* \in M$ satisfying i) and ii) above must be a proper initial segment of $\tilde{\sigma}(\mathrm{Ult}(\tilde{M}||\tilde{\nu}, E_{\tilde{\nu}}^{\tilde{M}})) = \mathrm{Ult}(M||\nu, E_\nu^M)$. In particular, this holds of $W^* = W_M^\nu = \bar{M}$, which obviously satisfies i) and ii).

From now on assume that M is sound. We adopt the strategy from Case 2 in the proof of the solidity theorem. Since M is sound, to each $\alpha \in p_M - \nu$ there is a $\xi_\alpha < \omega\varrho_M^{n+1}$ such that $W_M^\alpha = \tilde{h}_M^{n+1}(\xi_\alpha, p_M)$. Let $\xi < \omega\varrho_M^{n+1}$ be an upper bound on all such ξ_α, let M' be the transitive collapse of $\tilde{h}_M^{n+1}(\xi \cup \{p_M\})$ and let $\sigma' : M' \to M$ be the associated uncollapsing map. The map σ' is $\Sigma_1^{(n)}$-preserving, so the structure M' is a premouse by Lemma 9.1.7. Moreover, $\tilde{h}_{M'}^{n+1}(\xi \cup \{p'\}) = M'$ where p' and ν' is the preimage of $p_M - \nu$ and ν under σ', respectively. Lemma 1.12.5 then guarantees that $p' = p_{M'} - \nu'$, since each α has a generalized witness w.r.t. M' and p' in M', namely the σ'-preimage of $W_M^{\sigma(\alpha)}$. Notice also that $\omega\varrho_{M'}^{n+1} \leq \xi$ and $\sigma' \upharpoonright \xi = \mathrm{id}$, so M' has an $(\omega_1 + 1, \omega_1)$-iteration strategy by Corollary 9.2.13. Also, M' can be encoded in a $\Sigma_1^{(n)}(M)$ subset of $\xi < \omega\varrho_M^{n+1}$ which must be an element of M, so M', being reconstructible from this set inside M, is an element of M as well. In particular, $\mathrm{ht}(M') < \mathrm{ht}(M)$. By (9.3.20), the condensation lemma holds for any appropriate M'' and σ'' where $\sigma'' : M'' \to M'$.

In the following we will apply the previous observation with $W_{M'}^{\nu'}$ and the associated witness map in place of M'' and σ''. Before we go to this, we state a general fact about acceptable structures. Consider acceptable structures Q and $\bar{Q}$, finite sets of ordinals $p \subset Q$ and $\bar{p} \subset \bar{Q}$ and an ordinal $\alpha < \min(p), \min(\bar{p})$ such that $\omega\varrho_Q^{n+1} \leq \alpha < \omega\varrho_Q^n$. If $\bar{Q} = \tilde{h}_{\bar{Q}}^{n+1}(\alpha \cup \{\bar{p}\})$ and the statement

$$(\forall \xi_1, \ldots, \xi_\ell < \alpha)[Q \models \varphi(\xi_1, \ldots, \xi_\ell, p) \longleftrightarrow \bar{Q} \models \varphi(\xi_1, \ldots, \xi_\ell, \bar{p})] \quad (9.3.29)$$

holds for any $\Sigma_1^{(n)}$-formula $\varphi(v_1, \ldots, v_\ell)$, then $\bar{Q}$ is the *standard* witness for α w.r.t. M and p. To see this, notice that (9.3.29) guarantees that the value $\tilde{h}_{\bar{Q}}^{n+1}(\xi, \bar{p})$ is defined iff $\tilde{h}_Q^{n+1}(\xi, p)$ is and that $\tilde{h}_{\bar{Q}}^{n+1}(\xi, \bar{p}) = \tilde{h}_{\bar{Q}}^{n+1}(\eta, \bar{p})$ iff $\tilde{h}_Q^{n+1}(\xi, p) = \tilde{h}_Q^{n+1}(\eta, p)$, which enables us to define a map $\sigma_{\bar{Q},Q} : \bar{Q} \to Q$ by $\tilde{h}_{\bar{Q}}^{n+1}(\xi, \bar{p}) \mapsto \tilde{h}_Q^{n+1}(\xi, p)$. One more application of (9.3.29) shows that this map is $\Sigma_1^{(n)}$-preserving, so its range is precisely $\tilde{h}_Q^{n+1}(\alpha \cup \{p\})$ and $\bar{Q}$ is the transitive collapse of $\mathrm{rng}(\sigma_{\bar{Q},Q})$. Thus, $\bar{Q} = W_Q^{\alpha,p}$.

We now apply the above fact to M and $W = W_M^\nu$. Letting $p = p_M - (\nu + 1)$ and $\bar{p}$ be the preimage of p under the canonical witness map associated with W, the statement (9.3.29) can be expressed over M as

$$(\forall \xi_1, \ldots, \xi_\ell < \nu)[\varphi(\xi_1, \ldots, \xi_\ell, p) \longleftrightarrow W \models \varphi(\xi_1, \ldots, \xi_\ell, \bar{p})],$$

i.e. in a $\Pi_2^{(n)}$-fashion. But we know that σ' is $\Sigma_1^{(n)}$-preserving and has W_M^ν in its range, so the above formula is downward preserved under σ' and therefore holds with ν', p', W'

and $\bar{p}'$ in place of ν, p, W and $\bar{p}$ where $\sigma'(W', \bar{p}') = (W_M^\nu, \bar{p})$. We thus conclude that $W' = W_{M'}^{\nu',p'} = W_{M'}^{\nu'}$. Moreover, W' is sound and $\omega\varrho_{W'}^{n+1} \leq \nu'$, as the same holds of W and ν. Now we apply the condensation lemma to M', W' and the associated witness map; a discussion of the same sort as the one used above to draw the conclusion on W^* then shows that W' is a proper initial segment of $\mathrm{Ult}(M'||\nu', E_{\nu'}^{M'})$. This statement is preserved upward under σ', so W_M^ν is a proper initial segment of $\mathrm{Ult}(M||\nu, E_\nu^M)$. This completes the discussion in Case 4.4 and thus the proof of Case 4. □(Case 4)

We have verified the condensation lemma in all special cases; the proof is now complete. □(Theorem 9.3.2)

A minor modification of the condensation lemma holds even for maps σ which are not cardinal preserving. Notice that if σ is such a map, then $\bar{M}$ is passive, $n = 0$ and σ is non-cofinal, as σ maps the largest cardinal of $\bar{M}$ to the largest cardinal of M if the former fails and is Σ_1-preserving if one of the latter two fails. Thus, we have the following corollary:

Corollary 9.3.3. *Let M be as in the condensation lemma, let $\bar{M}$ be a passive premouse and let $\sigma : \bar{M} \to M$ be Σ_0-preserving, but not cardinal preserving. Assume that $\omega\varrho_{\bar{M}}^1 \leq \nu = \mathrm{cr}(\sigma)$. Then $\bar{M}$ is weakly iterable, and therefore solid. Moreover, granting that $\bar{M}$ is sound above ν, either*

- *$\bar{M} = J_\zeta^E$ for some $\zeta < \mathrm{ht}(M)$ (here $E = E^M$) or else*
- *one of* b) – d) *of the condensation lemma holds.*

Proof. Recall that $\bar{M}$ must be passive, $n = 0$ and σ must be non-cofinal. Let $\bar{\zeta} = \mathrm{ht}(\bar{M})$ and $\omega\zeta = \sup(\sigma''\omega\bar{\zeta}) < \omega\zeta$. Letting $M' = J_\zeta^E$ where $E = E^M$, the embedding $\sigma : \bar{M} \to M'$ is Σ_0 cofinal, and therefore cardinal preserving. Hence we can apply the condensation lemma to M', $\bar{M}$ and σ. If a) holds, then $\bar{M} = \mathrm{core}_\nu(M') = M'$, as M' is sound, so the former conclusion of the corollary is the case. If one of b) – d) holds, it is easy to verify that the same is the case if we replace M' by M, so the latter conclusion of the corollary holds. □(Corollary 9.3.3)

9.4 Uniqueness of Well-Founded Branches

So far, we have studied fine structural properties of premice on the assumption that we can come up with iteration strategies. Finding such strategies turns out to be a really deep and difficult task, and we will merely present a first step towards the general solution – a criterion for the existence of the uniqueness strategy. It is a discovery of Martin and Steel that the existence of unique well-founded branches is a smallness condition. They proved, roughly speaking, that any premouse, whose iteration has two cofinal well-founded branches, must contain a "large" transitive model with a Woodin cardinal (see below for the exact formulation). Originally, they established the result for coarse iterations and the fine structural refinement of the method appeared only later in [MiS94]. Our presentation of the method, however, significantly differs from that one; the reason has its roots in our alternative way of indexing extenders. Mitchell and Steel use the "shortest" possible indexing, i.e. they index each extender by the cardinal successor of the supremum of its generators in the corresponding ultrapower, which makes the argument simpler due to the very strong initial segment condition they obtain. We, instead, use the "longest" possible indexing and therefore have to deal with cutbacks of top extenders which need not be elements of the respective premice.

We recommend reviewing the section on strong and Woodin cardinals from the second chapter. To start with, we state a general fact about Woodinness in acceptable structures, whose proof is left to the reader as an exercise.

Fact. Let $N = J_\alpha^A$ be acceptable and $\delta \in N$. Then δ is Woodin in N iff for every $B \in \mathcal{P}(\delta) \cap N$ there is a cardinal $\kappa < \delta$ which is B-strong in $\langle J_\delta^A, B\rangle$. More precisely, given any $\beta < \delta$, there is an extender F at (κ, β) in J_δ^A such that $F(B \cap \kappa) = B \cap \beta$.

Now assume that $\mathfrak{I}$ is a normal iteration of a premouse M with two distinct cofinal well-founded branches b and c. Let $\zeta_0 \in b - c$ be such that $b - \zeta_0$ and $c - \zeta_0$ are disjoint and neither of them contains truncation points. For $n > 0$ we define ζ_n as follows:

$$\begin{aligned} \zeta_{2n+1} &= \min(c - \zeta_{2n}) \\ \zeta_{2n+2} &= \min(b - \zeta_{2n+1}). \end{aligned}$$

It follows that the sequence $\langle \zeta_n;\ i \in \omega\rangle$ is strictly increasing and, by the minimality of ζ_n, each ζ_n is a successor ordinal. Letting $T = T^{\mathfrak{I}}$, we fix the following notation:

$$\gamma_n = \zeta_n - 1 \quad \text{and} \quad \zeta_n^* = T(\zeta_n) = \xi_{\gamma_n}.$$

We now state a list of facts regarding various objects related to $\mathfrak{I}$.

a) $\zeta_{n+1}^* < \zeta_n < \zeta_{n+1}$.

b) $\kappa_{\gamma_{n+1}} < \lambda_{\zeta_{n+1}^*} \le \lambda_{\gamma_n} \le \kappa_{\gamma_{n+2}}$.

c) $\pi_{\zeta_n^*,\zeta_n} : M_{\zeta_n^*} \xrightarrow[E^n]{*} M_{\zeta_n}$ where $E^n = E_{\nu_{\gamma_n}}^{M_{\gamma_n}}$.

d) $\sup\{\zeta_n;\ n \in \omega\} = \theta = \text{length}(\mathfrak{I})$.

e) τ_{γ_n} is a cardinal in $M_{\zeta'}$ whenever $\zeta' \ge \zeta_n^*$. (Recall that τ_{γ_n} is a cardinal successor of κ_{γ_n} in $M_{\gamma_n} \| \nu_{\gamma_n}$.)

Let us give more details on the conclusions which are not completely trivial. The left inequality in a) follows from the minimality of ζ_{n+1}. Notice that it implies that $\zeta_{n+1}^* \le \gamma_n$, which yields the middle inequality in b). To see the right inequality thereof, recall that $\kappa_{\gamma_{n+2}}$ is larger of equal than λ_ζ whenever $\zeta < \zeta_{n+2}^*$ and that $\gamma_n < \zeta_n \le \zeta_{n+2}^*$.

Clause d) is a consequence of the fact that branches are closed sets of ordinals. Assuming that the supremum of all ζ_n is some $\bar{\theta} < \theta$, then $\bar{\theta}$, being a limit point of both $b \cap \bar{\theta}$ and $c \cap \bar{\theta}$, would be in $b \cap c$. This is impossible, as $\bar{\theta} > \zeta_0$. Finally, e) follows from general properties of normal iterations and from the fact that ζ_n is not a truncation point.

Definition. Given a normal iteration $\Im$ of M as above, we define

$$\delta(\Im) = \sup_{n \in \omega} \kappa_{\gamma_n} = \sup_{n \in \omega} \nu_{\gamma_n}$$
$$N(\Im) = J_\delta^E = \bigcup_{n \in \omega} J_{\kappa_{\gamma_n}}^{E^{M_n}} = \bigcup_{n \in \omega} J_{\nu_{\gamma_n}}^{E^{M_n}}.$$

Set $\delta = \delta(\Im)$ and $N = N(\Im)$. For all but finitely many n, we construct a sequence extenders $\langle G_i^n;\ i \in \omega\rangle$ with the following properties:

(P1) $G_i^n \in N$ and N is extendible by G_i^n, i.e. $\mathrm{Ult}(N, G_i^n)$ exists.

(P2) $\mathrm{cr}(G_i^n) = \kappa_n^*$ for all $i \in \omega$ where $\kappa_{\gamma_n} \le \kappa_n^* < \kappa_{\gamma_{n+1}}$.

(P3) G_i^n is at $(\kappa_n^*, \kappa_{\gamma_{n+1+i}})$.

(P4) $G_i^n = G_j^n | \kappa_{\gamma_{n+1+i}}$ whenever $i \le j$.

(P5) G_i^n is coherent with E. In other words, letting $N' = \mathrm{Ult}(N, G_i^n)$, we have $E^{N'} | \kappa_{\gamma_{n+1+i}} = E | \kappa_{\gamma_{n+1+i}}$.

We shall later see that this sequence of extenders have even better coherency properties – they witness, inside N, that each κ_n^* is A-strong for a fairly large class of sets $A \subset \delta$. One of the crucial points of the whole construction is the choice of G_0^n. Since b) guarantees that $\kappa_{\gamma_{n+1}} < \lambda_{\gamma_n}$, a natural candidate for G_0^n would be $E^n | \kappa_{\gamma_{n+1}}$ (see c) for the definition of E^n). This strategy works in the Mitchell–Steel setting, but need not work here, as P1 might fail if E^n is the top extender of M_{γ_n}. For the construction of G_0^n in this case we will need the next series of lemmata.

In the following, we shall consider pairs of ordinals (γ, κ) such that:

(Q1) κ is a cardinal in $M_\gamma || \nu_\gamma$ and $\kappa_\gamma < \kappa < \lambda_\gamma$.

(Q2) $E_{\nu_\gamma}^{M_\gamma} | \kappa \notin M_\gamma$.

Clearly, Q2 implies that $E_{\kappa_\gamma}^{M_\gamma} = E_{\mathrm{top}}^{M_\gamma}$ and $\omega\varrho_{M_\gamma}^1 \le \kappa$.

Lemma 9.4.1. *Let (γ, κ) be as above and let $\beta + 1 \le_T \gamma$ be such that $\kappa < \lambda_\beta$. Then*

a) $\mathrm{cr}(\pi_{\beta+1,\gamma}) \ge \kappa$.

b) *No $\beta' + 1$ satisfying $\beta + 1 <_T \beta' + 1 \le \gamma$ is a truncation point, i.e. the embedding $\pi_{\beta+1,\gamma}$ is total.*

Proof. Assume w.l.o.g. that $\beta + 1 <_T \gamma$. We leave the easy verification of a) to the reader and proceed directly to b). Granting that b) fails, let $\beta' + 1$ be the last truncation point on the branch $(\beta + 1, \gamma]$. Then $\pi_{\xi_{\beta'},\gamma} : M^*_{\beta'} \to M_\gamma$ is total and its critical point is at least κ. As $\omega\varrho^1_{M_\gamma} \le \kappa$ by the remarks preceding this lemma, the general properties of normal iterations guarantee that $\omega\varrho^1_{M^*_{\beta'}} = \omega\varrho^1_{M_\gamma} \le \kappa$. Since $M^*_{\beta'}$ is a proper initial segment of $M_{\beta'}$ and $\lambda_\beta > \kappa$ is an element of $M^*_{\beta'}$, λ_β is collapsed in $M_{\beta'}$. On the other hand, $\beta < \beta'$, so λ_β must be a cardinal in $M_{\beta'}$ by the general properties of normal iterations. Contradiction. □(Lemma 9.4.1)

Definition. To each pair or ordinals (γ, κ) as above we define an ordinal $\gamma^*(\kappa) < \gamma$ by

$$\gamma^*(\kappa) \simeq \text{the unique } \beta \text{ satisfying } \beta + 1 \le_T \gamma \text{ and } \kappa_\beta < \kappa < \lambda_\beta.$$

For all other pairs (γ, κ) we leave $\gamma^*(\kappa)$ undefined.

Definition. Given a pair of ordinals (γ, κ) as above, we define a descending sequence of ordinals $\gamma^i(\kappa)$ inductively as follows:

$$\gamma^0 = \gamma$$
$$\gamma^{i+1} \simeq (\gamma^i)^*(\kappa).$$

and set

$$p(\gamma, \kappa) = \text{the largest } i \text{ such that } \gamma^i(\kappa) \text{ is defined.}$$

Clearly, $p(\gamma, \kappa)$ always exists, as $\langle \gamma^i(\kappa) \rangle_i$ is a descending sequence of ordinals. Finally, for any ordinal κ set

$$\beta(\kappa) \simeq \text{the least } \beta \text{ satisfying } \kappa < \lambda_\beta.$$

Lemma 9.4.2. *Assume (γ, κ) as in the above definitions, $p = p(\gamma, \kappa)$ and $\bar\gamma = \gamma^p(\kappa)$. Then one of the following holds:*

a) $E^{M_{\bar\gamma}}_{\nu_{\bar\gamma}} | \kappa \in M_{\bar\gamma}$

b) $\beta(\kappa) \le_T \bar\gamma$.

Proof. Granting that a) fails, Q1 and Q2 hold with $(\bar\gamma, \kappa)$ in place of (γ, κ). Q1 can be easily verified by induction on $i \le p$. We now derive a contradiction from the failure of b). Let α' be the least ordinal $\le_T \bar\gamma$ which is larger than $\beta(\kappa)$. Clearly, $\alpha' = \alpha + 1$ for some α. By the minimality of α' and the failure of b), $\xi_\alpha = T(\alpha') < \beta(\alpha)$. It follows that $\lambda_{\xi_\alpha} \le \kappa$. From this and the fact that $\kappa_\alpha < \lambda_{\xi_\alpha}$ we obtain $\kappa_\alpha < \kappa < \lambda_\alpha$, which says that $\alpha = (\bar\gamma)^*(\kappa)$. However, the value on the right cannot be defined, as $\bar\gamma = \gamma^p(\kappa)$. Contradiction. □(Lemma 9.4.2)

We now specify the set Ω of those n for which G^n_i is *not* defined:

$$\Omega = \{n \in \omega;\ \kappa_{\gamma_n} \le \operatorname{cr}(E^{M_{\zeta_{n+1}}}_{\text{top}}) < \kappa_{\gamma_{n+1}}\}$$

To see that $\mathfrak{Q}$ is finite, notice that if $\hat{\kappa} = \mathrm{cr}(E_{\mathrm{top}}^{M_{\zeta_{n+1}}}) < \kappa_{\gamma_{n+1}}$, then $\mathrm{cr}(E_{\mathrm{top}}^{M_{\zeta_m}}) = \hat{\kappa}$ whenever $\zeta_{n+1} \leq_T \zeta_m$, as $\mathrm{cr}(\pi_{\zeta_{n+1},\zeta_m}) \geq \lambda_{\gamma_{n+1}} > \kappa_{\gamma_{n+1}} > \hat{\kappa}$. Since both

$$\langle \zeta_{n+1+2k};\ k \in \omega \rangle \quad \text{and} \quad \langle \zeta_{n+2k};\ k \in \omega \rangle$$

are strictly increasing sequences converging to θ, $\hat{\kappa}$ can happen to fall into $[\kappa_{\gamma_m}, \kappa_{\gamma_{m+1}})$ only for finitely many such m. This proves that both $\mathfrak{Q} \cap \{m;\ \zeta_m \in b\}$ and $\mathfrak{Q} \cap \{m;\ \zeta_m \in c\}$ are finite, so $\mathfrak{Q}$ is finite as well.

Lemma 9.4.3. *Assume $n \notin \mathfrak{Q}$. Let $p = p(\gamma_n, \kappa_{\gamma_{n+1}})$, and $\bar{\gamma} = (\gamma_n)^p(\kappa_{\gamma_{n+1}})$. Then $E_{\nu_{\bar{\gamma}}}^{M_{\bar{\gamma}}} | \kappa_{\gamma_{n+1}} \in M_{\bar{\gamma}}$.*

Proof. First of all we inductively verify that for all $i < p$ we have:

$$\kappa_{\gamma_n} \leq \kappa_{\gamma^i} < \kappa_{\gamma^{i+1}} < \kappa_{\gamma_{n+1}}. \tag{9.4.1}$$

Clearly, $\kappa_{\gamma^0} = \kappa_{\gamma_n}$ and each κ_{γ^i} is smaller than $\kappa_{\gamma_{n+1}}$ by the definition of γ^i. Thus, it suffices to show that the sequence $\langle \kappa_{\gamma^i} \rangle_i$ is ascending. Since $i+1$ is defined, Q1 and Q2 hold of $(\gamma, \kappa) = (\gamma^i, \kappa_{\gamma_{n+1}})$, so letting $\xi = \xi_{\gamma^{i+1}}$, b) of Lemma 9.4.1 guarantees that π_{ξ,γ^i} is total. Notice that $\pi_{\xi,\gamma^i}(\kappa_{\gamma^{i+1}}) \geq \lambda_{\gamma^{i+1}} > \kappa_{\gamma^i}$ and that $\kappa_{\gamma^{i+1}}$ is the critical point of π_{ξ,γ^i}; the latter follows from a) of the above lemma, which tells us that the critical point of $\pi_{\gamma^{i+1}+1,\gamma^i}$ is at least $\lambda_{\gamma^{i+1}}$. Now κ_{γ^i}, being the critical point of the top extender of M_{γ^i}, must be in the range of π_{ξ,γ^i} and it follows easily from the above relationship between κ_{γ^i} and $\pi_{\xi,\gamma^i}(\kappa_{\gamma^{i+1}})$ that this is possible only if $\kappa_{\gamma^i} < \mathrm{cr}(\pi_{\xi,\gamma^i}) = \kappa_{\gamma^{i+1}}$. This proves (9.4.1).

Towards the proof of the lemma, assume that $E_{\nu_{\bar{\gamma}}}^{M_{\bar{\gamma}}} \notin M_{\bar{\gamma}}$. Then $\beta = \beta(\kappa_{\gamma_{n+1}}) \leq_T \bar{\gamma}$ by Lemma 9.4.2. Notice also that $\beta = \zeta_{n+1}^*$. We now distinguish two cases.

If $\beta = \bar{\gamma}$, then actually $\bar{\gamma} = \zeta_{n+1}^*$, so $\mathrm{cr}(E_{\mathrm{top}}^{M_{\zeta_{n+1}^*}}) = \kappa_{\bar{\gamma}}$. Since $\mathrm{cr}(\pi_{\zeta_{n+1}^*,\zeta_{n+1}}) = \kappa_{\gamma_{n+1}}$ and ζ_{n+1} is not a truncation point, $\mathrm{cr}(E_{\mathrm{top}}^{M_{\zeta_{n+1}}}) = \kappa_{\bar{\gamma}}$, as $\kappa_{\bar{\gamma}} \in [\kappa_{\gamma_n}, \kappa_{\gamma_{n+1}})$ by (9.4.1). It follows that $n \in \mathfrak{Q}$. Contradiction.

Now assume that $\beta <_T \bar{\gamma}$. Then $\pi_{\beta,\bar{\gamma}}$ is total by Lemma 9.4.1 and, if α is such that $\beta = T(\alpha+1)$, we have $\kappa_\alpha \geq \kappa_{\gamma_{n+1}}$, as $(\bar{\gamma})^*(\kappa_{\gamma_{n+1}})$ is undefined. Thus, $\mathrm{cr}(\pi_{\beta,\bar{\gamma}}) = \kappa_\alpha \geq \kappa_{\gamma_{n+1}}$. In the following we observe that $\pi_{\beta,\bar{\gamma}}$ is defined on the entire M_β. Suppose that this fails, in other words, that M_α^* is a proper initial segment of M_β. Recall that $\omega\varrho_{M_{\bar{\gamma}}}^1 \leq \kappa_{\gamma_{n+1}}$; this is a consequence of the assumption that $E_{\mathrm{top}}^{M_{\bar{\gamma}}} | \kappa_{\gamma_{n+1}} \notin M_{\bar{\gamma}}$. So $\omega\varrho_{M_\beta}^1 \leq \kappa_{\gamma_{n+1}}$, as we have seen that $\mathrm{cr}(\pi_{\beta,\bar{\gamma}}) \geq \kappa_{\gamma_{n+1}}$. Pick a set $a \in \mathcal{P}(\kappa_{\gamma_{n+1}}) \cap \Sigma_1(M_\alpha^*)$ which is not an element of M_α^*. However, $a \in M_\beta = M_{\zeta_{n+1}^*}$ and, as $\mathrm{cr}(\pi_{\zeta_{n+1}^*,\zeta_{n+1}}) = \kappa_{\gamma_{n+1}}$, we also have $a \in M_{\zeta_{n+1}}$. Now $\beta = \zeta_{n+1}^* < \zeta_{n+1}$, so λ_β is a cardinal in $M_{\zeta_{n+1}}$. It follows that $a \in M_{\zeta_{n+1}} || \lambda_\beta = M_\beta || \lambda_\beta = M_\alpha^* || \lambda_\beta$, as $\lambda_\beta > \kappa_{\gamma_{n+1}}$. This contradicts our choice of a, so $M_\alpha^* = M_\beta$ after all. But then $\mathrm{cr}(E_{\mathrm{top}}^{M_\beta}) = \kappa_{\bar{\gamma}}$ and we can proceed exactly as in the previous paragraph to infer that $\kappa_{\bar{\gamma}} = \mathrm{cr}(E_{\mathrm{top}}^{M_{\zeta_{n+1}}})$ and that $n \in \mathfrak{Q}$. We have obtained a contradiction again, so necessarily $E_{\nu_{\bar{\gamma}}}^{M_{\bar{\gamma}}} | \kappa_{\gamma_{n+1}} \in M_{\bar{\gamma}}$. □(Lemma 9.4.3)

The previous lemma makes it clear how to define G_i^n for $n \notin \mathcal{Q}$. Letting $p = p(\gamma_n, \kappa_{\gamma_{n+1}})$ and $\bar{\gamma} = (\gamma_n)^p(\kappa_{\gamma_{n+1}})$, we set

$$G_0^n = E_{\nu_{\bar{\gamma}}}^{M_{\bar{\gamma}}} | \kappa_{\gamma_{n+1}}$$
$$\kappa_n^* = \mathrm{cr}(G_0^n) = \kappa_{\bar{\gamma}}.$$

Then

$$G_0^n \in M_{\zeta_{n+1}^*}. \text{ In fact, } G_0^n \in M_{\zeta_{n+1}^*} || \lambda_{\zeta_{n+1}^*}. \tag{9.4.2}$$

The former is merely the conclusion of the previous lemma. To see the latter, we recall that $\kappa_{\gamma_{n+1}} < \lambda_{\bar{\gamma}}$, so $\zeta_{n+1}^* \le \bar{\gamma}$. If $\zeta_{n+1}^* < \bar{\gamma}$, the conclusion follows immediately, as $\lambda_{\zeta_{n+1}^*}$ is a cardinal in $M_{\bar{\gamma}}$. If $\zeta_{n+1}^* = \bar{\gamma}$, then $G_0^n = \pi_{\zeta_{n+1}^*, \zeta_{n+1}}(G_0^n) | \kappa_{\gamma_{n+1}} \in M_{\zeta_{n+1}}$, so $G_0^n \in M_{\zeta_{n+1}} || \lambda_{\zeta_{n+1}^*} = M_{\zeta_{n+1}^*} || \lambda_{\zeta_{n+1}^*}$, as $\lambda_{\zeta_{n+1}^*}$ is a cardinal in $M_{\zeta_{n+1}}$.

Next we will need a means for verifying that N is extendible by G_i^n. Here one can use various arguments, which all turn out to be equivalent. Since all G_i^n will be elements of N, the ultrapowers $\mathrm{Ult}(N, G_i^n)$ are internal, and their well-foundedness can be decided inside N. Thus, it is convenient to apply Lemma 2.5.3 on the equivalence of the extendibility and the ω-completeness for internal extenders. Notice also that ω-completeness is a local property in the sense that it can be decided in any structure which contains the power set of the length of the extender in question. In the context of acceptable structures this means that we only need to know the structure up to the local cardinal successor of the length of the extender. Thus, the following observation will be useful:

$$G_0^n \text{ is } \omega\text{-complete in } M_{\zeta_{n+1}^*} || \lambda_{\zeta_{n+1}^*}. \tag{9.4.3}$$

Indeed, $M_{\zeta_{n+1}^*} || \lambda_{\zeta_{n+1}^*}$, being an initial segment of $M_{\bar{\gamma}}$ (see the above proof of (9.4.2)), is extendible by $E_{\nu_{\bar{\gamma}}}^{M_{\bar{\gamma}}}$ and so by any cutback thereof, in particular by G_0^n. (9.4.3) then follows from the above mentioned Lemma 2.5.3.

Lemma 9.4.4. *The extender G_0^n has the properties* P1 – P5.

Proof. Property P1 follows from (9.4.2) and (9.4.3); the latter yields that G_0^n is ω-complete in N. P2 follows from (9.4.1) and both P3 and P4 are trivial. To see P5, recall that the extender sequences of $M_{\bar{\gamma}}$ and N agree below $\kappa_{\zeta_{n+1}^*}$ (in fact, they agree below $\nu_{\bar{\gamma}}$). So it suffices to see that the extender sequences of $M_{\bar{\gamma}}$ and $\mathrm{Ult}(M_{\bar{\gamma}}, G_0^n)$ agree below $\kappa_{\zeta_{n+1}^*}$. But this is certainly true with $E_{\nu_{\bar{\gamma}}}^{M_{\bar{\gamma}}}$ in place of G_0^n; the conclusion then follows from the fact that $\mathrm{Ult}(M_{\bar{\gamma}}, G_0^n)$ and $\mathrm{Ult}(M_{\bar{\gamma}}, E_{\nu_{\bar{\gamma}}}^{M_{\bar{\gamma}}})$ agree below $\kappa_{\zeta_{n+1}^*}$. □(Lemma 9.4.4)

Having established the basic properties of G_0^n, we define G_i^n for all i:

$$\tilde{G}_i^n = \pi_{\zeta_{n+1+i}^*, \zeta_{n+1+i}^*}(G_i^n)$$
$$G_{i+1}^n = \tilde{G}_i^n | \kappa_{\gamma_{n+2+i}}.$$

Of course, this definition makes only sense if we know that $\tilde{G}_i^n$ is defined. We shall verify this together with P1 – P5 in the following lemma.

Lemma 9.4.5. *Each extender G_i^n is an element of $M_{\zeta^*_{n+1+i}}\|\lambda_{\zeta^*_{n+1+i}}$, is ω-complete in $M_{\zeta^*_{n+1+i}}\|\lambda_{\zeta^*_{n+1+i}}$ and has properties* P1 – P5.

Proof. We shall proceed by induction on i. The case $i = 0$ was done in the previous lemma, so let us focus on the induction step. Notice that P2 – P4 are immediate consequences of the definition of G_{i+1}^n and of the fact that $\kappa_n^* < \kappa_{\gamma_{n+1}}$. Also, P4 guarantees that the lengths of G_i^n ascend; to see this recall that the sequence $\langle\kappa_{\gamma_m}\rangle_m$ is strictly monotone, as follows by b).

Property P1 follows from the first part of the lemma, which we prove now. The induction hypothesis implies $\tilde{G}_i^n \in M_{\zeta_{n+1+i}}$. Now it suffices to recall that $\lambda_{\zeta^*_{n+2+i}}$ is a cardinal in $M_{\zeta_{n+1+i}}$ and that $M_{\zeta^*_{n+2+i}}$ agrees with $M_{\zeta_{n+1+i}}$ below $\lambda_{\zeta^*_{n+2+i}}$. To see that N is extendible by G_{i+1}^n we use (9.4.3), which together with the preservation properties of $\pi_{\zeta^*_{n+1+i},\zeta_{n+1+i}}$ yields the ω-completeness of $\tilde{G}_i^n$ in $M_{\zeta_{n+1+i}}\|\pi_{\zeta^*_{n+1+i},\zeta_{n+1+i}}(\lambda_{\zeta^*_{n+1}})$ and thus the ω-completeness of G_{i+1}^n in $M_{\zeta_{n+1+i}}\|\lambda_{\zeta^*_{n+2+i}} = M_{\zeta^*_{n+2+i}}\|\lambda_{\zeta^*_{n+2+i}}$.

It remains to verify P5. This is very much like the argument in the previous paragraph. By the agreement between N and the models $M_{\zeta^*_{n+1+j}}$, it suffices to prove P5 with $M_{\zeta^*_{n+1+j}}\|\lambda_{\zeta^*_{n+1+j}}$ and G_j^n in place of N and G_j^n for all j. We know from the induction hypothesis that

$$M_{\zeta^*_{n+1+i}}\|\lambda_{\zeta^*_{n+1+i}}$$

agrees with the ultrapower

$$\mathrm{Ult}(M_{\zeta^*_{n+1+i}}\|\lambda_{\zeta^*_{n+1+i}}, G_i^n)$$

below $\kappa_{\gamma_{n+1+i}}$. Applying the map $\pi_{\zeta^*_{n+1+i},\zeta_{n+1+i}}$, we obtain that

$$M_{\zeta_{n+1+i}}\|\pi_{\zeta^*_{n+1+i},\zeta_{n+1+i}}(\lambda_{\zeta^*_{n+1+i}})$$

agrees with the ultrapower

$$\mathrm{Ult}(M_{\zeta^*_{n+1+i}}\|\pi_{\zeta^*_{n+1+i},\zeta_{n+1+i}}(\lambda_{\zeta^*_{n+1+i}}), \tilde{G}_i^n)$$

below $\lambda_{\gamma_{n+1+i}}$. Now cutting $M_{\zeta_{n+1+i}}$ at $\lambda_{\zeta^*_{n+2+i}}$ and $\tilde{G}_i^n$ at $\kappa_{\gamma_{n+2+i}}$ yields the desired conclusion. □(Lemma 9.4.5)

We are now ready to prove that the sequences of extenders $\langle G_i^n;\ i \in \omega\rangle$ witness Woodinness of δ in certain carefully chosen end-extensions of N. We fix the following notation.

- M_b and M_c is the direct limit along the branch b and c, respectively;
- $\pi_{\beta,b} : M_\beta \to M_b$ and $\pi_{\gamma,c} : M_\gamma \to M_c$ are the associated iteration maps (we assume that $\beta, \gamma \geq \zeta_0$, so these maps are total).

Notice that by the agreement between the premice in the iteration, N is an initial segment of both M_b and M_c.

Lemma 9.4.6. *Let $a \subset \delta$ be in* $\mathrm{rng}(\pi_{\zeta_n^*,b}) \cap \mathrm{rng}(\pi_{\zeta_{n+1}^*,c})$. *Then the extenders*

$$\{G_i^n;\ i \in \omega\}$$

witness that κ_n^* is a-strong in $\langle N, a\rangle$.

Proof. We first verify that

$$G_0^n(a \cap \kappa_n^*) = a \cap \kappa_{\gamma_{n+1}} = a \cap \mathrm{length}(G_0^n). \tag{9.4.4}$$

Recall that $G_0^n = E_{\nu_{\bar\gamma}}^{M_{\bar\gamma}}$ where $\bar\gamma = (\gamma_n)^p(\kappa_{\gamma_{n+1}})$ and $p = p(\gamma_n, \kappa_{\gamma_{n+1}})$. By induction on h we prove that

$$E_{\nu_{\gamma^h}}^{M_{\gamma^h}}(a \cap \kappa_{\gamma^h}) \cap \kappa_{\gamma_{n+1}} = a \cap \kappa_{\gamma_{n+1}}, \tag{9.4.5}$$

which for $h = p$ yields (9.4.4).

If $h = 0$ then $\gamma^h = \gamma_n$, so letting $\bar a$ be the preimage of a under $\pi_{\zeta_n^*,b}$, the conclusion is immediate from the following computation:

$$\begin{aligned}
E_{\nu_{\gamma_n}}^{M_{\gamma_n}}(a \cap \kappa_{\gamma_n}) \cap \kappa_{\gamma_{n+1}} &= \pi_{\zeta_n^*,\zeta_n}(a \cap \kappa_{\gamma_n}) \cap \kappa_{\gamma_{n+1}} \\
&= \pi_{\zeta_n,b}(\pi_{\zeta_n^*,\zeta_n}(a \cap \kappa_{\gamma_n}) \cap \kappa_{\gamma_{n+1}}) \\
&= \pi_{\zeta_n,b} \circ \pi_{\zeta_n^*,\zeta_n}(a \cap \kappa_{\gamma_n}) \cap \pi_{\zeta_n,b}(\kappa_{\gamma_{n+1}}) \\
&= \pi_{\zeta_n^*,b}(a \cap \kappa_{\gamma_n}) \cap \kappa_{\gamma_{n+1}} \\
&= \pi_{\zeta_n^*,b}(\bar a \cap \kappa_{\gamma_n}) \cap \kappa_{\gamma_{n+1}} \\
&= a \cap \pi_{\zeta_n^*,b}(\kappa_{\gamma_n}) \cap \kappa_{\gamma_{n+1}} \\
&= a \cap \kappa_{\gamma_{n+1}}.
\end{aligned}$$

The second equality follows from the fact that $\mathrm{cr}(\pi_{\zeta_n,b}) \geq \lambda_{\gamma_n} > \kappa_{\gamma_{n+1}}$; the fifth one from the fact that $\bar a \cap \kappa_{\gamma_n} = a \cap \kappa_{\gamma_n}$, as $\kappa_{\gamma_n} = \mathrm{cr}(\pi_{\zeta_n^*,b})$.

Now assume that (9.4.5) holds for h. Recall that $\kappa_{\gamma^{h+1}} > \kappa_{\gamma^h}$ (see (9.4.5)). Furthermore, setting $\xi = \xi_{\gamma^{h+1}}$, the map $\pi_{\xi,\gamma^h} : M^*_{\gamma^{h+1}} \to M_{\gamma^h}$ is total and has critical point $\kappa_{\gamma^{h+1}}$; the former follows from Lemma 9.4.1. From the preservation properties of π_{ξ,γ^h} we obtain

$$E_{\mathrm{top}}^{M^*_{\gamma^{h+1}}} \restriction \kappa_{\gamma^{h+1}} = E_{\nu_{\gamma^h}}^{M_{\gamma^h}} | \kappa_{\gamma^{h+1}}; \tag{9.4.6}$$

here we recall that $E_{\nu_{\gamma^h}}^{M_{\gamma^h}}$ is the top extender of M_{γ^h}. These facts allow us to carry out the following computation:

$$
\begin{aligned}
E_{\nu_{\gamma^{h+1}}}^{M_{\gamma^{h+1}}}(a \cap \kappa_{\gamma^{h+1}}) \cap \kappa_{\gamma_{n+1}} &= \pi_{\xi,\gamma^{h+1}+1}(a \cap \kappa_{\gamma^{h+1}}) \cap \kappa_{\gamma_{n+1}} \\
&= \pi_{\gamma^{h+1}+1,\gamma^h}(\pi_{\xi,\gamma^{h+1}+1}(a \cap \kappa_{\gamma^{h+1}}) \cap \kappa_{\gamma_{n+1}}) \\
&= \pi_{\xi,\gamma^h}(a \cap \kappa_{\gamma^{h+1}}) \cap \kappa_{\gamma_{n+1}} \\
&= \pi_{\xi,\gamma^h}(E_{\mathrm{top}}^{M^*_{\gamma^{h+1}}}(a \cap \kappa_{\gamma^h}) \cap \kappa_{\gamma^{h+1}}) \cap \kappa_{\gamma_{n+1}} \\
&= E_{\mathrm{top}}^{M_{\gamma^h}}(a \cap \kappa_{\gamma^h}) \cap \pi_{\xi,\gamma^n}(\kappa_{\gamma^{h+1}}) \cap \kappa_{\gamma_{n+1}} \\
&= E_{\nu_{\gamma^h}}^{M_{\gamma^h}}(a \cap \kappa_{\gamma^h}) \cap \kappa_{\gamma_{n+1}} \\
&= a \cap \kappa_{\gamma_{n+1}}.
\end{aligned}
$$

The second equality follows from a) of Lemma 9.4.1, which guarantees that $\mathrm{cr}(\pi_{\gamma^{h+1}+1,\gamma^h}) > \kappa_{\gamma_{n+1}}$, the middle one from the induction hypothesis and (9.4.6) and the next to the last one from the fact that $\pi_{\xi,\gamma^{h+1}}(\kappa_{\gamma^{h+1}}) \geq \lambda_{\gamma^{h+1}} > \kappa_{\gamma_{n+1}}$. The last inequality is just the induction hypothesis. The rest is either clear from the discussion preceding the computation or standard. This completes the verification of (9.4.4).

In the following we verify by induction on i that

$$G_i^n(a \cap \kappa_n^*) = a \cap \kappa_{\gamma_{n+1+i}} = a \cap \mathrm{length}(G_i^n). \tag{9.4.7}$$

The case $i = 0$ is just (9.4.4) which we verified above, so we proceed directly to the induction step. Suppose that (9.4.7) holds for i. By (9.4.2), $G_i^n \in M_{\zeta^*_{n+1+i}}$. Let $\bar{a} \in M_{\zeta^*_{n+1+i}}$ be such that $\pi_{\zeta^*_{n+1+i},c}(\bar{a}) = a$. Then $\bar{a} \cap \kappa_{\gamma_{n+1+i}} = a \cap \kappa_{\gamma_{n+1+i}}$, which together with the induction hypothesis yields

$$G_i^n(\bar{a} \cap \kappa_n^*) = \bar{a} \cap \kappa_{\gamma_{n+1+i}}.$$

By applying $\pi_{\zeta^*_{n+1+i},\zeta_{n+1+i}}$ to this equality and an additional truncation to $\kappa_{\gamma_{n+2+i}}$ we obtain

$$
\begin{aligned}
\tilde{G}_i^n(\pi_{\zeta^*_{n+1+i},\zeta_{n+1+i}}(\bar{a}) \cap \kappa_n^*) &= \pi_{\zeta^*_{n+1+i},\zeta_{n+1+i}}(\bar{a}) \cap \lambda_{\gamma_{n+1+i}}, \\
G_{i+1}^n(\pi_{\zeta^*_{n+1+i},\zeta_{n+1+i}}(\bar{a}) \cap \kappa_n^*) &= \pi_{\zeta^*_{n+1+i},\zeta_{n+1+i}}(\bar{a}) \cap \kappa_{\gamma_{n+2+i}}.
\end{aligned}
$$

Now we apply $\pi_{\zeta_{n+1+i},c}$ and recall that $\mathrm{cr}(\pi_{\zeta_{n+1+i},c}) \geq \lambda_{\gamma_{n+1+i}} > \kappa_{\gamma_{n+2+i}}$, so

$$G_{i+1}^n(\pi_{\zeta^*_{n+1+i},c}(\bar{a}) \cap \kappa_n^*) = \pi_{\zeta^*_{n+1+i},c}(\bar{a}) \cap \kappa_{\gamma_{n+2+i}},$$

that is,

$$G_{i+1}^n(a \cap \kappa_n^*) = a \cap \kappa_{\gamma_{n+2+i}}.$$

This completes the proof of the induction step and thus that of (9.4.7). □(Lemma 9.4.6)

Corollary 9.4.7. *Let $\Im$ be a normal iteration of M of limit length with two cofinal well-founded branches b and c and let J_α^E be an end-extension of $N = N(\Im)$ such that $J_\alpha^E \subset M_b \cap M_c$. Then $\delta = \delta(\Im)$ is Woodin in J_α^E.*

Proof. If $a \subset \delta$ is in J_α^E, then there are β, γ such that $a \in \operatorname{rng}(\pi_{\beta,b}) \cap \operatorname{rng}(\pi_{\gamma,c})$, so for sufficiently large n we have $a \in \operatorname{rng}(\pi_{\zeta_n^*,b}) \cap \operatorname{rng}(\pi_{\zeta_{n+1}^*,c})$. By the previous lemma, the extenders $\{G_i^n;\ i \in \omega\}$ witness that κ_n^* is a-strong in $\langle N, a\rangle$, and therefore also in J_α^E. □(Corollary 9.4.7)

The following lemma and corollary can be applied to premice satisfying suitable smallness condition. The simplest condition of this sort is 1-*smallness.*

Definition. A premouse M is 1-*small* iff there is no $\nu \le \operatorname{ht}(M)$ such that, letting $\kappa = \operatorname{cr}(E_\nu^M)$, we have

$$J_\nu^{E^M} \models \text{There is a Woodin cardinal.}$$

Lemma 9.4.8. *Let $\Im$ be a normal iteration of a 1-small premouse M and let b and c be two cofinal well-founded branches. Assume that $E_\nu^{M_c} \neq \varnothing$ for some $\nu > \delta$. Then $M_b = M_c||\beta = J_\beta^E$ for some $\beta < \nu$ where $E = E^N$. Consequently, if $\operatorname{ht}(M_b) = \operatorname{ht}(M_c)$, then $M_b = J_\alpha^E = M_c$ where $\operatorname{ht}(M_b) = \alpha = \operatorname{ht}(M_c)$.*

Proof. Assume for a contradiction that $E_\nu^{M_b} \neq \varnothing$ for some $\nu \le \operatorname{ht}(M_b), \operatorname{ht}(M_c)$. Then J_ν^E satisfies the assumptions of the previous corollary with ν in place of α, so δ is Woodin in J_ν^E. Letting $\kappa = \operatorname{cr}(E_\nu^{M_b})$, $\tau = \kappa^{+M_b||\nu}$, $E^b = E^{M_b}$ and $\pi : J_\tau^{E_b} \to J_\nu^{E_b}$ be the ultrapower map associated with $\operatorname{Ult}(J_\tau^{E_b}, E_\nu^{M_b})$, we conclude that $J_\tau^{E^b}$ contains a Woodin cardinal $\le \kappa$, call it $\bar\delta$. If $\delta < \kappa$, we have a contradiction to 1-smallness of M, as $\bar\delta$ is obviously Woodin in $J_\kappa^{E_b}$. Since τ is a cardinal in $M_b||\nu$, it follows that $\bar\delta$ is Woodin in $J_\nu^{E^b}$, so if $\bar\delta = \kappa$, one more downward application of π yields a $\bar\delta' < \kappa$ which is Woodin in $J_\kappa^{E_b}$. Contradiction again. This proves the first part from the lemma; the second part then follows immediately. □(Lemma 9.4.8)

Corollary 9.4.9. *Let M be an active presolid 1-small premouse. Then any normal iteration of M of limit length has at most one cofinal well-founded branch.*

Proof. Suppose for a contradiction that $\Im$ is a normal iteration of M with two distinct cofinal well-founded branches, b and c. The previous lemma then implies that there must be truncation points on both b and c. Indeed, if both b and c were without truncation points, then both M_b and M_c would be active, and of heights larger than δ (because they are active!) which is impossible. If there were a truncation point say on c but not on b, then M_c would fail to be sound, so M_b would be an active initial segment of M_c of height larger than δ, which is impossible again. Now as both b and c contain a truncation point, neither of M_b, M_c is sound, so neither of them can be a proper initial segment of the other. The previous lemma then yields $M_b = M_c$.

Let $\beta + 1$ and $\gamma + 1$ be the last truncation point on b and c, respectively. Using the properties of iteration maps, as in the proof of Lemma 9.1.14 we conclude that $M_\beta^* = \operatorname{core}(M_b) = \operatorname{core}(M_c) = M_\gamma^*$ and that $\pi_{\xi_\beta,b}$ and $\pi_{\xi_\gamma,c}$, being the associated core maps, agree. From this we infer that the extenders $E_{\nu_\beta}^{M_\beta}$ and $E_{\nu_\gamma}^{M_\gamma}$ are compatible, so $\beta = \gamma$. It follows that $\xi_\beta = \xi_\gamma \leq_T \zeta$ where $\zeta = \max(b \cap c)$. Since M^*_β is sound, M_ζ is sound above $\tilde{\lambda} = \sup\{\lambda_\alpha;\ \alpha < \zeta\}$ and both iteration maps $\pi_{\zeta,b}$ and $\pi_{\zeta,c}$ are above $\tilde{\lambda}$. It follows then as before that $M_\zeta = \text{core}_{\tilde{\lambda}}(M_b) = \text{core}_{\tilde{\lambda}}(M_c)$ and that $\pi_{\zeta,b}$ and $\pi_{\zeta,c}$, being the corresponding core maps, agree. Again, we have that $E^{M_{\beta'}}_{\nu_{\beta'}}$ and $E^{M_{\gamma'}}_{\nu_{\gamma'}}$ are compatible, where β' and γ' are such that $\beta' + 1 \in b$, $\gamma' + 1 \in c$ and $\zeta = T(\beta' + 1) = T(\gamma' + 1)$. This would mean that $\beta' = \gamma'$, which is impossible, as $\beta' + 1 \in b - \zeta$ and $\gamma' + 1 \in c - \zeta$. □(Corollary 9.4.9)

An argument of the same sort yields the unique iterability above the ultimate projectum:

Corollary 9.4.10. *Let M be a sound presolid 1-small premouse. Then every normal iteration of M above $\omega\varrho^\omega_M$ of limit length has at most one cofinal well-founded branch.*

Proof. This follows by the arguments from the proof of the previous lemma. □(Corollary 9.4.10)

An elaboration on the above methods yields a slightly stronger result:

Corollary 9.4.11. *Let M be a sound presolid 1-small premouse which has an iteration strategy and such that $\omega\varrho^\omega_M \leq \nu$ for some ν satisfying $E^M_\nu \neq \varnothing$. Then every normal iteration of M of limit length has at most one cofinal well-founded branch.*

Proof. Exercise. □(Corollary 9.4.11)

Results along these lines are used in the proofs that K^c-constructions generate models whose initial segments are iterable (or, in the case of larger K^c-models, weakly iterable) premice.

9.5 Towards the Ultimate Model K^c

Time and space do not permit us to develop K^c-models for overlapping extenders, and in this section we merely point out the problems connected with building these models and give some references for further study. Let us recall the K^c-construction in a somewhat more general setting.

The construction consists of two sequences of premice,

$$\langle N_\alpha;\ \alpha \in \mathbf{On}\rangle \quad \text{and} \quad \langle M_\alpha;\ \alpha \in \mathbf{On}\rangle,$$

such that $M_\alpha = \text{core}(N_\alpha)$, inductively defined as follows.

$$N_0 = \varnothing.$$

Setting $N_\alpha = \langle J_\zeta^{E^\alpha}, E^\alpha_{\omega\zeta}\rangle$ and $M_\alpha = \langle J_{\bar\zeta}^{\bar E^\alpha}, \bar E^\alpha_{\omega\bar\zeta}\rangle$,

$$N_{\alpha+1} = \begin{cases} \langle J_{\bar\zeta}^{\bar E^\alpha}, F\rangle & \text{if } \bar E^\alpha_{\omega\bar\zeta} = \varnothing \text{ and } F \text{ is the unique } \textit{properly backgrounded} \\ & \text{extender such that } \langle J_{\bar\zeta}^{\bar E^\alpha}, F\rangle \text{ is a premouse;} \\ \langle J_{\bar\zeta+1}^{\bar E^\alpha}, \varnothing\rangle & \text{otherwise.} \end{cases}$$

Let ϱ_ξ denote $\varrho^\omega_{N_\xi} = \varrho^\omega_{M_\xi}$ and, given α such that $\xi \le \alpha \le \infty$ (we exclude the case $\xi = \alpha = \infty$), set

$$\begin{aligned} \varrho_{\xi\alpha} &= \min\{\varrho_\eta;\ \xi \le \eta < \alpha\} \\ \tau_{\xi\alpha} &= (\omega\varrho_{\xi\alpha})^{+N_\xi} \\ \zeta &= \sup\{\tau_{\xi\alpha};\ \xi < \alpha\}. \end{aligned}$$

For limit α we set

$$N_\alpha = \langle J_\zeta^{E^\alpha}, \varnothing\rangle \quad \text{where} \quad E^\alpha = \bigcup_{\xi<\alpha} E^\xi \restriction \tau_{\xi\alpha}.$$

This defines a hierarchy of premice which, granting that the construction does not break down, stops at ∞. We set

$$\boldsymbol{K}^c = N_\infty = \bigcup_{\xi<\infty} J^{E^\xi}_{\tau_{\xi\infty}}.$$

The construction can be carried out provided that at each step α we can form the appropriate core and that there is at most one extender which can be put on top of the premouse M_α. Both requirements can be ensured by sufficient iterability of $N_{\alpha+1}$, and we purposely used the vague expression "properly backgrounded", as there are various conditions which can be imposed on F to guarantee iterability. Let us mention three of them, which are most frequently used in the literature:

- F is *fully backgrounded*, i.e. there is an extender F' on V such that:
  - $F(x) = F'(x) \cap \lambda_F$ for all $x \in \mathrm{dom}(F)$;
  - $V_{\lambda_F+2} \subset \mathrm{Ult}(V, F')$.
- F has *background certificates*, i.e. for every $a \subset \kappa = \mathrm{cr}(F)$ there is a pair (N, G) satisfying:
  - N is a transitive ZFC^--model and $V_\kappa \in N$;
  - G is an extender on N such that $\lambda_F \le \lambda_G$ and $V_{\lambda+1} \subset \mathrm{Ult}(N, G)$;
  - $F(x) = G(x) \cap \lambda_F$ for all $x \in N \cap \mathrm{dom}(F)$.
- F is ω-complete.

These conditions are descending in strength. We used ω-completeness to construct iterable models containing up to one strong cardinal. This was originally done by Jensen in [Je1]. One can go a bit further and construct models with a proper class of strong cardinals [Sch1] or a strong limit of strong cardinals [Je01]. In [Sch1] a full core model theory at the level of a proper class of strong cardinals is developed. At present, it is not clear how far exactly one can go with ω-completeness; it is known, however, that one needs a stronger condition in order to build larger models, i.e. models with Woodin cardinals. This was first done by Mitchell and Steel in [MiS94], where a fully background certified model is built. We have already mentioned that the indexing of extenders used by Mitchell and Steel differs from that used in our approach, and the corresponding model in our setting was constructed by Jensen in [Je97]. The full backgrounding is, however a very strong condition, and the outcome of the construction, although being iterable, need not meet our expectations regarding the reflection of large cardinal properties. Thus, it might happen that there are inner models with Woodin cardinals, but at the same time there need not be any extender on V. In this case, the fully background certified model is nothing other than L. In order to obtain models with better reflection properties we have to relax the background condition, and this leads to the notion of a background certificate, first designed by Steel in [St96]. He called this model "K^c" and this terminology is also used in the literature. In our book, we took liberty to also use this terminology in connection with models constructed from ω-complete extenders (and now even in connection with fully background certified models). Again, the version of background certificate stated above was used by Jensen in [Je97]. There are other notions of background certificate, which seem to be somewhat weaker [Sch2], but in all known cases, K^c-constructions of this type seem to yield "essentially" the same model. Under a suitable smallness condition, K^c-models can be used to develop a full core model theory, i.e. to construct K and prove weak covering for it. This is done in [St96] and [MSS94]. Steel [St96] also showed the rigidity theorem for K and the fact that K embeds into any universal weasel. If our smallness condition is not too restrictive, it is not be true anymore that any universal weasel is an iterate of K; however in the case of K^c this *is* true. The same also applies to a natural broad class of weasels, as was proved by Schindler [Sch3]. Let us stress that there is one important difference between building K^c and K. While the weak iterability of the levels of K^c-construction alone is sufficient for building K^c, in order to build K we need the *full* iterability of K^c. In general, it is more difficult to construct K^c-models than fully background certified models, and there is no general method for constructing either of them at present. Recall that the main open question in connection with building these models is weak iterability of the levels of the construction, and that the larger the model we aim at, the more difficult the iterability proof. The largest known fully backgrounded models were constructed by Neeman [N1] and hover at the level of Woodin limit of Woodin cardinals. In the case of the K^c-models this level is considerably lower, and corresponds to, roughly speaking, Con($AD_{\mathbb{R}}$), see [ANS].

Under the bibliography we also list literature with applications of the core model theory, which we did not touch on in our book at all (because of the lack of time and space). Our list is far from being complete, and we would like to illustrate this area by stating some typical results. One group of applications comprise the results on changing cofinalities and the singular cardinal hypothesis, which are due to Gitik and Mitchell; see [G91], [GM96], [Mi91], [Mi87], [Mi92] and [Mi1]. The second group is related to infinitary combinatorics, here belong e.g. [FMS], [Je2], [Je94], [JeZ00], [Scm96], [Scm99], [SZ01], [SZ1], [W79], [Z97], [Z00] and [SZ2]. The third large group consists of the applications to descriptive set theory; here we mention [HS95], [HS00], [N2], [SR], [St96], [St93], [St95], [St2] and [SW98].

A brief review of the core model theory with several applications can be found in [LS99], [St1] and [W98].

Notes

The notion of premouse as defined here is again due to Jensen [Je97]. Mitchell and Steel [MiS94] use a different notion of premouse, with a much denser indexing. The notion of iteration tree was originally developed for coarse iterations in [MS89] and [MS94] and only later adapted to fine structural iteration in [MiS94], where the method of comparing premice using phalanxes also appeared. The notion of the weak Dodd–Jensen property, and its applications are due to Neeman and Steel [NS99]. The proofs of the solidity theorem and the condensation lemma for mice with overlapping extenders are also due to Mitchell and Steel [MiS94]; our version of these proofs from weak iterability is due to Jensen [Je99], who uses a mix of the methods from [Je97] and [MiS94]. The proofs presented in this book are based on the pure $*$-theory and are due to the author. The method of obtaining extenders witnessing Woodinness from non-unique iterations was invented in by Martin and Steel [MS94] and the fine structural version of the method developed by Mitchell and Steel in [MiS94]. The adaptation to of these methods to our context requires several extra ideas, which are due to Jensen.

Bibliography

[ANS] Andretta, A, Neeman, I. and Steel, J. R., The domestic levels of K^c are iterable, Israel J. Math. 125 (2001), 157–201.

[De84] Devlin, K. J., *Constructibility*, Perspect. Math. Logic, Springer, Berlin 1984.

[DeJ75] Devlin, K. J. and Jensen, R. B., Marginalia to a theorem of Silver, in *Logic Conference Kiel* 1974, Müller, G. H., Oberschelp, A. and Potthoff, K. (eds.), Lecture Notes in Math. 499, Springer, Berlin 1975, 115–142.

[D82] Dodd, A. J., *The Core Model,* London Math. Soc. Lecture Note Ser. 61, Cambridge Univ. Press, Cambridge 1982.

[DJ81] Dodd, A. J. and Jensen, R. B., The core model, Ann. of Math. Logic 20 (1981), 43–75.

[DJ81] Dodd, A. J. and Jensen, R. B., The covering lemma for **K**, Ann. of Math. Logic 22 (1982), 1–30.

[DJ81a] Dodd, A. J. and Jensen, R. B., The covering lemma for L^U, Ann. of Math. Logic 22 (1982), 127–155.

[DJK81] Donder, H. D., Jensen, R. B. and Koppelberg, B. J., Some application of the core model, in *Set Theory and Model Theory* (*Bonn* 1979), *Jensen, R. B. and Prestel, A.(eds.),* Lecture Notes in Math. 872, Springer, Berlin 1981, 55–97.

[FMS] Foreman, M., Magidor, M. and Schindler, R. D., The consistency strength of successive cardinals with the tree property, to appear in J. Symbolic Logic.

[FW] Friedman, S. D. and Wylie, D., Fragmented extenders in core models up to 0¶, handwritten notes.

[G91] Gitik, M., The strength of the failure of the singular cardinal hypothesis, Ann. Pure Appl. Logic 51 (3) (1991), 215–240.

[G93] Gitik, M., On measurable cardinals violating the continuum hypothesis, Ann. Pure Appl. Logic 63 (3) (1993) 227–240

[GM96] Gitik, M., and Mitchell, W. J., Indiscernible sequences for extenders and the singular cardinal hypothesis, Ann. Pure Appl. Logic 82 (3) (1996) 273–316.

[HS95] Hauser, K., The consistency strength of projective absoluteness, Ann. Pure Appl. Logic 74 (3) (1995), 245–295.

[HS00] Hauser, K. and Schindler, R. D., Projective uniformization revisited, Ann. Pure Appl. Logic 103 (1–3) (2000), 109–153.

[Jc78] Jech, T., *Set Theory*, Academic Press, New York 1978.

[Je72] Jensen, R. B., The fine structure of the constructible hierarchy, Ann. of Math. Logic 4 (1972), 229–308.

[Je88] Jensen, R. B., *Measures of Order 0*, handwritten notes, Oxford, 1988.

[Je1] Jensen, R. B., *Nonoverlapping Extenders.* handwritten notes, Oxford.

[Je2] Jensen, R. B., Some remarks on $\square$ below $0^{\P}$, handwritten notes, Oxford.

[Je94] Jensen, R. B., More on $\square$ below $0^{\P}$, handwritten notes, Oxford 1994.

[Je97] Jensen, R. B., A new fine structure theory for higher core models, handwritten notes, Berlin, 1997.

[Je98] Jensen, R. B., Addendum to "A new fine structure theory for higher core models", handwritten notes, Berlin, 1998.

[Je98] Jensen, R. B., More on iterability, handwritten notes, Berlin, 1998.

[Je99] Jensen, R. B., Corrections and remarks, handwritten notes, Berlin, 1999.

[Je01] Jensen, R. B., T-Mice, handwritten notes, Berlin, 2001.

[JeZ00] Jensen, R. B. and Zeman, M., Smooth categories and global square, Ann. Pure Appl. Logic 102 (1–2) (2000), 101–138.

[Ka94] Kanamori, A., *The Higher Infinite*, Perspect. Math. Logic, Springer, Berlin 1994.

[Ko89] Koepke, P., Fine structure for inner models up to one strong cardinal, Habilitation, Freiburg, 1989.

[Ku70] Kunen, K., Some applications of iterated ultrapowesr in set theory, Ann. of Math. Logic 1 (1970), 179–227.

[Ku80] Kunen, K., *Set theory. An introduction to independence proofs*, Stud. Logic Found. Math. 102, North-Holland, Amsterdam 1980.

[LS99] Löwe, B. and Steel, J. R., An introduction to core model theory, in *Sets and Proofs*, Logic Colloquium 1997 (Leeds, 1997), Vol. 1, Cooper, S. B. and Truss, J. (eds.), London Math. Soc. Lecture Note Ser. 258, Cambridge Univ. Press, Cambridge 1999, 103–157.

[MS89] Martin, D. A. and Steel, J. R., A Proof of Projective Determinancy, J. Amer. Math. Soc. 2 (1989), 71–125.

[MS94] Martin, D. A. and Steel, J. R., Iteration Trees, J. Amer. Math. Soc. 7 (1) (1994), 1–73.

[Mi74] Mitchell, W. J., Sets constructible from sequences of ultrafilters, J. Symbolic Logic 39 (1974), 57–66.

[Mi84] Mitchell, W. J., The core model for sequences of measures I, Math. Proc. Cambridge Philos. Soc. 95 (1984), 228–260.

[Mi1] Mitchell, W. J., The core model for sequences of measures II, unpublished.

[Mi84a] Mitchell, W. J., Indiscernibles, skies and ideals, in *Axiomatic Set Theory, Baumgartner,* J. E., Martin, D. A. and Shelah, S. (eds.), Contemp. Math. 31, Amer. Math. Soc., Providence, RI, 1984, 161–182.

[Mi87] Mitchell, W. J., Applications of the core model for sequences of measures, Trans. Amer. Math. Soc. 299 (1987), 41–58.

[Mi91] Mitchell, W. J., Definable singularity, Trans. Amer. Math. Soc. 327 (1991), 407–426.

[Mi92] Mitchell, W. J., On the singular cardinal hypothesis, Trans. Amer. Math. Soc. 329 (1992), 507–530.

[MiS95] Mitchell, W. J., and Schimmerling, E., Weak covering without countable closure, Math. Res. Lett. 2 (5) (1995), 595–609.

[MSS94] Mitchell, W. J., Schimmerling, E. and Steel, J. R., The covering lemma up to a Woodin cardinal, Ann. Pure Appl. Logic 84 (2) (1994), 219–255.

[MiS94] Mitchell, W. J. and Steel, J. R., *Fine structure and iteration trees,* Lecture Notes in Logic 3, Springer, Berlin 1994.

[Mi1] Mitchell, W. J., The covering lemma, in *Handbook of Set Theory,* Foreman, M., Kanamori, A. and Magidor, M. (eds.), to appear.

[N1] Neeman, I., Inner models in the region of Woodin limits of Woodin cardinals, to appear in the Ann. Pure Appl. Logic.

[N2] Neeman, I., Mice and Scales, preprint.

[NS99] Neeman, I. and Steel, J. R., A weak Dodd-Jensen lemma, J. Symbolic Logic 64 (3) (1999), 1285–1294.

[SR] Rudominer, M. and Steel, J. R., Inner models with wellorders in $L(\mathbb{R})$, preliminary draft.

[Sch99] Schindler, R. D., Successive weakly compact or singular cardinals, J. Symbolic Logic 64 (1999), 139–146.

[Sch1] Schindler, R. D., The core model for almost linear iterations, to appear in Ann. Pure Appl. Logic.

[Sch2] Schindler, R. D., K^c without large cardinals in V, submitted to Arch. Math. Logic.

[Sch3] Schindler, R. D., Iterates of the core model, preprint.

[SSZ] Schindler, R. D., Steel, J. R. and Zeman, M., Deconstructing inner model theory, preprint.

[Scm96] Schimmerling, E., Combinatorial principles in the core model for one Woodin cardinal, Ann. Pure Appl. Logic 74 (2) (1995), 153–201.

[Scm98] Schimmerling, E., Combinatorial set theory and inner models, in *Set theory, Techniques and Applications, Curaçao* 1995 *and Barcelona* 1996 *Conferences*, Di Prisco, C. A., Larson J. A., Bagaria, J. and Mathias, A. R. D. (eds.), Kluwer, 1998, 207–212.

[Scm99] Schimmerling, E., *A finite family weak square principle*, J. Symbolic Logic 64 (3) (1999), 1087–1110.

[Scm99a] Schimmerling, E., Covering properties of core models, in *Sets and Proofs*, S. B. Cooper and J. K. Truss (eds.), London Math. Soc. Lecture Note Ser. 258, Cambridge Univ. Press, Cambridge 1999, 281–299.

[SS96] Schimmerling, E. and Steel, J. R., Fine structure for tame inner models, J. Symbolic Logic 61 (2) (1996), 621–639.

[SS99] Schimmerling, E and Steel, J. R., The maximality of the core model, Trans. Amer. Math. Society 351 (8) (1999), 3119–3141.

[SW] Schimmerling, E. and Woodin, W. H., The Jensen covering property, to appear in J. Symbolic Logic.

[SZ01] Schimmerling, E. and Zeman, M., Square in core models, Bull. Symbolic Logic 7 (3) (2001), 305–314

[SZ1] Schimmerling, E. and Zeman, M., A characterization of $\Box_\kappa$ in core models, preprint.

[SZ2] Schimmerling, E. and Zeman, M., Cardinal transfer theorems in extender models, in preparation.

[St96] Steel, J. R., *The core model iterability problem*, Lecture Notes in Logic 8, Springer, Berlin 1996.

[St93] Steel, J. R., Inner models with many Woodin cardinals, Ann, Pure Appl, Logic 65 (2) (1993), 185–209.

[St95] Steel, J. R., Projectively well-ordered inner models, Ann. Pure Appl. Logic 74 (1) (1995), 77–104.

[St1] Steel, J. R., An Outline of Inner Model Theory, in *Handbook of Set Theory*, Foreman, M., Kanamori, A. and Magidor, M. (eds.), to appear.

[St2] Steel, J. R., Woodin's analysis of $\mathrm{HOD}^{L(\mathbb{R})}$, unpublished.

[St3] Steel, J. R., Theorem of Woodin on mouse sets, unpublished.

[St3] Steel, J. R., Core models with more Woodin cardinals, preliminary draft.

[SW98] Steel, J. R. and Welch, P. D., Σ^1_3-absoluteness and the second uniform indiscernible, Israel J. Math. 104 (1998), 157–190.

[V93] Vickers, J., Doctoral dissertation, Bristol, 1993.

[W79] Welch, P. D., Combinatorial principles in the core model, D. Phil. thesis, Oxford 1979.

[W99] Welch, P. D., Some remarks on the maximality of inner models, in *Logic Colloquium'98 (Prague, 1998)*, Buss, S. R., Hájek, P. and Pudlák, P. (eds.), Lecture Notes in Logic 13, Association for Symbolic Logic and A. K. Peters, Ltd., Natick, MA, 1999, 516–540.

[W98] Welch, P. D., Doing without determinancy-aspects or inner models, in *Proceedings of the Hull Logic Colloquium*, Drake, F. and Truss, J. (eds.), North-Holland Stud. Logic Ser., Amsterdam 1998, 333–342.

[Wy90] Wylie, D., Ph. D. thesis, MIT 1990.

[Z97] Zeman, M., The core models for non-overlapping extenders and its applications, Doctoral dissertation, Berlin 1997.

[Z00] Zeman, M., $\Diamond$ at Mahlo cardinals, J. Symbolic Logic 65 (4) (2000), 1813–1822.

[Z1] Zeman, M., Dodd solidity in extender models with long indexing, preprint.

[Z2] Zeman, M., Global square in extender models, preprint.

Index

Bei Fragen zur Produktsicherheit wenden Sie sich bitte an:
If you have any questions regarding product safety,
please contact:

Walter de Gruyter GmbH
Genthiner Straße 13
10785 Berlin
productsafety@degruyterbrill.com